Antioxidants from Food and Food Wastes for Nutraceutical, Pharmaceutical and Cosmetic Fields

Antioxidants from Food and Food Wastes for Nutraceutical, Pharmaceutical and Cosmetic Fields

Guest Editors

Monica Locatelli
Vincenzo Disca

MDPI

Basel • Beijing • Wuhan • Barcelona • Belgrade • Novi Sad • Cluj • Manchester

Guest Editors

Monica Locatelli
Department of
Pharmaceutical Sciences
Università del Piemonte
Orientale
Novara
Italy

Vincenzo Disca
Department of
Pharmaceutical Sciences
Università del Piemonte
Orientale
Novara
Italy

Editorial Office
MDPI AG
Grosspeteranlage 5
4052 Basel, Switzerland

This is a reprint of the Special Issue, published open access by the journal *Antioxidants* (ISSN 2076-3921), freely accessible at: https://www.mdpi.com/journal/antioxidants/special_issues/XP53Z2415V.

For citation purposes, cite each article independently as indicated on the article page online and as indicated below:

Lastname, A.A.; Lastname, B.B. Article Title. *Journal Name* **Year**, *Volume Number*, Page Range.

ISBN 978-3-7258-7452-1 (Hbk)
ISBN 978-3-7258-7453-8 (PDF)
https://doi.org/10.3390/books978-3-7258-7453-8

Contents

About the Editors

Monica Locatelli

Monica Locatelli holds a PhD in Science of Bioactive Substances and is Associate Professor of Food Chemistry at the Department of Pharmaceutical Sciences, Università del Piemonte Orientale (Italy). Her research activity is focused on the chemical characterization and the study of bioactive compounds of foods and food by-products, as well as on the evaluation of the effect of food processing (both at the industrial and domestic level). She is an expert in the application of extraction/separation methods (also considering green approaches), characterization of complex phytoextracts (spectrophotometric and chromatographic methods, such as HPLC and HPLC-MS), and evaluation of bioactivity and antioxidant properties. Part of her research activity is dedicated to the study of bioaccessibility of bioactive compounds through simulated in vitro digestion and on the formulation/stabilization of functional ingredients and functional foods. Additional expertise is in the use of multivariate statistical analysis techniques and artificial neural networks for the processing and interpretation of chemical data.

Vincenzo Disca

Vincenzo Disca is a Postdoctoral Research Fellow at the Food Chemistry, Biotechnology and Nutrition unit of the Department of Pharmaceutical Sciences, Università del Piemonte Orientale (Italy). He is Co-founder and Chief Research Officer of nutractiva S.r.l., an academic start-up focused on the development of innovative functional ingredients for the food, nutraceutical, and cosmetic sectors. His research activity integrates food chemistry, biotechnology, and nutrition, with particular emphasis on dietary fiber modification, polyphenol extraction, gut microbiota-related bioactivity, and sustainable processing technologies, especially in the context of food by-product valorization. He has gained international research experience at Wageningen University & Research (the Netherlands), Universidad de Córdoba (Spain), and the University of Patras (Greece).

antioxidants

Editorial

Editorial: Antioxidants from Food and Food Wastes for Nutraceutical, Pharmaceutical and Cosmetic Fields

Vincenzo Disca [1,2,*] and Monica Locatelli [1,*]

1 Department of Pharmaceutical Sciences, Università del Piemonte Orientale, Largo Donegani 2, 28100 Novara, Italy
2 nutractiva Srl, Largo Donegani 2, 28100 Novara, Italy
* Correspondence: vincenzo.disca@uniupo.it (V.D.); monica.locatelli@uniupo.it (M.L.)

Antioxidants from foods and agri-food by-products are becoming increasingly recognized as strategically important for the transition toward greener, circular value chains in the nutraceutical, pharmaceutical, and cosmetic sectors. The scientific challenge is no longer limited to demonstrating antioxidant capacity, per se, but rather to building robust pipelines that connect (i) sustainable recovery and "green" extraction, (ii) advanced chemical characterization and standardization, and (iii) formulation strategies that preserve stability and improve bioavailability, enabling applications that are clinically and industrially relevant. The ten contributions published in this Special Issue collectively address these priorities, spanning from marine and terrestrial matrices to extraction and process intensification, chemical fingerprinting, and translational endpoints in metabolic health and skin aging.

A core pillar of this Special Issue is the concern with sustainable extraction and process optimization, with several studies explicitly tackling how recovery technologies can shape yield, composition, and antioxidant performance. Tapia-Quirós et al. (2025) focus their study on olive leaves, an abundant agricultural residue, and illustrate how the valorization of this widely available by-product can be aligned with functional ingredient design (Contribution 1).

In the same "process-to-function" logic, Nabiyeva et al. (2025) optimize conventional and ultrasound-assisted extraction to maximize the total phenolic recovery and in vitro antioxidant activity from *Crataegus almaatensis* leaves, exemplifying how extraction intensification can be used as a lever to increase bioactivity while supporting more sustainable processing routes (Contribution 2).

Krivošija et al. (2025) similarly address extraction, in their case through using subcritical water and pressurized solvents to explore orange peel "herbal dust" as a source of antioxidant compounds (Contribution 3).

Together, these papers reinforce one of the key messages of this Special Issue: the efficacy of antioxidant ingredients is directly correlated to the technologies used to recover them, and so optimized and scalable extraction methods are prerequisites for reproducible nutraceutical and cosmetic development.

A second pillar of this Special Issue is advanced characterization and analytical rigor, which are essential to overcome the variability inherent in by-products and complex natural matrices. Rivera-Tovar et al. (2025) offer a strong contribution to this objective through an integrated extraction–purification workflow for recovering phlorotannins from *Durvillaea incurvata*, coupled with advanced characterization, an approach that directly addresses the Special Issue's call for analytical methods that enable meaningful quality control and application-driven purification (Contribution 4).

Extending the omics dimension even further, Martakos et al. (2026) provide a lipidomics-centered case study on marine by-product oils, showing how biological origin

 https://doi.org/10.3390/antiox15020226

and extraction technology can reshape lipid class distribution and antioxidant molecule retention, supporting traceability, standardization, and product development for high-value applications (Contribution 5).

Importantly, the marine focus expands the scope of this Special Issue beyond polyphenol-centric narratives, illustrating how antioxidant innovation from by-products also includes lipid-soluble antioxidants (e.g., tocopherols, carotenoids, squalene) and PUFA-rich matrices that require oxidation-aware processing.

A third major theme is formulation and application-readiness in the cosmetic field, where antioxidant value must translate into measurable skin health outcomes which are relevant to photoaging and inflammation. Chiu et al. (2025) assess djulis (*Chenopodium formosanum*) leaf extract as a cosmetic antioxidant ingredient, representing the pathway from the botanical source to skin-directed application (Contribution 6).

Complementarily, Hyun et al. (2025) move into a higher translational level by validating a synergistic marine nutricosmetic ingredient (GABALAGEN®) through randomized human trials, aligning closely with the Special Issue's emphasis on efficacy and delivery (Contribution 7).

Taken together, these contributions highlight two routes for cosmetic innovation, topical cosmeceutical ingredients derived from plant-based resources and ingestible nutricosmetic approaches derived from marine resources, both of which are dependent on robust characterization and stability-preserving processing.

A fourth pillar involves nutraceutical and health-oriented positioning, particularly for metabolic disorders where oxidative stress and inflammation intersect with diet. Buonaiuto et al. (2025) examine betaine supplementation and its biological relevance in vivo, contributing to the broader discussion of antioxidant nutraceutical strategies that influence redox, inflammatory, and metabolic homeostasis (Contribution 8).

At a broader level, Yuksek et al. (2025) review the use of dietary supplements derived from food by-products in diabetes management, explicitly integrating sustainability with mechanistic plausibility, safety, toxicology, and regulatory considerations, which are precisely the translational bottlenecks highlighted in this Special Issue's scope (Contribution 9). This review is of particular importance because it frames valorized by-products not only as sources of bioactive compounds, but as candidates that must meet real-world constraints: standardization, contaminant control, dosing reproducibility, and regulatory compliance.

Finally, the Special Issue is anchored by a wide-angle perspective on the polyphenols derived from by-products. Demir et al. (2026) review the polyphenols recovered from by-products, discussing their applications and health effects, and thereby connecting extraction technologies, compound classes, and downstream uses across the nutraceutical, pharmaceutical, and cosmetic spectra (Contribution 10).

This contribution provides a conceptual backbone for the Special Issue: polyphenols remain dominant targets for waste-valorization strategies, but their successful deployment depends on matching recovery methods to compound stability, ensuring compositional control, and designing formulations that preserve activity through processing and storage.

Across all the published works, a cohesive narrative emerges that is well-aligned with the stated aims of this Special Issue. Firstly, by-products are treated as legitimate, high-value feedstocks, not as marginal residues, regardless of whether they are olive leaves, hawthorn leaves, orange peels, macroalgae, or marine processing side streams. Secondly, the methodological choices (extraction, purification, and analytical workflows) are shown to be decisive, often being used to determine whether an "antioxidant ingredient" is reproducible, stable, and application-ready. Third, the collection demonstrates a welcome move toward translation, as is exemplified by the clinical evaluation in nutricosmetics and by the reviews that foreground safety, regulatory frameworks, and real formulation constraints.

https://doi.org/10.3390/antiox15020226

Overall, this Special Issue emphasizes the need for harmonized methodologies, improved quality control, and sustainability-oriented strategies that can support the effective translation of antioxidant research. On this basis, a second edition will expand the discussion by concentrating on advanced extraction methods, novel formulation approaches, and sustainable technologies, with the aim of further strengthening the practical implementation of antioxidant systems.

Author Contributions: Conceptualization, V.D. and M.L.; writing—original draft preparation, V.D.; writing—review and editing, V.D. and M.L. All authors have read and agreed to the published version of the manuscript.

Funding: This research received no external funding.

Conflicts of Interest: Vincenzo Disca is co-founder of the company nutractiva Srl. The remaining author declares that the research was conducted in the absence of any commercial or financial relationships that could be construed as a potential conflict of interest.

List of Contributions:

1. Tapia-Quirós, P.; Mir-Cerdà, A.; Granados, M.; Sentellas, S.; Saurina, J. From Waste to Resource: Exploring Green Approaches for Phenolics Recovery from Olive Leaves. *Antioxidants* **2025**, *14*, 136. https://doi.org/10.3390/antiox14020136.
2. Nabiyeva, Z.; Kulaipbekova, A.; Carpentieri, S.; Pronina, Y.; Samadun, A.; Assembayeva, E.; Ferrari, G. Optimization of Conventional and Ultrasound-Assisted Extraction to Maximize Recovery of Total Phenolic Content and In Vitro Antioxidant Activity from *Crataegus almaatensis* Leaves. *Antioxidants* **2025**, *14*, 1003. https://doi.org/10.3390/antiox14081003.
3. Krivošija, S.; Ballesteros-Gómez, A.; Zloh, M.; Milić, N.; Popović, A.; Nastić, N.; Vidović, S. Subcritical Water and Pressurised Ethanol Extractions for Maximum Recovery of Antioxidants from Orange Peel Herbal Dust with Evaluation of Its Pharmacological Potential Using in Silico and In Vitro Analysis. *Antioxidants* **2025**, *14*, 638. https://doi.org/10.3390/antiox14060638.
4. Rivera-Tovar, P.R.; Contreras-Contreras, G.; Rivas-Reyes, P.I.; Pérez-Jiménez, J.; Martínez-Cifuentes, M.; Pérez-Correa, J.R.; Mariotti-Celis, M.S. Sustainable Recovery of Phlorotannins from *Durvillaea incurvata*: Integrated Extraction and Purification with Advanced Characterization. *Antioxidants* **2025**, *14*, 250. https://doi.org/10.3390/antiox14030250.
5. Martakos, I.C.; Tzika, P.; Dasenaki, M.E.; Kalogianni, E.P.; Thomaidis, N.S. Lipidomic Characterization of Marine By-Product Oils: Impact of Species and Extraction Methods on Lipid Profile and Antioxidant Potential. *Antioxidants* **2026**, *15*, 95. https://doi.org/10.3390/antiox15010095.
6. Chiu, C.-S.; Chan, Y.-J.; Wu, Y.-Z.; Lu, W.-C.; Chiang, P.-Y.; Li, P.-H. Bioactive Compounds and Antioxidant Efficacy of Djulis (*Chenopodium formosanum*) Leaves: Implications for Sustainable Cosmeceutical Development. *Antioxidants* **2025**, *14*, 202. https://doi.org/10.3390/antiox14020202.
7. Hyun, J.; Rheu, K.-M.; Lee, B.-J.; Ryu, B. Targeting Aging Skin with GABALAGEN®: A Synergistic Marine Nutricosmetic Ingredient Validated Through Human Randomized Trials. *Antioxidants* **2025**, *14*, 245. https://doi.org/10.3390/antiox14030245.
8. Buonaiuto, G.; Federiconi, A.; Vecchiato, C.G.; Benini, E.; Mordenti, A.L. Betaine Dietary Supplementation: Healthy Aspects in Human and Animal Nutrition. *Antioxidants* **2025**, *14*, 771. https://doi.org/10.3390/antiox14070771.
9. Yuksek, E.N.; Pereira, A.G.; Prieto, M.A. Dietary Supplements Derived from Food By-Products for the Management of Diabetes Mellitus. *Antioxidants* **2025**, *14*, 1176. https://doi.org/10.3390/antiox14101176.
10. Demir, R.; Sarıtaş, S.; Bechelany, M.; Karav, S. Polyphenols from Byproducts: Their Applications and Health Effects. *Antioxidants* **2026**, *15*, 87. https://doi.org/10.3390/antiox15010087.

https://doi.org/10.3390/antiox15020226

antioxidants

MDPI

Article

From Waste to Resource: Exploring Green Approaches for Phenolics Recovery from Olive Leaves

Paulina Tapia-Quirós [1,2], Aina Mir-Cerdà [1,3], Mercè Granados [1,3], Sonia Sentellas [1,3,4] and Javier Saurina [1,3,*]

1. Department of Chemical Engineering and Analytical Chemistry, Universitat de Barcelona, Martí i Franquès 1-11, E08028 Barcelona, Spain; paulina.tapia@ub.edu (P.T.-Q.); ainamir@ub.edu (A.M.-C.); mgranados@ub.edu (M.G.); sonia.sentellas@ub.edu (S.S.)
2. Chemical Engineering Department, Escola d'Enginyeria de Barcelona Est (EEBE), Universitat Politècnica de Catalunya (UPC)-BarcelonaTECH, Eduard Maristany 10-14, Campus Diagonal Besòs, E08930 Barcelona, Spain
3. Research Institute in Food Nutrition and Food Safety, Universitat de Barcelona, Av. Prat de la Riba 171, Edifici Recerca (Gaudí), E08921 Santa Coloma de Gramenet, Spain
4. Serra Húnter Fellow, Generalitat de Catalunya, E08003 Barcelona, Spain

* Correspondence: xavi.saurina@ub.edu

Abstract: Waste management presents a significant challenge for agri-food industries, but also an opportunity to recover valuable bioactive compounds, particularly phenolics, aligning with circular economy principles. This study compares the efficacy of conventional solvents and natural deep eutectic solvents (NADES) in extracting polyphenols from olive leaves using the scalable techniques of solid–liquid extraction (SLE) with mechanical stirring and microwave-assisted extraction (MAE). Key factors affecting extraction yield, including solvent composition, time, and temperature were investigated. Extraction efficiency was evaluated by measuring total polyphenol content (TPC) by high-performance liquid chromatography (HPLC), antioxidant capacity (FRAP assay), and individual phenolic compounds, also quantified using HPLC. Additionally, differential pulse voltammetry (DPV) was employed to evaluate the antioxidant quality of the extracts. NADES demonstrated superior extraction performance compared to conventional solvents, with the choline–glycerol system showing the highest efficiency. The combination of MAE and NADES emerged as a green and sustainable alternative to traditional methods, offering improved yield and speed. In contrast, SLE with water or ethanol/water mixtures required extended times or higher temperatures for comparable results but offered easier scalability for industrial applications.

Keywords: microwave-assisted extraction; phenolic compounds; natural deep eutectic solvents; agri-food waste valorization; circular economy

1. Introduction

Olive oil production is a major economic activity in Southern Europe. It is estimated that 2.55 million metric tons of olive oil are produced per year. Therefore, this industry generates large amounts of solid and liquid waste, about 10 million tons of waste per year, including olive pomace, wastewater, and olive leaves as the main ones [1]. Leaves from *Olea europaea* contain a higher quantity of phenolic compounds compared to fruit or virgin olive oil [2] (ca. 1450 mg of total phenols/100 g of fresh leaves [3] in contrast to 110 mg/100 g of fruit [4], and 23 mg/100 mL of extra virgin olive oil [2,5]).

Phenolic compounds are an extensive group of secondary plant metabolites, synthesized as protective agents against external environmental or parasitic aggressions, with powerful activities as antioxidant and antiradical agents to scavenge free radicals, as structural molecules for the formation of protective parts in plants against insects, or as antibiotics to combat microorganisms that can affect the plant [6–8]. Under the generic framework of polyphenols, various families of compounds widespread throughout the plant kingdom, such as phenolic acids and flavonoids, are included [9,10]. However, phenylethanoids also stand out in olives as abundant and strategically remarkable chemicals. In general, phenolics arouse special interest in the food, cosmetic, and pharmaceutical industries because of their noticeable attributes and benefits for human health [8]. Olive leaves have been used from ancient times in traditional medicine to make infusions for a wide range of ailments [11,12]. Olive leaves contain a large number of phenolic compounds such as oleuropein, tyrosol, hydroxytyrosol, luteolin, coumaric acid, caffeic acid, ferulic acid, vanillic acid, ligstroside, rutin glucosides, etc. [13–17]. In recent times, there has been scientific evidence about their potential health benefits, with anti-inflammatory, anticarcinogenic, antihypertensive, antiviral, antimicrobial, hypoglycemic, antithrombotic, and hypocholesterolemic effects [11,12].

Phenolic compounds are commonly extracted with organic solvents, such as methanol, acetone, or ethyl acetate [18–21]. Nevertheless, these solvents are toxic and unsuitable (even incompatible) for food, cosmetic, and pharmaceutical purposes. For this type of application, the most appropriate solvent is water, but the extraction yields may be lower than other options, especially for apolar compounds such as flavonoids and phenylethanoid derivatives (e.g., oleuropein, with apolar moieties in their structures). Ethanol is an excellent option to improve the analyte affinity towards the solvent without incurring incompatibilities with food or nutraceutical applications, especially when used in hydro-organic mixtures.

Natural deep eutectic solvents (NADES) provide a green alternative to conventional solvents because of their enhanced extraction performance [18,22–24]. NADES consist of a mixture of two or more components, a hydrogen bond acceptor (e.g., choline chloride) and a hydrogen bond donor (e.g., glycerol), at a particular molar ratio to formulate eutectic mixtures. NADES have distinctive physicochemical properties, such as adjustable viscosity, low volatility, water solubility, strong solubilization capability for a wide range of compounds, and pharmaceutically acceptable toxicity [18,22–24].

Several extraction techniques have been proposed to recover polyphenols from olive mill waste, ranging from solid–liquid extraction (SLE) by simple mechanical stirring to advanced techniques (e.g., ultrasound-assisted extraction (UAE), pressurized liquid extraction (PLE), or microwave-assisted extraction (MAE)) [25,26]. SLE is the most widely used technique for the extraction of polyphenols from agri-food residues due to its ease of implementation and reduced operating costs. Beyond SLE, which is readily scalable for pilot plant or industrial applications, UAE and MAE are also implementable with reasonable ease. In particular, MAE proves to be a highly efficient technique, providing high extraction yields in a short processing time. Microwave radiation causes cell walls in the waste matrix to swell and rupture, which enhances the release of target compounds [24,27]. Additionally, MAE offers simplified handling, low energy requirements, and reduced solvent consumption.

This paper addresses the recovery of bioactive phenolics from the olive leaf matrix using scalable extraction techniques and green solvents, including some conventional solvent mixtures and NADES. The study evaluates the extraction capacity of different techniques and solvents to identify the most appropriate approach based on diverse outcomes such as the antioxidant capacity estimated by the ferric reducing antioxidant power (FRAP) assay, the global phenolic content determined from the chromatographic profile recorded at 280 nm, and the concentration of quantitatively remarkable compounds, such as luteolin, hydroxytyrosol, and their derivatives. Additionally, an electrochemical study of the antioxidant properties of the extract has been carried out to establish the compounds with the greatest reducing capacity, i.e., those that are easily oxidized at lower potentials and are expected to be the most powerful or proactive.

2. Materials and Methods

2.1. Chemicals

Polyphenol standards were purchased from the following suppliers: gallic acid, *p*-coumaric acid, oleoside, luteolin, and luteolin-7-O-glucoside were from Sigma Aldrich (St. Louis, MO, USA); rutin and 3-hydroxytyrosol were from TCI (Tokyo, Japan); oleuropein and oleuropein aglycone were from Biosynth Carbosynth (Berkshire, UK). Stock solutions of each of the compounds were prepared in methanol at a concentration of ca. 5000 mg L^{-1} and were stored in the freezer at −20 °C until use. The standard solutions for calibration were prepared from these stock solutions.

Reagents used for spectrophotometric assays were: 2,4,6-tripyridyl-S-triazine (TPTZ) from Alfa Aesar (Kandel, Germany), Fe(III) chloride from Merck (Darmstadt, Germany), Trolox from Carbosynth (Berkshire, UK).

Chemicals for the preparation of extraction solvents and the mobile phase were as follows: choline chloride (ChCl), glycerol (Gly), and urea from Thermo Fisher (Kandel, Germany); lactic acid (Lac) from Acros Organics (Geel, Belgium); ethylene glycol (EG) from PanReac (Barcelona, Spain); ethanol (EtOH) from Honeywell, Riedel-de Haën™ (Offenbach, Germany); acetonitrile (ACN) from Fisher Scientific (Leicestershire, UK); dimethylsulfoxide (DMSO), formic acid (FA) 98–100% *w*/*w*, and hydrochloric acid (HCl) 32% *w*/*w* from Merck (Darmstradt, Germany). Ultrapure water was generated with a Milli-Q system (Merck Millipore, Bedford, MA, USA).

2.2. Samples and Sample Treatment

Leaves from various olive tree (*Olea europaea* L.) cultivars, such as Arbequina and Verdiella, were obtained during extra virgin olive oil production in November 2021 in Albelda (Huesca, Spain). The olive harvest was sifted to separate the leaves from the fruit, and the leaf samples were then collected. The fresh leaves were processed in the laboratory with a domestic grinder for 1 min and then sieved through a 2 mm mesh. The resulting powder was stored at −18 °C until being used.

The samples were extracted with various solvents and techniques to evaluate the effectiveness of each procedure in the recovery of phenolic compounds. Extracts were filtered with nylon syringe filters (13 mm, 0.45 µm) from Agilent Technologies (Waldbronn, Germany). All conditions were tested in duplicate to have information on the variability of the results and their statistical significance.

2.3. NADES Preparation

Several NADES compositions based on ChCl and various donors, such as Gly, urea, Lac, and EG (see Section 2.1), were studied to extract phenolic compounds. The acceptor and donor components were mixed in a molar ratio of 1:5 and heated with constant stirring at 60 °C in a water bath until a colorless and homogeneous liquid was obtained. Then, 30% water was added and mixed [26]. Also, conventional solvents like ethanol/water solutions and water were compared.

2.4. Extraction Procedures

2.4.1. Microwave-Assisted Extraction (MAE)

The Milestone Microwave Labstation system (Ethos E, Milestone, Shelton, CT, USA) was used for MAE with NADES. The procedure used one gram of fresh sample placed in a PTFE vessel with 20 mL of the corresponding NADES system. Stirring was set at 50% with a magnet inside each vessel and power at 500 W. The influence of solvent composition, temperature (80–180 °C), and time (5–15 min) on the phenolic recovery was assessed by a factorial experimental design. After extraction, samples were transferred to a Falcon tube, centrifuged for 15 min at 3500 rpm with the Labofuge 400 centrifuge, filtered with nylon syringe filters, and stored at 4 °C until analysis. Experiments were performed in duplicate.

2.4.2. Solid–Liquid Extraction (SLE)

SLE with NADES and conventional solvents relied on a previously optimized methodology [26] using an RCT basic hot plate stirrer with temperature controller (IKA®, Staufen, Germany). Briefly, 1 g of fresh sample was mixed with 20 mL of the corresponding solvent. The extraction was at 80 °C in a water bath with constant stirring using a magnet (450 rpm) with the extraction time varying from 5 min to 2 h. After each batch, the extracts were centrifuged for 15 min at 3500 rpm with a Labofuge 400 centrifuge (Heraeus, Hanau, Germany), filtered with nylon syringe filters (13 mm, 0.45 μm; Agilent Technologies, Waldbronn, Germany), and stored at 4 °C until analysis. Experiments were performed in duplicate.

2.5. Analysis Methods

2.5.1. Ferric Reducing Antioxidant Power Assay (FRAP)

The antioxidant activity of samples was estimated spectrophotometrically by the FRAP method adapted from the publication by Alcalde et al. [27] using an Agilent 8453 UV-Vis spectrophotometer (Santa Clara, CA, USA) with QS quartz glass high-performance cuvettes (10 mm optical path) from Hellma Analytics (Jena, Germany). The FRAP reagent consisted of 20 mmol L^{-1} FeCl3 solution, 10 mmol L^{-1} TPTZ solution (50 mmol L^{-1} HCl), and 50 mmol L^{-1} FA buffer solution, mixed in a 1:2:10 ($v/v/v$) proportion. A volume of 300 μL of FRAP was mixed with appropriate sample volumes and diluted with Milli-Q water to a final volume of 2.5 mL. The reaction was developed for 5 min, and the absorbance was measured at 595 nm using a reagent blank as the reference. Trolox standards in the concentration range of 0.2 to 10 mg L^{-1} were used for calibration. Antioxidant capacity was expressed as Trolox equivalent units per gram of fresh sample (mg TE g^{-1}). Analyses were performed in duplicate.

2.5.2. Liquid Chromatography with UV Detection (HPLC-UV)

HPLC-UV was used to determine the global content of phenolic compounds and quantify relevant compounds such as luteolin glucoside, luteolin, *p*-coumaric acid, or rutin using method conditions optimized elsewhere [26]. The instruments used consisted of an Agilent Series 1200 system (Agilent Technologies, Palo Alto, CA, USA) equipped with a quaternary pump, an automatic injection system, and a diode array detector (DAD). Phenolic compound separated under the reversed-phase mode with a Kinetex C18 column (100 mm × 4.6 mm × 2.6 μm, Phenomenex, Torrance, CA, USA) using 0.1% formic acid aqueous solution (A) and ACN (B) as the mobile phase components. The gradient program was as follows: 0 min, 3%B; 0–18 min, 30%B; 18–28 min, 65%B; 28–30 min, 90%B; 30–32.5 min, 90%B; 32.5–33 min, 3%B; 33–40 min, 3%B. The flow rate was 0.5 mL min^{-1} and the injection volume was 10 μL. Chromatograms were recorded at 280, 325, and 370 nm. The total polyphenol content (TPC) was estimated from the total peak area at 280 nm, in the time range between 5 and 30 min. TPC was expressed in terms of mg of gallic acid equivalent (GAE) per gram of fresh sample (mg GAE g^{-1}). Analogously, specific phenolics were determined from the corresponding peak area at suitable wavelengths of 280 nm for 3-hydroxytyrosol, oleoside, oleuropein, and oleuropein aglycone; 325 nm for *p*-coumaric acid; and 370 nm for luteolin, luteolin-7-O-glucoside, and rutin. Calibration curves were built for each analyte using the corresponding standards, in the concentration range of 0.5 to 20 mg L^{-1}.

2.5.3. Differential Pulse Voltammetry (DPV)

The redox features of the most significant compounds and the ethanolic or NADES extracts were evaluated electrochemically with a µAutolab system Type (III) instrument (EcoChemie, Utrech, Netherlands) attached to a 663VA Stand (Metrohm, Herisau, Switzerland), using a screen-printed carbon DRP-110 device (Dropsens, Oviedo, Spain) as the working, reference, and auxiliary electrode. DPV was used for measurements according to the conditions by Alcalde et al. [27]. Briefly, an appropriate volume of sample or standard was added to 10 mL of supporting electrolyte (0.1 M sodium acetate–acetic acid buffer at pH 5) in the electrochemical cell. Differential pulse voltammograms were recorded from −0.5 to +0.1 V. Other conditions were scan rate 0.25 mV s^{-1}, potential step 10 mV, pulse potential 100 mV, and pulse time 0.01 s. Data was acquired with the GPES 4.9 software.

2.5.4. Statistical Analysis

The significance of factors was evaluated by ANOVA and Students' *t*-test using Microsoft Excel 2016. The significance level was 0.05.

2.5.5. AGREE Score Calculation

The AGREEprep calculator [28], based on the principles of green chemistry, was used to assign Analytical Greenness Metric (AGREE) scores to the extraction procedures.

3. Results and Discussion

This paper evaluates the effectiveness of phenolic compound extraction from olive leaves, considered a waste matrix from the extra virgin oil production process. The extraction capacity of various green solvents was compared, including water, ethanol/water solutions, and NADES formulated with ChCl as acceptor and donor agents such as Gly, EG, urea, or Lac. These donors have demonstrated their efficiency in other similar cases [26].

The first preliminary study aimed to compare three green solvent systems using MAE (10 min at 80 °C) as the extraction technique, as follows: water, a 20:80 (*v*/*v*) ethanol/water solution, and ChCl/Gly. The FRAP assay reveals significant differences ($p < 0.001$) in antioxidant capacity among the three solvents (see orange bars in Figure 1), with the NADES demonstrating superior performance compared to the other solvents. This result agrees with previous studies by other authors who compared NADES with conventional organic solvents and water [2,22]. Water is less effective for extracting less polar compounds, with the antioxidant capacity of the aqueous extract being 65% lower than that of the NADES extract. However, water-based systems remain a viable option for industrial applications due to their environmental benefits and low cost. Adding 20% ethanol, a conventional green organic solvent, increases extraction efficiency by 1.5 times without considerably raising costs. In line with this, Jurmanović et al. [29] proposed a 20:80 (*v*/*v*) ethanol/water mixture for extracting polyphenols from olive pomace using MAE.

The TPC results determined by HPLC (see blue bars in Figure 1) led to similar conclusions, indicating that NADES was the most efficient option ($p < 0.001$). The concentration of individual compounds in the extracts (Figure S1) provided a more detailed picture of the behavior of the different extraction media. In particular, flavonoid glycosides (luteolin glucoside and rutin) and hydroxycinnamic acids (*p*-coumaric acid) follow a pattern similar to that set out above, but aglycones (e.g., luteolin) are hardly extracted by water or the hydroalcoholic mixture because of their lower polarity.

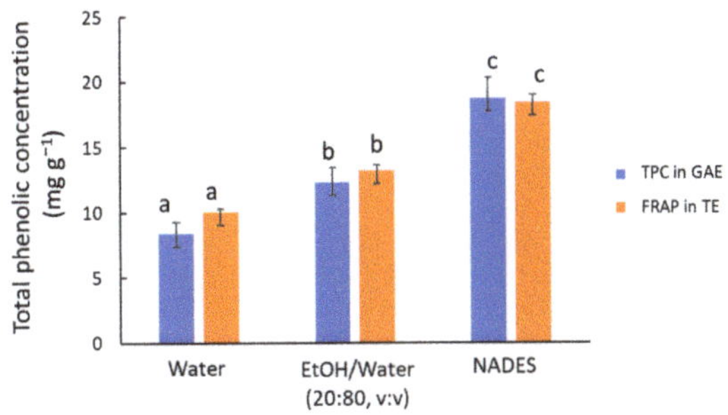

Figure 1. Recovery of phenolic compounds from olive leaves using water, ethanol/water mixture (20:80 v/v), and ChCl/Gly NADES (1:5 m/m, 30% water) using MAE as an extraction technique. Blue bars, TPC values by HPLC-UV in mg GAE g^{-1}; orange bars, antioxidant capacity values by FRAP in mg TE g^{-1}. Comparing TPC or TE values, different letters mean significant differences and vice versa.

The extraction performance of conventional solvents can be significantly improved when working at higher temperatures, reaching yields comparable to NADES. For instance, in the MAE assay for 10 min at temperatures ranging from 80 °C to 180 °C (Figure S2), ANOVA showed no significant differences in FRAP activity between ethanol/water (20:80 v/v) and ChCl/Gly. With water at 80 °C, the FRAP index was considerably lower than for the other systems; it increased progressively with temperature (up to 3.8 times above 160 °C, reaching ca. 85% of the antioxidant capacity achieved by the other solvents). In the literature, various authors proposed MAE extraction temperatures of 70–80 °C for olive leaves using NADES [2,18,22].

Based on their promising performance in similar studies [2,22], the extraction efficiency of four NADES, composed of ChCl with Gly, Lac, urea, or EG as donors, was compared using a 1:5 ChCl-to-donor ratio and 30% water. In addition to the MAE technique (10 min at 80 °C), conventional solid–liquid extraction (SLE) with mechanical stirring (2 h at 80 °C) was also considered. The outcomes of global antioxidant capacity by FRAP (Figure 2) indicate that Gly, EG, and urea provide high activities, ca. 15 to 20 mg g^{-1} TE, with urea possibly being the most effective donor. In contrast, Lac was found to be a less effective agent. The behavior of Gly and EG is similar, in line with the structural and functional similarity of these two molecules. Except for the ChCl/urea system, Figure 2 shows that MAE provides FRAP values for global antioxidant capacity that are ca. 20% higher than those of SLE. However, there are noticeable differences in process time (2 h for SLE vs. 10 min for MAE). It is important to note that shorter process times in SLE provided significantly lower extraction yields (Figure S3). Chanioti et al. [24,30] compared MAE, SLE, UAE, and high hydrostatic pressure extraction, and found that SLE provided the highest extraction efficiency for recovering phenolic compounds from olive pomace matrix. In that study, the authors selected ChCl/citric acid (1:2) with 20% of water or ChCl/caffeic acid (1:2), 60 °C, and 30 min of extraction time as the working conditions.

When examining the chromatographic results (Figure S4), the behaviors of luteolin glycoside and rutin under the different conditions are similar, with slightly higher values observed in the case of SLE and NADES containing urea. Luteolin results are more erratic, likely due to potential degradation processes. Overall, TPC is ca. 20 mg g^{-1} GAE in the extracts for both MAE and SLE, highlighting the potential of this matrix as a source of these phytochemicals.

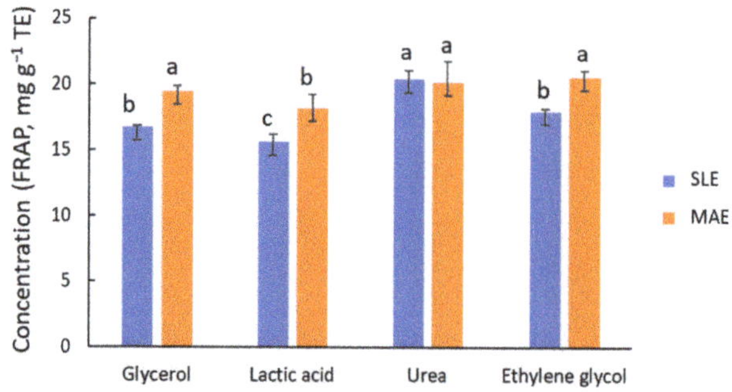

Figure 2. Influence of NADES composition on the recovery of phenolic compounds, estimated by FRAP, from olive leaves using MAE (orange bars) and SLE (blue bars) as extraction techniques. Different letters mean significant differences and vice versa.

The extraction performance of MAE was assessed with more detail by an experimental design of two factors at three levels: time (5, 10, and 15 min) and temperature (80, 100, 120 °C). NADES systems formulated with Gly and urea were chosen as representative cases. The trends are similar in both systems, and the antioxidant capacity of urea and Gly extracts is similar (on average ca. 5% higher for urea) (Figure 3). The significance of extraction time and temperature factors was evaluated by ANOVA using antioxidant activity and TPC as the data. The antioxidant capacity from FRAP (Figure 3a) showed that increasing temperature or time slightly improved the antioxidant capacity of the extracts for the ChCl/Gly system (temperature, p-value = 0.0031; time, p-value = 0.035), while the interaction was irrelevant. Similarly, these factors were also relevant for the urea-based NADES (temperature, p-value = 3.4×10^{-6}; time, p-value = 5.8×10^{-10}; interaction, p-value = 0.0015) The chromatographic results lead to similar conclusions on the influence of temperature and time on the TPC; the response increases with these factors, although the effect of temperature is quantitatively more notable. ChCl/urea outcomes (Figure S5) were ca. 20% higher than ChCl/Gly ones (Figure S6), with the differences becoming accentuated at elevated temperatures. However, chromatographic profiles denoted that these differences are not attributable to polyphenols but rather to degradation products associated with urea that interfered with the chromatograms; this process contributes to the overall response, though the antioxidant capacity of these degradation products is null. Therefore, urea-based NADES should be avoided under more rigorous conditions, e.g., high temperatures or prolonged processing times, as this decomposition also entails an unpleasant odor. Consequently, urea-based NADES are not recommended for applications in food, nutraceuticals, or cosmetics. Besides, when focusing on given polyphenols (luteolin glucoside, rutin, and luteolin), the extraction improves slightly with temperature or time (Figure S6), but the results with urea are not superior to those with glycerol.

In light of all the previous results, a final study was carried out to compare the options considered most relevant for various reasons. The selected conditions include MAE (80 °C, 10 min) with Gly and urea NADES, and SLE (80 °C, 2 h) with Gly NADES and an ethanol/water mixture. Figure 4 shows the profiling of the most quantitatively significant polyphenols, such as 3-hydroxytyrosol, luteolin-7-O-glucoside, oleuropein, luteolin-4-glucoside, oleuropein aglycone, and oleoside, which were found at concentrations above 1 mg g^{-1} in the sample, representing ca. 90% of the total polyphenols. On the whole, the MAE/NADES Gly system provides better recoveries for flavonoids; this finding is in accordance with the results of Gao et al. [31] dealing with the flavonoid extraction (rutin, isoquercetin, astragalin) from Mulberry leaves using MAE (ChCl/Gly (1:2) with 20% of water). When comparing the performance of SLE and MAE (Figure 4), MAE is generally superior. However, SLE with either ChCl/Gly NADES or the hydroalcoholic mixture (20% EtOH) is more suitable for hydroxytyrosol derivatives (e.g., 3-hydroxytyrosol and oleuropein). This outcome is in line with Boli et al. [18], who showed that ethanol is an effective solvent for oleuropein recovery from olive leaves by SLE, particularly at high temperatures (70 °C). For compounds occurring at concentrations below 0.5 mg g^{-1}, such as rutin, coumaric acid, and luteolin, the general behavior seems more complex, and their extraction yield may depend on unknown affinity issues. Overall, since the quantitative differences between the options are not remarkable, it suggests that all approaches are viable. The choice between them will depend on the specific applications of the extracts and the scaling requirements.

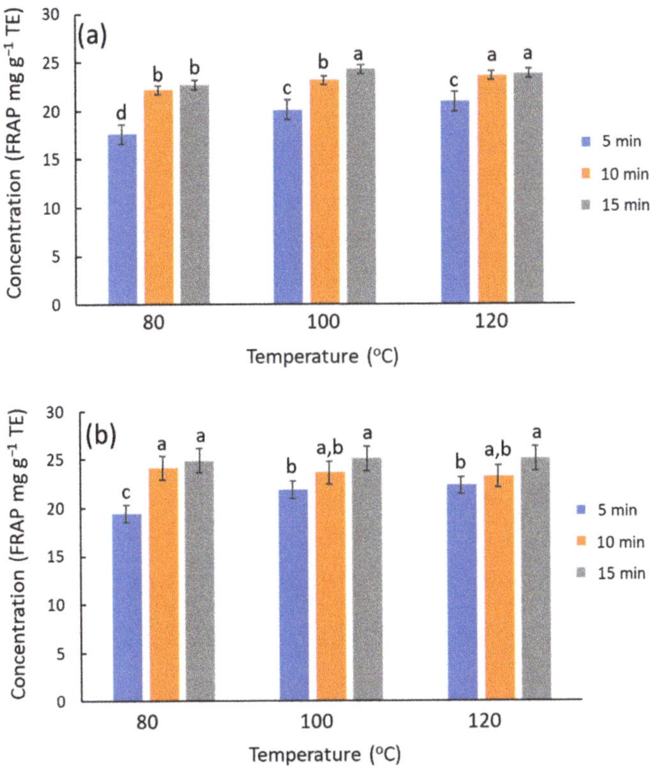

Figure 3. Influence of time and temperature on the recovery of phenolic compounds from olive leaves using MAE, estimated by FRAP. (**a**) ChCl/Gly, (**b**) ChCl/urea. For each NADES system, different letters mean significant differences and vice versa.

Electrochemical studies were also performed to estimate the antioxidant quality of the extracts. On the one hand, the intensity of the DPV signals is a quantitative measurement of the overall antioxidant rate, i.e., providing analogous information to the FRAP or TPC index. On the other hand, the reduction potential provides a qualitative idea of the potency to neutralize or combat different types of oxidizing compounds in the medium. Hence, DPV peaks at lower potentials correspond to more active reducing extracts/compounds as excellent protectors against oxidation processes. Figure 5 depicts a representative voltammogram of an ethanol/water extract. As mentioned, the peak at 0.2 V corresponds to the most electroactive compounds, such as hydroxytyrosol, luteolin, and their derivatives, which contain dihydroxyphenyl moieties (hydroxyphenyls with OH- groups in ortho- or para-positions). The band at 0.6 V is associated with less active phenolic groups, including moieties with simple hydroxyphenyl or dihydroxyphenyls with OH- groups in meta. DPV analysis showed the Gly-based extracts are 20% more active than the urea ones. The effect of the extraction conditions on the antioxidant quality was also confirmed, observing that when working at elevated temperatures, the most labile compounds (peak at ca. 0.2 V, corresponding to dihydroxyphenyl moieties) decreased their intensity with respect to extracts under milder conditions; conversely, the least labile ones (peak at ca. 0.6 V) remained practically constant even under strong experimental conditions.

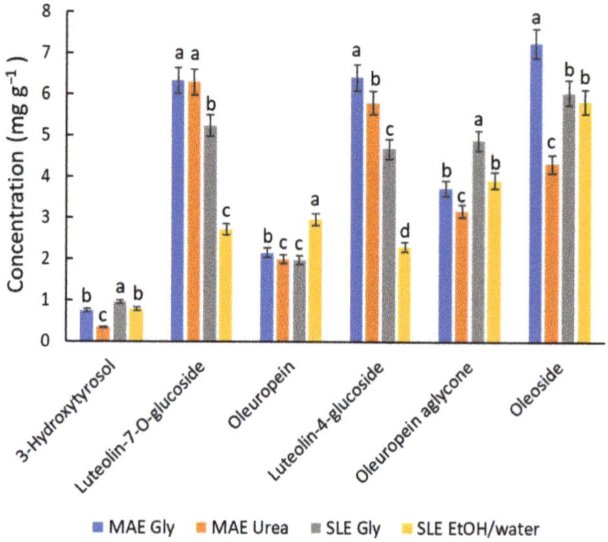

Figure 4. Recovery of the main phenolic compounds, expressed in mg g^{-1}, determined by HPLC depending on the solvents and extraction techniques under optimal conditions. For each compound, different letters mean significant differences and vice versa.

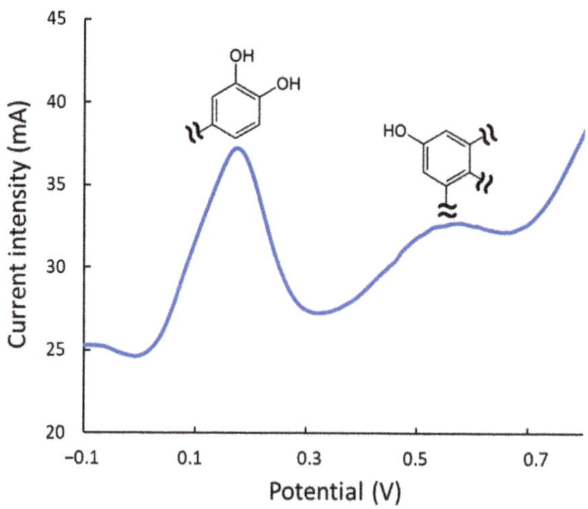

Figure 5. DPV of an ethanolic extract (ethanol/water 20:80, v/v) of olive leaves with the peaks of the two main types of antioxidant moieties.

In a general comparison of our results with other recent studies published in the literature, Musolino et al. also reported that oleuropein and luteolin-7-O-glucoside are the two main polyphenolic compounds in olive leaf extract [32]. Beyond the exhaustive characterization of olive leaf extracts obtained with conventional and eutectic solvents by Mir-Cerdà et al. [26], other authors explored various types of waste generated in the production of olive oil, including leaves, to evaluate their possibilities as a source of phenolic and terpenic compounds [33].

As of now, NADES have scarcely been used to recover phenolics from this waste type [18,22,24,26,34]. Beyond the above-commented examples, Mitar and Kardum relied on NADES containing aluminum oxide nanoparticles to facilitate the transfer of polyphenols from olive leaves [34]. Although extraction with NADES is fast and efficient, solvent removal and further extract purification may be an unsolved challenge. In this scenario, resin-based procedures are sometimes applied to isolate the bioactive compounds from the solvent and obtain enriched by-products [35]. Conversely, most authors preferred traditional solvents, such as methanol, ethanol, and water, or their mixtures [19–21,36], also obtaining good performances regarding the recovery of polyphenols or overall antioxidant capacity; additionally, EtOH/water mixtures also guaranteed the greenness of the method [36]. For instance, another study compared the extraction ability of MeOH and water, demonstrating that results were better when working with hydroalcoholic mixtures, especially when trying to extract some less polar compounds such as oleuropein [37].

Regarding extraction techniques, traditional options, including maceration or decoction, have an evident green nature, although they are methodologically slow [38]. In this context, homogenization-assisted extraction, UAE, and MAE seem to be very popular among the scientific community; anyway, other innovative options, such as pulsed electric fields and high-pressure extraction, also demonstrate high performance, becoming highly competitive in terms of cost and scaling possibilities [26,39,40].

Finally, a greenness assessment of the different approaches was conducted to evaluate the sustainability of the extraction process. The AGREEprep calculator [28], which assigns a so-called normalized AGREE score based on the principles of green chemistry, was used. The score ranges from 0 to 1, with values closer to 1 indicating a greener process. Both MAE and SLE techniques were evaluated using water, 20% ethanol, and ChCl/Gly as the solvents. For all solvents tested, the AGREE score for MAE was consistently higher (0.74, 0.71, 0.74) compared to the corresponding SLE system (0.63, 0.60, 0.63) (see Figure 6). The main reason for the difference in scores between the two extraction techniques is the sample throughput, with MAE being more versatile than SLE. The lower scores observed for the ethanol/water system, whether using MAE or SLE, are mostly linked to operator safety concerns. This assessment was performed on a laboratory scale, and factors like sample throughput may differ when considering upscaling.

In summary, ChCl/Gly NADES is the best option considering antioxidant activity and extraction of individual bioactive compounds. While MAE (80 °C, 10 min) provides slightly better recovery than SLE (80 °C, 2 h), along with a higher AGREE score and shorter processing time, SLE remains interesting for large-scale applications due to its simplicity and ease of scaling.

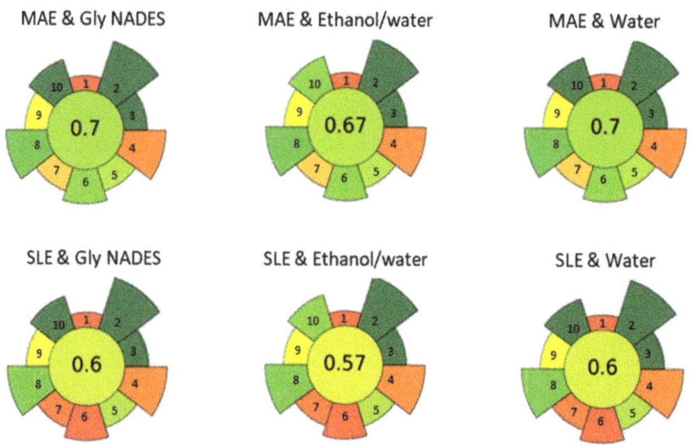

Figure 6. Evaluation of the greenness and sustainability of the sample treatment using the AGREE approach.

4. Conclusions

This study compared various green solvents, including NADES, ethanol/water mixtures, and pure water, for recovering antioxidant compounds from olive leaf waste generated during extra virgin oil production. The performance of SLE was evaluated via mechanical stirring and MAE, analyzing the extracts using FRAP antioxidant capacity, HPLC, and voltammetry. The optimization of working conditions yielded comparable results for most solvents in terms of polyphenolic content and antioxidant capacity. This equivalence is probably due to achieving near-quantitative extraction, limiting further improvements. It should be noted, however, that when less efficient solvents are used (in the case of water or the ethanol/water mixture), the working conditions to achieve similar performances become more extreme, requiring longer times or higher temperatures.

The MAE/NADES combination demonstrated high efficiency with short process times (10 min) and reasonable temperatures (80 °C). Among the NADES tested, choline chloride-glycerol was the most effective, matching the urea-based NADES, without degradation or unpleasant odors. Nonetheless, while MAE/NADES being promising, scalability considerations may impact its cost-effectiveness. SLE using water or ethanol/water mixtures provided excellent results at 80 °C but with longer processing times (120 min). Combining SLE with an ethanol/water mixture appears practical for large-scale operations, but its AGREEprep score is lower compared to MAE options.

For industrial applications, balancing efficiency, scalability, cost, and sustainability is crucial. Nonetheless, developing sustainable and efficient processes for valorizing olive leaf waste will require pilot-scale studies to validate laboratory results, as well as investigating solvent recycling to enhance sustainability.

Supplementary Materials: The following supporting information can be downloaded at: https://www.mdpi.com/article/10.3390/antiox14020136/s1, Figure S1. Recovery of individual phenolic compounds from olive leaves using water, 20/80 ethanol/water mixture (*v*/*v*) and ChCl/Gly NADES (1:5 m:m, 30% water) using MAE as extraction technique, analyzed by HPLC-UV (mg g^{-1}), showing (a) luteolin glucoside, (b) rutin, (c) luteolin, (d) *p*-coumaric acid. Figure S2. MAE study of temperature (80–180 °C) at 10 min of extraction time by FRAP (mg TE g^{-1}). Figure S3. SLE study of time (5–120 min) at 80 °C of extraction temperature. Analysis of (a) TPC by HPLC-UV (mg GAE g^{-1}), (b) TPC by FRAP (mg TE g^{-1}), (c) luteolin glucoside (mg g^{-1}), (d) rutin (mg g^{-1}), (e) luteolin (mg g^{-1}). Figure S4. Influence of NADES composition on the recovery of phenolic compounds, estimated by HPLC-UV, from olive leaves using MAE (blue bars) and SLE (orange bars) as extraction techniques, showing

(a) TPC (mg GAE g^{-1}), (b) luteolin glucoside (mg g^{-1}), (c) rutin (mg g^{-1}), (d) luteolin (mg g^{-1}). Figure S5. Influence of extraction temperature and time on the recovery of phenolic compounds, estimated by HPLC-UV, from olive leaves using MAE as extraction technique and Gly-based NADES, showing (a) TPC (mg GAE g^{-1}), (b) luteolin glucoside (mg g^{-1}), (c) rutin (mg g^{-1}), (d) luteolin (mg g^{-1}). Figure S6. Influence of extraction temperature and time on the recovery of phenolic compounds, estimated by HPLC-UV, from olive leaves using MAE as extraction technique and urea-based NADES, showing (a) TPC (mg GAE g^{-1}), (b) luteolin glucoside (mg g^{-1}), (c) rutin (mg g^{-1}), (d) luteolin (mg g^{-1}).

Author Contributions: Conceptualization, M.G., S.S. and J.S.; methodology, P.T.-Q. and A.M.-C.; formal analysis, P.T.-Q. and J.S.; investigation, P.T.-Q. and A.M.-C.; writing—original draft preparation, P.T.-Q. and J.S.; writing—review and editing, P.T.-Q., M.G., S.S. and J.S.; supervision, M.G., S.S. and J.S.; funding acquisition, M.G. and J.S. All authors have read and agreed to the published version of the manuscript.

Funding: This research was supported by the project PID2023-147160OB-C22, financed by the Agencia Estatal de Investigación (Ministerio de Ciencia, Innovación y Universidades, España). Paulina Tapia-Quirós thanks to her Margarita Salas postdoctoral fellowship (agreement CG/2022/03/27, of 5 April 2022) from the Ministerio de Universidades (MIU), funded by the European Union-Next Generation EU. We also thank INSA-UB Maria de Maeztu Unit of Excellence (Grant CEX2021-001234-M) funded by MICIN/AEI/FEDER, UE.

Institutional Review Board Statement: Not applicable.

Informed Consent Statement: Not applicable.

Data Availability Statement: All the data is in the paper or in the Supplementary Material.

Conflicts of Interest: The authors declare no conflicts of interest.

References

1. Tapia-Quirós, P.; Granados, M.; Sentellas, S.; Saurina, J. Microwave-Assisted Extraction with Natural Deep Eutectic Solvents for Polyphenol Recovery from Agrifood Waste: Mature for Scaling-Up? *Sci. Total Environ.* **2024**, *912*, 168716. [PubMed]
2. Bonacci, S.; Di Gioia, M.L.; Costanzo, P.; Maiuolo, L.; Tallarico, S.; Nardi, M. Natural Deep Eutectic Solvent as Extraction Media for the Main Phenolic Compounds from Olive Oil Processing Wastes. *Antioxidants* **2020**, *9*, 513. [CrossRef] [PubMed]
3. Silva, S.; Gomes, L.; Leitão, F.; Coelho, A.V.; Boas, L.V. Phenolic Compounds and Antioxidant Activity of *Olea europaea* L. Fruits and Leaves. *Food Sci. Technol. Int.* **2006**, *12*, 385–396.
4. Kountouri, A.M.; Mylona, A.; Kaliora, A.C.; Andrikopoulos, N.K. Bioavailability of the Phenolic Compounds of the Fruits (Drupes) of *Olea europaea* (Olives): Impact on Plasma Antioxidant Status in Humans. *Phytomedicine* **2007**, *14*, 659–667. [CrossRef]
5. Owen, R.W.; Mier, W.; Giacosa, A.; Hull, W.E.; Spiegelhalder, B.; Bartsch, H. Identification of Lignans as Major Components in the Phenolic Fraction of Olive Oil. *Clin. Chem.* **2000**, *46*, 976–988.
6. Montenegro-Landívar, M.F.; Tapia-Quirós, P.; Vecino, X.; Reig, M.; Valderrama, C.; Granados, M.; Cortina, J.L.; Saurina, J. Polyphenols and Their Potential Role to Fight Viral Diseases: An Overview. *Sci. Total Environ.* **2021**, *801*, 149719. [PubMed]
7. Shahidi, F.; Ambigaipalan, P. Phenolics and Polyphenolics in Foods, Beverages and Spices: Antioxidant Activity and Health Effects—A Review. *J. Funct. Foods* **2015**, *18*, 820–897.
8. Zhang, Y.; Cai, P.; Cheng, G.; Zhang, Y. A Brief Review of Phenolic Compounds Identified from Plants: Their Extraction, Analysis, and Biological Activity. *Nat. Prod. Commun.* **2022**, *17*, 1934578X2110697.
9. Rana, A.; Samtiya, M.; Dhewa, T.; Mishra, V.; Aluko, R.E. Health Benefits of Polyphenols: A Concise Review. *J. Food Biochem.* **2022**, *46*, e14264. [CrossRef]
10. Cassano, A.; De Luca, G.; Conidi, C.; Drioli, E. Effect of Polyphenols-Membrane Interactions on the Performance of Membrane-Based Processes. A Review. *Coord. Chem. Rev.* **2017**, *351*, 45–75.
11. Romani, A.; Ieri, F.; Urciuoli, S.; Noce, A.; Marrone, G.; Nediani, C.; Bernini, R. Health Effects of Phenolic Compounds Found in Extra-Virgin Olive Oil, By-Products, and Leaf of *Olea europaea* L. *Nutrients* **2019**, *11*, 1776. [CrossRef] [PubMed]
12. Talhaoui, N.; Taamalli, A.; Gómez-Caravaca, A.M.; Fernández-Gutiérrez, A.; Segura-Carretero, A. Phenolic Compounds in Olive Leaves: Analytical Determination, Biotic and Abiotic Influence, and Health Benefits. *Food Res. Int.* **2015**, *77*, 92–108.

13. Rufino-Palomares, E.E.; Pérez-Jiménez, A.; García-Salguero, L.; Mokhtari, K.; Reyes-Zurita, F.J.; Peragón-Sánchez, J.; Lupiáñez, J.A. Nutraceutical Role of Polyphenols and Triterpenes Present in the Extracts of Fruits and Leaves of *Olea europaea* as Antioxidants, Anti-Infectives and Anticancer Agents on Healthy Growth. *Molecules* **2022**, *27*, 2341. [CrossRef] [PubMed]
14. Martín-García, B.; De Montijo-Prieto, S.; Jiménez-Valera, M.; Carrasco-Pancorbo, A.; Ruiz-Bravo, A.; Verardo, V.; Gómez-Caravaca, A.M. Comparative Extraction of Phenolic Compounds from Olive Leaves Using a Sonotrode and an Ultrasonic Bath and the Evaluation of Both Antioxidant and Antimicrobial Activity. *Antioxidants* **2022**, *11*, 558. [CrossRef] [PubMed]
15. Araújo, M.; Pimentel, F.B.; Alves, R.C.; Oliveira, M.B.P.P. Phenolic Compounds from Olive Mill Wastes: Health Effects, Analytical Approach and Application as Food Antioxidants. *Trends Food Sci. Technol.* **2015**, *45*, 200–211.
16. Dermeche, S.; Nadour, M.; Larroche, C.; Moulti-Mati, F.; Michaud, P. Olive Mill Wastes: Biochemical Characterizations and Valorization Strategies. *Process Biochem.* **2013**, *48*, 1532–1552.
17. Gouvinhas, I.; Machado, N.; Sobreira, C.; Domínguez-Perles, R.; Gomes, S.; Rosa, E.; Barros, A.I.R.N.A. Critical Review on the Significance of Olive Phytochemicals in Plant Physiology and Human Health. *Molecules* **2017**, *22*, 1986. [CrossRef]
18. Boli, E.; Prinos, N.; Louli, V.; Pappa, G.; Stamatis, H.; Magoulas, K.; Voutsas, E. Recovery of Bioactive Extracts from Olive Leaves Using Conventional and Microwave-Assisted Extraction with Classical and Deep Eutectic Solvents. *Separations* **2022**, *9*, 255. [CrossRef]
19. Aissa, I.; Kharrat, N.; Aloui, F.; Sellami, M.; Bouaziz, M.; Gargouri, Y. Valorization of Antioxidants Extracted from Olive Mill Wastewater. *Biotechnol. Appl. Biochem.* **2017**, *64*, 579–589.
20. Panzella, L.; Moccia, F.; Nasti, R.; Marzorati, S.; Verotta, L.; Napolitano, A. Bioactive Phenolic Compounds From Agri-Food Wastes: An Update on Green and Sustainable Extraction Methodologies. *Front. Nutr.* **2020**, *7*, 60.
21. Elkacmi, R.; Kamil, N.; Bennajah, M. Separation and Purification of High Purity Products from Three Different Olive Mill Wastewater Samples. *J. Environ. Chem. Eng.* **2017**, *5*, 829–837. [CrossRef]
22. Alañón, M.E.; Ivanović, M.; Gómez-Caravaca, A.M.; Arráez-Román, D.; Segura-Carretero, A. Choline Chloride Derivative-Based Deep Eutectic Liquids as Novel Green Alternative Solvents for Extraction of Phenolic Compounds from Olive Leaf. *Arab. J. Chem.* **2020**, *13*, 1685–1701. [CrossRef]
23. Isci, A.; Kaltschmitt, M. Recovery and Recycling of Deep Eutectic Solvents in Biomass Conversions: A Review. *Biomass Convers. Biorefin.* **2022**, *12*, 197–226. [CrossRef]
24. Chanioti, S.; Tzia, C. Extraction of Phenolic Compounds from Olive Pomace by Using Natural Deep Eutectic Solvents and Innovative Extraction Techniques. *Innov. Food Sci. Emerg. Technol.* **2018**, *48*, 228–239. [CrossRef]
25. Sridhar, A.; Ponnuchamy, M.; Kumar, P.S.; Kapoor, A.; Vo, D.V.N.; Prabhakar, S. Techniques and Modeling of Polyphenol Extraction from Food: A Review. *Env. Chem. Lett.* **2021**, *19*, 3409–3443. [CrossRef]
26. Mir-Cerdà, A.; Granados, M.; Saurina, J.; Sentellas, S. Green Extraction of Antioxidant Compounds from Olive Tree Leaves Based on Natural Deep Eutectic Solvents. *Antioxidants* **2023**, *12*, 995. [CrossRef] [PubMed]
27. Alcalde, B.; Granados, M.; Saurina, J. Exploring the Antioxidant Features of Polyphenols by Spectroscopic and Electrochemical Methods. *Antioxidants* **2019**, *8*, 523. [CrossRef]
28. Wojnowski, W.; Tobiszewski, M.; Pena-Pereira, F.; Psillakis, E. AGREEprep—Analytical GREEnness Metric for sample preparation. *Trend Anal. Chem.* **2022**, *149*, 116553. [CrossRef]
29. Jurmanović, S.; Jug, M.; Safner, T.; Radić, K.; Domijan, A.M.; Pedisić, S.; Šimić, S.; Jablan, J.; Čepo, D.V. Utilization of Olive Pomace as a Source of Polyphenols: Optimization of Microwave-Assisted Extraction and Characterization of Spray-Dried Extract. *J. Food Nutr. Res.* **2019**, *58*, 51–62.
30. Chanioti, S.; Katsouli, M.; Tzia, C. Novel Processes for the Extraction of Phenolic Compounds from Olive Pomace and Their Protection by Encapsulation. *Molecules* **2021**, *26*, 1781. [CrossRef] [PubMed]
31. Gao, M.Z.; Cui, Q.; Wang, L.T.; Meng, Y.; Yu, L.; Li, Y.Y.; Fu, Y.J. A Green and Integrated Strategy for Enhanced Phenolic Compounds Extraction from Mulberry (*Morus alba* L.) Leaves by Deep Eutectic Solvent. *Microchem. J.* **2020**, *154*, 104598. [CrossRef]
32. Musolino, V.; Macrì, R.; Cardamone, A.; Serra, M.; Coppoletta, A.R.; Tucci, L.; Maiuolo, J.; Lupia, C.; Scarano, F.; Carresi, C. Nocellara Del Belice (*Olea europaea* L. Cultivar): Leaf Extract Concentrated in Phenolic Compounds and Its Anti-Inflammatory and Radical Scavenging Activity. *Plants* **2023**, *12*, 27. [CrossRef] [PubMed]
33. López-Salas, L.; Díaz-Moreno, J.; Ciulu, M.; Borrás-Linares, I.; Quirantes-Piné, R.; Lozano-Sánchez, J. Monitoring the Phenolic and Terpenic Profile of Olives, Olive Oils and By-Products throughout the Production Process. *Foods* **2024**, *12*, 1555. [CrossRef]
34. Mitar, A.; Kardum, J.P. Intensification of Mass Transfer in the Extraction Process with a Nanofluid Prepared in a Natural Deep Eutectic Solvent. *Chem. Eng. Technol.* **2020**, *43*, 2286–2294. [CrossRef]
35. Mir-Cerdà, A.; Ramos-Calero, V.; Beltrán, J.L.; Cortina, J.L.; Saurina, J.; Sentellas, S.; Granados, M. Towards a sustainable recovery of polyphenols from agrifood waste: Performance of polymeric sorbents with natural deep eutectic solvent. *LWT Food Sci. Technol.* **2024**, *207*, 116632. [CrossRef]

36. Ronca, C.L.; Marques, S.S.; Ritieni, A.; Giménez-Martínez, R.; Barreiros, L.; Segundo, M.A. Olive Oil Waste as a Source of Functional Food Ingredients: Assessing Polyphenolic Content and Antioxidant Activity in Olive Leaves. *Foods* **2024**, *13*, 189. [CrossRef] [PubMed]
37. Hernández-Fernández, A.; Garrido, Y.; Iniesta-López, E.; de los Ríos, A.P.; Quesada-Medina, J.; Hernández-Fernández, F.J. Recovering Polyphenols in Aqueous Solutions from Olive Mill Wastewater and Olive Leaf for Biological Applications. *Processes* **2023**, *11*, 2668. [CrossRef]
38. Ezzarrouqy, K.; Sbahi, S.; Hejjaj, A.; Idlimam, A.; Mandi, L. A green extraction method of phenolic compounds from olive leaves (*Olea europaea* L.): Evaluation and prediction using multiple linear regression. *Int. J. Environ. Sci. Technol.* **2024**, *21*, 3761–3774.
39. Marquez, K.; Marquez, N.; Avila, F.; Cruz, N.; Burgos-Edwards, A.; Pardo, X.; Carrasco, B. Oleuropein Enriched Extract from Olive Mill Leaves by Homogenizer-Assisted Extraction and Its Antioxidant and Antiglycating Activities. *Front. Nutr.* **2022**, *9*, 895070. [CrossRef]
40. Tsevdou, M.; Ntzimani, A.; Katsouli, M.; Dimopoulos, G.; Tsimogiannis, D.; Taoukis, P. High Pressure Processing on the Extraction of Antioxidants from Olive Pomace. *Molecules* **2024**, *29*, 2303. [CrossRef] [PubMed]

antioxidants

MDPI

Article

Optimization of Conventional and Ultrasound-Assisted Extraction to Maximize Recovery of Total Phenolic Content and In Vitro Antioxidant Activity from *Crataegus almaatensis* Leaves

Zhanar Nabiyeva [1,2], Akerke Kulaipbekova [2,*], Serena Carpentieri [3,*], Yuliya Pronina [2], Abdyssemat Samadun [1], Elmira Assembayeva [2] and Giovanna Ferrari [3,4]

1 Research Institute of Food Safety, Almaty Technological University, Almaty 050012, Kazakhstan; atu_nabiyeva@mail.ru (Z.N.); abdu.93_93@mail.ru (A.S.)
2 Department of Food Technology, Almaty Technological University, Almaty 050012, Kazakhstan; medvezhonok_87@inbox.ru (Y.P.); elmiraasembaeva@mail.ru (E.A.)
3 Department of Industrial Engineering, University of Salerno, 84084 Fisciano, Italy; gferrari@unisa.it
4 ProdAl Scarl, University of Salerno, 84084 Fisciano, Italy
* Correspondence: a.kulaipbekova@atu.edu.kz (A.K.); scarpentieri@unisa.it (S.C.)

Abstract

Background: *Crataegus almaatensis*, an endemic hawthorn species from Kazakhstan, is known for its rich content of phenolic compounds and flavonoids with significant pharmacological potential. This study aimed to optimize and compare conventional solid–liquid extraction (SLE) and ultrasound-assisted extraction (UAE) processes for maximizing the extractability of bioactive compounds from hawthorn leaves powder. Methods: The effects of temperature, extraction time, ethanol concentration, and solid-to-liquid ratio (or ultrasound power in the case of UAE) on total phenolic content (TPC), total flavonoid content (TFC), and antioxidant activity (FRAP, DPPH, and ABTS assays) were systematically evaluated. Results: The UAE method yielded higher concentrations of TPC and TFC, with up to 16% improvement in TPC and reduced ethanol usage (40% (v/v)) compared to SLE (75% (v/v)), demonstrating its efficiency and sustainability. Optimal extraction conditions were identified as 70 °C, 75% ethanol, 34 min, and an S/L ratio of 0.05 g/mL for SLE, 70 °C, 40% ethanol, 44 min, and 100 W US power for UAE. High-resolution HPLC-DAD and LC-Q/TOF-MS analyses confirmed the presence of key phenolic acids and flavonoid glycosides, including chlorogenic acid and apigenin-8-C-glucoside-2′-rhamnoside as the most abundant compounds identified. Conclusions: These findings validate UAE as an innovative, eco-friendly method for extracting bioactive compounds from hawthorn leaves and highlight its potential for developing natural antioxidants for pharmaceutical and nutraceutical applications.

Keywords: hawthorn leaves; extraction; ultrasound; bioactive compounds; optimization; HPLC-DAD; LC-Q/TOF-MS

1. Introduction

Hawthorn (*Crataegus* spp.) is a flowering plant in the rose family (Rosaceae), which includes various species native to temperate latitudes of the Northern Hemisphere, including Europe, Asia, and North America.

The main morphological characteristics of hawthorn are its shrubby or tree-like form, resistance to various climatic conditions, and ability to grow on a wide range of soils, including stony and rubbly substrates [1]. In South Kazakhstan, particularly in the Zailiyskiy

Alatau, *Crataegus altaica*, *Crataegus almaatensis*, and *Crataegus ambigua* species are found growing in mountain areas, gorges, and river valleys, preferring well-drained soils [2]. Hawthorn berries are the main edible part of hawthorn, traditionally used in foods (juice, jams, teas, supplements). The leaves are byproducts derived from hawthorn juice and jam production processes, often harvested alongside the berries and subsequently used in herbal infusions and teas, extracts for supplements (cardiovascular support), flavorings, or functional ingredients. Biologically active compounds, such as flavonoids (e.g., hyperoside, vitexin), oligomeric procyanidins, and phenolic acids, contained in the fruits, leaves, and flowers of hawthorn have antioxidant properties, contributing to the strengthening of blood vessels, improving blood circulation, and maintaining cardiovascular system health. *Crataegus almaatensis* has been previously characterized by determining its chemical composition, biological activity, and adaptive properties. Several studies [3–5] confirmed the presence of many biologically active components in *Crataegus almaatensis* leaves, which can be used for the prevention and treatment of various diseases, including liver diseases. Phytochemical studies conducted by Soares et al. (2019) [2] demonstrated that *Crataegus almaatensis* leaf extracts contain key flavonoids, including hyperoside, quercitrin, and afzelin. Pharmacological studies showed that the aqueous extract (APCa) had vasodilatory activity, whereas the ethanolic extract (CECa) promoted an increase in vascular tone. In addition, APCa showed marked anti-inflammatory properties. Studies conducted by Bekbolatova et al. (2018) [6] provided a detailed analysis of the phenolic composition and antioxidant potential of different parts of *Crataegus almaatensis*. Compared with the European species *Crataegus oxyacantha* L., *Crataegus almaatensis* revealed the presence of 22 phenolic compounds, including flavonoids and phenolic acids. Among them, hyperoside is one of the major flavonoids, while chlorogenic acid is the main phenolic compound in leaves and flowers. Moreover, leaf extracts showed higher antioxidant activity compared to that detected in the fruit extracts. Therefore, this species can be considered to be a promising source for the extraction and isolation of bioactive substances for potential pharmacological and food applications. Conventional extraction processes of bioactive compounds from plant materials face significant limitations due to the presence of plant cell walls and membranes, which act as a physical barrier, impeding the efficient recovery of targeted intracellular compounds and often requiring longer extraction times, higher solvent volumes, and temperatures. This challenge has prompted researchers to investigate the application of various cell disruption techniques aimed at weakening or breaking the plant cell envelope. Ultrasound-assisted extraction (UAE) [7,8], extraction with supercritical CO_2 [9] microwave-assisted extraction [10] were recently investigated to recover antioxidant compounds from hawthorn leaves and fruits. Such approaches facilitate the release of intracellular compounds, thereby improving extraction efficiency while reducing solvent usage and processing time. Among the cell disruption technologies, ultrasound (US) has emerged as a promising technology for enhancing the recovery of intracellular compounds from plant matrices. The application of ultrasonic waves generates cavitation effects that disrupt plant cell walls, thereby accelerating the release of valuable phytochemicals into the extraction medium. UAE has demonstrated significant potential in enhancing the recovery of phenolic compounds from hawthorn leaves [11–14], offering a green and efficient alternative to traditional extraction methods. However, the effective application of UAE requires a well-defined optimization process to fully realize its advantages over conventional solid–liquid extraction (SLE) processes. Parameters such as US power, extraction time, solvent-to-solid ratio, temperature, and solvent composition must be finely tuned to achieve maximum extractability and quality of the extracted compounds. Despite the growing interest in hawthorn leaves, a limited number of studies focused on the optimization of UAE parameters to recover bioactive compounds from different species of Crataegus leaves, such as *Crataegus pinnatifida* [11,12] and *Crataegus monogyna* [13]. Moreover, not all relevant factors were optimized comprehensively, and a direct comparison between UAE and conventional extraction methods, which would be essential to assess the benefits of using US technology, was not reported [11,13,14].

The aim of this study was to optimize SLE and UAE processing conditions, investigating the effect of temperature, extraction time, ethanol concentration, and solid-to-liquid ratio (or ultrasound power in the case of UAE) on the phenolic content, flavonoid content, and antioxidant activity of the extracts from *Crataegus almaatensis* leaves. The phenolic composition of the obtained extracts was assessed using HPLC-DAD and LC-Q/TOF-MS analyses.

2. Materials and Methods

2.1. *Raw Materials and Chemicals*

As part of the IRN AP23489874 project, "Development of gentle technology for production of natural extracts on processing of local plant raw materials for enrichment of sweets" from 5 August 2024, to 24 August 2024, aerial parts of the hawthorn (*Crataegus almaatensis*)—including leaves, stems, and berries—were collected from the mountainous area of Ile-Alatau National Park. To determine the optimal extraction conditions for bioactive compounds, experiments were conducted on ultrasonic extraction from hawthorn (*Crataegus almaatensis*) leaves. Before extraction, hawthorn leaves were pre-sorted, debris was removed, and damaged or diseased leaves were discarded. The leaves were then dried using the shade-drying method in well-ventilated rooms on silk screens with a layer thickness of 2–3 cm, away from direct sunlight, at a temperature of 25–30 °C for 12–15 days until the moisture content reached approximately 8% (AACC 44-15.02 method). During the first two days, the leaves were turned over 1–2 times daily; thereafter, they were turned every other day to ensure uniform drying.

Once dried, the leaves were sorted and ground using a GM 200 knife mill (RETSCH, Haan, Germany) to a particle size of 0.5–1 mm to increase the contact surface area with the extractant. The crushed raw material was then packed into clean, airtight glass containers and stored in well-ventilated rooms at temperatures between 20 and 25 °C, away from sunlight. Under these storage conditions, the shelf life of the crushed raw material without loss of quality is up to 7 months.

Ethanol $\geq$ 99.8%, methanol $\geq$ 99.9%, Folin–Ciocalteu, sodium carbonate $\geq$ 99.5%, aluminum chloride hexahydrate 99%, sodium hydroxide $\geq$ 97%, TPTZ (2,4,6-Tris(2-pyridyl)-s-triazine) $\geq$ 98%, sodium acetate $\geq$ 99%, acetic acid $\geq$ 99.99%, iron(III) chloride hexahydrate $\geq$ 98%, DPPH (1,1-diphenyl-2-picrylhydrazyl) $\geq$ 95%, gallic acid $\geq$ 98%, quercetin $\geq$ 99%, ascorbic acid $\geq$ 99%, chlorogenic acid $\geq$ 95%, apigenin-8-C-glucoside-2′-rhamnoside $\geq$ 99%, epicatechin $\geq$ 99%, and catechin $\geq$ 99% were purchased from Sigma Aldrich (Steinheim, Germany).

2.2. *Ultrasound (US) Equipment*

US treatments of hawthorn leaves powder were conducted using a US processor UP 400S (Hielscher GmbH, Chamerau, Germany) with a maximum power of 400 W, a constant frequency of 24 kHz, a Tip H3 (titanium, 3 mm in diameter) probe, and an acoustic power density of 460 W cm^{-2}.

2.3. *Solid–Liquid Extraction (SLE)*

The conventional SLE process was carried out using simple maceration on a magnetic stirring and heating plate (IKA® C-MAG HS 7, IKA-Werke GmbH & Co. KG, Staufen im Breisgau, Germany) set to 300 rpm, under the processing conditions described in the following Section 2.4.

2.4. Experimental Design

Response surface methodology was used to determine the optimal conditions of the conventional solid–liquid extraction process (SLE) and ultrasound-assisted extraction (UAE), which maximize the extractability of total phenolic compounds, flavonoids, and antioxidant activity (FRAP and DPPH) of hawthorn leaves powder extracts.

A four-factor face-centered central composite design (FC-CCD) was employed to optimize both solid–liquid extraction (SLE) and ultrasound-assisted extraction (UAE) processes. For the SLE process, the following parameters and ranges were selected based on preliminary experiments: extraction temperature (X_1, 25–70 °C), extraction time (X_2, 10–90 min), ethanol concentration (X_3, 0–80%, v/v) in a water–ethanol mixture, and solid-to-liquid ratio (X_4, 0.05–0.1 g/mL). The response variables were total phenolic content (Y_1), flavonoid content (Y_2), and antioxidant activity assessed by FRAP (Y_3), DPPH (Y_4), and ABTS (Y_5) assays. The experimental design used is shown in Table 1.

Table 1. Effect of the processing variables investigated on the response variables (TPC, TFC, FRAP, DPPH, and ABTS) in extracts from hawthorn leaves powder obtained after the SLE process.

Run	Variables				SLE				
	T (°C)	t (min)	EtOH (%)	S/L (g/mL)	TPC	TFC	FRAP	DPPH	ABTS
1	25	10	0	0.1	12.6 ± 0.1 $_{ab}$	11.6 ± 0.3 $_{ab}$	16.8 ± 0.0 $_{a}$	54.3 ± 2.1 $_{de}$	21.8 ± 0.5 $_{a}$
2	25	90	0	0.1	13.2 ± 0.4 $_{abc}$	12.3 ± 2.9 $_{ab}$	22.7 ± 1.6 $_{b}$	64.5 ± 1.3 $_{fgh}$	27.5 ± 1.1 $_{b}$
3	25	90	0	0.05	11.8 ± 0.2 $_{a}$	12 ± 0.2 $_{ab}$	24.2 ±1.6 $_{bc}$	64.2 ± 3.1 $_{fgh}$	30.5 ± 0.9 $_{bc}$
4	25	10	80	0.1	31.9 ± 0.4 $_{ghi}$	28.7 ± 0.1 $_{ghi}$	58.9 ± 3.4 $_{ghi}$	63.2 ± 3.7 $_{fgh}$	77.6 ± 2.4 $_{op}$
5	25	10	80	0.05	34.2 ± 0.7 $_{hil}$	30.7 ± 1 $_{hil}$	70.3 ± 1.3 $_{lm}$	49.4 ± 0.4 $_{cd}$	89.2 ± 2.7 $_{q}$
6	25	50	40	0.075	32.5 ± 0.5 $_{hi}$	25.6 ± 1.4 $_{gh}$	64.9 ± 1.4 $_{lmn}$	60.7 ± 2.4 $_{efgh}$	80.4 ± 1.8 $_{p}$
7	25	90	80	0.1	35.4 ± 0.4 $_{mn}$	35.4 ± 3 $_{lmn}$	56.9 ± 1.9 $_{gh}$	66.6 ± 3 $_{h}$	69.8 ± 1.5 $_{mn}$
8	25	90	80	0.05	41.8 ± 0.5 $_{o}$	38 ± 1.5 $_{def}$	71.0 ± 1 $_{n}$	53.4 ± 2.2 $_{de}$	92.3 ± 1.3 $_{q}$
9	25	10	0	0.05	18.9 ± 0.8 $_{e}$	20 ± 0.9 $_{no}$	28.7 ± 0.2 $_{cd}$	54.5 ± 2.9 $_{de}$	38.1 ± 0.8 $_{de}$
10	47.5	50	40	0.1	24.2 ± 0.9 $_{f}$	24.5 ± 1 $_{fg}$	48.5 ± 1 $_{f}$	66.5 ± 2.5 $_{h}$	53.1 ± 1.3 $_{gh}$
11	47.5	10	40	0.075	31.2 ± 0.3 $_{gh}$	29.8 ± 1.3 $_{hi}$	62.4 ± 1.8 $_{il}$	64.9 ± 1.6 $_{gh}$	77.1 ± 0.7 $_{op}$
12	47.5	50	80	0.075	34 ± 0.2 $_{il}$	30.7 ± 1 $_{hil}$	56.0 ± 0.8 $_{gh}$	67.6 ± 0.8 $_{h}$	72.8 ± 1.0 $_{no}$
13	47.5	50	40	0.05	31.7 ± 0.7 $_{ghi}$	28.3 1.8 $_{ghi}$	63.7 ± 2.4 $_{il}$	67.1 ± 1.1 $_{h}$	80.8 ± 2.1 $_{p}$
14	47.5	50	0	0.075	12.1 ± 0.5 $_{a}$	19.3 ± 0.4 $_{cde}$	25.4 ± 1.3 $_{bcd}$	58.2 ± 1.7 $_{efg}$	33.0 ± 1.7 $_{c}$
15	47.5	50	40	0.075	26 ± 1.3 $_{f}$	28.3 ± 1 $_{ghi}$	49.9 ± 0.5 $_{f}$	65.9 ± 2.4 $_{h}$	68.9 ± 0.9 $_{mn}$
16	47.5	50	40	0.075	26.1 ± 1.5 $_{f}$	26.5 ± 2 $_{gh}$	49.2 ± 1.2 $_{f}$	65.5 ± 2.9 $_{gh}$	65.6 ± 1.5 $_{lm}$
17	47.5	90	40	0.075	24.2 ± 0.4 $_{f}$	23.5 ± 3 $_{efg}$	46.7 ± 2.7 $_{f}$	66.2 ± 3.8 $_{h}$	55.7 ± 1.5 $_{hi}$
18	70	90	80	0.1	29.2 ± 0.8 $_{g}$	32.5 ± 1.2 $_{ilm}$	48.9 ± 1 $_{f}$	64.4 ± 2.1 $_{fgh}$	60.6 ± 1.3 $_{il}$
19	70	10	80	0.05	39.6 ± 0.2 $_{no}$	42.3 ± 1.3 $_{o}$	91.9 ± 3.4 $_{p}$	63.3 ± 2.4 $_{fgh}$	109.5 ± 2.4 $_{s}$
20	70	10	80	0.1	33.8 ± 0.2 $_{hil}$	37.9 ± 2.4 $_{no}$	66.7 ± 2 $_{lmn}$	63 ± 1.9 $_{fgh}$	88.1 ± 2.6 $_{q}$
21	70	50	40	0.075	41.7 ± 0.9 $_{o}$	37 ± 1.4 $_{no}$	85.2 ± 2.6 $_{o}$	65.6 ± 1.8 $_{gh}$	112.8 ± 2.4 $_{s}$
22	70	90	80	0.05	36.9 ± 0.7 $_{mn}$	38.9 ± 0.2 $_{no}$	80.2 ± 0.4 $_{o}$	64.1 ± 0.1 $_{fgh}$	98.7 ± 1.5 $_{r}$
23	70	10	0	0.05	15.9 ± 1 $_{cd}$	14.5 ± 2.8 $_{bc}$	30.1 ± 3.2 $_{d}$	57.1 ± 3.9 $_{ef}$	35.1 ± 1.2 $_{cd}$
24	70	10	0	0.1	14.9 ± 1.1 $_{bc}$	14.4 ± 0.5 $_{bc}$	35.9 ± 2.5 $_{e}$	44.2 ± 2.2 $_{bc}$	45.7 ± 1.3 $_{f}$
25	70	90	0	0.05	17.7 ± 0.2 $_{de}$	15.4 ± 0.5 $_{bcd}$	37.9 ± 0.3 $_{e}$	40.2 ± 1.9 $_{ab}$	50.6 ± 0.8 $_{fg}$
26	70	90	0	0.1	11.8 ± 2 $_{a}$	8.9 ± 2.7 $_{a}$	27.4 ± 1.7 $_{bcd}$	33.5 ± 1.9 $_{a}$	40.6 ± 1.2 $_{e}$

TPC (mg GAE/g_{DM}), TFC (mg QE/g_{DM}), FRAP (mg AAE/g_{DM}), DPPH (%), ABTS (mg AAE/g_{DM}). The results are expressed as mean ± standard deviation (n = 2 for factorial and axial points, n = 5 for central point). T (extraction temperature, °C), t (extraction time, min), EtOH (ethanol concentration in ethanol–water mixture, %, v/v); S/L (solid–liquid ratio, g/mL). Values with different lowercase letters within the same column are significantly different ($p \leq 0.05$).

Based on the optimized SLE conditions, the same FC-CCD structure (25 runs; Table 2) was applied to the UAE process with adjusted factors appropriate for ultrasonic treatment:

input power (X_1, 100–400 W), extraction temperature (X_2, 40–70 °C), extraction time (X_3, 30–90 min), and ethanol concentration (X_4, 40–70%, v/v). The solid-to-liquid ratio was fixed at 0.05 g/mL to reduce variability from this factor during ultrasonic treatment. The same response variables (Y_1–Y_5) were measured for comparison. During the extraction process, the flask containing the sample was placed in a thermostatic water bath to maintain a constant temperature. The sample temperature was monitored at regular intervals throughout the extraction process using a digital thermometer.

Table 2. Effect of the processing variables investigated on the response variables (TPC, TFC, FRAP, DPPH, and ABTS) in extracts from hawthorn leaves powder obtained after the UAE process.

Run	Variables				UAE				
	T (°C)	t (min)	EtOH (%)	P (W)	TPC	TFC	FRAP	DPPH	ABTS
1	40	60	40	100	39.6 ± 1.4 bcdef	39.7 ± 2.1 bcdef	75.1 ± 1.1 fg	59.8 ± 0.2 efg	96.7 ± 2.4 gh
2	40	30	70	400	35.6 ± 0.4 ab	35.9 ± 2.5 ab	68.2 ± 2.4 cde	59.1 ± 1.4 ef	80.7 ± 2.1 ab
3	40	60	70	400	36.5 ± 2.4 abc	36 ± 1 ab	63.3 ± 0.0 a	54.4 ± 0.4 ab	80.3 ± 1.0 ab
4	40	30	40	400	40.6 ± 0.1 cdefhi	38.5 ± 1.7 abcde	71.6 ± 1.7 def	58.6 ± 0.5 cde	89.1 ± 2.2 def
5	40	45	55	250	38.6 ± 0.8 abcde	38.3 ± 1.2 abcde	66.4 ± 0.4 abc	61.6 ± 0.5 fghi	80.3 ± 0.5 ab
6	40	60	70	100	34.9 ± 1.8 a	34.4 ± 0.0 a	64.8 ± 2 abc	60.1 ± 1 fgh	87.2 ± 1.6 cde
7	40	30	70	100	35.7 ± 0.7 a	37.1 ± 2.2 abcd	68.5 ± 0.9 cde	59 ± 0.9 def	89.0 ± 0.9 def
8	40	60	40	400	36.7 ± 1.9 abcd	35.6 ± 2.6 abc	63.3 ± 1 a	53.9 ± 0.8 ab	78.3 ± 1.3 a
9	40	30	40	100	37.8 ± 0.5 abcd	37.4 ± 1.8 abcd	64 ± 0.6 ab	62.9 ± 1.2 hil	83.2 ± 1.8 bc
10	55	45	55	250	40.2 ± 1.5 cdefg	38.9 ± 2.8 abcde	71.9 ± 1.4 ef	56.9 ± 0.6 bcde	90.5 ± 1.8 ef
11	55	30	55	250	40.8 ± 2.3 cdefgh	39.8 ± 1 bcdef	72.5 ± 1.6 fg	55.4 ± 0.9 b	92.3 ± 1.1 fg
12	55	45	55	100	44.1 ± 0.3 ghi	43.5 ± 1.5 fgh	76 ± 0.0 gh	56 ± 0.6 bcd	98.8 ± 1.0 hi
13	55	45	70	250	37.8 ± 0.3 abcd	37 ± 1.8 abcd	72.9 ± 0.3 fg	55.7 ± 1.2 bc	84.8 ± 0.8 bcd
14	55	45	40	250	42 ± 2.4 efghi	42.1 ± 1.3 defgh	79.8 ± 1.1 hi	54.2 ± 1.3 ab	100.7 ± 1.4 hil
15	55	45	55	400	43.1 ± 1 fghi	42.8 ± 0.9 efgh	80.7 ± 0.7 il	59.2 ± 1 efg	104.9 ± 1.2 lm
16	55	60	55	250	41 ± 2.9 defgh	38.1 ± 0.6 abcd	67.7 ± 0.5 bcd	55.2 ± 0.8 b	88.0 ± 0.9 def
17	70	60	40	400	51 ± 2.1 m	47 ± 2.5 h	89.6 ± 1 n	62.2 ± 1.6 ghil	106.5 ± 1.3 mn
18	70	30	40	100	44.2 ± 1 ghi	45.6 ± 2 gh	94.5 ± 0.8 o	61.1 ± 1.7 fghi	112.9 ± 1.0 op
19	70	30	70	100	40.1 ± 0.5 cdefg	38.4 ± 2 abcde	80.9 ± 1.3 il	51.5 ± 0.5 a	115.2 ± 0.7 opq
20	70	30	40	400	45.7 ± 2 il	40.6 ± 1.5 bcdefg	88 ± 1.2 mn	60.1 ± 1 fgh	104.4 ± 2.6 lm
21	70	60	70	100	41 ± 0.2 fgh	34 ± 0.5 a	84.5 ± 1 lm	59.8 ± 0.9 efg	110.8 ± 0.8 no
22	70	45	55	250	44.9 ± 1.4 hil	41.5 ± 0.4 cdefg	88.9 ± 0.9 n	61.3 ± 0.9 fghi	115.6 ± 1.2 pq
23	70	60	70	400	40.5 ± 0.9 cdefg	38.9 ± 0.3 abcde	87 ± 0.7 mn	63.6 ± 0.5 il	103.1 ± 1.1 ilm
24	70	60	40	100	48.9 ± 0.9 lm	45.5 ± 1.7 gh	95.3 ± 2.8 o	65.2 ± 1 l	117.9 ± 1.6 q
25	70	30	70	400	39.3 ± 1.5 abcdef	37.1 ± 0.2 abcd	80.1 ± 1.5 i	59.2 ± 0.7 efg	104.1 ± 2.0 lm

TPC (mg GAE/g_{DM}), TFC (mg QE/g_{DM}), FRAP (mg AAE/g_{DM}), DPPH (%), ABTS (mg AAE/g_{DM}). The results are expressed as mean ± standard deviation (n = 2 for factorial and axial points, n = 5 for central point). T (extraction temperature, °C), t (extraction time, min), EtOH (ethanol concentration in ethanol–water mixture, %, v/v); P (US power, W). Values with different lowercase letters within the same column are significantly different ($p \leq 0.05$).

For both SLE and UAE processes, a second-order polynomial model, reported in Equation (1), was reduced by stepwise elimination and utilized to fit the experimental data. Specifically, terms with p-values greater than 0.05 were systematically excluded using a stepwise backward elimination approach, as implemented by the Design Expert version 12 software.

$$Y_k = \alpha_0 + \sum_{i=1}^{5} \alpha_i X_i + \sum_{i=1}^{5} \alpha_{ii} X_i^2 + \sum_{i=1}^{5} \sum_{j=i+1}^{5} \alpha_{ij} X_i X_j \quad (1)$$

where Y_k is the predicted response variable; X_i and X_j are the independent variables; α_0, α_i, α_{ii}, and α_{ij} are the intercept, regression coefficients of the linear, quadratic, and interaction terms of the model, respectively. The number of candidate terms initially considered was 14 for both SLE and UAE, and after the elimination of nonsignificant terms, 5, 2, 6, 3, 5 terms were retained in the final reduced model for TPC, TFC, FRAP, DPPH, ABTS, in the case of SLE, and 6, 3, 2, 4, 2 terms were retained in the final reduced model for TPC, TFC, FRAP, DPPH, ABTS, in the case of UAE.

To determine the optimal extraction conditions, a multi-response optimization was conducted using the desirability function available in Design Expert software. The goal was to maximize all the responses and identify the most favorable combination of extraction parameters.

2.5. *Analysis of the Extracts*

2.5.1. Extraction Yield

At the end of the extraction processes, the extracts were centrifuged at 5700× *g* for 20 min (PK121R model, ALC International, Cologno Monzese, Milan, Italy) to obtain clear supernatants. The supernatants obtained were further concentrated in a R-200/205 Rotavapor (BÜCHI Labortechnik AG, Flawil, Switzerland) until a decrease in volume of up to 90% was achieved. The concentrated extracts were freeze-dried using a 25 L VirTis Genesis freeze-drier (SP Scientific, Gardiner, NY, USA). The dry extract was packed under vacuum in plastic–aluminum pouches and stored under refrigerated conditions (8 °C).

2.5.2. Total Phenolic Content (TPC)

The Folin–Ciocalteau assay was used to evaluate the TPC of hawthorn leaves powder extracts as reported by Carpentieri et al. (2022) [15]. A 1 mL portion of the diluted supernatant was combined with 5 mL of 10% (*v*/*v*) Folin–Ciocalteu reagent and left at room temperature for 5 min. Then, 4 mL of 7.5% (*w*/*v*) sodium carbonate was added. After mixing, the solution was incubated in the dark at room temperature for one hour. The absorbance was measured at 765 nm using a V-650 spectrophotometer (Jasco Inc., Easton, MD, USA). Calibration curves were prepared using gallic acid standards (1–100 mg/L). The TPC was expressed as milligrams of gallic acid equivalents (GAE) per g of dry weight (g_{DM}) of hawthorn leaves powder.

2.5.3. Total Flavonoid Content (TFC)

Aluminum chloride assay was used to determine the TFC of hawthorn leaves powder extracts as previously reported by Carpentieri et al. (2022) [15]. A total of 1 mL of the supernatant was combined with 4 mL of distilled water, followed by the addition of 0.3 mL of 5% (*w*/*v*) sodium nitrite. The mixture was kept in the dark for 5 min, after which 0.3 mL of 10% (*w*/*v*) aluminum chloride hexahydrate ($AlCl_3 \cdot 6H_2O$) in water was added. After 5 min of incubation, 2 mL of 1.0 M sodium hydroxide was introduced, and the final volume was brought up to 10 mL with distilled water. Standard quercetin solutions (20–100 μg/mL) were prepared using the same procedure. Absorbance of both samples and standards was measured at 510 nm. The TFC was expressed as mg of quercetin equivalent (QE) per g_{DM} of hawthorn leaves powder.

2.5.4. Ferric Reducing Antioxidant Power (FRAP)

The FRAP assay for hawthorn leaves powder extracts was performed following the method described by Benzie and Strain (1996) [16]. In brief, 0.5 mL of the diluted supernatant was mixed with 2.5 mL of FRAP reagent and incubated for 10 min. The absorbance was then recorded at 593 nm. Ascorbic acid, in concentrations ranging from 0 to 2 mmol/L, was used to construct the standard calibration curve. The FRAP values were expressed as mg of ascorbic acid equivalents (mg AAE) per g_{DM} of hawthorn leaves powder.

2.5.5. Free-Radical-Scavenging Capacity (DPPH)

The DPPH assay was used to evaluate the free-radical-scavenging capacity of the extracts [17]. Briefly, 3.9 mL of a 25 ppm DPPH solution in methanol was mixed with 0.1 mL of supernatant and incubated in the dark for 5 min. The absorbance of the mixture was read at a wavelength of 515 nm. The inactivation level of extracts was evaluated as reported in the following equation (Equation (2)):

$$\%I = 100 - \left(\frac{ABS_S}{ABS_B}\right) * 100\% \tag{2}$$

where ABS_s is the absorbance of the sample and ABS_B is the absorbance of the blank.

The antioxidant activity was expressed as both percentage of DPPH radical inhibition ($\%I_{DPPH}$) and IC_{50} values, which were estimated by linear interpolation between two concentrations bracketing 50% inhibition, using data across a concentration range of the extract of 0.05 to 1.00 mg/mL.

2.5.6. ABTS Scavenging Activity

The ABTS (2,2-azo-bis (3-ethyl benzothiazoline-6-sulfonic acid)) assay was used to evaluate the scavenging activity of extracts following the methodology of Olędzki et al. (2022) [18]. The absorbance of the reacting mixture was recorded at 734 nm. Ascorbic acid was used as a standard to generate the calibration curve (100–800 μmol/L). The antiradical capacity was expressed as mg of ascorbic acid equivalents (mg AAE) per g_{DM} of hawthorn leaves powder.

2.5.7. HPLC-DAD Analysis

The High-Performance Liquid Chromatography—Diode Array Detection (HPLC-DAD) analyses of the hawthorn leaves powder extracts obtained at the optimal extraction conditions were performed as reported by Rrucaj et al. (2024) [19]. The separation of the extracts was carried out using a Vanquish HPLC System (Thermo Fisher Scientific, Waltham, MA, USA) equipped with a Luna C18 reverse phase column, 250 mm × 4.6 mm with a 5 μm particle size (Phenomenex, Torrance, CA, USA), at 37 °C. The following gradient of the solvent B (ACN/0.1% TFA) was used to carry out the separations at a 1.0 mL min^{-1} constant flow rate: 0–4 min: 0% B; 4–14 min 0–14% B; 14–30 min 14–28% B; 30–34 min 28% B; 34–42 min 28–60% B; 42–45 min 60–80% B; 45–50 min 80–100% B. Solvent A consisted of 0.1% TFA in HPLC grade water. The injection volume of the extract was 50 μL. A diode array detector (DAD) was used to record the UV-vis spectra every 2 s in the 190–650 nm range. The HPLC separations were monitored by recording the λ = 520, 360, 320, and 280 nm wavelengths. Data were processed using the Chromeleon™ Chromatography Data System (CDS) Software, version 7 (Thermo Fisher Scientific). Standard solutions, in the range of 5–250 mg kg^{-1} in methanol and ten-fold diluted with 0.1% TFA in water prior to injection, were used to build calibration curves to quantify the phenolic compounds ($R^2 > 0.99$).

2.5.8. LC-Q/TOF-MS Analysis

The chromatographic analysis was carried out using an Agilent Technologies (Santa Clara, CA, USA) 1260 Infinity Series LC coupled with an Agilent Technologies 6530 UHD Accurate Mass LC-Q/TOF-MS, equipped with an electrospray ionization Agilent Technologies Dual Jet Stream ion source (Dual AJS ESI). Chromatographic separation was performed with a 250 mm × 2.1 mm i.d. C18 reversed-phase Aeris core-shell column, 3.6 μm particle diameter (Phenomenex, Torrance, CA, USA), and 50 μL was the volume of injection. The column temperature was maintained at 37 °C during the LC analyses. The solvents were 0.1% formic acid in water, Milli-Q (A), and 0.1% formic acid in acetonitrile (B), at a flow rate of 0.2 mL min^{-1}, with the following gradient: 5–60% B in 60 min after 5 min isocratic elution at 5% B. The Q/TOF-MS conditions were the following: drying gas flow (N_2), 12.0 L min^{-1}; nebulizer pressure, 45 psi; gas drying temperature, 350 °C; capillary voltage,

3500 V; fragmentor voltage, 150 V; skimmer voltage 65 V and octopole RF peak, 750 V. Dual AJS ESI interface was used in negative ionization mode and negative ions were acquired in the range of 100–1200 m/z for MS scans and auto MS/MS scans, at a scan rate of 2 scans s^{-1} for MS and for MS/MS, respectively. Automatic acquisition mode MS/MS was performed using collision energy automatically calculated. Internal mass correction was enabled, using two reference masses at 121.0509 and 922.0098 m/z. Instrument control and data acquisition were carried out using Agilent MassHunter Workstation software B10.00. All the MS and MS/MS data of the validation standards were integrated by MassHunter Quantitative Analysis B.10.0 (Agilent Technologies). For the putative identification of metabolites in extract, raw MS and MS/MS data were converted into CEF files and processed using the SIRIUS software tool (vers. 6.1.0, Jena University, Germany). Molecular formulas were predicted based on accurate mass and isotope patterns, while MS/MS fragmentation patterns were used for structural elucidation and compound annotation. Identifications were assigned by matching experimental spectra to in silico fragmentation trees and reference compound databases integrated within SIRIUS.

2.5.9. Statistical Analysis

All the analyses of the obtained extracts were performed in triplicate, and the results were reported as means ± standard deviations. Differences among mean values were analyzed by one-way variance (ANOVA) using SPSS 20 (SPSS IBM, Chicago, IL, USA) statistical package. Tukey test was carried out as a post hoc test to determine statistically significant differences ($p < 0.05$). The software package Design Expert Version 12 (Minneapolis, MN, USA) was used to define the experimental design (FC-CCD), as well as to perform the analysis of data including the determination of ANOVA parameters, such as R^2, adjusted R^2, predicted R^2, F-values, lack of fit, p-values, adequate precision, sum of squares, degree of freedom, mean square, model term levels of significance, and the determination and validation of the optimal processing conditions using desirability functions. The contribution percentages of each significant term were also calculated by dividing the sum of squares of the individual factor by the total model sum of squares.

3. Results and Discussion

3.1. *Effect of SLE and UAE Processing Conditions on the Extractability of Bioactive Compounds from Hawthorn Leaves Powder*

3.1.1. Model Fitting of the Experimental Data (SLE)

Various factors, including type of solvent, extraction temperature, diffusion time, and solid-to-liquid (S/L) ratio, are pivotal in affecting the recovery of valuable intracellular compounds from herbal matrices [20–22]. These factors significantly influence the disruption of cellular structures and the solubilization of intracellular constituents, thereby impacting overall extractability. Optimizing these conditions is essential for maximizing the recovery of phenolic compounds, flavonoids, and other antioxidants from herbal matrices.

In this study, the effect of four independent factors, such as extraction temperature, diffusion time, ethanol concentration, and S/L ratio, on TPC, TFC, and antioxidant activity (FRAP, DPPH, and ABTS) of extracts from hawthorn leaves powder achieved from conventional SLE was evaluated to optimize the processing conditions. The values and significance of the regression coefficients of the model and the corresponding p-values, adjusted and predicted R^2, lack of fit, and F-value of the model are reported in Table 3. Equations of reduced models, including significant terms, determination coefficients (R^2), are reported in Table S1 of the Supplementary Material. The coefficients and p-values of the unrefined model have been reported for transparency in Table S3 of the Supplementary Material.

Table 3. Coefficients and ANOVA parameters of the reduced second-order polynomial model for TPC, TFC, and antioxidant activity (FRAP, DPPH, and ABTS) in extracts from hawthorn leaves powder obtained after the SLE process.

	SLE						
	TPC (mgGAE/g_{DM})						
	Coefficients	**Sum of Squares**	**df**	**Mean Square**	**F-Value**	***p*-Value**	**Contribution (%)**
α_0	38.73962	-	-	-	-	*	-
α_1 (T)	−1.07297	38.03	1	38.03	5.41	*	1.74
α_3 (EtOH)	0.74447	1736.46	1	1736.46	247.04	***	79.52
α_4 (S/L ratio)	−85.44843	82.14	1	82.14	11.69	**	3.76
α_{11} (T × T)	0.011974	120.45	1	120.45	17.14	***	5.52
α_{33} (EtOH × EtOH)	−0.006237	326.37	1	326.37	46.43	***	14.95
Model	-	2183.54	5	436.71	62.13	***	-
Lack of fit	-	140.58	19	7.40	1.45	ns	-
R^2				0.940			
Adjusted R^2				0.924			
Predicted R^2				0.887			
Adequate precision				23.002			
	TFC (mgQE/g_{DM})						
α_0	5.69088	-	-	-	-	*	-
α_3 (EtOH)	0.298550	2567.00	1	2567.00	85.88	***	94.14
α_{12} (T × t)	−0.003347	145.19	1	145.19	4.86	*	5.32
Model	-	2726.71	4	681.68	22.81	**	-
Lack of fit	-	626.02	20	31.30	1.48	ns	-
R^2				0.883			
Adjusted R^2				0.878			
Predicted R^2				0.788			
Adequate precision				19.169			

Table 3. *Cont.*

SLE							
TPC (mgGAE/g_{DM})							
	Coefficients	Sum of Squares	df	Mean Square	F-Value	*p*-Value	Contribution (%)
FRAP (mgAAE/g_{DM})							
α_0	73.91293	-	-	-	-	**	-
α_1 (T)	−2.38291	446.29	1	446.29	9.45	**	4.6
α_3 (EtOH)	1.81099	6884.20	1	6884.20	4.86	***	70.98
α_4 (S/L ratio)	−96.66875	738.95	1	738.95	15.64	**	7.62
α_{34} (EtOH × S/L ratio)	−3.99053	254.79	1	254.79	5.39	*	2.63
α_{11} (T × T)	0.02741	631.27	1	631.27	13.36	**	6.50
α_{33} (EtOH × EtOH)	−0.012785	1371.54	1	1371.54	29.03	***	14.14
Model	-	9699.39	6	1616.57	34.21	***	-
Lack of fit	-	897.40	18	49.86	0.42	ns	-
R^2				0.915			
Adjusted R^2				0.889			
Predicted R^2				0.835			
Adequate precision				20.890			
DPPH (%)							
α_0	50.31451	-	-	-	-	***	-
α_3 (EtOH)	−0.38078	392.62	1	392.62	8.34	***	36.19
α_{12} (T × t)	−0.00366	173.65	1	173.65	7.69	*	16
α_{13} (T × EtOH)	0.005884	448.65	1	448.65	12.53	***	41.36
Model	-	1084.82	5	216.96	4.61	***	-
Lack of fit	-	941.73	19	49.56	2.80	ns	-
R^2				0.870			
Adjusted R^2				0.820			
Predicted R^2				0.795			
Adequate precision				13.187			

Table 3. *Cont.*

	SLE						
	TPC (mgGAE/g_{DM})						
	Coefficients	**Sum of Squares**	**df**	**Mean Square**	**F-Value**	***p*-Value**	**Contribution (%)**
	ABTS (mgAAE/g_{DM})						
α_0	109.9680	-	-	-	-	**	-
α_1 (T)	−3.093153	728.35	1	728.35	6.83	*	5
α_3 (EtOH)	1.890647	10,546.36	1	105,46.36	98.92	***	72.51
α_4 (S/L ratio)	−311.11111	1088.89	1	1088.89	10.21	**	7.49
α_{11} (T × T)	0.035535	1060.80	1	1060.80	9.95	**	7.29
α_{33} (EtOH × EtOH)	−0.016069	2166.65	1	2166.65	20.32	**	14.90
Model	-	14,543.80	5	2908.76	27.28	***	-
Lack of fit	-	2126.88	19	111.94	1.56	ns	-
R^2				0.942			
Adjusted R^2				0.910			
Predicted R^2				0.838			
Adequate precision				16.237			

ns not significant for $p > 0.05$. * Significant for $p \leq 0.05$; ** significant for $p \leq 0.01$; *** significant for $p \leq 0.001$.

Results show that, for the extracts obtained after SLE of bioactives from hawthorn leaves powder, the extraction temperature resulted in statistically significant linear and quadratic effects on TPC, FRAP, and ABTS values, with a nonsignificant effect detected only for TFC and DPPH.

Although the magnitude of the quadratic term is relatively small, its statistical significance indicates that temperature influences the extraction process in a U-shaped manner. Specifically, the responses initially declined slightly with increasing temperature before rising again at higher temperatures, suggesting a moderate thermal sensitivity of phenolic compounds, balanced by enhanced extractability at elevated temperatures.

This trend is consistent with previous findings across various plant matrices. For instance, Rittisak et al. (2022) [23] observed a mild U-shaped relationship between temperature and antioxidants recovered from roasted rice germ herbal tea, attributing the initial decrease to the thermal degradation of heat-labile phenolics, followed by increased recovery at higher temperatures due to cell wall disruption and improved solubility. Similarly, Dar et al. (2024) [24] identified a significant but low-magnitude positive quadratic term for temperature in the ultrasound-assisted extraction of antioxidants from Nigella sativa seeds.

Flavonoid content was not significantly affected by temperature, implying that flavonoid structures might be more stable under the studied temperature range (25–70 °C) compared to simple phenolic acids.

Ethanol concentration in water was the most significant factor affecting all responses. It showed a positive linear effect and a significant negative quadratic effect on TPC, TFC, FRAP, and ABTS values ($p < 0.001$) of extracts obtained from SLE. This trend indicates an inverted U-shaped relationship, where increasing ethanol percentage initially enhanced extraction efficiency, due to improved solubility and polarity, then, beyond an optimal ethanol concentration, further increases led to a decline in response values.

This type of ethanol-dependent behavior is well-documented in studies investigating the extractability of phenolic compounds from plant-based biomasses, where a mixed-polarity solvent maximized extraction yield [15,25].

The interaction between ethanol percentage and solid-to-liquid ratio showed a notable negative effect ($p < 0.01$) on FRAP values, implying that beyond certain concentrations, increasing both parameters simultaneously reduces the extractability of antioxidant compounds.

Extraction time did not exert significant linear and quadratic effects on the response variables investigated, suggesting that this factor played a secondary role in the recovery of valuable intracellular compounds from hawthorn leaves powder. Indeed, prolonged extraction processes did not enhance the extractability of the responses investigated. In some cases, prolonged extraction can lead to degradation of antioxidants or unnecessary exposure to oxidative conditions [13]. It is worth highlighting that in the case of SLE, the temperature—time interaction coefficient is negative and significant for TFC and DPPH, indicating that increasing both temperature and time does not provide additional benefits to the extraction efficiency.

The S/L ratio exerts highly negative linear effects on TPC, FRAP, and ABTS, indicating a strong inverse relationship between the amount of solid and phenolic compounds and antioxidant activity in the extracts obtained. Interestingly, the S/L ratio had no significant impact on TFC and DPPH values, suggesting that while TPC, FRAP, and ABTS decreased, the flavonoids and specific antioxidant compounds contributing to scavenging activities might not scale proportionally with the amount of plant material in the solvent.

3.1.2. Model Fitting of the Experimental Data (UAE)

The same design was utilized to investigate the effect of extraction temperature, diffusion time, ethanol concentration, and US power on bioactive compounds of extracts from hawthorn leaves powder achieved from the UAE process at a fixed S/L ratio of 0.05 g/mL, selected according to the optimized SLE processing conditions.

The values and significance of the regression coefficients of the model and the corresponding *p*-values, adjusted and predicted R^2, lack of fit, and F-value of the model are reported in Table 4. Equations of reduced models, including significant terms, determination coefficients (R^2), are reported in Table S2 of the Supplementary Material. The coefficients and *p*-values of the unrefined model have been reported for transparency in Table S4 of the Supplementary Material.

Table 4. Coefficients and ANOVA parameters of the reduced second-order polynomial model for TPC, TFC, and antioxidant activity (FRAP, DPPH, and ABTS) in extracts from hawthorn leaves powder obtained after the UAE process.

			UAE				
			TPC (mgGAE/g_{DM})				
	Coefficients	**Sum of Squares**	**df**	**Mean Square**	**F-Value**	***p*-Value**	**Contribution %**
α_0	3.81775	-	-	-	-	**	-
α_1 (T)	0.297404	191.07	1	191.07	195.33	***	54.75
α_3 (EtOH)	1.25735	93.51	1	93.51	95.59	***	28.26
α_{12} (T × t)	0.003581	10.38	1	10.38	10.62	**	3.14
α_{13} (T × EtOH)	−0.004388	15.59	1	15.59	15.94	**	4.71
α_{33} (EtOH × EtOH)	−0.010618	18.30	1	18.30	18.71	**	5.56
α_{44} (Power × Power)	0.000060	5.92	1	5.92	6.05	*	1.79
Model	-	330.87	8	41.36	42.48	***	-
Lack of fit	-	94.5	10	9.45	1.05	ns	-
R^2				0.955			
Adjusted R^2				0.932			
Predicted R^2				0.881			
Adequate precision				21.866			
			TFC (mgQE/g_{DM})				
α_0	22.50699	-	-	-	-	**	-
α_1 (T)	0.476451	71.87	1	71.87	11.09	**	30.62
α_3 (EtOH)	0.164074	130.01	1	130.01	20.07	***	55.39
α_{13} (T × EtOH)	−0.006241	31.55	1	31.55	4.87	*	13.44
Model	-	234.73	4	58.68	12.48	***	-
Lack of fit	-	16.42	6	2.74	1.87	ns	-
R^2				0.910			
Adjusted R^2				0.875			
Predicted R^2				0.793			
Adequate precision				13.445			

Table 4. *Cont.*

UAE							
TPC (mgGAE/g_{DM})							
	Coefficients	**Sum of Squares**	**df**	**Mean Square**	**F-Value**	***p*-Value**	**Contribution %**
FRAP (mgAAE/g_{DM})							
α_0	48.79112	-	-	-	-	***	-
α_1 (T)	0.68029	1874.34	1	1874.34	97.75	***	92.79
α_3 (EtOH)	−0.189417	145.31	1	145.31	7.58	**	7.19
Model	-	2019.94	4	504.99	26.33	***	-
Lack of fit	-	74.5	10	7.45	1.250	ns	-
R^2				0.940			
Adjusted R^2				0.909			
Predicted R^2				0.857			
Adequate precision				17.279			
DPPH (%)							
α_0	136.43547	-	-	-	-	**	-
α_{12} (T × t)	0.00842	57.43	1	57.43	12.91	***	27.38
α_{14} (T × Power)	0.000657	34.50	1	34.50	7.76	**	14.45
α_{34} (EtOH × Power)	0.000559	25.33	1	25.33	5.69	*	12.08
α_{11} (T × T)	0.015732	63.15	1	63.15	14.20	***	30.10
Model	-	209.77	8	26.22	6.12	**	-
Lack of fit	-	20.23	6	3.37	0.101	ns	-
R^2				0.886			
Adjusted R^2				0.857			
Predicted R^2				0.778			
Adequate precision				18.105			
ABTS (mgAAE/g_{DM})							
α_0	56.21941	-	-	-	-	***	-
α_1 (T)	0.83593	2830.03	1	2830.03	87.23	***	93.34
α_4(Power)	−0.02233	202.00	1	202.00	6.23	*	6.67
Model	-	3032.03	2	1516.02	46.73	***	-
Lack of fit	-	187.18	6	31.20	0.065	ns	-
R^2				0.890			
Adjusted R^2				0.822			
Predicted R^2				0.797			
Adequate precision				16.105			

ns not significant for $p > 0.05$. * Significant for $p \leq 0.05$; ** significant for $p \leq 0.01$; *** significant for $p \leq 0.001$.

In the case of the UAE process, temperature had a significant linear positive effect on TPC, TFC, and FRAP ($p < 0.001$, $p < 0.01$, $p < 0.001$, respectively), and a quadratic effect on DPPH ($p < 0.001$) showing that an increase in temperature enhanced phenolic compounds, flavonoids extraction and antioxidant activity.

Ethanol showed a strong positive effect on TPC and TFC ($p < 0.001$), similar to SLE, confirming that ethanol is crucial for phenolic solubilization.

However, it negatively affected FRAP values. Regarding the linear dependence of the responses on ethanol concentration, it is interesting to note that a reduced significance was detected for FRAP, DPPH, and ABTS values upon US application compared to SLE. This implies that, in the investigated variable domain, US treatment decreased the effect of ethanol concentration on the extractability of antioxidants. These findings suggest that while ethanol concentration is a critical factor in the extraction of antioxidants, the application of ultrasound may modify its impact, potentially reducing its significance compared to other extraction parameters.

Moreover, the interaction between ethanol percentage and temperature showed a notable negative effect ($p < 0.01$) for TPC and TFC, implying that increasing both parameters simultaneously reduces the recovery of these compounds.

Extraction time did not exert significant linear and quadratic effects on the response variables investigated. Nonetheless, in contrast to the SLE process, the temperature—time interaction coefficient is positive and significant for TPC and DPPH, suggesting a synergistic effect between temperature and time, which justifies the selection of a moderately longer extraction time under optimal conditions (Table 5).

Table 5. TPC, TFC, FRAP, DPPH, and ABTS predicted and experimental values of hawthorn leaves powder extracts obtained after SLE and UAE processes at the optimized conditions.

	SLE		**UAE**		**% Deviation**	
Optimal processing conditions	T: 70 °C t: 34 min EtOH%: 75% S/L ratio: 0.05 g/mL		T: 70 °C t: 44 min EtOH%: 40% S/L ratio: 0.05 g/mL US power: 100 W			
	Predicted value	Experimental value	Predicted value	Experimental value	SLE	UAE
TPC (mgGAE/g_{DM})	39.75 ± 0.30 $_a$	41.28 ± 0.90 $_a$	47.75 ± 0.85 $_b$	48.00 ± 0.85 $_b$	3.85	0.52
TFC (mgQE/g_{DM})	39.74 ± 0.10 $_a$	40.64 ± 0.63 $_a$	46.82 ± 0.85 $_b$	43.68 ± 0.77 $_b$	2.26	6.71
FRAP (mgAAE/g_{DM})	86.97 ± 1.20 $_a$	91.98 ± 1.48 $_a$	89.77 ± 0.85 $_b$	98.61 ± 1.15 $_b$	5.76	9.84
DPPH (%)	67.83 ± 0.95 $_b$	64.26 ± 1.25 $_a$	64.49 ± 0.85 $_a$	66.73 ± 1.10 $_a$	5.27	3.47
ABTS (mgAAE/g_{DM})	106.08 ± 1.30 $_a$	109.08 ± 1.70 $_a$	112.50 ± 0.85 $_b$	118.50 ± 1.35 $_b$	2.83	5.33

Values with different lowercase letters within the same line for predicted values and experimental values are significantly different ($p \leq 0.05$).

As regards the US power, no significant linear effect on the responses was observed. It showed a slight positive quadratic effect on TPC; however, the small absolute values of the coefficient suggested that most of the bioactive compounds were efficiently extracted at the lowest US power studied (100 W). Increasing US power, no significant enhancements in the extractability of bioactive compounds were detected. These findings align with previous research on the optimization of UAE parameters for extracting phenolic compounds from *Centella asiatica* [26]. These authors found that ethanol concentration and S/L ratio significantly influenced TPC and TFC, while US power had nonsignificant effects.

In addition, the interaction between ethanol concentration and temperature with US power had a positive effect on DPPH values, indicating that these variables synergistically affect the scavenging capacity of extracts.

3.1.3. Model Validation and Accuracy

The Analysis of Variance (ANOVA) revealed that the relationship between the response variables and extraction parameters had determination coefficients (R^2) ranging from 0.870 to 0.942 for SLE and from 0.886 to 0.955 for UAE, indicating a strong correlation between the experimental data and the values predicted by the model. For both SLE and UAE, the regression models show strong statistical significance across all responses, as indicated by high model F-values (all $p < 0.01$ or 0.001) and nonsignificant lack of fit, confirming good model adequacy. The models demonstrate good fit and predictive power, predicted R^2 values are in reasonable agreement with the adjusted R^2 values, with differences lower than 0.08.

This is further supported by the correlation plots of predicted vs. actual values reported in Figures S1 and S2 of the Supplementary material, where a strong alignment along the diagonal demonstrates the accuracy of the model in estimating experimental data. Normal probability plots of residuals are used to assess whether the residuals (errors) from the models follow a normal distribution, which is one of the assumptions for valid regression analysis. These plots are reported in Figures S3 and S4. For TPC, TFC, FRAP, and ABTS, the residuals exhibit an approximately linear trend along the reference line, indicating that the errors are normally distributed. This suggests that the models for these responses are statistically sound and meet the underlying assumptions of ANOVA and regression analysis. DPPH, in the case of SLE, shows slight deviations from linearity, which may indicate minor non-normality or potential heteroscedasticity.

3.1.4. Optimization of SLE and UAE Processing Conditions

The 3D response surface graphs showing the interaction between the extraction temperature (25–70 °C), ethanol concentration (0–80%, *v*/*v*) at a S/L ratio of 0.05 g/mL and 0.1 g/mL on TPC, TFC, FRAP, DPPH, and ABTS of the extracts obtained after SLE of hawthorn leaves powder are reported in Figures 1 and 2. Figures 3 and 4 depict the 3D response surface graphs demonstrating the interacting effect between the extraction temperature (25–70 °C), ethanol concentration (40–70%, *v*/*v*) at a US power of 100 W and 400 W, and a S/L ratio of 0.05 g/mL, on TPC, TFC, FRAP, DPPH, and ABTS of the extracts obtained after UAE of hawthorn leaves powder.

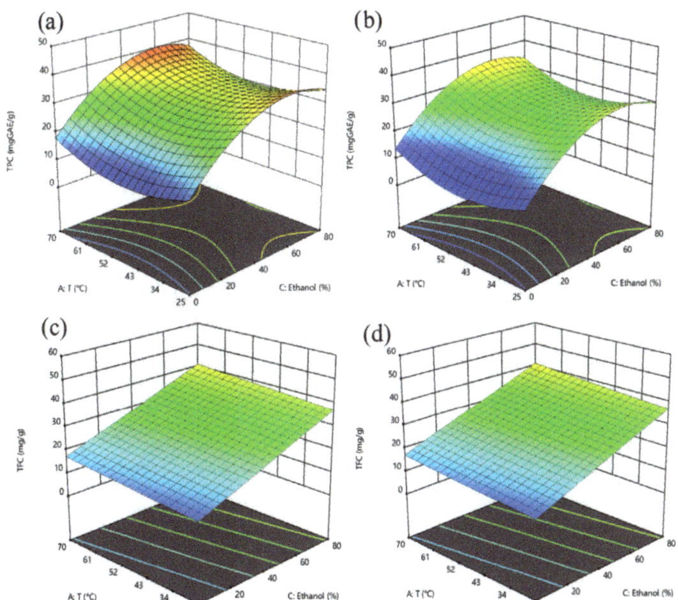

Figure 1. Response surface graphs for TPC (mg GAE/g_{DM}) (**a**,**b**) and TFC (mg QE/g_{DM}) (**c**,**d**) of extracts from hawthorn leaves powder after SLE as a function of temperature and ethanol concentration, at S/L ratio of 0.05 g/mL (**a**,**c**), and 0.1 g/mL (**b**,**d**). The surface color gradient represents the magnitude of the response: cooler colors (e.g., blue) indicate lower response values, while warmer colors (e.g., yellow to red) correspond to higher response values.

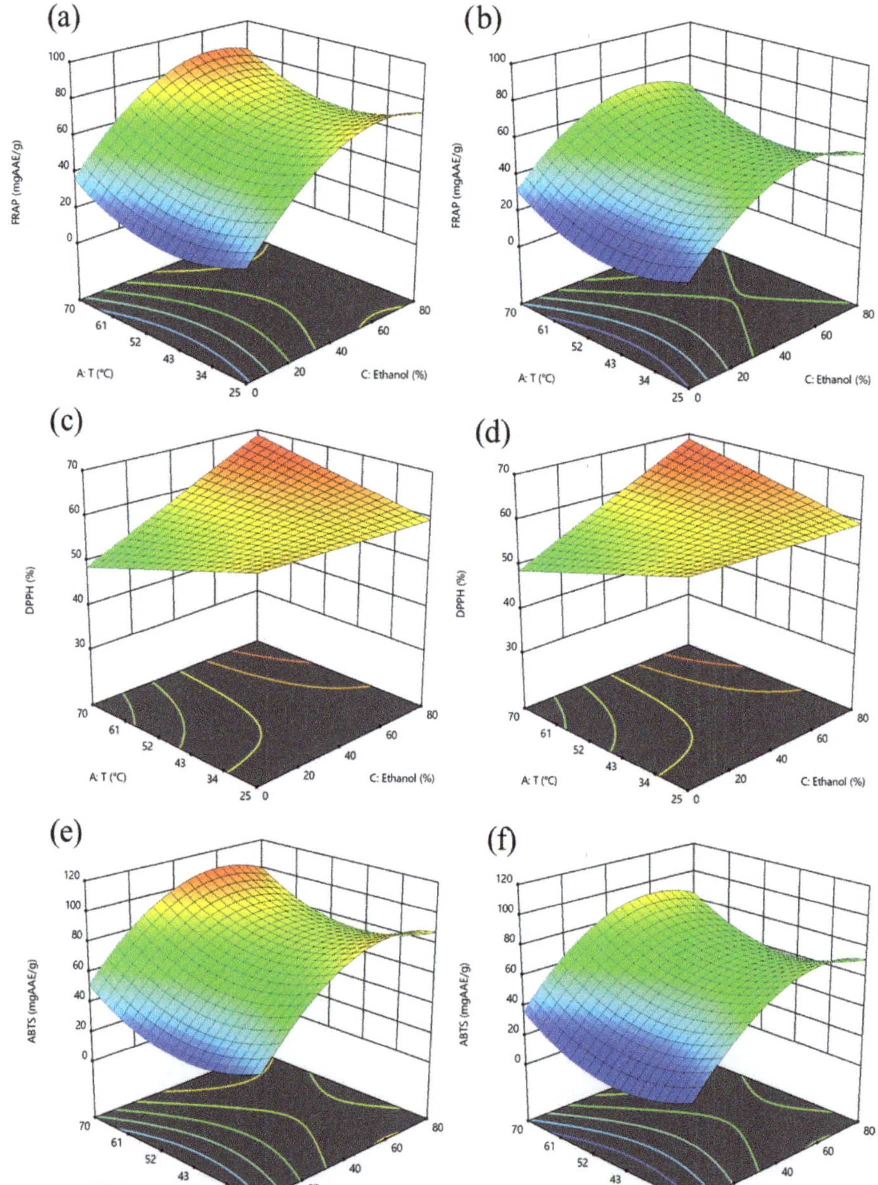

Figure 2. Response surface graphs for FRAP (mg AAE/g_{DM}) (**a**,**b**), DPPH (%) (**c**,**d**) and ABTS (mg AAE/g_{DM}) (**e**,**f**) of extracts from hawthorn leaves powder after SLE as a function of temperature and ethanol concentration, at S/L ratio of 0.05 g/mL (**a**,**c**,**e**), and 0.1 g/mL (**b**,**d**,**f**). The surface color gradient represents the magnitude of the response: cooler colors (e.g., blue) indicate lower response values, while warmer colors (e.g., yellow to red) correspond to higher response values.

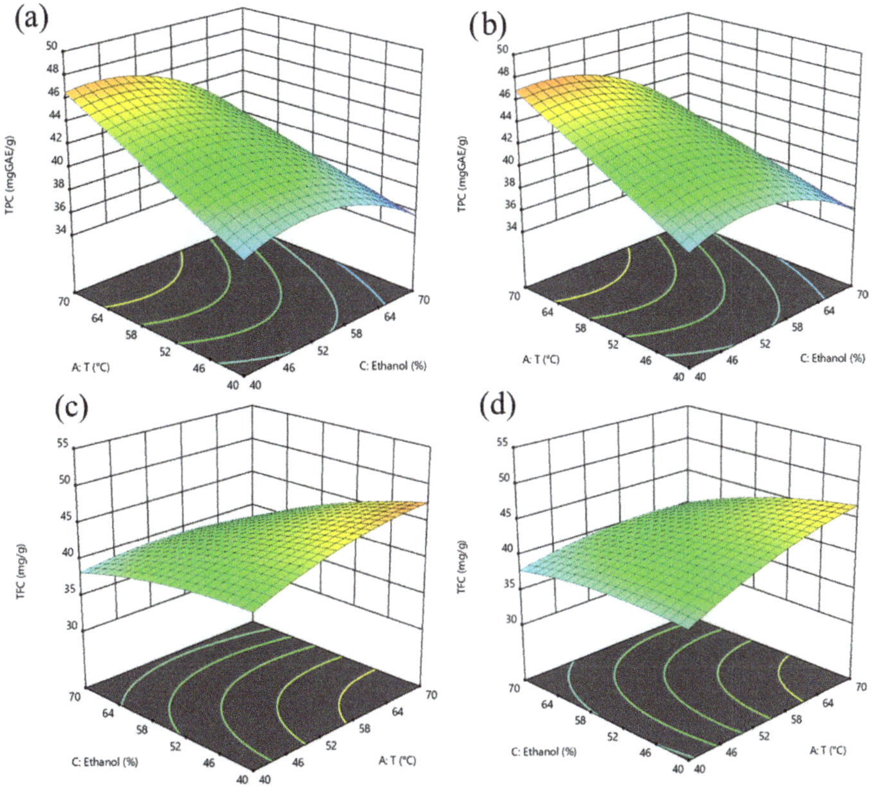

Figure 3. Response surface graphs for TPC (mg GAE/g_{DM}) (**a**,**b**) and TFC (mg QE/g_{DM}) (**c**,**d**) of extracts from hawthorn leaves powder after UAE as a function of temperature, ethanol concentration, at US power of 100 W (**a**,**c**), and 400 W (**b**,**d**). The surface color gradient represents the magnitude of the response: cooler colors (e.g., blue) indicate lower response values, while warmer colors (e.g., yellow to red) correspond to higher response values.

Throughout the entire investigated domain, the response variables exhibited a similar trend. The UAE process applied to hawthorn leaves powder resulted in a higher amount of bioactive compounds in the extracts compared to the SLE process. It can be observed that the ethanol concentration in water appeared as the factor that most influenced the investigated responses, as confirmed by the significance of the coefficients (Tables S1 and S2).

Moreover, the influence of ethanol concentration on the extractability of bioactive compounds showed a parabolic trend, which appeared less pronounced when applying UAE compared to the conventional SLE process.

The concentration of ethanol used in the extraction mixture significantly influences the recovery of phenolic compounds from plant materials. Ethanol–water mixtures are widely used for extracting phenolic compounds due to their ability to solubilize a broad spectrum of molecules with varying polarities. By combining both solvents, the resulting mixture offers an intermediate polarity that enhances the overall extraction efficiency of diverse phenolic constituents.

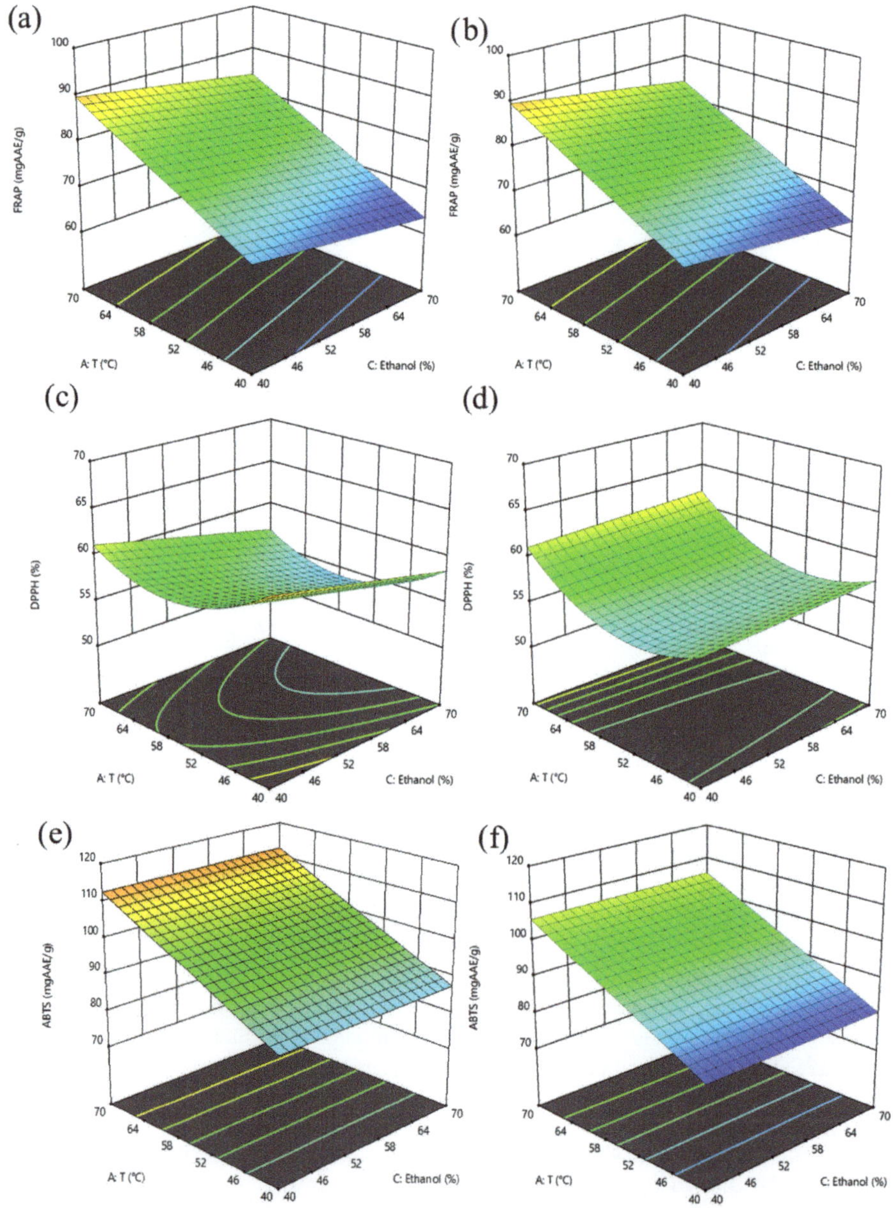

Figure 4. Response surface graphs for FRAP (mg AAE/g_{DM}) (**a**,**b**), DPPH (%) (**c**,**d**) and ABTS (mg AAE/g_{DM}) (**e**,**f**) of extracts from hawthorn leaves powder after UAE as a function of temperature and ethanol concentration, at US power of 100 W (**a**,**c**,**e**), and 400 W (**b**,**d**,**f**). The surface color gradient represents the magnitude of the response: cooler colors (e.g., blue) indicate lower response values, while warmer colors (e.g., yellow to red) correspond to higher response values.

Studies on hawthorn leaves and other medicinal herbs consistently report a parabolic relationship between ethanol concentration and total phenolic content. As ethanol concentration increases, phenolic extraction initially improves due to enhanced solubility of less polar compounds and increased disruption of plant cell membranes. However, beyond a certain ethanol concentration, the concentration begins to decline. This decrease is typically attributed to the reduced ability of high-ethanol mixtures to solubilize more polar phenolics, along with reduced swelling of plant tissues, which hinders solvent penetration [27].

This trend has also been observed in similar herbs and plant materials. For example, in rosemary, optimal phenolic extraction occurred at ethanol concentrations ranging from 30% to 50% (*v*/*v*), depending on the extraction technique used; beyond this range, extraction efficiency decreased [28].

When employing UAE, the optimal ethanol concentration tends to be lower compared to conventional methods. In the case of *Crataegus pinnatifida* (a species of hawthorn) leaves, 39% ethanol (*v*/*v*) has been identified as the optimal concentration for maximizing total phenolic content [11].

Ultrasound enhances extraction by inducing cavitation, which disrupts plant cell walls and facilitates solvent diffusion. This mechanical effect can compensate for the reduced solvating power of lower ethanol concentrations. Likewise, Martin-Garcia et al. (2021) [13], who investigated the optimization of UAE of phenolic compounds from *Crataegus monogyna* leaves, found that the optimal acetone concentration in water (*v*/*v*), ranging from 20% to 80% was 50% (*v*/*v*).

Nevertheless, increasing the US power from 100 W to 400 W did not induce any significant increases in TPC, TFC, FRAP, DPPH, and ABTS values in the extracts (Figures 3 and 4). The quadratic coefficient of US power (Table 4) was positive for TPC with a slight significance and a small absolute value, indicating that extreme power levels are not necessary to enhance the extractability of target phenolic compounds.

Together with the type of solvent, the extraction temperature plays a critical role in the efficiency of phenolic compound recovery from plant matrices. From the 3D response surface graphs, it is possible to observe, for TPC, FRAP, and ABTS (Figures 1 and 2), in the case of SLE, and for DPPH (Figure 4), in the case of UAE, the U-shaped trend, in line with the positive quadratic coefficients of temperature. Higher temperatures can disrupt the structural integrity of plant tissues, significantly enhancing the mass transfer process by increasing the solubility and diffusivity of phenolic compounds in the solvent. As the temperature rises, the surface tension and viscosity of the solvent decrease, allowing it to effectively penetrate plant cell walls and facilitate the release of intracellular bioactive compounds [27]. Several studies have demonstrated that moderate to elevated extraction temperatures improved the extractability of phenolic compounds from various medicinal and aromatic herbs, including hawthorn leaves. For instance, the extraction of phenolics from white tea leaves [29] and rosemary [30,31] showed peak efficiency at extraction temperatures of 65–70 °C. Similarly, Liang et al. (2022) [14] and Pan et al. (2012) [32], who investigated the optimization of UAE of flavonoids from hawthorn leaves and seeds, respectively, found an optimal temperature of 65 °C. However, it is important to note that excessively high temperatures may lead to the breakdown of certain sensitive phenolic compounds, thus lowering the overall antioxidant capacity of the extract, and seeking a compromise between enhanced extractability and compound stability.

Regarding the extraction time, the 3D surface graphs revealed that, regardless of the application of US treatment, this factor scarcely affected TPC, TFC, FRAP, DPPH, and ABTS values in the extracts.

Liang et al. (2022) [14] found a relatively short optimal extraction time of 40 min when investigating the optimization of ultrasonic-assisted aqueous two-phase extraction of flavonoids from hawthorn leaves.

Similarly, other authors observed optimal extraction times ranging from 31 min to 55 min when optimizing the UAE process of phenolic compounds from *Crataegus pinnatifida* and *Crataegus monogyna* leaves [11,13].

Moreover, from the graphs reported in Figures 1 and 2, the lowest S/L ratio investigated (0.05 g/mL) enhanced the extractability of phenolic compounds and flavonoids from hawthorn leaves powder compared to higher S/L ratios. A lower solid content, in the range of 0.03–0.07 g/mL, provides enhanced solvent saturation capacity, while a higher solvent volume increases the contact surface area between the solid and the solvent, reducing viscosity and clumping of powdery leaves and improving solubility and diffusion of solutes [11,14].

Based on the findings obtained and the use of desirability function, the optimal values of the independent variables that maximized the investigated responses were shown by the adopted model to be 70 °C, 75% ethanol–water mixture, 34 min, and S/L ratio 0.05 g/mL for extracts from SLE, and 70 °C, 40% ethanol–water mixture, 44 min, and US power of 100 W for extracts from UAE of hawthorn leaves powder, with desirability values of 0.93 and 0.91, respectively. Under these optimized conditions, the TPC, TFC, FRAP, DPPH, and ABTS predicted and experimental values of the extracts achieved from SLE and UAE processes are reported in Table 5. To validate the model predictions, experimental values were measured under the optimized extraction conditions and compared with the predicted responses. The % deviations between predicted and actual values for TPC, TFC, FRAP, DPPH, and ABTS ranged from 0.52% to 9.84%, supporting the adequacy and predictive capability of the response surface models, indicating a good agreement between experimental and predicted data.

Even though the amount of bioactives extracted from herbal matrices depends on several factors such as variety, growing conditions, harvesting time, post-harvest handling, equipment and experimental protocols, these results appeared consistent with those observed by other authors who found comparable concentrations of phenolic compounds (12.41–82.74 mg GAE/g_{DM}) [13,14,33,34], flavonoids (0.97–63.34 mg QE/g_{DM}) [33,35], and antioxidant activity (49.62–134.68 mg Trolox/g_{DM}) [13] in extracts from different species of hawthorn leaves.

Interestingly, the application of optimized UAE process can be successfully used to intensify the extractability of phenolic compounds (16%), flavonoids (7%), antioxidant activity (FRAP, 7%, DPPH, 4%, and ABTS, 9%) from hawthorn leaves powder with a significantly reduced ethanol concentration compared to conventional SLE process.

This reduction is made possible by the mechanical effects of ultrasound, which enhance cell disruption, mass transfer, and solvent interaction, reducing the dependence on high solvent polarity, making UAE a more sustainable and economical alternative for plant bioactive extraction.

Ultrasound generates high-frequency sound waves that create microbubbles in the solvent. When these bubbles collapse, they produce intense localized energy that disrupts plant cell walls, improves solvent penetration into cellular structures, and effectively releases intracellular phenolic compounds [36].

The results obtained are consistent with those reported by Martin-Garcia et al. [13], who demonstrated that optimized UAE conditions led to a significant enhancement in phenolic compound recovery from *C. monogyna* leaves. Similarly, Luo et al. [11] reported an increase in TPC by approximately 13–18% in *C. pinnatifida* using UAE compared to conventional methods, confirming the high efficiency and versatility of UAE across different *Crataegus* species. The extraction yield, expressed as % dry weight of extract per gram of raw material, was 18% for UAE and 15% for SLE (Solid–Liquid Extraction). This 20% increase in yield using UAE compared to SLE highlights the enhanced efficiency of ultrasound in disrupting plant cell walls and improving mass transfer.

Moreover, the optimal hawthorn leaf extract exhibited an IC_{50} value of 0.35 ± 0.02 mg/mL for SLE and 0.3 ± 0.01 mg/mL for UAE using the DPPH assay, indicating a moderate antioxidant activity. This value is consistent with that found for leaf extracts from commonly used plants for their antioxidant properties, including *Crataegus oxyacantha* [37] and *Crataegus pinnatifida* leaves [11], which have IC_{50} values ranging between 0.04 and 0.69 mg/mL, widely depending on the type of solvent and extraction method used.

The optimized extraction conditions developed in this study offer promising advantages in terms of sustainability and potential scalability for industrial applications. The reduced ethanol concentration not only minimizes solvent consumption and associated environmental impact but also lowers costs related to solvent recovery and handling. Furthermore, UAE operates efficiently at moderate temperatures and extraction times, with relatively low power applied (100 W), contributing to the overall energy efficiency of the process. These features, combined with the ability to enhance extraction yields without the need for harsh conditions or large solvent volumes, support the feasibility of UAE as a more sustainable and economically viable alternative for large-scale production of plant-based bioactives.

3.1.5. Phenolic Composition of the Extracts

HPLC analysis and mass spectroscopy aimed at investigating the phenolic composition of the extracts from hawthorn leaves powder obtained after SLE (70 °C, 75% ethanol–water mixture, 34 min, S/L ratio of 0.05 g/mL) and UAE (70 °C, 40% ethanol–water mixture, 44 min, S/L ratio of 0.05 g/mL, US power of 100 W) under optimized conditions.

The HPLC-DAD profiles of extracts obtained after SLE and UAE were comparable with concentrations of chlorogenic acid, apigenin-8-C-glucoside-2′-rhamnoside, epicatechin, and catechin reported in Table 6. Notably, the application of ultrasound did not result in any significant alteration or degradation of the HPLC profile compared to the extract obtained via SLE, confirming the structural integrity and preservation of the phenolic composition. The obtained HPLC-DAD profile of the extract is reported in Figure 5, and the identified phenolic compounds are listed in Table 7.

Table 6. Concentrations (mg/g_{DM}) of catechin, chlorogenic acid, epicatechin, and apigenin-8-C-glucoside-2′-rhamnoside identified in hawthorn leaf powder extracts obtained after SLE and UAE processes at the optimized conditions.

N.	Compound	Concentration	
		SLE	UAE
3	Catechin	$0.04 \pm 0_a$	$0.04 \pm 0_b$
4	Chlorogenic acid	$1.68 \pm 0.2_a$	$2.42 \pm 0.3_b$
6	Epicatechin	$0.06 \pm 0_a$	$0.08 \pm 0_b$
11	Apigenin-8-C-glucoside-2′-rhamnoside	$1.28 \pm 0.1_a$	$1.75 \pm 0.2_b$

Values with different lowercase letters within the same line are significantly different ($p \leq 0.05$).

Figure 5 and Table 7 show that the dominant peaks are associated with chlorogenic acid and apigenin-8-C-glucoside-2′-rhamnoside, suggesting that these are major bioactive constituents of hawthorn leaves powder extract. Consistently with previous findings [3,38,39], several classes of compounds were putatively identified in hawthorn leaves extracts, such as phenolic acids (neochlorogenic, chlorogenic, cryptochlorogenic acids), flavan-3-ols (catechin, epicatechin, procyanidin C1), flavone/flavonol glycosides (apigenin, luteolin, quercetin derivatives), dihydrochalcones (nothofagin), and lignan (cinchonain 1A).

Chlorogenic acid is a significant phenolic compound found in hawthorn leaves; however, its concentration can vary based on species, extraction methods, and environmental factors. The accumulation of phenolics in leaves, indeed, is commonly suggested to be sensitive to the environment [40]. For instance, a study analyzing methanolic extracts of

Crataegus pinnatifida reported chlorogenic acid content at approximately 0.84 mg/g_{DW} [41]. Research focusing on the phenolic composition of hawthorn leaves identified chlorogenic acid as one of the prevalent compounds, with contents in the leaves varying from 3 to 11 mg/g_{DW} [40].

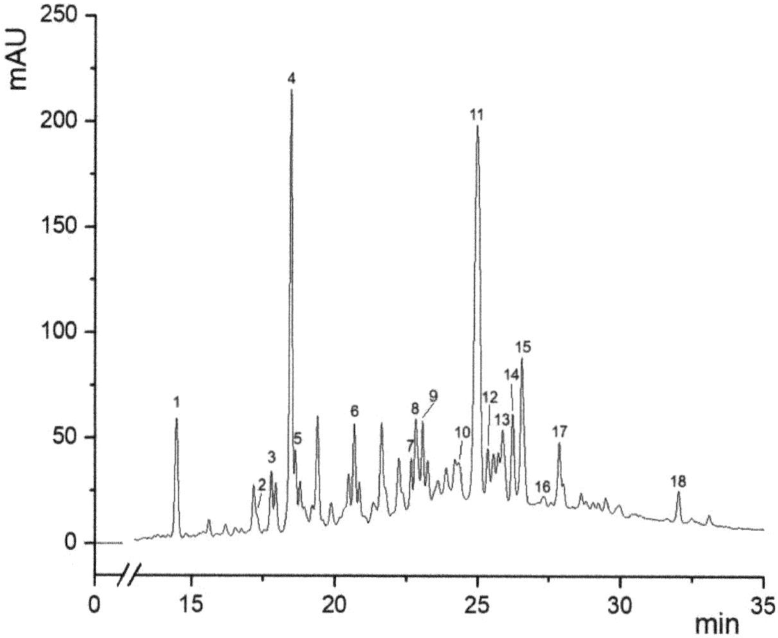

Figure 5. HPLC-DAD profile (280 nm) of the extract obtained after SLE under optimized conditions (temperature of 70 °C, solvent 75% ethanol–water mixture, time of 34 min, and S/L ratio of 0.05 g/mL).

Table 7. High-resolution mass spectrometry (LC-Q/TOF)—based assignment of chromatographic peaks in the extract obtained after SLE under optimized conditions (temperature of 70 °C, solvent 75% ethanol–water mixture, time of 34 min, and S/L ratio of 0.05 g/mL).

N.	Compound	Formula	[M-H]- Theoretical (*m*/*z*)	[M-H]- Measured (*m*/*z*)	Error (ppm)
1	Neochlorogenic acid	$C_{16}H_{18}O_9$	353.0872	353.0867	−1.42
2	1-*O*-galloyl-l-rhamnose	$C_{13}H_{16}O_9$	315.0716	315.0714	−0.63
3	Catechin	$C_{15}H_{14}O_6$	289.0712	289.0714	0.69
4	Chlorogenic acid	$C_{16}H_{18}O_9$	353.0872	353.0869	−0.85
5	Cryptochlorogenic acid	$C_{16}H_{18}O_9$	353.0872	353.0881	2.55
6	Epicatechin	$C_{15}H_{14}O_6$	289.0712	289.0719	2.42
7	Procyanidin C1	$C_{45}H_{38}O_{18}$	865.1979	865.1990	1.27
8	Apigenin-6-C-glucoside-8-*C*-arabinoside	$C_{26}H_{28}O_{14}$	563.1400	563.1411	1.95
9	Luteolin-7-*O*-glucoside	$C_{21}H_{20}O_{11}$	447.0927	447.0932	1.12
10	Isovitexin-2″-*O*-arabinoside	$C_{26}H_{28}O_{14}$	563.1400	563.1398	−0.36
11	Apigenin-8-*C*-glucoside-2′-rhamnoside	$C_{27}H_{30}O_{14}$	577.1557	577.1565	1.39
12	Isovitexin (apigenin-6-*C*-glucoside)	$C_{21}H_{20}O_{10}$	431.0978	431.0975	−0.70
13	Vitexin (apigenin-8-*C*-glucoside)	$C_{21}H_{20}O_{10}$	431.0978	431.0982	0.93
14	Hyperoside (quercetin-3-*O*-galactoside)	$C_{21}H_{20}O_{12}$	463.0876	463.0871	−1.08
15	Isoquercitrin (quercetin-3-*O*-glucoside)	$C_{21}H_{19}O_{12}$	463.0876	463.0877	0.22
16	Nothofagin	$C_{21}H_{24}O_{10}$	435.1291	435.1297	1.38
17	Quercetin 3-*O*-(6″-*O*-malonyl)-glucoside	$C_{24}H_{22}O_{15}$	549.0880	549.0888	1.46
18	Cinchonain 1A	$C_{24}H_{20}O_9$	451.1029	451.1036	1.55

Moreover, a comprehensive review of the chemical composition of hawthorn leaves reported that flavonoids are the main components of hawthorn leaves and that over 60 flavonoids and flavonoid glycosides have been identified, including vitexin derivatives [3].

Apigenin 8-*C*-glucoside (vitexin) derivatives, including *C*-glycosides and their rhamnosylated forms such as vitexin 2″-*O*-rhamnoside and isovitexin 2″-*O*-rhamnoside, are well-known compounds in *Crataegus* species. Vitexin and vitexin-2″-*O*-rhamnoside were found among the most abundant phenolic compounds in hawthorn leaves [33].

4. Conclusions

This study successfully optimized both solid–liquid extraction (SLE) and ultrasound-assisted extraction (UAE) methods for recovering phenolic compounds, flavonoids, and antioxidants from *Crataegus almaatensis* (hawthorn) leaves. Response surface methodology identified ethanol concentration as the most critical factor affecting extraction efficiency. UAE outperformed SLE by increasing total phenolic content by up to 16%, while requiring significantly less ethanol (40% vs. 75%), highlighting its effectiveness and sustainability. Temperature showed a significant non-linear effect, particularly in SLE, while ultrasound power and extraction time had minimal impact, supporting the potential for energy-efficient processing.

Phytochemical analysis (HPLC-DAD, LC-Q/TOF-MS) confirmed the presence of diverse bioactives, most notably chlorogenic acid and apigenin-8-C-glucoside-2′-rhamnoside, which contribute to the antioxidant and anti-inflammatory potential of hawthorn leaves. These findings establish UAE as a greener and more efficient technique for producing high-value extracts and support the future development of functional food, nutraceutical, and pharmaceutical applications. Overall, this research validates the utilization of the UAE process as a superior extraction method for the valorization of hawthorn leaves, offering both higher extraction efficiency and a greener approach through reduced solvent use compared to SLE. The results pave the way for the future development of standardized extracts and formulations from hawthorn leaves, contributing to a broader utilization of endemic plant species in food and health-promoting products. Further investigation into anti-inflammatory, cytoprotective effects on human cell lines will help guide the potential application of the optimized extracts.

Supplementary Materials: The following supporting information can be downloaded at: https://www.mdpi.com/article/10.3390/antiox14081003/s1. Table S1. Reduced second-order polynomial model equations for TPC, TFC, and antioxidant activity (FRAP, DPPH, and ABTS) in extracts from hawthorn leaves powder obtained after SLE process. Table S2. Reduced second-order polynomial model equations for TPC, TFC, and antioxidant activity (FRAP, DPPH, and ABTS) in extracts from hawthorn leaves powder obtained after UAE process. Table S3. Coefficients of the full second-order polynomial model for TPC, TFC, and antioxidant activity (FRAP and DPPH) in extracts from hawthorn leaves powder obtained after SLE process. Table S4. Coefficients of the full second-order polynomial model for TPC, TFC, and antioxidant activity (FRAP and DPPH) in extracts from hawthorn leaves powder obtained after UAE process. Figure S1. Correlation of predicted vs. actual values of TPC, TFC, FRAP, DPPH, and ABTS of extracts obtained after SLE process of hawthorn leaves powder. Figure S2. Correlation of predicted vs. actual values of TPC, TFC, FRAP, DPPH, and ABTS of extracts obtained after UAE process of hawthorn leaves powder. Figure S3. Normal plots of residuals for TPC, TFC, FRAP, DPPH, and ABTS of extracts obtained after SLE process of hawthorn leaves powder. Figure S4. Normal plots of residuals for TPC, TFC, FRAP, DPPH, and ABTS of extracts obtained after UAE process of hawthorn leaves powder.

Author Contributions: Conceptualization, G.F. and Z.N.; methodology, A.K. and S.C.; software, S.C.; validation, G.F.; investigation, A.K., S.C. and A.S.; resources, E.A. and A.S.; data curation, G.F. and

Z.N.; writing—original draft preparation, A.K., S.C. and E.A.; writing—review and editing, G.F. and Y.P.; visualization, S.C.; supervision, G.F.; project administration, Z.N.; funding acquisition, Z.N. All authors have read and agreed to the published version of the manuscript.

Funding: This research has been funded by the Science Committee of the Ministry of Science and Higher Education of the Republic of Kazakhstan, Grant No. AP23489874, the project "Development of gentle technology for production of natural extracts on processing of local plant raw materials for enrichment of sweets".

Institutional Review Board Statement: Not applicable.

Informed Consent Statement: Not applicable.

Data Availability Statement: The data presented in this study are available within the article. The original data that support the findings of this study are available from the corresponding author upon reasonable request.

Acknowledgments: The authors would like to thank Francesco Siano and Gianluca Picariello for their invaluable help with LC-Q/TOF-MS analysis and data interpretation.

Conflicts of Interest: The authors declare no conflicts of interest.

References

1. Zhao, M.; Mu, Y.; Shi, Z.; Wang, X.; Liu, W.; Zhou, Y.; Yi, H.; Zhang, L.; Zhang, Z. Effects of Different Lactic Acid Bacteria on the Physicochemical Properties, Functional Characteristics and Metabolic Characteristics of Fermented Hawthorn Juice. *Food Chem.* **2025**, *470*, 142672. [CrossRef]
2. Soares, S.S.; Bekbolatova, E.; Cotrim, M.D.; Sakipova, Z.; Ibragimova, L.; Kukula-Koch, W.; Giorno, T.B.S.; Fernandes, P.D.; Fonseca, D.A.; Boylan, F. Chemistry and Pharmacology of the Kazakh Crataegus Almaatensis Pojark: An Asian Herbal Medicine. *Antioxidants* **2019**, *8*, 300. [CrossRef] [PubMed]
3. Guo, W.; Shao, T.; Peng, Y.; Wang, H.; Chen, Z.S.; Su, H. Chemical Composition, Biological Activities, and Quality Standards of Hawthorn Leaves Used in Traditional Chinese Medicine: A Comprehensive Review. *Front. Pharmacol.* **2023**, *14*, 1275244. [CrossRef] [PubMed]
4. Chow, J.; Yang, X.; Hu, J.; Yu, Q.; Zhong, Y.; Hu, X.; Liang, J.; Zhu, C.; Yan, S.; Li, L.; et al. Gastrointestinal Absorption and Its Regulation of Hawthorn Leaves Flavonoids. *Sci. Rep.* **2025**, *15*, 658. [CrossRef]
5. Dai, H.; Wang, J.; Li, Y.; Lv, Z. Hawthorn-Leaf Flavonoid Alleviate Intestinal Health and Microbial Dysbiosis Problems Induced by Glyphosate. *Ecotoxicol. Env. Saf.* **2024**, *284*, 116901. [CrossRef]
6. Bekbolatova, E.; Kukula-Koch, W.; Baj, T.; Stasiak, N.; Ibadullayeva, N.; Koch, W.; Głowniak, K.; Tulemissov, S.; Sakipova, Z.; Boylan, F. Phenolic composition and antioxidant potential of different organs of Kazakh Crataegus almaatensis Pojark: A comparison with the European *Crataegus oxyacantha* L. flowers). *Open Chem.* **2018**, *16*, 415–426. [CrossRef]
7. Koohestani, H.; Nabilu, M.; Balooch, A. Biosynthesis and Investigation of Antibacterial Properties of Green Silver Nanoparticles Using Fruit Extracts of Wild Barberry, Medlar (*Mespilus Germanica* L.), and Hawthorn. *Food Chem. Adv.* **2025**, *6*, 100850. [CrossRef]
8. Mousa, A.M.; Allemailem, K.S. Could Hawthorn Have a Cardioprotective Impact against Obesity-Induced Cardiac Injury in Rats via Antioxidant, Hypolipidemic, Anti-Inflammatory, Antiapoptotic, and Antifibrotic Properties? *Tissue Cell* **2025**, *93*, 102673. [CrossRef]
9. Cvetković, K.; Đorđević, N.; Karabegović, I.; Danilović, B.; Dordevic, D.; Dordevic, S.; Kushkevych, I. Production and Testing of Carrageenan-Based Films Enriched with Chinese Hawthorn Extract in Strawberry Packaging. *Processes* **2025**, *13*, 379. [CrossRef]
10. Ma, J.X.; Yang, W.; Ng, C.Y.J.; Tang, X.D.; Wong, S.; Gan, R.Y.; Zhong, L. The Hawthorn (Crataegus Pinnatifida Bge.) Fruit as a New Dietary Source of Bioactive Ingredients with Multiple Beneficial Functions. *Food Front.* **2024**, *5*, 1534–1558. [CrossRef]
11. Luo, M.; Hu, J.Y.; Song, Z.Y.; Jiao, J.; Mu, F.S.; Ruan, X.; Gai, Q.Y.; Qiao, Q.; Zu, Y.G.; Fu, Y.J. Optimization of Ultrasound-Assisted Extraction (UAE) of Phenolic Compounds from Crataegus Pinnatifida Leaves and Evaluation of Antioxidant Activities of Extracts. *RSC Adv.* **2015**, *5*, 67532–67540. [CrossRef]
12. Jiao, P.; Ma, S.; He, X.; Wang, S.; Wang, P.; Yang, P.; Niu, Q. Enhanced Green Extraction of Phenolic Antioxidants from Hawthorn Leaves Based on Deep Eutectic Solvent and Ultrasonic Wave. *Waste Biomass Valorization* **2025**, *16*, 1–14. [CrossRef]
13. Martín-García, B.; Razola-Díaz, M.D.C.; Gómez-Caravaca, A.M.; Benítez, G.; Verardo, V. Setup of an Ultrasonic-Assisted Extraction to Obtain High Phenolic Recovery in Crataegus Monogyna Leaves. *Molecules* **2021**, *26*, 4536. [CrossRef]
14. Liang, J.; Zhang, H.; Zhu, C. Optimization of Ultrasonic-Assisted Aqueous Two-Phase Extraction of Flavonoids from Hawthorn Leaves Using Response Surface Methodology. *J. Phys. Conf. Ser. Inst. Phys.* **2022**, *2329*, 012041. [CrossRef]

15. Carpentieri, S.; Ferrari, G.; Pataro, G. Optimization of Pulsed Electric Fields-Assisted Extraction of Phenolic Compounds From White Grape Pomace Using Response Surface Methodology. *Front. Sustain. Food Syst.* **2022**, *6*, 854968. [CrossRef]
16. Benzie, I.F.F.; Strain, J.J. The Ferric Reducing Ability of Plasma (FRAP) as a Measure of "Antioxidant Power": The FRAP Assay. *Anal. Biochem.* **1996**, *239*, 70–76. [CrossRef] [PubMed]
17. Rapisarda, P.; Tomaino, A.; Lo Cascio, R.; Bonina, F.; De Pasquale, A.; Saija, A. Antioxidant Effectiveness as Influenced by Phenolic Content of Fresh Orange Juices. *J. Agric. Food. Chem.* **1999**, *47*, 4718–4723. [CrossRef]
18. Olędzki, R.; Lutosławski, K.; Nowicka, P.; Wojdyło, A.; Harasym, J. Non-commercial grapevines hybrids fruits as a novel food of high antioxidant activity. *Foods* **2022**, *11*, 2216. [CrossRef] [PubMed]
19. Rrucaj, E.; Carpentieri, S.; Scognamiglio, M.; Siano, F.; Ferrari, G.; Pataro, G. Sustainable Valorization of Industrial Cherry Pomace: A Novel Cascade Approach Using Pulsed Electric Fields and Ultrasound Assisted-Extraction. *Foods* **2024**, *13*, 1043. [CrossRef]
20. Mungwari, C.P.; King'ondu, C.K.; Sigauke, P.; Obadele, B.A. Conventional and Modern Techniques for Bioactive Compounds Recovery from Plants: Review. *Sci. Afr.* **2025**, *27*, e02509. [CrossRef]
21. Carpentieri, S.; Soltanipour, F.; Ferrari, G.; Pataro, G.; Donsì, F. Emerging Green Techniques for the Extraction of Antioxidants from Agri-Food By-Products as Promising Ingredients for the Food Industry. *Antioxidants* **2021**, *10*, 1417. [CrossRef]
22. Weremfo, A.; Abassah-Oppong, S.; Adulley, F.; Dabie, K.; Seidu-Larry, S. Response Surface Methodology as a Tool to Optimize the Extraction of Bioactive Compounds from Plant Sources. *J. Sci. Food Agric.* **2023**, *103*, 26–36. [CrossRef] [PubMed]
23. Rittisak, S.; Charoen, R.; Choosuk, N.; Savedboworn, W.; Riansa-ngawong, W. Response Surface Optimization for Antioxidant Extraction and Attributes Liking from Roasted Rice Germ Flavored Herbal Tea. *Processes* **2022**, *10*, 125. [CrossRef]
24. Dar, I.H.; Junaid, P.M.; Ahmad, S.; Shams, R.; Dash, K.K.; Shaikh, A.M.; Béla, K. Optimization of ultrasound-assisted extraction of Nigella sativa seed oil for enhancement of yield and antioxidant activity. *Appl. Sci.* **2023**, *6*, 104. [CrossRef]
25. Frontuto, D.; Carullo, D.; Harrison, S.M.; Brunton, N.P.; Ferrari, G.; Lyng, J.G.; Pataro, G. Optimization of Pulsed Electric Fields-Assisted Extraction of Polyphenols from Potato Peels Using Response Surface Methodology. *Food Bioproc. Tech.* **2019**, *12*, 1708–1720. [CrossRef]
26. Hiranpradith, V.; Therdthai, N.; Soontrunnarudrungsri, A.; Rungsuriyawiboon, O. Optimisation of Ultrasound-Assisted Extraction of Total Phenolics and Flavonoids Content from Centella Asiatica. *Foods* **2025**, *14*, 291. [CrossRef]
27. Dai, J.; Mumper, R.J. Plant Phenolics: Extraction, Analysis and Their Antioxidant and Anticancer Properties. *Molecules* **2010**, *15*, 7313–7352. [CrossRef]
28. Psarrou, I.; Oreopoulou, A.; Tsimogiannis, D.; Oreopoulou, V. Extraction Kinetics of Phenolic Antioxidants from the Hydro Distillation Residues of Rosemary and Effect of Pretreatment and Extraction Parameters. *Molecules* **2020**, *25*, 4520. [CrossRef]
29. Peiró, S.; Gordon, M.H.; Blanco, M.; Pérez-Llamas, F.; Segovia, F.; Almajano, M.P. Modelling Extraction of White Tea Polyphenols: The Influence of Temperature and Ethanol Concentration. *Antioxidants* **2014**, *3*, 684–699. [CrossRef]
30. Nguyen-Kim, M.T.; Truong, Q.C.; Nguyen, M.T.; Cao-Thi, B.H.; Tong, T.D.; Dao, T.P.; Tran, T.H.; Van Tan, L.; Le, X.T. Optimized Extraction of Polyphenols from Leaves of Rosemary (*Rosmarinus officinalis* L.) Grown in Lam Dong Province, Vietnam, and Evaluation of Their Antioxidant Capacity. *Open Chem.* **2021**, *19*, 1043–1051. [CrossRef]
31. Hosseini, H.; Bolourian, S.; Yaghoubi Hamgini, E.; Ghanuni Mahababadi, E. Optimization of Heat- and Ultrasound-Assisted Extraction of Polyphenols from Dried Rosemary Leaves Using Response Surface Methodology. *J. Food Process. Preserv.* **2018**, *42*, e13778. [CrossRef]
32. Pan, G.; Yu, G.; Zhu, C.; Qiao, J. Optimization of Ultrasound-Assisted Extraction (UAE) of Flavonoids Compounds (FC) from Hawthorn Seed (HS). *Ultrason. Sonochem.* **2012**, *19*, 486–490. [CrossRef]
33. Alirezalu, A.; Salehi, P.; Ahmadi, N.; Sonboli, A.; Aceto, S.; Maleki, H.H.; Ayyari, M. Flavonoids Profile and Antioxidant Activity in Flowers and Leaves of Hawthorn Species (*Crataegus* Spp.) from Different Regions of Iran. *Int. J. Food Prop.* **2018**, *21*, 452–470. [CrossRef]
34. Turkmen, D.; Dursun, A.; Caliskan, O.; Koksal Kavrak, M.; Guler, Z. Volatile Compounds, Phenolic Content, and Antioxidant Capacity in Sultan Hawthorn (*Crataegus azarolus* L.) Leaves. *J. Agric. Sci. Technol.* **2023**, *25*, 1089–1099. [CrossRef]
35. Renda, G.; Özel, A.; Barut, B.; Korkmaz, B.; Yayli, N. In Vitro Protection by Crataegus Microphylla Extracts against Oxidative Damage and Enzyme Inhibition Effects. *Turk. J. Pharm. Sci.* **2018**, *15*, 77–84. [CrossRef]
36. Chemat, F.; Rombaut, N.; Sicaire, A.G.; Meullemiestre, A.; Fabiano-Tixier, A.S.; Abert-Vian, M. Ultrasound Assisted Extraction of Food and Natural Products. Mechanisms, Techniques, Combinations, Protocols and Applications. A Review. *Ultrason. Sonochem.* **2017**, *34*, 540–560. [CrossRef]
37. Loubna, A.; Belhattab, R. Antioxidant activity of aqueous extracts from Crataegus oxyacantha leaves. *Pharmacogn. Commn.* **2015**, *5*, 229–232. [CrossRef]
38. Cui, M.; Cheng, L.; Zhou, Z.; Zhu, Z.; Liu, Y.; Li, C.; Liao, B.; Fan, M.; Duan, B. Traditional Uses, Phytochemistry, Pharmacology, and Safety Concerns of Hawthorn (Crataegus Genus): A Comprehensive Review. *J. Ethnopharmacol.* **2024**, *319*, 117229. [CrossRef]

39. Elsadig Karar, M.G.; Kuhnert, N. UPLC-ESI-Q-TOF-MS/MS Characterization of Phenolics from Crataegus Monogyna and Crataegus Laevigata (Hawthorn) Leaves, Fruits and Their Herbal Derived Drops (Crataegutt Tropfen). *J. Chem. Biol. Ther.* **2016**, *01*, 2572-0406. [CrossRef]
40. Liu, P.; Kallio, H.; Yang, B. Phenolic Compounds in Hawthorn (Crataegus Grayana) Fruits and Leaves and Changes during Fruit Ripening. *J. Agric. Food Chem.* **2011**, *59*, 11141–11149. [CrossRef]
41. Zhang, L.L.; Zhang, L.F.; Xu, J.G. Chemical Composition, Antibacterial Activity and Action Mechanism of Different Extracts from Hawthorn (Crataegus Pinnatifida Bge.). *Sci. Rep.* **2020**, *10*, 8876. [CrossRef] [PubMed]

antioxidants

MDPI

Article

Subcritical Water and Pressurised Ethanol Extractions for Maximum Recovery of Antioxidants from Orange Peel Herbal Dust with Evaluation of Its Pharmacological Potential Using In Silico and In Vitro Analysis

Slađana Krivošija [1], Ana Ballesteros-Gómez [2], Mire Zloh [3], Nataša Milić [4], Aleksandra Popović [5], Nataša Nastić [1] and Senka Vidović [1,*]

1 Department of Pharmaceutical Engineering, Faculty of Technology Novi Sad, University of Novi Sad, Boulevard Cara Lazara 1, 21000 Novi Sad, Serbia; sladjana.krivosija@uns.ac.rs (S.K.); natasa.nastic@uns.ac.rs (N.N.)
2 Department of Analytical Chemistry, Institute of Chemistry for Energy and the Environment, University of Córdoba, 14071 Cordoba, Spain; ana.ballesteros@uco.es
3 UCL School of Pharmacy, University College London, 29/39 Brunswick Square, London WC1N 1AX, UK; mire.zloh@ffns.ac.rs
4 Department of Pharmacy, Faculty of Medicine, University of Novi Sad, Hajduk Veljkova 3, 21000 Novi Sad, Serbia; natasa.milic@mf.uns.ac.rs
5 Department of Physiology, Faculty of Medicine, University of Novi Sad, Hajduk Veljkova 3, 21000 Novi Sad, Serbia; aleksandra.rakovac@mf.uns.ac.rs
* Correspondence: senka.vidovic@uns.ac.rs

Abstract: This research explored the potential of pressurised liquid extraction techniques for valorising herbal orange peel dust (OPD) waste from the filter tea industry. A series of experiments were conducted, varying the temperature (120–220 °C) and solvent (water and 50% (v/v) ethanol), while pressure and time were kept constant. Afterward, the obtained extracts were analysed by LC-ESI-MS/MS for determining the chemical composition. The highest concentrations of the most dominant compounds, the antioxidants hesperidin (662.82 ± 22.11 mg/L) and naringin (62.37 ± 2.05 mg/L), were found at specific temperatures using subcritical water extraction. In silico studies indicated that these compounds could interact with sirtuin-1 and growth factor beta receptors, suggesting potential anti-ageing benefits for skin. In vitro experiments on rat hepatoma cells (H4IIE) revealed that OPD extracts had antitumor potential, inhibiting cell proliferation and altering cell morphology. These findings underscore the importance of temperature and extraction technique in obtaining antioxidant-rich extracts with pharmacological potential. The resulting extracts, obtained using green solvents, show promise for cosmetic applications, though further in vivo studies are needed to confirm their therapeutic efficacy.

Keywords: pressurised liquid extraction; subcritical water extraction; pressurised ethanol extraction; orange peel herbal dust; waste valorisation; antitumor activity

1. Introduction

For decades, traditional herbal medicine has played a vital role in the treatment of numerous diseases. High-value bioactive compounds found in herbs represent the basis for the prevention and treatment of various diseases [1]. The thousands of years old extraction process has been improved to such an extent that not only has the maximum use of bioactive principles from plant material been achieved, but on the other hand, the emphasis is now placed on the ecology, economic profitability, and sustainability of the processes themselves. The application of such technologies not only reduces or eliminates the use of organic solvents but also contributes to the safety, quality, and applicability of plant extracts [2].

Among the various new green extraction technologies that have been developed recently, pressurised liquid extraction (PLE) enjoys great scientific interest. This extraction technique is based on the use of solvents under pressurised conditions and at temperatures between their boiling and critical points. Various liquids are applicable in the PLE technique. Among them, subcritical water extraction (SWE) is considered the most promising [3]. This attractive technique involves the use of water in a subcritical state, which is achieved at temperatures between 100 and 374.15 °C and pressures high enough to keep water in a liquid state [4]. Water, in this case, replaces conventional organic, easily volatile solvents, and also represents an easily available, cheap, safe, and effective extractant. In SWE, the pressure does not have such a significant effect on the extraction process efficiency, as long as it is high enough to keep the water in a liquid state [3]. On the other hand, temperature is the most significant influencing factor, because with increasing temperature, there is better and easier mass transfer between medium and matrix, a higher solubility of dissolved substances, and a diffusion speed at lower solvent viscosity and surface tension [5,6]. Namely, with an increase in temperature, the physical and chemical properties of water can vary significantly, especially its dielectric constant [7,8]. Also, with the increase in this parameter, water changes from a polar to a non-polar solvent, which makes it a potential multifunctional solvent for extraction. What can also greatly affect the quality of SWE extracts is the extraction time. Long-term exposure to high temperatures can lead to the degradation of thermosensitive compounds or the creation of new antioxidants [6,9]. Consequently, in order to ensure the most efficient process possible, it is necessary to optimise the process conditions in terms of minimum energy consumption and operating costs, with maximum yield of bioactive compounds.

In PLE, it is also possible to use a mixture of water/ethanol and ethanol as solvents because they are also generally recognised as safe solvents [10]. Subcritical ethanol or ethanol under pressure is heated and maintained in a subcritical state under temperatures between 78.37 °C and 240.8 °C and pressures between 1 and 6.13 MPa [11]. The chemical properties of ethanol also change under the influence of temperature, as the viscosity, dielectric constant, and surface tension decrease with increasing temperature, while diffusivity increases. The effects of pressurised ethanol extraction (PEE) on product yield and quality may be different depending on operating conditions, and the obtained bioactive products represent a potentially important source of resources for the sectors of pharmaceuticals, food, etc. [12]. A study conducted by Mufari et al. [13] demonstrated the efficacy of subcritical extraction using an ethanol/water mixture to isolate the bioactive principles of germinated quinoa. Subcritical extraction using aqueous ethanol was also studied on defatted rice bran [14], and it improved the cold-water swelling ability of oat starch [15]. Moreover, subcritical treatment using ethanol and aqueous ethanol solution showed the ability to achieve a high total phenol content and antioxidant activity of rice stem, as well as *Arctium lappa* leaves [16,17]. However, scarce information has been reported on the extraction of bioactive compounds from orange peel using this type of extraction. Therefore, PEE, as another green technique, was chosen in order to investigate the influence of ethanol's addition to SWE, as well as to compare both the yield and the profile of the total content of polyphenols from the herbal dust of orange peel.

Orange fruit (*Citrus sinensis* L.) is the most common citrus fruit grown worldwide, even being grown in over 130 countries. Orange peel waste obtained during the production of either juice or canned food is one of the most abundant wastes in the food industry.

Annually, more than 76 thousand tons of oranges are produced worldwide [18], and consequently, due to large production or processing, a large amount of waste is created, even up to 50% of the total fruit weight [19], which further causes large economic–industrial costs. This valuable biowaste is underutilised and is mostly either dumped in landfills or used as compost [20,21].

However, considering the compositional profile of the peel, its transformation into value-added products is considered a promising option from an economic and ecological perspective [22]. Therefore, more and more work is being performed on the analysis of such raw material in an attempt to valorise it and reuse it. Orange peel is a chemically complex substrate and can be used for the extraction and isolation of high-value compounds like proteins, fibres, polysaccharides, phenolic compounds, aromatic compounds, vitamins, and carotenoids. Naringin and hesperidin stand out as the most abundant flavonoids in orange peel [23]. The bioactivity shown by the polyphenolic compounds of orange peel has long been known [24], and, in addition to the dominant antioxidant effect, may include chemopreventive, anti-inflammatory, and antimicrobial effects [25]. The citrus peel extracts have also shown considerable skin anti-ageing activity [26], with their components explored for the development of topical applications, including hesperidin as nanocrystals [27] and naringin in microemulsion formulations [28].

Orange peel is important in the production of herbal teas as an aromatic ingredient. In the process of filter tea production, orange peel dust is generated as food waste. The particle size of this powder is smaller than the pore size of the filter paper (<0.315 mm), and accordingly, it cannot be used for the production of filter tea and is considered waste. Although a certain loss of bioactivity of these materials has been observed, generally, herbal dusts still represent an exceptional source of bioactive compounds. This has already been confirmed in several studies conducted using different techniques for the valorisation of such waste [29–31].

The effectiveness of PLE for the extraction of bioactive compounds from orange peel herbal dust using a combination of pressurised water and ethanol has not, to our knowledge, been evaluated so far. Therefore, this study was designed to gain new insight into this topic and contribute to the development of new uses of herbal orange peel dust using more environmentally friendly techniques to reduce the amount of waste in the filter tea industry. Therefore, the effect of extraction temperature, as the most important parameter affecting the efficiency of SWEs and PEEs was evaluated in terms of its impact on the recovery, stability, and degradation of phenolic constituents. The obtained extracts were analysed using liquid chromatography–high resolution tandem mass spectrometry (LC-ESI-MS/MS) in order to identify and quantify the main polyphenolic compounds that also contribute to the use of extracts in one of the many branches of industry. Further, molecular docking was used as a tool that would highlight the obtained extract's potential to affect the sirtuin-1 and growth factor beta (TGF-β) receptors, protein targets involved in pathways related to skin ageing. Finally, the antitumor potential of the extracts was evaluated using an in vitro model based on rat hepatoma cells (H4IIE) to further explore their pharmaceutical relevance.

2. Materials and Methods

2.1. Plant Material and Chemicals

The investigated material, orange peel herbal dust (*Citrus sinensis* L.), was obtained from the local filter tea factory Fructus d.o.o. (Bačka Palanka, Serbia). The orange peel dust (OPD) was created as waste during the process of cutting, grinding, and fractionating the raw material in the filter tea factory. The particle size of the material obtained in this way was <0.315 mm, while the moisture content in the obtained material was 3.36 ± 0.63%.

All reagents were of analytical grade. Gallic acid and Folin–Ciocalteu reagent were supplied by Sigma-Aldrich (Steinheim, Germany). Methanol was supplied by Riedel-de Haën (Seelze, Germany), while sodium carbonate was purchased from Aldrich (St. Louis, MO, USA). Ultrapure water was produced in an Elix® Essential 3 water purification system from Merck Millipore (Madrid, Spain). Formic acid, ammonium formiate, and standards for phenolic compounds (hesperidin, naringin, p-coumaric acid, caffeic acid, and gallic acid) were acquired from Aldrich (St. Louis, MO, USA).

2.2. Pressurised Liquid Extraction

Both types of extraction—subcritical water extraction (SWE) and pressurised ethanol extraction (PEE)—were performed using a high-pressure serial extractor (Parr Instrument Company, Moline, IL, USA). The extractor was filled with 7 g of OPD plant material and 140 mL double-distilled water for SWE, while for PEE, 50% (*v*/*v*) ethanol was used as an extraction solvent. The independent variable for both types of extraction (SWE and PEE) was temperature (120–220 °C), while pressure (20 bar) and time (15 min) were kept constant. The obtained extracts were filtered through filter paper under a vacuum and stored at 4 °C until further analysis.

During further analyses, OPD extracts were named as SWE and PEE depending on the type of extraction by which they were obtained, as well as given numbers from 1 to 6 depending on the applied process temperature (from the lowest to the highest).

2.3. Determination of Extraction Yield and Total Phenol Content in the OPD Extracts

The extraction yield (EY) was determined by evaporating a known volume of solvent extract under vacuum. After that, drying was carried out at 105 °C until a constant mass was reached. All experiments were performed in triplicate and the results were expressed as mean values in percentage (%), i.e., the mass of the dry extract obtained (g) per 100 g of dry plant material.

The total phenol content (TPC) in OPD extracts, obtained by SWE or PEE, was determined by the Folin–Ciocalteu method. For TPC quantification, the SWE and PEE extracts were diluted 1:10 *v*/*v* with an adequate solvent (solvent used for extraction). After that, 50 μL of the diluted sample, 1.5 mL of distilled water, 100 μL of 0.1N Folin–Ciocalteu reagent, and 300 μL of sodium carbonate solution (200 g/L) were added to a 2 mL Eppendorf tube. After 90 min of incubation in the dark, absorbance was measured at 760 nm using a Thermo Spectronic Helios ε spectrophotometer from Lubbock (Madrid, Spain). Quantitative analysis was carried out using standard solutions of gallic acid prepared in ultrapure water in concentration range (5–1000 mg/L), which were then processed as reported before [32]. The results were expressed as gallic acid equivalents per gram of dry weight (mg GAE/g dw). All experiments were performed in triplicate and the results were expressed as mean values.

2.4. LC-ESI-MS/MS Analysis of OPD Extracts

OPD extracts obtained by SWE or PEE were analysed by liquid chromatography coupled with high-resolution mass spectrometry for both the quantitation of representative phenolic compounds (hesperidin, naringin, p-coumaric acid, caffeic acid, and gallic acid) and for the further screening of other compounds of interest (other phenolic compounds, fatty acids, etc.). For this purpose, we employed a Bruker ELUTE UHPLC (Bruker Daltonics, Bremen, Germany) device connected to a hybrid triple quadrupole/TOF of ion mobility (TimsTOF, Q-TOF) and equipped with an ESI source operating in a positive and negative mode from Bruker Daltonics (Bremen, Germany).

The stationary phase, chromatographic conditions, and ESI source parameters were reported by Sánchez-Vallejo et al. [32]. Acquisition for quantitation was performed with the bbCID (broadband collision induced dissociation) mode and data were processed with the programme TASQ Client 2.1 (Bruker Daltonics). For the identification of other compounds of interest, the acquisition was performed by the autoMSMS mode and data treatment was performed with Metaboscape 6.0.2 (Bruker Daltonics) and open libraries (EU Mass Bank), together with those administered by the equipment vendor for plant metabolites.

OPD extracts were obtained, as described in Section 2.2, by being diluted with methanol, centrifuged (5 min, 10,000 rpm) to remove particles, and then 100 μL of the supernatant was transferred to LC vials with an insert and analysed. Calibration curves were prepared in methanol at concentrations in the ranges of 0.1–1 mg/L for p-coumaric acid, 0.1–0.5 mg/L for caffeic and gallic acids, and 2–25 mg/L for hesperidin and naringin. For the screening of other compounds of interest, identification criteria were set on the basis of exact mass ($\leq$10 ppm), isotopic pattern fit (mSigma $\leq$ 200), and MS/MS fragmentation score (>900).

2.5. In Silico Studies

A set of protein targets relevant to skin health and ageing was identified and the crystal structures were downloaded from the Protein Data Bank (PDB): 2TCL (catalytic domain of fibroblast collagenase), 1ELB (elastase), 4I5I (catalytic domain of NAD-dependent deacetylase sirtuin-1), and 1VJY (the growth factor beta receptor type). All selected protein targets were prepared by removing water molecules and adding missing hydrogen atoms. The co-crystalized ligands in the binding site were removed from structures prior to docking and their centre of mass was used to position a box with a size of 20 Å × 20 Å × 20 Å. The docking of two selected components was conducted using Gold v. 2024.1.0 software and the DockThor web server platform, both using genetic algorithms to generate ligand poses. The docking performed by GOLD software was conducted with GA run for each compound. In a single GA run, 125,000 operations were performed, with the rest of the settings set to autoscale level 2. GoldScore was selected for a scoring function which was used to rank the docking positions of the molecules. The genetic algorithm parameters for DockThor were set to its default values: number of evaluations = 500,000, population size = 750, number of runs = 12, and seed at run #1 = −1985.

2.6. Cell Culture

A commercially available rat hepatoma H4IIE cell line was purchased from the American Type Culture Collection (Manassas, VA, USA) (ATCC cat. no. CRL-1548). The cells were cultured and maintained in minimum essential medium (MEM) supplemented with 10% foetal bovine serum (FBS), 1% penicillin/streptomycin (100 U/mL penicillin and 100 μg/mL streptomycin), and 1% L-glutamine in a humidified atmosphere of 5% CO_2 at 37 °C. The logarithmic growth phase (approximately 70–80% confluence) was used for the following tests. All chemicals were purchased from Capricorn Scientific GmbH (Ebsdorfergrund, Germany).

2.7. Preparation of Extract Solution for Treatment

The cells were treated for 24 and 48 h with different concentrations of OPD and its extracts (SWE 3 and PEE 4) ranging from 1 to 300 μg/mL (a stock solution concentration was 3 mg/mL), which were dissolved in fresh culture medium MEM and filtered two times through a bacteriological filter (JCAS022025K, Filter-lab CA syringe filter, 0.22 μm, Filtros Anoia, Barcelona, Spain) to reduce the risk of the contamination of extract solutions. Control cells were treated with fresh medium.

2.8. H4IIE Cell Viability Measurement

Cell viability was determined using the MTT [3-(4, 5-dimethylthiazol-2-yl)-2, 5-diphenyltetrazoli-um bromide], as described earlier [33]. H4IIE cells were seeded into a 96-well plate at a density of 2×10^4 cells/well and cultured overnight. Different concentrations of SWE 3 and PEE 4 extracts (0–300 μg/mL) were added and incubated for 24 and 48 h. Following the treatment, 20 μL of the MTT solution (5 mg/mL in MEM) was added and further incubated for 3 h at 37 °C. The insoluble blue-coloured formazan product was dissolved with 100 μL of 0.04 M HCl/isopropanol and incubated at room temperature for 10 min. Absorbance was measured using a 96-well plate reader (Multiscan MCC340, Labsystems, Helsinki, Finland) at a wavelength of 540/690 nm. Measurements were performed two times in quadruplicate. The cell viability was calculated as a percentage using the underlying equation:

$$\text{Cell viability (\%)} = \frac{\text{the absorbance of viable cells per well}}{\text{the mean absorbance of the viable cells in control wells}} \times 100\%$$

2.9. Morphological Observation

H4IIE cells were plated into six-well plates at a density of 1.5×10^4 cells/well and cultured overnight. Then, the cells were exposed to treatment for 24 and 48 h. The morphology and shape of the H4IIE cells were observed and photographed under an inverted phase contrast microscope (Leica DMIL, LED microscope, Wetzlar, Germany) with the camera Leica MC 190 at 200× magnification. Analysis was performed in a population of 1000 cells.

2.10. Colony Formation Assay (Long-Term Cell Survival Assay)

Cells (2×10^3 cells/well) were initially grown in six-well plates overnight. The cells were then treated with prepared extract solutions with different concentrations (0–300 μg/mL) for 24 h at 37 °C. Subsequently, the treatment was discarded, the cells were washed with PBS and a fresh complete medium was added. The medium was changed twice a week. After 14 days, colonies were fixed with 100% methanol for 20 min at room temperature and stained with 0.5% crystal violet solution [34]. Due to a large number of colonies, the results were interpreted only qualitatively using a stereomicroscope (OZR 564, KERN Optics, Kern & Sohn GmbH, Balingen-Frommern, Germany).

2.11. Statistical Analysis

Results related to EY, TPC, and LC-ESI-MS/MS quantification were presented as mean ± standard deviation (SD). Statistical differences between samples were assessed using one-way analysis of variance (ANOVA), followed by Tukey's post hoc test, with significance considered at $p < 0.05$.

For cytotoxicity assays, the results were also presented as mean ± SD. Variations between groups were statistically analysed in SPSS v26.0 (IBM Corp., Chicago, IL, USA) using one-way ANOVA with Tukey's post hoc test. The graphical presentation of the results was carried out with GraphPad Prism software package version 7.05 (San Diego, CA, USA). Statistical significance was considered as follows: * $p < 0.05$ and # $p < 0.001$.

3. Results and Discussion

3.1. Extraction Yield and Total Phenolic Content of Obtained OPD Extracts

Various factors are known to influence the presence of biologically active substances in orange peel, including cultivation practices, climatic conditions, morphology, soil composition, and variety, but also the type and parameters of extraction. As a strategy for OPD valorisation and the recovery of biologically active compounds from this kind of waste material, two green techniques (SWE and PEE) were applied under different process set-ups in order to evaluate the impact of process parameters and to define the extraction protocol which would result in the production of OPD extracts of the highest quality. Extraction conditions and results are given in Table 1.

Table 1. Extraction yield (EY) and total phenolic content (TPC) obtained using pressurised liquid extractions (PLEs).

Samples	Pressure [bar]	Temperature [°C]	EY [%]	TPC [mg GAE/g Dry Weight]
SWE orange peel 1	20	120	60.22 ± 0.86 [a, 2–4]	13.83 ± 0.001 [b, 6]
SWE orange peel 2	20	140	66.67 ± 0.45 [a, 1–2]	20.51 ± 2.85 [b, 5]
SWE orange peel 3	20	160	64.43 ± 1.50 [a, 1–3]	32.39 ± 2.85 [a, 4]
SWE orange peel 4	20	180	52.37 ± 0.76 [ab, 4–5]	36.52 ± 0.30 [a, 4]
SWE orange peel 5	20	200	42.07 ± 2.62 [bc, 5–6]	35.99 ± 3.15 [a, 4]
SWE orange peel 6	20	220	30.03 ± 0.64 [c, 7]	35.67 ± 1.50 [a, 4]
PEE orange peel 1	20	120	54.48 ± 0.90 [D, 3–4]	16.59 ± 1.20 [E, 5–6]
PEE orange peel 2	20	140	63.03 ± 1.36 [BC, 1–4]	18.92 ± 2.40 [E, 5–6]
PEE orange peel 3	20	160	72.60 ± 2.07 [A, 1]	32.17 ± 1.05 [D, 4]
PEE orange peel 4	20	180	66.52 ± 0.77 [B, 1–2]	46.07 ± 2.10 [C, 3]
PEE orange peel 5	20	200	60.70 ± 0.22 [C, 2–4]	56.88 ± 3.60 [B, 2]
PEE orange peel 6	20	220	38.93 ± 1.43 [E, 6–7]	70.56 ± 5.85 [A, 1]

Different letters within a column indicate significant differences between samples at $p < 0.05$: lowercase letters are related to SWE extracts, uppercase letters are associated with PEE extracts, and numbers represent the difference between SWE and PEE extracts. SWE—subcritical water extraction; PEE—pressurised ethanol extraction.

The EY values, obtained in the case of the SWE of OPD, ranged from 30.03 to 66.67%, showing a slight dependence of the yield on the applied process parameters. EY increased as the temperature increased from 120 °C to 140 °C, then a slight decreasing trend began with a further increase in the applied temperature. The maximum EY value of SWE extracts (66.67 ± 0.45%) was obtained at a temperature of 140 °C, an extraction time of 15 min, and a process pressure of 20 bar, while the lowest yield (30.03 ± 0.64%) was obtained at the highest temperature and was 2.2 times lower compared to the maximum. In the case of the PEE of OPD, the yield increase showed a slightly slower trend compared to SWE, and EY grew and reached the maximum yield at a temperature of 160 °C (72.60 ± 2.07%), after which the yield also began to decrease, and it was much more noticeable in the interval between 200 °C and 220 °C. If we compare these two techniques in terms of yield, it can be concluded that the maximum yield with PEE was achieved at a slightly higher temperature and that it was 1.09 times higher than the maximum yield achieved using SWE.

Thanks to a series of studies conducted so far, it is now well established that high-value polyphenolic compounds in citrus fruits play a very important role both in improving human health and delaying/preventing the onset of a large number of chronic diseases [35]. They have shown exceptional activity in removing free radicals to protect fruits and vegetables from infection by various pathogens, preventing photo-oxidative damage by UV radiation, etc. [36]. Polyphenolic compounds in citrus peels possess a wide range of bioactivities that improve human health, including anti-inflammatory, anti-allergic, antiproliferative, antioxidant, antiviral, neuroprotective, anti-carcinogenic, and antimicrobial properties that protect against oxidative stress-related diseases [35,37]. It was previously demonstrated by Guimarães et al. [38] that citrus peels contained higher levels of TPC than other parts of citrus fruits.

In this work, TPC in different SWE OPD extracts ranged from 13.83 to 36.52 mg GAE/g dw. A significant increase in the phenol content was observed from 120 to 180 °C reaching a maximum at 180 °C (36.52 ± 0.30 mg GAE/g dw). With a further increase in temperature, the phenol content decreased, and the SWE OPD extract obtained at the highest temperature of 220 °C had 1.02 times lower phenol content compared to the highest value, while the lowest TPC was obtained at the lowest temperature of 120 °C and was 2.64 times lower compared to the highest value. The increase in TPC originated from the increased solubility of this group of compounds in subcritical water due to the increase in process temperature. Water at room temperature has a relatively high dielectric constant, which decreases due to the application of high temperatures and therefore increases the extraction power of water. Moreover, the application of high temperatures favours the kinetics of mass transfer by disrupting the interaction of the matrix and the analyte, especially dipole–dipole forces and hydrogen bonding, which greatly facilitates the initial desorption of the analyte from the sample matrix. Also, there is an acceleration of diffusion and a change in solubility [39]. However, higher temperatures, in this case higher than 180 °C, lead to the decomposition of phenolics. Regarding PEE, the TPC ranged from 16.59 to 70.56 mg GAE/g dw. In relation to SWE, in this type of extraction, TPC constantly increased with the increase in temperature, and the maximum was obtained at 220 °C (70.56 ± 5.85 mg GAE/g dw). The obtained TPC was almost 2 times higher compared to the maximum TPC obtained using subcritical water. The increase in TPC with the applied temperature was recorded by Benito-Román et al. [40], who isolated the highest TPC from onion skin waste using SWE at the highest applied temperature (180 °C).

In a study conducted by Brezo-Borjan et al. [41], the influence of temperature (120–200 °C) and time (5–60 min) on TPC from orange peel obtained using SWE was investigated. The TPC ranged from 27.58 mg GAE/g dw to 45.45 mg GAE/g dw, which is in agreement with the results obtained in this study.

3.2. LC-ESI-MS/MS Profiling of Liquid OPD Extracts

The LC-ESI-MS/MS quantitative analysis of OPD extracts, obtained in SWE and PEE, confirmed the presence of all the targeted bioactive compounds, namely hesperidin and naringin, followed by p-coumaric, gallic, and caffeic acids. According to the literature, the two most dominant flavonoids in the fresh and dried orange by-products were hesperidin and narirutin [42]. This was also confirmed in our previous study where ultrasound-assisted extraction was applied for the recovery of hesperidin and narirutin, as well as naringin and rutin, and the results showed that these compounds could be extracted in significant amounts from OPD [43]. M'hiri et al. [44] identified 10 phenolic compounds in ethanol extracts of Maltese orange peel obtained by different methods, namely flavanones (eriocitrin, narirutin, naringin, hesperidin, neohesperidin, didimine) and polymethoxylated flavones (sinensetin, hexamethoxyflavone, nobiloxyflavone). Also, hesperidin was one of the most dominant compounds in the study conducted by Londono-Londono et al. [45].

As already mentioned, the most common flavonoid in orange peel is hesperidin. According to the results, the same flavonoid is the common flavonoid of OPD, which is meaningful, as OPD is produced during the orange peel processing; therefore, it is the same natural material. Based on Table 2, it is obvious that hesperidin is the most dominant flavonoid in all tested OPD extracts. Naringin was present as the second most dominant compound, while phenolic acids were identified as supporting elements for which the concentration was quite low.

Table 2. Compounds quantified by liquid chromatography–high resolution tandem mass spectrometry (LC-ESI-MS/MS) in orange peel dust (OPD) extracts.

Samples	Temperature [°C]	Hesperidin [mg/L]	Naringin [mg/L]	p-Coumaric Acid [mg/L]	Gallic Acid [mg/L]	Caffeic Acid [mg/L]
SWE orange peel 1	120	$180.28 \pm 20.05^{c, 5}$	$62.37 \pm 2.05^{a, 1}$	$0.32 \pm 0.01^{b, 3–4}$	$0.02 \pm 0.001^{d, 5}$	$0.07 \pm 0.001^{b, 4–5}$
SWE orange peel 2	140	$401.02 \pm 21.21^{b, 3–4}$	$57.65 \pm 8.41^{b, 1–3}$	$0.31 \pm 0.06^{b, 3–4}$	$0.05 \pm 0.01^{c, 3–4}$	$0.18 \pm 0.01^{a, 2–3}$
SWE orange peel 3	160	$662.82 \pm 22.11^{a, 1}$	$59.99 \pm 2.65^{ab, 1–2}$	$0.40 \pm 0.05^{a, 3}$	$0.12 \pm 0.001^{a, 1}$	$0.19 \pm 0.01^{a, 1–2}$
SWE orange peel 4	180	$173.43 \pm 13.80^{c, 5}$	$12.17 \pm 1.09^{c, 7}$	$0.21 \pm 0.16^{c, 4}$	$0.12 \pm 0.001^{a, 1}$	$0.06 \pm 0.001^{b, 4–6}$
SWE orange peel 5	200	$6.39 \pm 8.99^{d, 6}$	$0.83 \pm 0.42^{d, 8}$	$0.04 \pm 0.03^{d, 5}$	$0.12 \pm 0.001^{a, 1}$	$0.04 \pm 0.02^{c, 6–7}$
SWE orange peel 6	220	$0.09 \pm 0.10^{d, 6}$	$0.14 \pm 0.20^{d, 8}$	$0.05 \pm 0.05^{d, 5}$	$0.07 \pm 0.01^{b, 2}$	-
PEE orange peel 1	120	$366.42 \pm 12.75^{BC, 3–4}$	$49.49 \pm 1.07^{AB, 3–4}$	$0.38 \pm 0.07^{C, 3}$	$0.01 \pm 0.001^{C, 5}$	$0.03 \pm 0.01^{E, 7}$
PEE orange peel 2	140	$325.50 \pm 40.26^{C, 4}$	$46.56 \pm 3.61^{AB, 4–5}$	$0.76 \pm 0.05^{B, 2}$	$0.02 \pm 0.01^{C, 5}$	$0.08 \pm 0.01^{C, 4}$
PEE orange peel 3	160	$441.76 \pm 58.88^{AB, 2–3}$	$51.49 \pm 3.57^{A, 2–4}$	$1.19 \pm 0.05^{A, 1}$	$0.04 \pm 0.01^{B, 4}$	$0.21 \pm 0.01^{A, 1}$
PEE orange peel 4	180	$491.77 \pm 28.89^{A, 2}$	$40.56 \pm 2.38^{B, 5}$	$0.66 \pm 0.09^{B, 2}$	$0.07 \pm 0.001^{A, 2}$	$0.16 \pm 0.02^{B, 3}$
PEE orange peel 5	200	$323.84 \pm 28.72^{C, 4}$	$25.15 \pm 3.04^{C, 6}$	$0.29 \pm 0.10^{C, 3–4}$	$0.07 \pm 0.001^{A, 2}$	$0.05 \pm 0.02^{DE, 5–7}$
PEE orange peel 6	220	$5.81 \pm 4.99^{D, 6}$	$0.66 \pm 0.26^{D, 8}$	$0.04 \pm 0.00^{D, 5}$	$0.06 \pm 0.01^{A, 2–3}$	$0.06 \pm 0.03^{CD, 4–6}$

Different letters within a column indicate significant differences between samples at $p < 0.05$: lowercase letters are related to SWE extracts, uppercase letters are associated with PEE extracts, and numbers represent the difference between SWE and PEE extracts. SWE—subcritical water extraction; PEE—pressurised ethanol extraction.

According to the literature, Lachos-Perez et al. [46] investigated the possibility of using SWE to obtain high-added-value products from orange peel (OP) as a first step in a biorefinery concept. Hesperidin and narirutin were identified as the most dominant flavonoids. The yield of hesperidin ranged from 9.82 ± 0.01 to 22.99 ± 0.7 mg/g OP, while the yield of narirutin was significantly lower (1.9 ± 0.2 mg/g OP), reaching the highest value at 150 °C using a solvent flow of 10 mL/min. A study by Barrales et al. [47] aimed to investigate and compare the possibility of extracting phenolic compounds from the orange peel with and without supercritical carbon dioxide pretreatment using three different techniques: PLE, UAE, and Soxhlet extraction. Among the applied extraction methods, PLE achieved the highest yields of glycosylated flavonoids such as hesperidin, naringin, and narirutin. Among the identified flavonoids, hesperidin was the most abundant, which corresponds to the typical composition of flavonoids in citrus peel [48]. The PLE solvents were absolute ethanol and mixtures of ethanol and water (75% and 50% ethanol, v/v) at temperatures of 45, 55, and 65 °C and a pressure of 10 MPa. The best PLE conditions for the isolation of hesperidin were the use of 75% ethanol at 65 °C; however, when this concentration was compared to the extraction conducted using 50% ethanol, it could be concluded that the concentration of hesperidin was not significantly different. Hwang et al. [49] proved in their research that the extraction efficiency of narirutin and hesperidin from *Citrus unshiu* peel was improved by the pulsed electric field (PEF). Therefore, the samples were first treated with a PEF of 3 kV/cm for 60 and 120 s. After that, SWE was performed at temperatures ranging from 110 to 190 °C for 3 to 15 min. The maximum concentration of hesperidin was 46.96 mg/g dry peel after PEF treatment for 120 s, with a temperature of 150 °C for 15 min, while the highest concentration of narirutin was 8.7 mg/g dry peel after PEF treatment for 120 s, with a temperature of 190 °C for 5 min. Narirutin and hesperidin increased with the duration of PEF treatment. PEF increased the amount of extracted hesperidin by 22.1% and narirutin by 33.6%, respectively.

The concentration of hesperidin in the liquid OPD extracts obtained in SWE ranged from 0.09 ± 0.10 to 662.82 ± 22.11 mg/L depending on the applied temperature. The obtained values are in accordance with the previously mentioned literature. The highest concentration was obtained at a temperature of 160 °C, and after that, there was a significant decrease in the concentration of this high-value constituent. This is consistent with research conducted by Lachos-Perez et al. [46] who found that SWE degradation of hesperidin from defatted orange peel occurred at temperatures greater than 160 °C. If the process was carried out using 50% (v/v) ethanol as a solvent, then liquid OPD extracts with a slightly lower hesperidin concentration (5.81 ± 4.99–491.77 ± 28.89 mg/L) could be obtained. However, since the concentration of TP was significantly higher in this case, a detailed analysis of the obtained samples should be carried out, due to the possibility of creating new antioxidants.

Another dominant flavonoid, naringin, known for its antioxidant abilities [50], ranged in liquid SWE OPD extracts between 0.14 ± 0.20 and 62.37 ± 2.05 mg/L, with the highest concentration obtained at the lowest temperature (120 °C). In the second investigated case, the concentration of this compound was in the range of 0.66 ± 0.26–51.49 ± 3.57 mg/L, where the highest concentration was obtained at a significantly higher temperature compared to SWE extracts (180 °C). As in the case of hesperidin, the degradation of the naringin was noticed in the case of both applied processes. For SWE, the rapid degradation of naringin started above 160 °C, and this compound was almost all degraded at 220 °C. For PEE, the degradation was not so rapid; at 200 °C, the concentration decreased to approximately 50% of the highest obtained in PEE, but then at 220 °C was almost completely degraded. The same pattern of slower degradation was noticed for hesperidin.

The other targeted compounds (phenolic acids, namely p-coumaric, caffeic, and gallic acids) were detected at much lower concentrations in all OPD liquid extracts, either obtained by SWE or by PEE. These acids were also identified by Castro-Vázquez et al. [51] in their research as being compounds present in orange peel.

To the best of our knowledge, the comprehensive elucidation of compounds of interest in this type of extract was performed for the first time and carried out by LC-ESI-MS/MS. In liquid SWE OPD extracts, a wide variety of compounds—a total of 101 compounds—were identified, while in the case of PEE OPD extracts, even more—131 compounds—were identified. In total, 153 different compounds were tentatively identified. Results are shown in Table S1. The samples are marked with SWE and PEE depending on the type of extraction by which they were obtained, as well as numbers from 1 to 6 depending on the applied process temperature (from the lowest to the highest). Together with the previously quantified hesperidin (hesperetin-7-O-rutinoside), its unconjugated form (hesperetin) was also detected in all the SWE and PPE extracts. The flavanone glycosides derived from naringin, namely naringenin-7-O-glucoside and naringenin-7-O-rutinoside or narirutin, were also detected, with the latter only being present in PEE extracts. Other less abundant flavonoids were also present in all the extracts, such as luteolin, a recognised potent antioxidant, and its derivatives (diosmin, 6-methoxyluteolin, and luteolin-6-C-glucoside, the latter only being present in SWE extracts). Further tentatively identified minor flavonoids included rutin, limocitrin, isosakuranetin, and catechin, among others. Other phenolic compound classes with antioxidant potential were also detected, such as hydroxyferulic acid, homovanillic acid, hydroxycinnamic acids (hydroxyferulic acid, 7,8-dihydroxycoumarin, etc.), and vanillin, among many others. Finally, other common plant metabolites were also identified, such as organic acids (malic acid, tartaric acid, citric acid, succinic acid, etc.), saccharides (D-galactose, D-ribose, etc.), and fatty acids and derivatives (palmitic acid, FA 18:2 + 1O, FA 18:2 + 2O, etc.).

Although SWE gave slightly higher concentrations of both hesperidin and naringin, this technique was more susceptible to spoilage and contamination by various microorganisms due to the nature of water extracts. In contrast, PEE extracts obtained using 50% ethanol as a solvent represent a significantly safer and more stable medium in terms of contamination. Furthermore, the results for TPC were significantly higher. The complex mixture of compounds with potential antioxidant properties tentatively identified by wide MS screening and possible synergistic effects could explain the difference. In fact, as mentioned before, a higher number of compounds of interest were detected in PEE extracts.

3.3. In Silico Evaluation of Skin Anti-Ageing Potential

The molecular docking of hesperidin and naringin against several protein targets implicated in skin ageing was performed using the GOLD software and binding affinities were predicted using the DockThor rescoring function (Table 3). The higher GOLDScore fitness values and more negative DockThor binding affinities (kcal/mol) indicated stronger predicted binding to the targets. This consensus of the molecular docking results suggested that hesperidin and naringin have the potential to interact with ageing-related proteins including collagenase, elastase [52], sirtuin-1 [53], and growth factor beta receptor [54]. Across the targets, hesperidin exhibited slightly higher predicted binding affinity based on the GOLDScore fits and DockThor scores. The consensus of the two methods suggested that the strongest interactions of two flavonoids were predicted with protein sirtuin-1 and growth factor beta receptor. For example, the flavonoidic backbone of the hesperidin could fit tightly into the hydrophobic cleft of the growth factor beta receptor and the glycoside moiety had additional affinity with the hydrophilic surface (Figure S1a), providing selectivity through the interaction with the Lys232 of the selectivity pocket [55] (Figure S1b). Similarly, naringin was predicted to adopt a comparable orientation and interaction pattern (Figure S2), suggesting its glycoside moiety may have contributed to selective binding akin to hesperidin.

Table 3. In silico prediction of binding interactions between most abundant flavonoids in the orange peel extract and selected skin ageing-associated proteins. The molecular docking was conducted using Gold software and the DockThor web platform.

Component	Hesperidin		Naringin	
Protein Target	**GoldScore Fitness Score**	**DockThor Docking Score (kcal/mol)**	**Gold Fitness Score**	**DockThor Docking Score (kcal/mol)**
Fibroblast collagenase	72.7	−0.8	66.2	−0.7
Elastase	63.8	−9.2	57.2	−8.7
Sirtuin-1	88.4	−10.2	89.6	−10.1
Growth factor beta receptor	86.5	−10.0	68.0	−10.2

Overall, the molecular modelling provided evidence that hesperidin and naringin can favourably interact with multiple biochemical targets linked to skin ageing, with hesperidin showing a slight edge over naringin. The findings provided insights into the molecular mechanisms by which these citrus flavonoids may mitigate skin ageing and support further investigation into their anti-ageing bioactivities.

While hesperidin and naringin could likely contribute to the bioactivity due to their abundance, the cosmetic benefits may also be derived from not only their synergistic interactions but also from their synergistic interaction with other compounds present in the extract via polypharmacology. Specifically, additional flavonoids like kaempferol [56] and quercetin [57] could provide complementary antioxidant and anti-inflammatory activities while organic acids like citric may enhance the absorption and bioavailability of the primary flavonoids [58] and limonene and menthol may improve skin permeation [59].

The multi-target activities predicted for the flavonoids suggest they may modulate interconnected pathways involved in various ageing phenotypes through network pharmacology effects. Cosmetic formulations incorporating citrus extracts like these, with bioactive profiles optimised based on the relative concentrations of key actives like hesperidin and naringin, may provide more comprehensive anti-ageing benefits by simultaneously supporting extracellular matrix remodelling, collagen production, and cell proliferation. However, some components like essential oils and simple phenols could potentially cause irritation or sensitization at high concentrations [60], while mycotoxins such as citrinin must be monitored and limited to safe levels [61]. Therefore, careful formulation and testing would be needed to fully evaluate synergies and safety, while further studies could explore the extraction conditions to OPD to optimise extracts to have the potential for application in multi-faceted and safe anti-ageing cosmetic products.

3.4. In Vitro Evaluation of Antitumor Potential

In recent years, more and more attention has been focused on research whose goal is the evaluation of anticancer bioprospecting on medicinal plants [62–65]. According to this burning problem, this study investigated the anticancer activity of OPD and different extracts prepared with different extraction methods on rat hepatoma cells.

To measure the viability of H4IIE cells after treatment with SWE 3 and PEE 4 extracts for different periods (24 h and 48 h), an MTT assay was used (Figure 1). After a 24 h incubation with sample extracts from water (SWE 3), there was no significant cytotoxicity at any tested concentrations (Figure 1A). The addition of the highest tested concentration, which was 300 μg/mL, for 48 h exhibited negligible cytotoxicity and a decrease in measured absorbance compared to other groups (survival rate 90%) (Figure 1B). Chen et al. [66] evaluated the influence of the dehydrated powder of the sweet orange peel (*Citrus sinensis* L.) suspended in water (WESP) on the cell viability of human hepatocellular carcinoma (HepG2). After cells were cultured with WESP in different concentrations ranging from 0 to 1000 μg/mL for 24 h, cell viability was determined by an MTT assay. They concluded that cell survival in the presence of all tested concentrations was >85%, so this extract had no cytotoxicity to HepG2 cells, which is consistent with our findings.

On the other hand, after H4IIE cells were cultured in a medium containing 300 μg/mL of PEE 4 extract, this treatment in both periods of time induced a significant decrease in tumour growth compared to other experimental groups (Figure 1C,D) and the control (Figure 1D) (* $p < 0.05$, # $p < 0.001$). The observed high toxicity of this ethanolic extract aligned with the findings reported by Banjerdpongchai et al. [67], who investigated the effect of bioflavonoids from citrus seeds against the HepG2 cell line. In this study, cells were treated with crude ethanolic extract from *Citrus* seeds and the commercially available active compounds hesperidin and naringin at various concentrations for 24 h. Results showed that IC_{50} values of more than 200 mg/μL from both the ethanolic crude extract of *Citrus* seed and the bioflavonoids from *Citrus* seeds had antitumor potential. The cytotoxic effect was attributed to active flavanone glycosides, hesperidin [68], and naringin [69]. These compounds demonstrated various biological activities through the regulation of several cellular signalling pathways, such as antioxidant [70], anti-inflammatory [71], and anticancer properties [72–76]. Studies have shown that a possible explanation is in the inhibition of cellular proliferation through different mechanisms such as growth suppression and development via apoptosis by inducing both extrinsic and intrinsic pathways [75,77–79], and cell cycle arrest at the G0/G1 phases in malignant cells [80].

Evaluation of morphological characteristics under an inverted microscope with phase contrast showed that there were no observable differences in cell morphology and structure at both 24 h and 48 h post-treatment with SWE 3 extract (1–300 μg/mL) compared to the control group of cells (Figure 2). Also, treatment with concentrations ranging from 1 to 100 μg/mL of PEE 4 extract produced no significant effect on the H4IIE cell morphology and cell death (Figure 3B–E,H–K), which was in agreement with the results of cell viability and showed that the tested extracts in the noted concentrations exhibited no cytotoxicity effect.

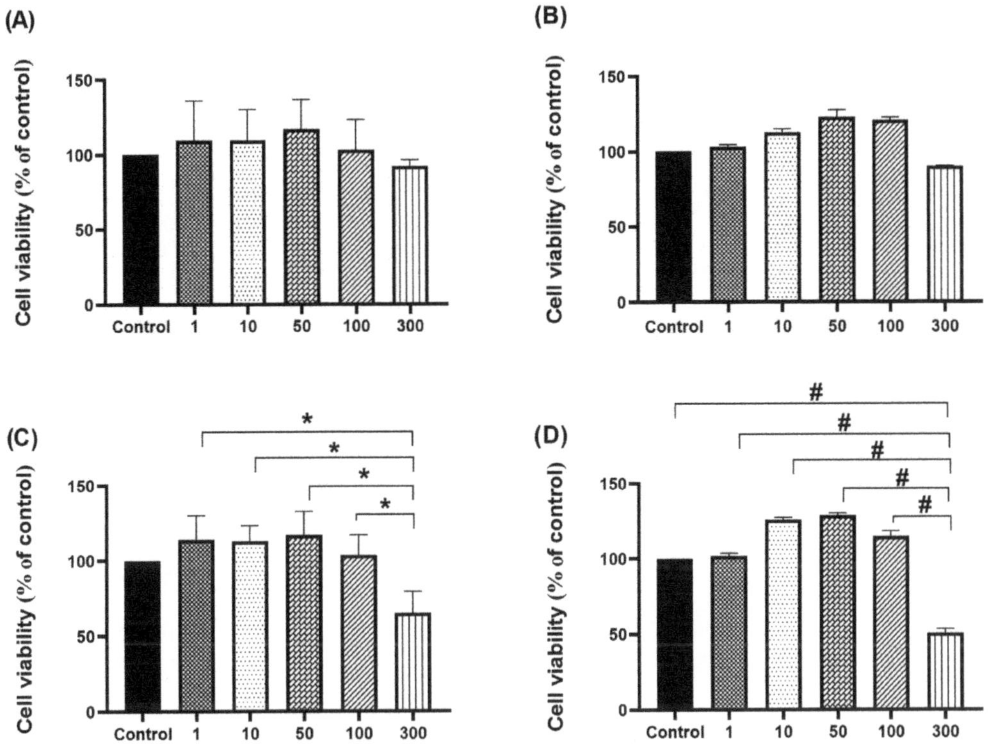

Figure 1. Effect of different concentrations (1 µg/mL, 10 µg/mL, 50 µg/mL, 100 µg/mL, and 300 µg/mL) of SWE 3 (extract obtained with subcritical water extraction procedure) (**A**,**B**) and PEE 4 (extract obtained with the pressurised ethanol extraction procedure) (**C**,**D**) extracts on the viability of H4IIE cells, measured by MTT assay. Cells were incubated with extracts for 24 h (**A**,**C**) and 48 h (**B**,**D**). Treatment with 300 µg/mL of the extract obtained with the pressurised ethanol extraction procedure in both periods significantly decreased the tumour growth rate compared to the other experimental groups (**C**,**D**) and the control (**D**). Results are expressed as viable cells (% of control) ± standard deviation (SD). Differences between groups were analysed where * $p < 0.05$ and # $p < 0.001$.

When cells were treated with a dose of 300 µg/mL of PEE 4 after 24 h (Figure 3F) and 48 h (Figure 3L), significant morphological changes in cell structure were observed in comparison to the control group (Figure 3A,G). Cells were more spherical and had smaller dimensions. After 48 h, extensive damage in the cellular structure could be seen. Nuclear disintegration and chromatin condensation of the treated cells was observed. In addition, the number of dead rounded cells and their debris was significantly higher 48 h post-treatment. These results indicate that the extract PEE 4 causes classic apoptotic cell death.

The ability of OPD extracts to inhibit cell growth was also determined by colony formation assay (Figures 4 and 5). No significant differences were observed in long-term cell survival upon treatment with extracts SWE 3 and PEE 4 in concentrations of 1–50 µg/mL compared to the control. However, higher concentrations of 100 µg/mL of PEE 4 extract (Figure 5E) or 300 µg/mL of SWE 3 and PEE 4 exerted a strong inhibitory effect on cell survival (Figures 4F and 5F).

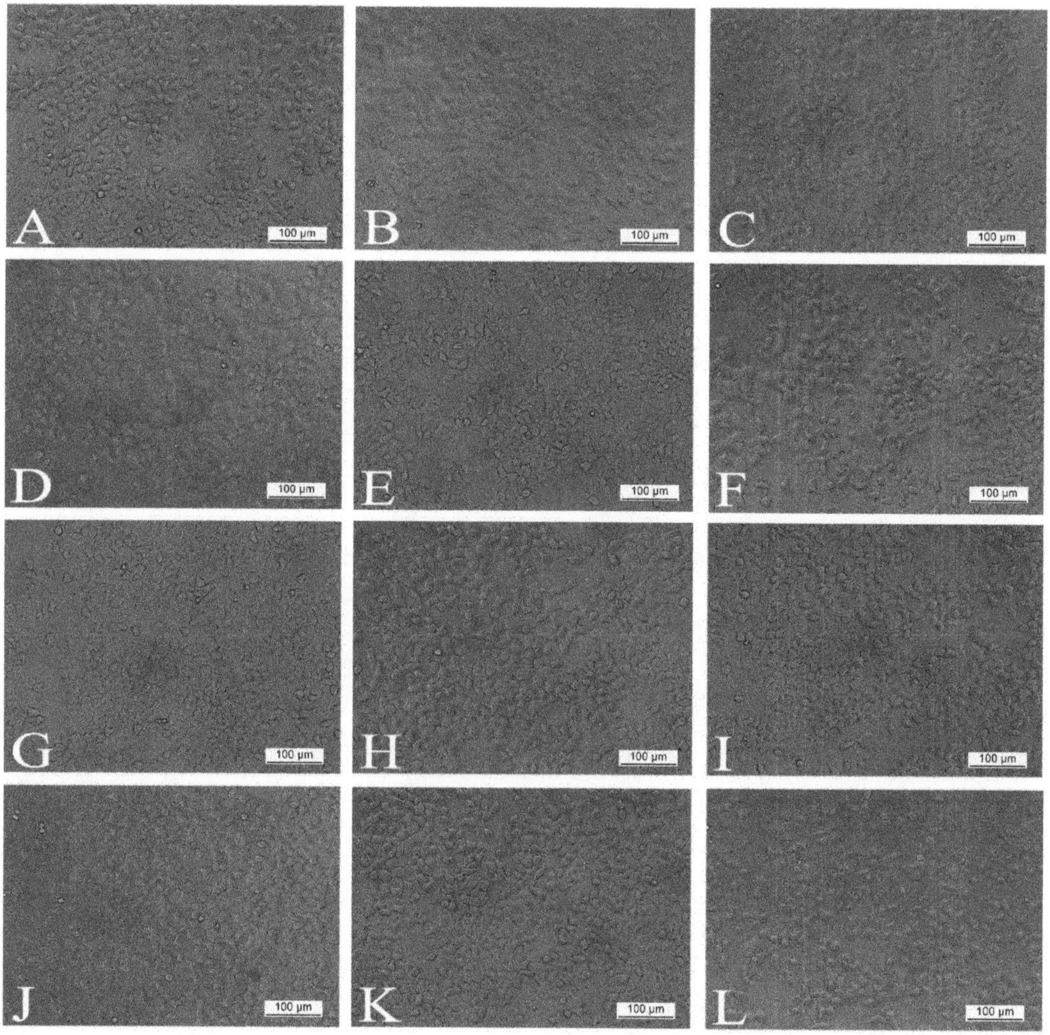

Figure 2. Morphology of the H4IIE cell line exposed to various concentrations of SWE 3 (extract obtained with a subcritical water extraction procedure): (**A**,**G**) control (non-treated), (**B**,**H**) treated with 1 µg/mL, (**C**,**I**) treated with 10 µg/mL, (**D**,**J**) treated with 50 µg/mL, (**E**,**K**) treated with 100 µg/mL, (**F**,**L**) treated with 300 µg/mL. Cellular morphological changes were analysed with phase contrast in different periods (**A**–**F**) after 24 h and (**G**–**L**) after 48 h. There were no observable differences in cell morphology and structure at both periods compared to the control group of cells. All photographs were taken at 200× magnification. Scale bars represent 100 µm.

These results from an in vitro study on rat hepatoma cells suggested that OPD extracts have antitumor potential. The quote activity of the ethanolic extract showed a higher potential compared to the water extract of orange peel. Similar results were reported by Gavamukulya et al. [81]. Analysis of the ethanolic and water leaf extracts of *A. muricata* leaves indicated that water extracts showed no effect throughout the range of tested concentrations (0, 250, 500, 750, 1000, and 1250 µg/mL) on the cell proliferation of human breast cancer cell lines MDA and SKBR3. However, the ethanolic extract had a very high cytotoxic potential with an IC_{50} of 248.77 µg/mL and 202.33 µg/mL against MDA and

SKBR3 cell lines, respectively. A possible explanation of our results is that the extracts contain other active biological compounds in addition to the isolated ones, but the methods for their identification are not readily available. Some of these compounds may have been present in very high quantities in the ethanolic extract, or the concentrations were lower or even absent in the water extract, leading to the discrepancy of anticancer potential. Lastly, more experimental in vitro studies are needed to elucidate the root cause of this difference.

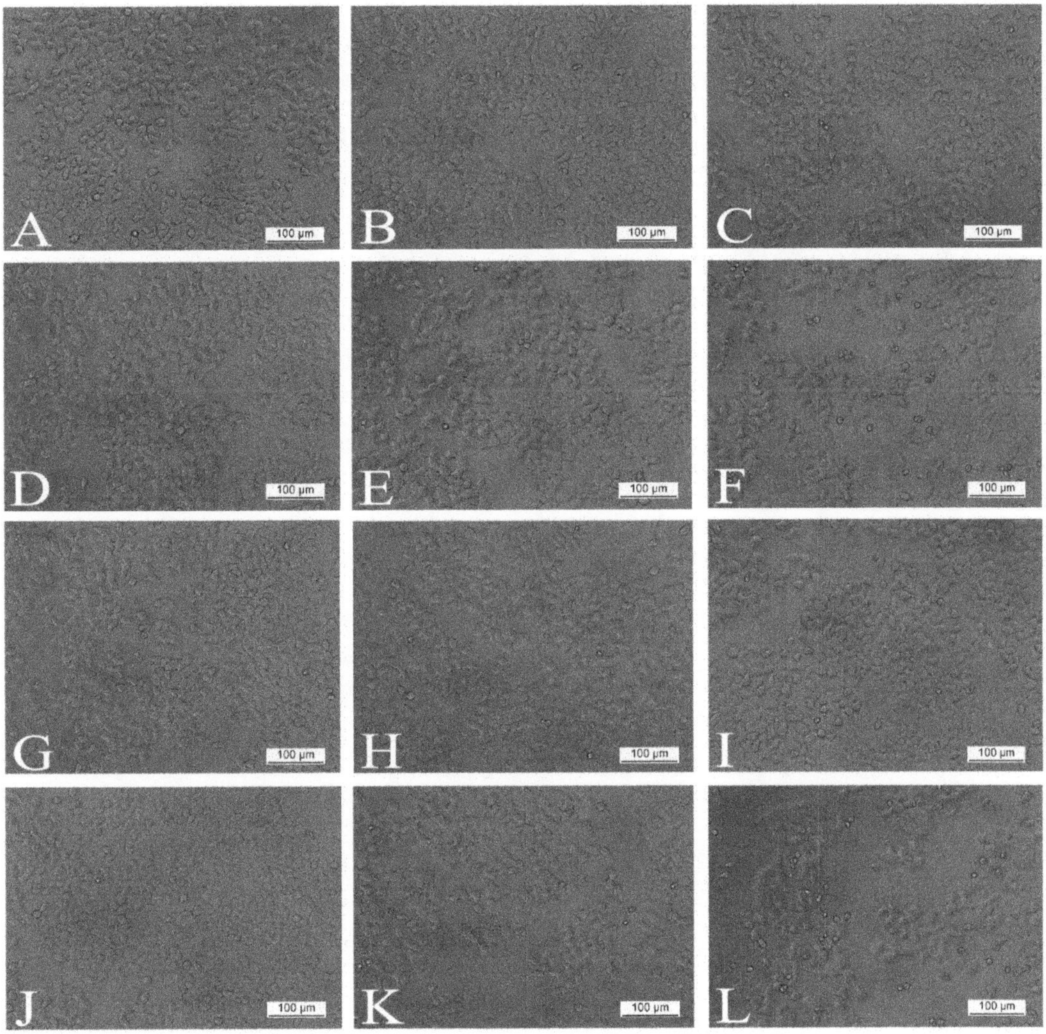

Figure 3. Morphology of the H4IIE cell line exposed to various concentrations of PEE 4 (extract obtained with the pressurised ethanol extraction procedure): (**A**,**G**) control (non-treated), (**B**,**H**) treated with 1 µg/mL, (**C**,**I**) treated with 10 µg/mL, (**D**,**J**) treated with 50 µg/mL, (**E**,**K**) treated with 100 µg/mL, (**F**,**L**) treated with 300 µg/mL. Cellular morphological changes were analysed with phase contrast in different periods: (**A–F**) after 24 h and (**G–L**) after 48 h. The extract obtained with the pressurised ethanol extraction procedure at 300 µg/mL concentration caused classic apoptotic cell death (rounded and smaller cells with nuclear disintegration and chromatin condensation). All photographs were taken at 200× magnification. Scale bars represent 100 µm.

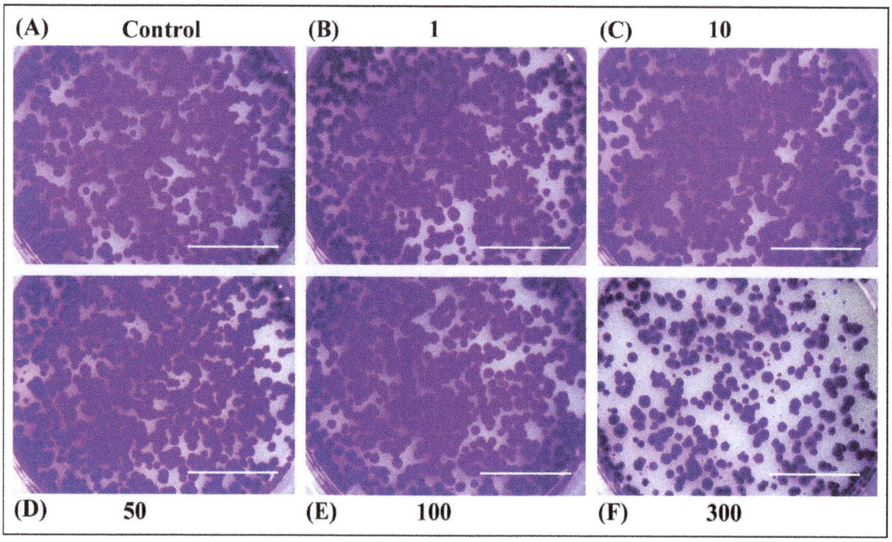

Figure 4. Rat hepatoma cells H4IIE were seeded in six-well plates (2000 cells/well) for long-term survival assay. After 24 h, the culture medium was replaced with (**A**) fresh medium or medium containing (**B**) 1 µg/mL, (**C**) 10 µg/mL, (**D**) 50 µg/mL, (**E**) 100 µg/mL, and (**F**) 300 µg/mL of SWE 3 extract (obtained with a subcritical water extraction procedure). Then, the cells were incubated for another 24 h. The culture medium was changed every three days for two weeks. Cells were fixed with methanol and stained with crystal violet. A 300 µg/mL concentration exerted a strong inhibitory effect on cell survival. Scale bars represent 1 cm.

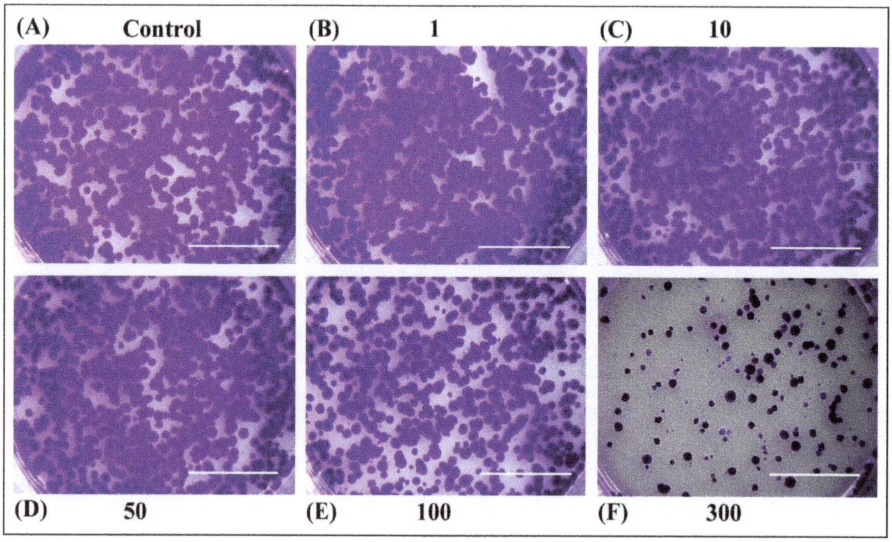

Figure 5. Rat hepatoma cells H4IIE were seeded in six-well plates (2000 cells/well) for long-term survival assay. After 24 h, the culture medium was replaced with (**A**) fresh medium, or medium containing (**B**) 1 µg/mL, (**C**) 10 µg/mL, (**D**) 50 µg/mL, (**E**) 100 µg/mL, and (**F**) 300 µg/mL of PEE 4 extract (obtained with the pressurised ethanol extraction procedure). Then, the cells were incubated for another 24 h. The culture medium was changed every three days for two weeks. Cells were fixed with methanol and stained with crystal violet. The higher concentrations of 100 µg/mL exerted a strong inhibitory effect on cell survival. Scale bars represent 1 cm.

4. Conclusions

In this study, different process parameters of green extraction methods, namely subcritical water extraction and ethanol extraction under pressure, were investigated, in order to determine the most adequate ones that ensured high-quality OPD extracts and maximum raw material exploitation without negative impacts on the environment. It was concluded that lower subcritical water extraction temperatures were more suitable for OPD valorisation because they provided a higher content of dominant antioxidants (hesperidin and naringin). However, pressurised ethanol extraction proved to be a more suitable method in terms of isolating TPC, as well as providing a higher extraction yield and a higher number of identified compounds of interest. The present findings from an in vitro study on rat hepatoma cell culture confirmed that PEE OPD extracts exerted stronger antitumor potential than SWE OPD extracts. Also, according to the in silico test results, and thanks to the bioactivity of the dominant compounds hesperidin and naringin, OPD extracts incorporated into cosmetic formulations could be beneficial against ageing by simultaneously supporting extracellular matrix remodelling, collagen production, and cell proliferation. These research results make a significant contribution to the concept of the circular economy considering that they propose sustainable solutions that include the reduction in waste/by-products from the filter tea industry, i.e., food, as well as the exploitation of OPD through making maximum use of natural resources.

Supplementary Materials: The following supporting information can be downloaded at: https://www.mdpi.com/article/10.3390/antiox14060638/s1. Table S1: LC-ESI-MS/MS qualitative analysis of liquid OPD extracts; Figure S1: A representative docked pose of hesperidin into the binding site of growth factor beta receptor (surface coloured according to hydrophobicity) shown as (a) a three-dimensional representation of a ligand (thick green sticks) and (b) a two-dimensional protein ligand plot; Figure S2: A representative docked pose of naringin into the binding site of growth factor beta receptor (surface coloured according to hydrophobicity) shown as (a) a three-dimensional representation of a ligand (thick green sticks) and (b) a two-dimensional protein ligand plot.

Author Contributions: Conceptualization, S.K. and S.V.; methodology, S.V., A.B.-G., N.M. and M.Z.; software, M.Z.; validation, S.K. and S.V.; formal analysis, S.K., A.P., N.M., A.B.-G. and S.V.; investigation, S.K.; resources, S.V., A.B.-G. and N.M.; data curation, S.V.; writing—original draft preparation, S.K.; writing—review and editing, S.V., A.B.-G., N.M., A.P., M.Z. and N.N.; visualisation, S.K., S.V. and N.N.; supervision, S.V.; project administration, S.V.; funding acquisition, S.K., S.V., A.B.-G. and M.Z. All authors have read and agreed to the published version of the manuscript.

Funding: This work has been supported by the Serbian Ministry of Science and Technological Development (451-03-66/2024-03/200134) and the Provincial Secretariat for Higher Education and Scientific Research, Autonomous Province of Vojvodina, Serbia (142-451-2683/2021-01). A. Ballesteros also acknowledges project PID2020-113743RB-I00, funded by MCIN/AEI/10.13039/501100011033.

Institutional Review Board Statement: Not applicable.

Informed Consent Statement: Not applicable.

Data Availability Statement: The data that support the findings of this study are available upon request from the corresponding author.

Acknowledgments: The authors sincerely thank Fructus d.o.o., Bačka Palanka, Serbia, for generously providing the raw material and supporting this study.

Conflicts of Interest: The authors declare no conflicts of interest.

References

1. Colegate, S.M.; Molyneux, R.J. *Bioactive Natural Products: Detection, Isolation, and Structural Determination*, 2nd ed.; CRC Press: Boca Raton, FL, USA, 2007.
2. Vidović, S.; Vladić, J.; Nastić, N.; Jokić, S. Subcritical and supercritical extraction in food by-product and food waste valorization. In *Innovative Food Processing Technologies: A Comprehensive Review*; Muthukumarappan, K., Knoerzer, K., Eds.; Elsevier: Amsterdam, The Netherlands, 2021; pp. 705–721. [CrossRef]
3. Cheng, Y.; Xue, F.; Yu, S.; Du, S.; Yang, Y. Subcritical Water Extraction of Natural Products. *Molecules* **2021**, *26*, 4004. [CrossRef] [PubMed]
4. Vladić, J.; Jakovljević, M.; Molnar, M.; Vidović, S.; Tomić, M.; Drinić, Z.; Jokić, S. Valorization of Yarrow (*Achillea millefolium* L.) by-Product through Application of Subcritical Water Extraction. *Molecules* **2020**, *25*, 1878. [CrossRef] [PubMed]
5. Vidović, S.; Nastić, N.; Gavarić, A.; Cindrić, M.; Vladić, J. Development of green extraction process to produce antioxidant-rich extracts from purple coneflower. *Sep. Sci. Technol.* **2019**, *54*, 1174–1181. [CrossRef]
6. Vladić, J.; Canli, O.; Pavlić, B.; Zeković, Z.; Vidović, S.; Kaplan, M. Optimization of Satureja Montana Subcritical Water Extraction Process and Chemical Characterization of Volatile Fraction of Extracts. *J. Supercrit. Fluids* **2017**, *120*, 86–94. [CrossRef]
7. Lachos-Perez, D.; Brown, A.B.; Mudhoo, A.; Timko, M.T.; Rostagno, M.A.; Forster-Carneiro, T. Applications of Subcritical and Supercritical Water Conditions for Extraction, Hydrolysis, Gasification, and Carbonization of Biomass: A Critical Review. *Biofuel Res. J.* **2017**, *4*, 611–626. [CrossRef]
8. Ciftci, D.; Saldaña, M.D.A. Hydrolysis of Sweet Blue Lupin Hull Using Subcritical Water Technology. *Bioresour. Technol.* **2015**, *194*, 75–82. [CrossRef]
9. Munir, M.T.; Kheirkhah, H.; Baroutian, S.; Quek, S.Y.; Young, B.R. Subcritical Water Extraction of Bioactive Compounds from Waste Onion Skin. *J. Clean. Prod.* **2018**, *183*, 487–494. [CrossRef]
10. Monrad, J.K.; Howard, L.R.; King, J.W.; Srinivas, K.; Mauromoustakos, A. Subcritical Solvent Extraction of Anthocyanins from Dried Red Grape Pomace. *J. Agric. Food Chem.* **2010**, *58*, 2862–2868. [CrossRef]
11. Marcus, Y. Extraction by Subcritical and Supercritical Water, Methanol, Ethanol and Their Mixtures. *Separations* **2018**, *5*, 4. [CrossRef]
12. Rahmana Putra, N.; Nur Rizkiyah, D.; Idham, Z.; Abbas Ahmad Zaini, M.; Azizi Che Yunus, M.; Hazim Abdul Aziz, A. Optimization and Solubilization of Interest Compounds from Roselle in Subcritical Ethanol Extraction (See). *Alex. Eng. J.* **2023**, *65*, 59–74. [CrossRef]
13. Mufari, J.R.; Rodríguez-Ruiz, A.C.; Bergesse, A.E.; Miranda-Villa, P.P.; Nepote, V.; Velez, A.R. Bioactive Compounds Extraction from Malted Quinoa Using Water-Ethanol Mixtures under Subcritical Conditions. *LWT* **2021**, *138*, 110574. [CrossRef]
14. Chiou, T.-Y.; Neoh, T.L.; Kobayashi, T.; Adachi, S. Properties of Extract Obtained from Defatted Rice Bran by Extraction with Aqueous Ethanol under Subcritical Conditions. *FSTR* **2012**, *18*, 37–45. [CrossRef]
15. Li, Y.; Qi, Y.; Li, H.; Chen, Z.; Xu, B. Improving the Cold Water Swelling Properties of Oat Starch by Subcritical Ethanol-Water Treatment. *Int. J. Biol. Macromol.* **2022**, *194*, 594–601. [CrossRef] [PubMed]
16. De Souza, A.R.C.; Stefanov, S.; Bombardelli, M.C.M.; Corazza, M.L.; Stateva, R.P. Assessment of Composition and Biological Activity of Arctium Lappa Leaves Extracts Obtained with Pressurized Liquid and Supercritical CO_2 Extraction. *J. Supercrit. Fluids* **2019**, *152*, 104573. [CrossRef]
17. Tangkhavanich, B.; Kobayashi, T.; Adachi, S. Effects of Repeated Treatment on the Properties of Rice Stem Extract Using Subcritical Water, Ethanol, and Their Mixture. *J. Ind. Eng. Chem.* **2014**, *20*, 2610–2614. [CrossRef]
18. FAO. *Citrus Fruit Statistical Compendium 2020*; FAO: Rome, Italy, 2021.
19. Kanaze, F.I.; Termentzi, A.; Gabrieli, C.; Niopas, I.; Georgarakis, M.; Kokkalou, E. The Phytochemical Analysis and Antioxidant Activity Assessment of Orange Peel (*Citrus sinensis*) Cultivated in Greece—Crete Indicates a New Commercial Source of Hesperidin. *Biomed. Chromatogr.* **2009**, *23*, 239–249. [CrossRef]
20. Satari, B.; Karimi, K. Citrus Processing Wastes: Environmental Impacts, Recent Advances, and Future Perspectives in Total Valorization. *Resour. Conserv. Recycl.* **2018**, *129*, 153–167. [CrossRef]
21. Ortiz-Sanchez, M.; Solarte-Toro, J.C.; Orrego-Alzate, C.E.; Acosta-Medina, C.D.; Cardona-Alzate, C.A. Integral Use of Orange Peel Waste through the Biorefinery Concept: An Experimental, Technical, Energy, and Economic Assessment. *Biomass Conv. Bioref.* **2021**, *11*, 645–659. [CrossRef]
22. Laufenberg, G.; Kunz, B.; Nystroem, M. Transformation of Vegetable Waste into Value Added Products. *Bioresour. Technol.* **2003**, *87*, 167–198. [CrossRef]
23. Sawalha, S.M.S.; Arráez-Román, D.; Segura-Carretero, A.; Fernández-Gutiérrez, A. Quantification of Main Phenolic Compounds in Sweet and Bitter Orange Peel Using CE–MS/MS. *Food Chem.* **2009**, *116*, 567–574. [CrossRef]
24. Tripoli, E.; Guardia, M.L.; Giammanco, S.; Majo, D.D.; Giammanco, M. Citrus Flavonoids: Molecular Structure, Biological Activity and Nutritional Properties: A Review. *Food Chem.* **2007**, *104*, 466–479. [CrossRef]

25. Yi, L.; Ma, S.; Ren, D. Phytochemistry and Bioactivity of Citrus Flavonoids: A Focus on Antioxidant, Anti-Inflammatory, Anticancer and Cardiovascular Protection Activities. *Phytochem. Rev.* **2017**, *16*, 479–511. [CrossRef]
26. Apraj, V.; Pandita, N. Evaluation of Skin Anti-Aging Potential of Citrus Reticulata Blanco Peel. *Pharmacogn. Res.* **2016**, *8*, 160–168. [CrossRef]
27. Stanisic, D.; Liu, L.H.B.; Dos Santos, R.V.; Costa, A.F.; Durán, N.; Tasic, L. New Sustainable Process for Hesperidin Isolation and Anti-Ageing Effects of Hesperidin Nanocrystals. *Molecules* **2020**, *25*, 4534. [CrossRef]
28. Şahin, D.; Çağlar, E.Ş.; Boran, T.; Karadağ, A.E.; Özhan, G.; Üstündağ Okur, N. Development, Characterization of Naringenin-Loaded Promising Microemulsion Formulations, and Demonstration of Anti-Aging Efficacy by in Vitro Enzyme Activity and Gene Expression. *J. Drug Deliv. Sci. Technol.* **2023**, *84*, 104422. [CrossRef]
29. Nastić, N.; Vasić, A.; Šoronja Simović, D.; Vladić, J.; Jokić, S.; Aladić, K.; Vidović, S. Underutilized Rosa Canina Herbal Dust as an Innovative Natural Functional and Health Promoting Ingredient: A Proposal of Two-Novel Approaches. *Waste Biomass Valor.* **2023**, *14*, 1207–1217. [CrossRef]
30. Živković, J.; Vladić, J.; Naffati, A.; Nastić, N.; Šavikin, K.; Tomić, M.; Vidović, S. Comparative Chemical Profiling of Underexploited Arctostaphylos Uva-Ursi l. Herbal Dust Extracts Obtained by Conventional, Ultrasound-Assisted and Subcritical Water Extractions. *Waste Biomass Valor.* **2022**, *13*, 4147–4155. [CrossRef]
31. Sulejmanović, M.; Milić, N.; Mourtzinos, I.; Nastić, N.; Kyriakoudi, A.; Drljača, J.; Vidović, S. Ultrasound-assisted and subcritical water extraction techniques for maximal recovery of phenolic compounds from raw ginger herbal dust toward in vitro biological activity investigation. *Food Chem.* **2024**, *437*, 137774. [CrossRef]
32. Sánchez-Vallejo, C.; Ballesteros-Gómez, A.; Rubio, S. Tailoring Composition and Nanostructures in Supramolecular Solvents: Impact on the Extraction Efficiency of Polyphenols from Vegetal Biomass. *Sep. Purif. Technol.* **2022**, *292*, 120991. [CrossRef]
33. Popovic, A.; Drljaca, J.; Popovic, M.; Miljkovic, D.; Marinovic, J.; Ljubkovic, M.; Kladar, N.; Capo, I. Mitochondrial Energy Metabolism in Baby Hamster Kidney (BHK-21/C13) Cells Treated with Karnozin Extra®. *Int. J. Morphol.* **2022**, *40*, 91–97. [CrossRef]
34. Crowley, L.C.; Christensen, M.E.; Waterhouse, N.J. Measuring Survival of Adherent Cells with the Colony-Forming Assay. *Cold Spring Harb. Protoc.* **2016**, *2016*, pdb.prot087171. [CrossRef]
35. Ma, G.; Zhang, L.; Sugiura, M.; Kato, M. Citrus and Health. In *The Genus Citrus*; Elsevier: Amsterdam, The Netherlands, 2020; pp. 495–511, ISBN 9780128121634.
36. Ignat, I.; Volf, I.; Popa, V.I. A Critical Review of Methods for Characterisation of Polyphenolic Compounds in Fruits and Vegetables. *Food Chem.* **2011**, *126*, 1821–1835. [CrossRef]
37. Oboh, G.; Ademosun, A.O. Characterization of the Antioxidant Properties of Phenolic Extracts from Some Citrus Peels. *J. Food Sci. Technol.* **2012**, *49*, 729–736. [CrossRef]
38. Guimarães, R.; Barros, L.; Barreira, J.C.M.; Sousa, M.J.; Carvalho, A.M.; Ferreira, I.C.F.R. Targeting Excessive Free Radicals with Peels and Juices of Citrus Fruits: Grapefruit, Lemon, Lime and Orange. *Food Chem. Toxicol.* **2010**, *48*, 99–106. [CrossRef]
39. Plaza, M.; Turner, C. Pressurized Hot Water Extraction of Bioactives. *TrAC Trends Anal. Chem.* **2015**, *71*, 39–54. [CrossRef]
40. Benito-Román, Ó.; Blanco, B.; Sanz, M.T.; Beltrán, S. Subcritical Water Extraction of Phenolic Compounds from Onion Skin Wastes (*Allium Cepa* Cv. Horcal): Effect of Temperature and Solvent Properties. *Antioxidants* **2020**, *9*, 1233. [CrossRef]
41. Brezo-Borjan, T.; Švarc-Gajić, J.; Morais, S.; Delerue-Matos, C.; Rodrigues, F.; Lončarević, I.; Pajin, B. Chemical and Biological Characterisation of Orange (*Citrus sinensis*) Peel Extracts Obtained by Subcritical Water. *Processes* **2023**, *11*, 1766. [CrossRef]
42. Razola-Díaz, M.D.C.; Guerra-Hernández, E.J.; Rodríguez-Pérez, C.; Gómez-Caravaca, A.M.; García-Villanova, B.; Verardo, V. Optimization of Ultrasound-Assisted Extraction via Sonotrode of Phenolic Compounds from Orange by-Products. *Foods* **2021**, *10*, 1120. [CrossRef]
43. Krivošija, S.; Jerković, I.; Nastić, N.; Zloh, M.; Jokić, S.; Banožić, M.; Aladić, K.; Vidović, S. Green Pathway for Utilisation of Orange Peel Dust and In Silico Evaluation of Pharmacological Potential. *Microchem. J.* **2023**, *193*, 109132. [CrossRef]
44. M'hiri, N.; Irina, I.; Cédric, P.; Ghoul, M.; Boudhrioua, N. Antioxidants of Maltease Orange Peel: Comparative Investigation of the Efficiency of Four Extraction Methods. *J. Appl. Pharm. Sci.* **2017**, *7*, 126–135. [CrossRef]
45. Londoño-Londoño, J.; Lima, V.R.D.; Lara, O.; Gil, A.; Pasa, T.B.C.; Arango, G.J.; Pineda, J.R.R. Clean Recovery of Antioxidant Flavonoids from Citrus Peel: Optimizing an Aqueous Ultrasound-Assisted Extraction Method. *Food Chem.* **2010**, *119*, 81–87. [CrossRef]
46. Lachos-Perez, D.; Baseggio, A.M.; Torres-Mayanga, P.C.; Ávila, P.F.; Tompsett, G.A.; Marostica, M.; Goldbeck, R.; Timko, M.T.; Rostagno, M.; Martinez, J.; et al. Sequential Subcritical Water Process Applied to Orange Peel for the Recovery Flavanones and Sugars. *J. Supercrit. Fluids* **2020**, *160*, 104789. [CrossRef]
47. Barrales, F.M.; Silveira, P.; Barbosa, P.D.P.M.; Ruviaro, A.R.; Paulino, B.N.; Pastore, G.M.; Macedo, G.A.; Martinez, J. Recovery of Phenolic Compounds from Citrus By-Products Using Pressurized Liquids—An Application to Orange Peel. *Food Bioprod. Process* **2018**, *112*, 9–21. [CrossRef]

48. Rafiq, S.; Kaul, R.; Sofi, S.A.; Bashir, N.; Nazir, F.; Ahmad Nayik, G. Citrus Peel as a Source of Functional Ingredient: A Review. *J. Saudi Soc. Agric. Sci.* **2018**, *17*, 351–358. [CrossRef]
49. Hwang, H.-J.; Kim, H.-J.; Ko, M.-J.; Chung, M.-S. Recovery of Hesperidin and Narirutin from Waste Citrus Unshiu Peel Using Subcritical Water Extraction Aided by Pulsed Electric Field Treatment. *Food Sci. Biotechnol.* **2021**, *30*, 217–226. [CrossRef]
50. Chen, R.; Qi, Q.-L.; Wang, M.-T.; Li, Q.-Y. Therapeutic Potential of Naringin: An Overview. *Phar Biol.* **2016**, *54*, 3203–3210. [CrossRef]
51. Castro-Vázquez, L.; Lozano, M.V.; Rodríguez-Robledo, V.; González-Fuentes, J.; Marcos, P.; Villaseca, N.; Arroyo-Jiménez, M.M.; Santander-Ortega, M.J. Pressurized Extraction as an Opportunity to Recover Antioxidants from Orange Peels: Heat Treatment and Nanoemulsion Design for Modulating Oxidative Stress. *Molecules* **2021**, *26*, 5928. [CrossRef]
52. Wittenauer, J.; Mäckle, S.; Sußmann, D.; Schweiggert-Weisz, U.; Carle, R. Inhibitory Effects of Polyphenols from Grape Pomace Extract on Collagenase and Elastase Activity. *Fitoterapia* **2015**, *101*, 179–187. [CrossRef]
53. Bielach-Bazyluk, A.; Zbroch, E.; Mysliwiec, H.; Rydzewska-Rosolowska, A.; Kakareko, K.; Flisiak, I.; Hryszko, T. Sirtuin 1 and Skin: Implications in Intrinsic and Extrinsic Aging—A Systematic Review. *Cells* **2021**, *10*, 813. [CrossRef]
54. Han, K.-H.; Choi, H.-R.; Won, C.-H.; Chung, J.-H.; Cho, K.-H.; Eun, H.-C.; Kim, K.-H. Alteration of the TGF-β/SMAD Pathway in Intrinsically and UV-Induced Skin Aging. *Mech. Ageing Dev.* **2005**, *126*, 560–567. [CrossRef]
55. Gellibert, F.; Woolven, J.; Fouchet, M.-H.; Mathews, N.; Goodland, H.; Lovegrove, V.; Laroze, A.; Nguyen, V.-L.; Sautet, S.; Wang, R.; et al. Identification of 1,5-Naphthyridine Derivatives as a Novel Series of Potent and Selective Tgf-β Type i Receptor Inhibitors. *J. Med. Chem.* **2004**, *47*, 4494–4506. [CrossRef]
56. Wang, J.; Fang, X.; Ge, L.; Cao, F.; Zhao, L.; Wang, Z.; Xiao, W. Antitumor, Antioxidant and Anti-Inflammatory Activities of Kaempferol and Its Corresponding Glycosides and the Enzymatic Preparation of Kaempferol. *PLoS ONE* **2018**, *13*, e0197563. [CrossRef]
57. Azeem, M.; Hanif, M.; Mahmood, K.; Ameer, N.; Chughtai, F.R.S.; Abid, U. An Insight into Anticancer, Antioxidant, Antimicrobial, Antidiabetic and Anti-Inflammatory Effects of Quercetin: A Review. *Polym. Bull.* **2023**, *80*, 241–262. [CrossRef]
58. Liu, L.; Wu, Y.; Zhao, Y.; Lu, C.; Zhao, R. Citric Acid Enhances the Activities of Astilbin on Psoriasis via Down-Regulation of *p* -Glycoprotein. *Mol. Pharm.* **2023**, *20*, 1964–1974. [CrossRef]
59. Yang, Z.; Teng, Y.; Wang, H.; Hou, H. Enhancement of Skin Permeation of Bufalin by Limonene via Reservoir Type Transdermal Patch: Formulation Design and Biopharmaceutical Evaluation. *Int. JPharm* **2013**, *447*, 231–240. [CrossRef]
60. Tisserand, R.; Young, R. *Essential Oil Safety: A Guide for Health Care Professionals*, 2nd ed.; Elsevier Ltd.: Amsterdam, The Netherlands, 2013.
61. Hamuel, J.D. The Occurrence, Properties and Significance of Citrinin Mycotoxin. *J. Plant Pathol. Microbiol.* **2015**, *6*, 2. [CrossRef]
62. Al Qaisi, Y.; Alfarrayeh, I.; Alsarayreh, A.; Khleifat, K.; Abu-Nwas, N. Assessment of Antioxidant Potential, Cytotoxicity, and Anticancer Activity of Methanolic Extracts from Selected Wild Medicinal Plants. *Phytomed. Plus* **2024**, *4*, 100534. [CrossRef]
63. Barzegarparay, F.; Najafzadehvarzi, H.; Pourbagher, R.; Parsian, H.; Ghoreishi, S.M.; Mortazavi-Derazkola, S. Green Synthesis of Novel Selenium Nanoparticles Using Crataegus Monogyna Extract (Senps@cm) and Investigation of Its Toxicity, Antioxidant Capacity, and Anticancer Activity against MCF-7 as a Breast Cancer Cell Line. *Biomass Conv. Bioref.* **2024**, *14*, 25369–25378. [CrossRef]
64. Jongrungraungchok, S.; Madaka, F.; Wunnakup, T.; Sudsai, T.; Pongphaew, C.; Songsak, T.; Pradubyat, N. In Vitro Antioxidant, Anti-Inflammatory, and Anticancer Activities of Mixture Thai Medicinal Plants. *BMC Complement. Med. Ther.* **2023**, *23*, 43. [CrossRef]
65. Rani, D.M.; Wongso, H.; Purwoko, R.Y.; Winarto, N.B.; Shalas, A.F.; Triatmoko, B.; Pratama, A.N.W.; Keller, P.A.; Nugraha, A.S. Anti-Cancer Bioprospecting on Medicinal Plants from Indonesia: A Review. *Phytochemistry* **2023**, *216*, 113881. [CrossRef]
66. Chen, Z.-T.; Chu, H.-L.; Chyau, C.-C.; Chu, C.-C.; Duh, P.-D. Protective Effects of Sweet Orange (*Citrus sinensis*) Peel and Their Bioactive Compounds on Oxidative Stress. *Food Chem.* **2012**, *135*, 2119–2127. [CrossRef]
67. Banjerdpongchai, R.; Wudtiwai, B.; Khaw-on, P.; Rachakhom, W.; Duangnil, N.; Kongtawelert, P. Hesperidin from Citrus Seed Induces Human Hepatocellular Carcinoma HepG2 Cell Apoptosis via Both Mitochondrial and Death Receptor Pathways. *Tumor Biol.* **2016**, *37*, 227–237. [CrossRef]
68. Bansal, K.; Bhati, H.; Vanshita; Bajpai, M. New Insights into Therapeutic Applications and Nanoformulation Approaches of Hesperidin: An Updated Review. *Pharmacol. Res.-Mod. Chin. Med.* **2024**, *10*, 100363. [CrossRef]
69. Zangade, S.B.; Dhulshette, B.S.; Patil, P.B. Flavonoid-Metal Ion Complexes as Potent Anticancer Metallodrugs: A Comprehensive Review. *Mini-Rev. Med. Chem.* **2024**, *24*, 1046–1060. [CrossRef]
70. Hasnat, H.; Shompa, S.A.; Islam, M.M.; Alam, S.; Richi, F.T.; Emon, N.U.; Ashrafi, S.; Ahmed, N.U.; Chowdhury, M.N.R.; Fatema, N.; et al. Flavonoids: A Treasure House of Prospective Pharmacological Potentials. *Heliyon* **2024**, *10*, e27533. [CrossRef]
71. Puteri, A.R.; Huda, H.B. Flavonoids and Their Role As Anti-Inflammatory Agents. *Biomed. J. Indones.* **2024**, *10*, 42–48. [CrossRef]
72. Lobo, C.L.; Prabhu, P.P.; Dubey, A.; Shetty, A.; Mahadev, M. Exploring the Potential of Hesperidin and Synergistic Formulations in Breast Cancer Management: A Comprehensive Review. *Int. J. Pharm. Investig.* **2024**, *14*, 1122–1130. [CrossRef]

73. Deng, R.; Liu, Y.; Wu, X.; Zhao, N.; Deng, J.; Pan, T.; Cao, L.; Zhan, F.; Qiao, X. Probing the Interaction of Hesperidin Showing Antiproliferative Activity in Colorectal Cancer Cells and Human Hemoglobin. *Int. J. Biol. Macromol.* **2024**, *281*, 136078. [CrossRef]
74. Bakhshan, M.A.; Sheikhzadeh, S.; Delirezh, N. Hesperidin Nanoparticles for Prostate Cancer Therapy: Preparation, Characterization and Cytotoxic Activity. *Biomed. Mater.* **2024**, *19*, 035044. [CrossRef]
75. Chang, T.-M.; Chi, M.-C.; Chiang, Y.-C.; Lin, C.-M.; Fang, M.-L.; Lee, C.-W.; Liu, J.-F.; Kou, Y.R. Promotion of ROS-Mediated Apoptosis, G2/M Arrest, and Autophagy by Naringenin in Non-Small Cell Lung Cancer. *Int. J. Biol. Sci.* **2024**, *20*, 1093–1109. [CrossRef]
76. Elsori, D.; Pandey, P.; Ramniwas, S.; Kumar, R.; Lakhanpal, S.; Rab, S.O.; Siddiqui, S.; Singh, A.; Saeed, M.; Khan, F. Naringenin as Potent Anticancer Phytocompound in Breast Carcinoma: From Mechanistic Approach to Nanoformulations Based Therapeutics. *Front. Pharmacol.* **2024**, *15*, 1406619. [CrossRef] [PubMed]
77. Ahmadi, A.; Shadboorestan, A. Oxidative Stress and Cancer; the Role of Hesperidin, a Citrus Natural Bioflavonoid, as a Cancer Chemoprotective Agent. *Nutr. Cancer* **2016**, *68*, 29–39. [CrossRef] [PubMed]
78. Bartoszewski, R.; Hering, A.; Marszałł, M.; Stefanowicz Hajduk, J.; Bartoszewska, S.; Kapoor, N.; Kochan, K.; Ochocka, R. Mangiferin Has an Additive Effect on the Apoptotic Properties of Hesperidin in Cyclopia Sp. Tea Extracts. *PLoS ONE* **2014**, *9*, e92128. [CrossRef] [PubMed]
79. Park, H.J.; Kim, M.-J.; Ha, E.; Chung, J.-H. Apoptotic Effect of Hesperidin through Caspase3 Activation in Human Colon Cancer Cells, SNU-C4. *Phytomedicine* **2008**, *15*, 147–151. [CrossRef]
80. Abroon, S.; Nouri, M.; Mahdavi, M. Hesperidin/Salinomycin Combination; a Natural Product for Deactivation of the Pi3k/Akt Signaling Pathway and Anti-Apoptotic Factors in Kg1a Cells. *J. Fluoresc.* **2024**. [CrossRef]
81. Gavamukulya, Y.; Abou-Elella, F.; Wamunyokoli, F.; AEl-Shemy, H. Phytochemical Screening, Anti-Oxidant Activity and in Vitro Anticancer Potential of Ethanolic and Water Leaves Extracts of Annona Muricata (Graviola). *Asian Pac. J. Trop. Med.* **2014**, *7*, S355–S363. [CrossRef]

antioxidants

MDPI

Article

Sustainable Recovery of Phlorotannins from *Durvillaea incurvata*: Integrated Extraction and Purification with Advanced Characterization

Pamela Raquel Rivera-Tovar [1], Gabriela Contreras-Contreras [1], Paulina Isabel Rivas-Reyes [2], Jara Pérez-Jiménez [3,4], Maximiliano Martínez-Cifuentes [5], José Ricardo Pérez-Correa [6] and María Salomé Mariotti-Celis [1,*]

1 Nutrition and Dietetics School, Faculty of Medicine, Universidad Finis Terrae, Pedro de Valdivia 1509, Providencia, Santiago 7501015, Chile; privera@uft.cl (P.R.R.-T.); gabriela.contreras@ug.uchile.cl (G.C.-C.)

2 Food Technology Section, Department of Analytical Chemistry, Nutrition and Food Science, School of Veterinary Sciences, University of Santiago de Compostela, 27002 Lugo, Spain; privasr@uft.edu

3 Department of Metabolism and Nutrition, Institute of Food Science, Technology and Nutrition (ICTAN-CSIC), 28040 Madrid, Spain; jara.perez@ictan.csic.es

4 CIBER de Diabetes y Enfermedades Metabólicas Asociadas (CIBERDEM), Instituto de Salud Carlos III (ISCIII), 28029 Madrid, Spain

5 Departamento de Química Orgánica, Facultad de Ciencias Químicas, Universidad de Concepción, Edmundo Larenas 129, Concepción 4070371, Chile; maxmartinez@udec.cl

6 Chemical and Bioprocess Engineering Department, School of Engineering, Pontificia Universidad Católica de Chile, Vicuña Mackenna 4860, Macul, Santiago 7820436, Chile; jperezc@uc.cl

* Correspondence: mmariotti@uft.cl

Abstract: The rising demand for bioactive compounds from marine resources highlights the need for sustainable separation technologies. This study introduces an integrated process combining ultrasound-assisted extraction (USAE) and resin purification (RP) to isolate phlorotannins from *Durvillaea incurvata*, a brown seaweed with significant biomedical potential. Using a 32.5% ethanol–water solvent system for USAE followed by RP on Diaion HP-20 resin, phlorotannins were enriched 2.4-fold, with simultaneous removal of interfering compounds such as mannitol (~100%), which was demonstrated by FTIR and HPLC-IR analysis. Advanced characterization using UHPLC-QToF-MS/MS identified five novel phlorotannins with polymerization degrees of 3 to 8 phloroglucinol units in both USAE extracts and post-RP. Mass balance based on spectrophotometric measurements indicated a purification factor of ~2, confirming process effectiveness. RP streams showed distinct phlorotannin profiles, with one phlorotannin exceeding 70% relative abundance. However, MS/MS results showed significantly lower recoveries than spectrophotometric data, revealing a novel insight into RP purification. These findings highlight the critical role of comprehensive chemical characterization in optimizing sustainable phlorotannin extraction from seaweed. They propose a framework for scalable, eco-efficient technologies for achieving high-purity phlorotannin recovery. This approach facilitates the development of phlorotannin-based applications in the nutraceutical and pharmaceutical industries.

Keywords: brown seaweed; integrated extraction–purification processes; spectrophotometric; UHPLC-QToF MS/MS; mass balance

1. Introduction

Researchers have described various forms of nutritional supplementation to prevent and control oxidative stress-induced pathologies [1]. Among various bioactive compounds, attention has also focused on polyphenols, a wide range of plant-based metabolites with multi-target biological functionality [2–6]. Phenolic compounds are not limited to the plant kingdom; they are also present in seaweeds—particularly brown seaweed, which contains phlorotannins [7].

Phlorotannins are phenolic compounds constituted by phloroglucinol (1,3,5-trihydroxybenzene) chains and net-like structures of diverse molecular weights. They are classified into six groups, according to the type of linkages between phloroglucinol units and their content of hydroxyl groups: (i) fucols, with aryl–aryl linkages; (ii) phlorethols, with aryl–ether linkages; (iii) fucophlorethols, with aryl–aryl and aryl–ether units; (iv) fuhalols, with aryl–ether linkages and additional OH groups in every third ring; (v) carmalols and (vi) eckols, both with dibenzodioxin linkages [8]. Eckols differ from carmalols by their low molecular weight and by the presence of a phenoxyl substitution at the C4 position, while in carmalols, the phenoxyl substitution occurs at the C3 position [8]. These highly reactive metabolites have been pointed out, in in vitro and preclinical studies, as potent modulators of various biochemical processes linked to the prevention and treatment of major chronic diseases such as cancer [9], diabetes type II [10], neurodegenerative disorders [11], among others. These conditions are frequently linked to increased oxidative stress and inflammatory responses [12,13]. The anti-inflammatory effects of phlorotannins have been reported in extracts rich in carmalols, fucols, and eckols, demonstrating efficacy in both cell-based and animal models [12,14,15]. And, notably, in a clinical trial involving healthy adults supplemented with capsules containing ethanolic extracts from brown seaweed, it was found that high-molecular-weight phlorotannins were metabolized and absorbed primarily in the large intestines, leading to a reduction in inflammation biomarkers [12].

Indeed, the magnitude of the activity of the brown seaweed extracts/fractions is significantly different from that of the individual phlorotannins contained therein, which makes fully characterized high-purity extracts novel candidates for developing natural therapeutic alternatives with potential medical applications [16]. Therefore, selective separation techniques are research areas of interest [10,17]. The sustainable extraction of phlorotannins from *Durvillaea incurvata*, a brown seaweed highly available on the Chilean coast, is an attractive alternative to scale up the production of phlorotannin-rich extracts [10,16,18]. The frond of this seaweed has shown a significantly higher total polyphenol content compared to those of other brown seaweed species, such as *Lessonia spicata* [10], making it a valuable source of bioactive compounds. Additionally, *Durvillaea incurvata* is highly abundant on the Chilean coastline (National Landing 2022: 11,179 tons, Sernapesca), where its excessive accumulation, exacerbated by climate change and oceanographic changes such as warmer temperatures and altered currents, has increasingly impacted coastal ecosystems. This accumulation negatively affects tourism and economic activities by altering beach aesthetics and emitting unpleasant odors as the biomass decomposes [19–21]. Sustainable valorization strategies, such as the extraction of phlorotannins, offer an opportunity to mitigate these negative effects while supporting the development of high-value products from this abundant marine resource.

Hot Pressurized Liquid Extraction (HPLE) has been one of the most employed environmentally friendly techniques for obtaining phlorotannin-rich extracts efficiently [10,18], although its scaling to commercial production is still challenging [22]. In turn, Ultrasound-Assisted Extraction (USAE) is a scalable technology that effectively recovers these bioactive compounds but, like HPLE, lacks selectivity [17,23]. Recent studies have shown that *Durvillaea incurvata* phlorotannin extracts obtained through HPLE are rich in mannitol. This sugar alcohol has been shown to preserve the antioxidant properties of various terrestrial plant extracts subjected to spray drying due to its thermoprotective effect [24,25]. However, this effect has been questioned because mannitol can also scavenge free radicals, leading to overestimating antioxidant capacity and total polyphenol measurements after spray drying [26].

Integrating selective purification methods following the extraction process appears to be an effective approach for obtaining phlorotannin-rich extracts with fewer interferents, thereby enhancing their bioactive potency [27]. Additionally, the analysis of purified extracts is simpler, leading to a deeper understanding of their chemical composition, which in turn enables more effective exploitation of their therapeutic potential.

Colorimetric techniques allow the determination of seaweed extracts' total phenolic compounds and antioxidant capacities. However, the complex polymeric structure and similar polarity of phlorotannins make them difficult to separate and quantify. In this sense, the continuous development of advanced instrumental techniques opens the possibility of unveiling the occurrence of moieties (bonds and functional groups). In this way, different mass spectrometry (MS) approaches, with or without coupling to liquid chromatography systems, have allowed the analysis of phlorotannin-rich extracts of diverse species of brown seaweed. Systems coupled with LC based on different analyzers are valid for detecting phlorotannins of low molecular weight [28]. At the same time, matrix-assisted laser desorption/ionization–time-of-flight (MALDI-ToF) can ionize and analyze larger compounds of up to 27 degrees of polymerization (DP) [28]. Also, Fourier transform infrared spectroscopy (FTIR) is a highly effective qualitative technique for identifying potential bonds and functional groups present in a sample, indicating the presence of any specific species of interest [29]. It has been suggested as a valid tool for characterizing polyphenol profiles in brown seaweed extracts [30]. However, due to the chemical complexity of phlorotannins, new analytical approaches need to combine the mentioned techniques, which allow their routine characterization, for instance, in the context of nutraceutical development.

The Chilean coast is rich in seaweed, and several studies have explored their phlorotannin profiles. Ethanolic extracts obtained under atmospheric conditions from three Chilean brown seaweed (*Durvillaea antarctica*, *Lessonia spicata*, and *Macrocystis integrifolia*) and analyzed by LC-DAD-ESI-MS/MS (Liquid Chromatography–Diode Array Detector–Electrospray Ionization–Tandem Mass Spectrometry) showed differential profiles regarding the DP of these compounds, e.g., trimers to octamers in *D. Antarctica* and trimers to tetramers in *L. spicata* [31]. Recently, phlorotannins from *Durvillaea incurvata* and *Lessonia spicata* extracts obtained by HPLE-RP (Hot Pressurized Liquid Extraction–Resin Purified Process) were characterized using liquid UHPLC-Orbitrap (Ultra-High-Performance Liquid Chromatography) and UHPLC-QoF-MSn (Ultra High-Performance Liquid Chromatography–Quadrupole Time of Flight–Tandem Mass Spectrometry with Multiple Stages) techniques [27]. Fucols, phlorethols, and fucophlorethols isomers up to 4 DP were the most representative phlorotannins in the extracts of both Chilean seaweeds. Eckols, carmalols, fuhalols, phenolic acids, and flavonoids were also detected.

Interestingly, the analysis of *Durvillaea incurvata* extracts showed low abundances of high molecular weight phlorotannins for the first time (11–21 DP) [27]. The differences in the metabolite profile of evaluated Chilean brown seaweed extracts indicate that achieving a deeper molecular and structural characterization of these compounds is essential to elucidate the mechanisms behind their selective separation. Moreover, this becomes even more crucial when considering scaling up their production using sustainable technologies.

Mass balances are essential for analyzing performance and designing integrated extraction and purification processes, particularly when dealing with complex streams [32]. However, few studies have utilized mass balances to assess the efficiency of integrated processes aimed at producing purified, polyphenol-rich extracts from natural sources. Mass balance analysis is especially challenging in the case of phlorotannin separation processes, as their identification and quantification are complicated by both the structural complexity of phlorotannins and the limited availability of analytical standards.

In the specific case of resin purification (RP) by adsorption, each outlet stream exhibits a distinct polyphenolic profile [33], resulting in a unique spectrum of bioactivities for each stream. Without a comprehensive mass balance analysis, it is difficult to accurately evaluate the overall performance of the integrated extraction and purification process or assess the potential of each outlet stream for use in natural therapeutic applications.

The objectives of this study were to develop an integrated water–ethanol-based USAE-RP process for the selective separation of phlorotannins from Durvillaea incurvata, characterize the extracted compounds using spectrophotometric analysis, HPLC-IR, FTIR, and UHPLC-QToF MS/MS, and assess the efficiency and selectivity of the process through a combination of parameters (mass balance analysis, evaluating yield, recovery rates, mannitol removal efficiency, purities, and purification factors). This integrative approach highlights the potential of marine resources for bioactive compound recovery and establishes a foundation for optimizing green separation technologies in natural alternatives with potential medical applications derived from *Durvillaea incurvata*.

2. Materials and Methods

2.1. Seaweed Material

Brown seaweed frond (*Durvillaea incurvata*) samples were collected from Concepción, Chile (36°49′37.2″ S 73°2′59.2″ W) in the fall of 2023. The samples were washed with water, cut, frozen, and freeze-dried. Then, they were ground and sieved (Tyler sieve n°25, <710 μm) and stored in dark polyester bags at 20 °C. The proximal composition of the raw material was determined using standard analytical procedures [34], among them moisture (10.90 ± 0.08%), ash (19.54 ± 0.03%), crude protein (0.82 ± 0.10%), crude fat (0.41 ± 0.03%), crude fiber (0.98 ± 0.25%), and carbohydrates by difference (67.36 ± 0.14%).

2.2. Chemicals

Analytical grade solvents (acetone, acetonitrile, and ethanol) and reagents (2,4-dimethoxybenzaldehyde [DMBA], Folin–Ciocalteu phenol reagent, HCl, H_3PO_4, NaOH, and the salts K_2HPO_4, KH_2PO_4, and Na_2CO_3) were acquired from Merck, Germany. Additional materials, including macroporous resins (Diaion HP-20), fluorescein, 2,2′-Azobis(2-amidinopropane) dihydrochloride (AAPH), and the standards Trolox, phloroglucinol, gallic acid, and mannitol were purchased from Sigma-Aldrich, Saint Louis, MI, USA.

2.3. Extraction Techniques

2.3.1. Atmospheric Solid–Liquid Extraction (ASLE)

For ASLE, 10 g of ground seaweed frond was mixed with 100 mL of 60% *v/v* acetone–water. The extraction was performed at 20 °C with constant magnetic stirring (IKA C-MAG HS, Stauffer, Germany) in a cooling tube system for 6 h. This extract was used as a reference extract (RE) since it is assumed to contain nearly 100% of the extractable polyphenols from the natural matrix [35]. The RE was used solely to calculate the recoveries of polyphenols and phlorotannins achieved by USAE and to estimate relative masses for the USAE mass balance.

2.3.2. Ultrasound-Assisted Extraction (USAE)

For USAE, 20 g of ground seaweed frond was mixed with 200 mL of 32.5% *v/v* ethanol–water. The extraction process was conducted using a 1500 W ultrasonic device (BIOBASE, Jinan, Shandong, China) operating at a constant frequency of 20 kHz, a controlled temperature of 30 °C, with a jacket cooling system, 40% power, and one cycle of 30 min.

Experimental conditions were established based on preliminary tests aimed at obtaining the highest phlorotannin content in the extracts. This extract was called crude extract (CE).

Following the extractions (ASLE and USAE), RE and CE were centrifuged for 5 min at 4.500 rpm (DM0412 DLAB centrifuge, Beijing, China) and subjected to rotary evaporation (RE-100 PRO DLAB, Beijing China) until dry. The resulting concentrated extracts were reconstituted with 26 mL of water and freeze-dried (BIOBASE, Jinan, Shandong, China) for 24 h. The solid powder extracts were stored in a bag in the dark until their physicochemical characterization.

2.4. *Resin Purification (RP) Process*

For RP, a column (Büchi C-690 26/100) was packed with macroporous resin (Diaion HP-20) as adsorbent, and 80% *v/v* ethanol–water was used as an elution agent [27]. In comparison to other macroporous resins, such as SP-850, XAD-2, XAD-4, and XAD-7HP, HP-20 resin has demonstrated the highest adsorption capacity for phlorotannins [36,37]. RP consisted of four stages: (i) adsorption of extract, (ii) washing with water, (iii) elution with ethanol, and (iv) regeneration of the resin with HCl and NaOH [38]. The CE and the solvents (water, ethanol, HCl, and NaOH) were pumped through the RP system at 3 mL/min using a peristaltic pump (PP-X575). The process followed this sequence: 20 mL of CE reconstituted in water (3 mg/mL), 40 mL of water, 40 mL of 80% EtOH, and finally, the regeneration consisted of passing 40 mL of water, 40 mL of HCl (2N), 40 mL NaOH (1N), and 40 mL of water [27]. Three main output streams were collected: the outlet stream during the column charging process (CS), the outlet washing stream (WS), and the elution stream (ES), which corresponds to the purified extract. These streams were subsequently freeze-dried for storage until their later analysis.

2.5. *Chemical Analyses*

2.5.1. Total Polyphenol Content (TPC)

The TPC of all samples was determined spectrophotometrically using the Folin–Ciocalteu assay [28]. For analysis, samples were reconstituted (20 mg in 10 mL of water) and analyzed in duplicate following the method described in [39], with some modifications for adaptation to a 96-well microplate reader (TECAN Infinite 200 PRO MPlex, Switzerland). Aliquots of each sample, standard, or water (20 µL) were mixed with Folin–Ciocalteu reagent diluted 1:2 with water (10 µL), 150 µL of water, and 10% *w/v* aqueous solution of Na_2CO_3 (20 µL). The mixture was shaken and allowed to react for 1 h at room temperature (20 °C) in darkness; then, the absorbance was measured at 765 nm. Sample absorbances were interpolated against a gallic acid calibration curve (30–150 mg/L) and a phloroglucinol calibration curve (50–250 mg/L). Results are expressed as milligrams of standard equivalent (gallic acid or phloroglucinol) per gram of dried extract or RP stream.

2.5.2. Total Phlorotannin Content (TPhC)

The samples were also analyzed using the DMBA method [40]. For analysis, samples were reconstituted (20 mg in 10 mL of water) and analyzed in duplicate following the method described in [40], with some modifications [41]. The reaction began with shaking for 30 min, followed by incubation at 25 °C in the dark for 60 min. Absorbances were measured at 515 nm using a 96-well microplate reader (TECAN Infinite 200 PRO MPlex, Männedorf, Switzerland). Sample absorbances were interpolated against a phloroglucinol calibration curve (1–40 mg/L). Results are expressed as milligrams of phloroglucinol equivalent (PhE) per gram of dried extract or RP stream.

2.5.3. Antioxidant Capacity (AC)/Oxygen Radical Absorbance Capacity Assay (ORAC)

The ORAC assay was performed as described in [42], with some modifications [10]. In a 96-well microplate, 250 μL of 55 nM fluorescein, diluted in phosphate buffer saline (PBS) at pH 7.4 (75 mM, prepared with K_2HPO_4 solution and KH_2PO_4 solution in nanopure water), was added to 25 μL of extract, Trolox standard, or PBS (blank). The microplate was incubated for 30 min at 37 °C, and the reaction was initiated by adding 25 μL of freshly prepared 153 mM AAPH solution in PBS using an automatic injector. Fluorescence was recorded every minute for 1 h at 37 °C using a microplate reader (TECAN Infinite 200 PRO MPlex, Männedorf, Switzerland). The reference calibration curve was constructed with Trolox solutions ranging from 2 to 8 mg/L; analyses were performed in duplicate. Results are reported as micromoles of Trolox equivalents (TEs) per gram of dried extract or RP stream.

2.5.4. Antioxidant Capacity (AC)/DPPH Assay

The DPPH method was performed as described by other authors [43], with modifications [10]. For analysis, samples were reconstituted (20 mg in 5 mL of water), diluted 1:6, and analyzed in duplicate. Absorbances were measured at 520 nm (Genesys 150 UV-Vis, Thermo Fisher Scientific, Waltham, MA, USA) and interpolated against a Trolox calibration curve (2–8 mg/L). Results are expressed as micromoles of Trolox equivalents (TEs) per gram of dried extract or RP stream.

2.5.5. Fourier Transform Infrared Spectroscopy Analysis of USAE and RP Streams

FTIR was performed in a Spectrum Two IR spectrophotometer (Perkin Elmer, Shelton, CT, USA) with attenuated total reflection (ATR) (Pike Instruments, Madison, WI, USA) in the range of 400–4000 cm^{-1} [44]. This analysis identified functional groups in the CE and the RP output streams (CS, WS, and ES). Furthermore, the FTIR spectra of phloroglucinol and mannitol were studied to determine the selectivity of the USAE-RP process.

2.5.6. Mannitol Content Determination

Mannitol was quantified in all samples using an HPLC-IR system [10]. Chromatographic analysis was performed on the Thermo Scientific Vanquish Flex HPLC system (Thermo Fisher Scientific, Waltham, MA, USA) coupled with a refractive index detector (RefractoMa 520 ERC, Thermo Fisher Scientific, Waltham, MA, USA). The autosampler was maintained at room temperature. The column used was an APS-2 Hypersil™ (150 mm × 4.6 mm, 5 μm) from Thermo Fisher Scientific (Waltham, MA, USA). The column temperature was set at 50 °C. The mobile phase consisted of 250 mM H_3PO_4 prepared in 80% acetonitrile. A sample volume of 20 μL was injected, and the flow rate was set at 0.3 mL/min. The total analysis time was 18 min. Mannitol quantification in the samples was carried out using a calibration curve (100 to 1000 mg/L). Analyses were performed in duplicate. Results are expressed as milligrams of mannitol (Ma) per gram of dried extract or RP stream.

2.5.7. Tentative Identification of Phlorotannins Using a UHPLC Coupled with a QToF Detector

A tentative identification of the phlorotannins in all USAE-RP streams was carried out using UHPLC-QToF. Mass spectrometry analysis was conducted using a previously described methodology, with slight modifications [27].

Sample Injection and Chromatograph and Detector Conditions

Briefly, UHPLC-MS^2 analysis was performed on a UHPLC-QToF system (Bruker Daltonics, Billerica, MA, USA). The autosampler was set at 4 °C, and a C18 column (2.1 mm × 100 mm, 1.7 μm) (Phenomenex, Torrance, CA, USA) was used with a temperature of 40 °C. A 5 μL sample was injected in a gradient mode, with a mobile phase A consisting of 0.1% formic acid and a mobile phase B consisting of 90% acetonitrile/0.1% formic acid. The gradient was set to start with 12% B for 1 min, then 99% B at 11 min, 99% B at 13.5 min, 12% B at 14 min, and 12% B at 15 min. The mass spectrometry data were acquired in the range 50 m/z to 1300 m/z in the negative mode. The capillary voltage was 4500 V, the nebulizer pressure was 2.5 bar, and the dry gas temperature was 250 °C, with a flow rate of 8 L/min. The collision energy was between 18 eV and 50 eV.

Identification of Phlorotannins

Signals corresponding to phlorotannins were identified using Compass Data Analysis 4.4 software (Bruker Daltonik, Billerica, MA, USA). MS^2 and MS^1 spectra were analyzed in this study to identify specific fragmentation patterns. Our analysis was focused on the MS^1 spectra, where characteristic fragmentation patterns were observed. Phlorotannins were identified based on their $[M-H]^-$ ions. This approach aligns with the methodology reported by Hyo Moon et al. [45], who successfully identified several phlorotannins using MS^1 spectra. After tentative identification, the phlorotannins were classified according to their DP. An error of up to ±10 ppm was considered, as typically considered for phenolic compounds on a UHPLC-QToF instrument [27,46].

Use of Areas

Once the tentative identification and classification were made, the relative areas for the different phlorotannins were determined. The distribution of phlorotannins in the CE and each of the three RP output streams was calculated. For this, the total area of phlorotannins identified in each sample was considered to represent 100 percent for each case.

2.6. Statistical Analysis

The ASLE, USAE, and RP processes were performed in triplicate, and chemical analyses were performed in duplicate. The experimental results obtained are expressed as means ± SD. Statistical analysis was carried out using Minitab® Statistical Software v.21. Analysis of variance (ANOVA) at a 95% confidence level ($p < 0.05$) was applied to compare all the obtained responses with Tukey's post-hoc test after verifying the normal distribution of the data and homogeneity of variances, respectively.

2.7. Relative Mass Balance and Selective Separation Performance Parameters

A complementary analysis was performed using the TPC, TPhC, mannitol measurements, and the relative areas from UHPLC to perform a relative mass balance in the USAE (Equation (1)) and the RP (Equation (2)).

$$m_{i_{\text{seaweed}}} = m_{i_{\text{CE}}} + m_{\text{extraction cake}} + e \quad (1)$$

$$m_{i_{\text{CE}}} = m_{i_{\text{CS}}} + m_{i_{\text{WS}}} + m_{i_{\text{ES}}} + e \quad (2)$$

where m is the mass (g or mg) and the subindex i refers to the evaluated components (extractable solids or ExS, total polyphenols, total phlorotannins, mannitol, specific phlorotannins); e represents the difference between the input and output relative mass. The relative masses of the seaweed or pre-USAE ($m_{i\ \text{seaweed}}$) are determined based on the reference extract (RE). The masses of ExS from CE and streams were determined by freeze-drying the respective samples. Masses of total polyphenols (TPs), total phlorotannins (TPhs), and mannitol were determined based on TPC, TPhC, and mannitol content values, respectively.

Selective separation performance parameters were calculated to determine (i) the yield achieved by USAE in terms of ExS (Equation (3)), (ii) the USAE polyphenol or phlorotannin recoveries (Equation (4)), (iii) the mannitol removal efficiency (Equation (5)),

(iv) the polyphenol and phlorotannin purities (Equation (6)), and (v) the purification factor in terms of TP and TPh (Equation (7)). These parameters were determined as follows:

$$\text{yield}_{\text{ExS}}\ (\%) = m_{\text{CE}}/m_{\text{seaweed}} \times 100 \quad (3)$$

where m_{CE} is the dry mass of the extract generated by USAE and $m_{seaweed}$ is the mass of dried seaweed used in USAE.

$$Rec_j\ (\%) = m_{j_{CE}}/m_{j_{\text{seaweed}}} \times 100 \quad (4)$$

where Rec_j is the recovery of component *j*, which can be TP or TPh, m_{TP} is an estimation of the TP mass as phloroglucinol equivalent from the Folin–Ciocalteu assay, and m_{TPh} is an estimation of the TPh mass as phloroglucinol equivalent from DMBA assay.

$$Eff_{\text{Ma}}\ (\%) = m_{\text{Ma removed}}/m_{\text{Ma}_{\text{CE}}} \times 100 \quad (5)$$

where Eff_{MA} is the efficiency of RP to remove the mannitol contained in CE, $m_{\text{Ma removed}}$ is the result of the difference between $m_{\text{Ma CE}}$ and $m_{\text{Ma purified extract}}$.

$$P_{\text{j}}\ (\%) = m_j/m_{\text{ExS}} \times 100 \quad (6)$$

where P_j is the purity of component *j*, which can be TP or TPh.

$$F_{\text{Purification}} = \text{P}_{\text{TP ó TPh}}\ \text{after RP}/\text{P}_{\text{TP ó TPh}}\ \text{before RP} \quad (7)$$

where $F_{\text{Purification}}$ is the purification factor, which represents the performance of the RP to increase the purity of the component of interest (TP and TPh).

During the preparation of this section, the authors used Grammarly, DeepL, and ChatGPT v.4 to enhance the readability of this text. After using these tools, the authors reviewed and edited the content as necessary; they take full responsibility for the content of the published article.

3. Results and Discussion

3.1. Spectrophotometric Characterization and Quantification: TPC, TPhC, AC_{ORAC}, and AC_{DPPH}

All samples, i.e., CE and the three RP streams (CS, WS, and ES), were characterized in terms of TPC, TPhC, and AC (by ORAC and DPPH assays), as shown in Figure 1. Overall, the values determined by the Folin–Ciocalteu method were one order of magnitude higher than those obtained by the DMBA method. This discrepancy could be due to an overestimation of the Folin–Ciocalteu values caused by interference by other non-phenolic reducing compounds present in the brown seaweed (sugar alcohols, polysaccharides, pigments, and other polyphenol–protein complexes [47]) or to an underestimation of the DMBA values due to the non-detection of branched phlorotannins or those with aryl bonds or fuhalols with additional hydroxyl groups in positions 2, 4, or 6, which differ structurally from those reacting with DMBA (1,3 and 1,3,5-trihydroxybenzenes) [28]. Considering both measurement methods could offer a better approximation of seaweed polyphenol and phlorotannin quantification.

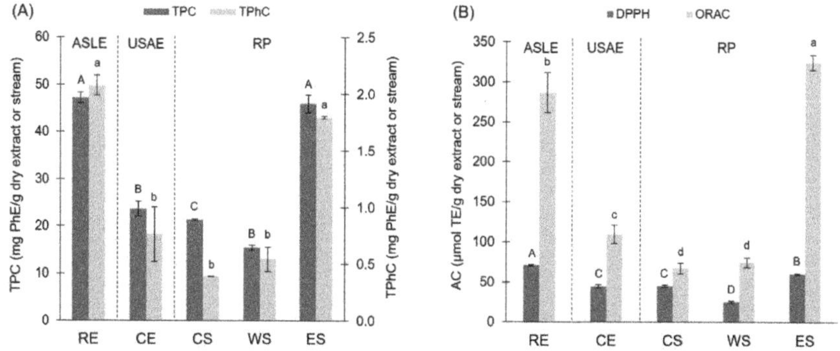

Figure 1. (**A**) TPC and TPhC, and (**B**) antioxidant capacity (DPPH and ORAC assays) of RE, CE, and RP streams (CS, WS, and ES) from *Durvillaea incurvata* frond. [(A–D; a–d)] Values that do not share a letter are significantly different.

TPC of CE (23.6 ± 1.6 mg PhE/g dry extract) was slightly higher than that obtained in a previous study of *Durvillaea incurvata* frond from "Las Cruces" by HPLE (15% *v/v* glycerol; 21.1 ± 9.2 mg PhE/g dry extract). However, the absolute recovery of total polyphenols in our extract (3.6 ± 0.2 mg PhE/g dried seaweed) was approximately half the amount found in that study [10]. Although both methods (USAE-EtOH and HPLE-glycerol) produced extracts with similar polyphenol concentrations, HPLE-glycerol was more effective for total polyphenol recovery. USAE-EtOH's lower performance may be related to the degradation of certain phlorotannins and the generation of highly reactive hydroxyl radicals in the air bubbles resulting from the acoustic cavitation of USAE [47]. Factors related to the raw material, such as harvest date, location, and pre-storage treatments, may also affect these values [48]. Nevertheless, in this case, this discrepancy in polyphenol recovery is mainly due to the extraction methods. In the previous study, the same seaweed species and anatomical part were used, and both reference extractions with ASLE achieved similar absolute yields (8.2 ± 0.2 PhE/g dry seaweed in the current study and 11.7 mg PhE/g dry seaweed in the previous study).

It is worth noting that the TPC of our CE sample expressed in GAE (12.7 ± 0.9 mg GAE/g dry extract) was practically the same as that of the optimal extract (12.8 ± 2.3 mg GAE/g extract in dry weight) obtained through USAE-EtOH from *Durvillaea incurvata* in another study, where a higher temperature (50 °C), a longer time (80.8 min), and a much higher ethanol concentration (70% *v*/*v*) were used [49]. It has been reported that phlorotannin extraction from brown seaweed begins to decrease with increasing ethanol concentration from 30 to 70% *v/v* [17]. Acoustic cavitation is the main extraction mechanism in USAE, so higher temperatures, higher co-solvent concentrations, and longer extraction times seem unnecessary [17,47].

Notably, the RP treatment resulted in a 1.9-fold and a 2.4-fold increase in TPC and TPhC, respectively (Figure 1A). These values for RE and ES were, on average, 2.3 times higher than those of CE (Figure 1A).

The antioxidant capacity was assessed by two distinct methodologies (DPPH and ORAC), considering that natural extracts present more than one antioxidant mechanism [50]. Overall, AC_{ORAC} showed higher values than AC_{DPPH} in all samples (Figure 1B). This difference may arise from several factors: the phenolic compounds extracted from *Durvillaea incurvata* frond could be more effective at scavenging peroxyl radicals than DPPH radicals; the methanolic medium utilized in the DPPH method might hinder the accessibility of phlorotannins to radicals; or the presence of certain compounds in the extract could interfere more with the DPPH method than the ORAC method [50–52]. The

AC of CE was equivalent to DPPH and ORAC values previously reported for *Durvillaea incurvata* subjected to extraction by HPLE–glycerol and about twice the DPPH obtained by USAE-EtOH 70% [49], showing that high ethanol concentrations are detrimental for extraction efficiency. Incorporating RP led to significant enhancement of the AC, with AC_{DPPH} and AC_{ORAC} ES values ~1.5 and ~3.0 times higher than those of CE. On average, the AC values for ES and RE were similar, where $DPPH_{ES}$ was lower than $DPPH_{RE}$, and $ORAC_{ES}$ was higher than $ORAC_{RE}$.

This suggests that RP selectively concentrated specific phlorotannins with higher reactivity toward the hydrogen atom transfer mechanism (DPPH assay) and peroxyl radicals (ORAC assay), leading to a significant increase in antioxidant capacity. Given the dose-dependent nature of phlorotannin bioactivity, it may be expected that this enhancement would translate into greater biological efficacy. In this sense, previous studies support this projection, as a purified *Durvillaea incurvata* extract exhibited a stronger α-glucosidase inhibitory effect than its crude counterpart [10]. Similarly, in diabetic mouse models, oral administration of 2,7′′-phloroglucinol-6,6′-bieckol delayed carbohydrate absorption by inhibiting the relevant enzyme in a dose-dependent manner [53]. Additionally, research on the anti-inflammatory properties of seaweed extracts in mouse models found that purified eckol significantly reduced proinflammatory cytokine levels at the highest doses [54].

Overall, these findings indicate that RP improves the bioactivity of phlorotannin CE, potentially allowing lower doses to achieve comparable or enhanced biological responses.

3.2. Fourier Transform Infrared Spectroscopy Analysis

CE, all RP output streams, and standard mannitol and phloroglucinol solutions were analyzed using FTIR Spectroscopy. Analyzing the characteristic bands of all streams contributed to a better understanding of the USAE-RP's selectivity (Figure 2).

In the six spectra, band 1 corresponding to O–H stretching was evident, which demonstrated the presence of groups characteristic of mannitol (aliphatic hydroxyl, Figure 2A,C–E) and polyphenols (aromatic hydroxyl, Figure 2B,F). Also, bands associated with aromatic rings (C–H stretching and C–H bending) were found in the six spectra. Specifically, FTIR spectra allowed the identification of several types of bonds commonly found in phlorotannins, including aryl ether linkages between aromatic rings that have been potentially associated with fucophloroethols [55]. Similarly, the FTIR spectrum of phloroglucinol enabled the detection of specific characteristic bands associated with this compound, and alterations in bands related to interferences, such as mannitol, were also evident for each analyzed sample.

The FTIR spectrum of CE (Figure 2C) showed an overlap of bands from mannitol and polyphenols, indicating the presence of both compounds. The low definition of a key band was attributed to residual water and ethanol from the USAE process. Additionally, a possible band of terrestrial polyphenols was detected (band 3 at 1619 cm^{-1}, typically associated with C=O), while CS spectra revealed a significant amount of mannitol, which was expected due to the selectivity offered by the HP-20 resin. However, it also contained unadsorbed phlorotannins, possibly due to insufficient residence time or resin saturation. WS stream was rich in mannitol, while the ES lacked mannitol and was rich in phenolic compounds.

An important point should be discussed about the RP streams (Figure 2D–F). They all have a new band not detected on the phloroglucinol spectra (Figure 2B), indicating a new vibration associated with the C–O–C linkage. This band is characteristic of aromatic rings connected via ether bonds [56]. For instance, a sharp band associated with a C–O–C linkage (band 5) was observed in the spectrum of the elution stream (Figure 2F). Along with the broad O–H stretching band, it can be concluded that the elution stream is rich in phlorotannins.

(a)

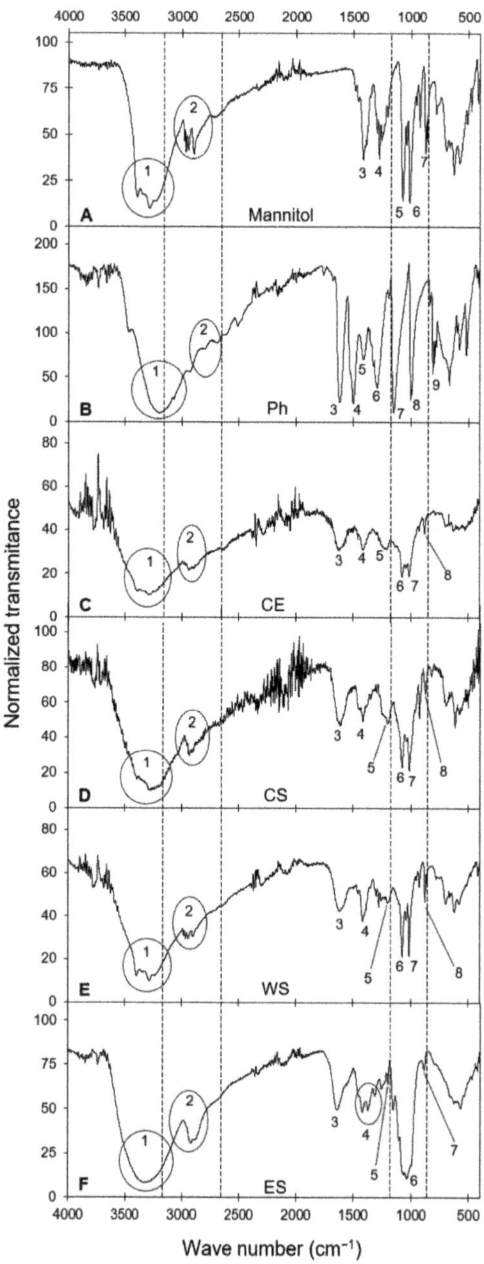

(b)

	Vibration type	Wavenumber (cm⁻¹)
A	1. ν O–H	3389
	2. ν C–H aliphatic	2869 – 2809
	3. δ O–H	1417
	4. δ C–H secondary	1281
	5. δ C–O secondary alcohol	1079
	6. ν C–O primary alcohol	1019
	7. δ C–H aliphatic	881
B	1. ν O–H	3191
	2. ν C–H aromatic	2931
	3. ν C=C aromatic	1622
	4. δ C=C aromatic	1502
	5. δ O–H	1414
	6. δ O–H	1296
	7. δ C–H in plane	1154
	8. δ C–O	1008
	9. δ C–H 1,3,5 substituted aromatic ring	814
C	1. ν O–H	3387 – 3284
	2. ν C–H aliphatic and aromatic	2946 – 2859
	3. ν C=C and possible ν C=O	1619
	4. δ C–OH aliphatic and aromatic	1414
	5. δ C–H aromatic	1212
	6. δ O–H secondary alcohol	1079
	7. ν O–H primary alcohol	1019
	8. δ C–H aliphatic	883
D	1. ν O–H aliphatic and aromatic	3481 – 3028
	2. ν C–H aliphatic and aromatic	2969 – 2809
	3. ν C=C and possible ν C=O	1614
	4. δ O–H aliphatic and aromatic	1418
	5. ν C–O–C aryl	1203
	6. δ C–OH aliphatic secondary alcohol	1082
	7. ν C–OH aliphatic primary alcohol	1018
	8. δ C–H aromatic	881
E	1. ν O–H aliphatic and aromatic	3359 – 3230
	2. ν C–H aliphatic and aromatic	2285 – 2903
	3. ν C=C and possible ν C=O	1611
	4. δ O–H aliphatic and aromatic	1417
	5. ν C–O–C aryl	1206
	6. δ C–OH aliphatic secondary alcohol	1079
	7. ν C–OH aliphatic primary alcohol	1019
	8. δ C–H aromatic	882
F	1. ν O–H aromatic	3328
	2. ν C–H aromatic	2920 – 2878
	3. ν C=C and possible ν C=O	1636
	4. δ O–H aliphatic and aromatic	1419, 1371, 1314
	5. ν C–O–C aryl	1202
	6. δ C–H aromatic and C–OH phenolic	1037
	7. δ C–H aromatic	889

Figure 2. (**a**) FTIR spectra of mannitol, phloroglucinol, CE, and RP streams. (**b**) Characteristic bands of each FTIR spectrum associated with specific types of vibrations.

3.3. Mannitol Content

Seaweed extracts contain undesirable compounds, such as mannitol, a predominant alcohol sugar of seaweeds [57]. Previously reported values for *Durvillaea incurvata* frond extracts ranged from 248 ± 39 to 310 ± 120 mg mannitol/g dried extract [10]. Here, the mannitol content of CE was 259 ± 24 mg mannitol/g dried extract, which falls within that range. However, the most interesting aspect is that the RP facilitated a substantial decrease in this constituent, with ES showing only 1% of the original mannitol value in CE. This low value indicates that applying USAE with a moderate concentration of EtOH, followed by RP, is quite effective for obtaining selective polyphenolic extracts from *Durvillaea incurvata* fronds with a reduced mannitol content [27]. Low mannitol content may enhance the value of the extract by increasing the concentration of polyphenols [58,59]. From a safety standpoint, the potential side effects associated with normal or elevated doses of mannitol (maximum dose: 1.5 g/kg/day, FDA) raise concerns regarding its use, as emphasized by relevant Food and Drug Administration guidelines. Therefore, future studies should focus on investigating the toxicity and side effects of sugar alcohols to assess their safety more accurately [60].

3.4. Tentative Identification of Phlorotannins and Determination of Their Relative Abundances by UHPLC-QToF Mass Spectrometry

The chemical identification of the phlorotannins present in the samples was initially based on a targeted analysis, which matched the identified phlorotannins with those previously reported [61]. Subsequently, a tentative identification of unreported phlorotannins was conducted. This tentative identification was based on matching the experimental molecular ions with the theoretical molecular weights of proposed phlorotannin structures, constructed according to the bonding types known for the different subfamilies (Fucols, Phloretols, Fucophlorethols, Fuhalols, Eckols) [62].

Five phlorotannins composed of 3 to 8 phloroglucinol units (PGUs) were proposed to be present in CE based on their $[M-H]^-$ ions (Mass spectra in supplementary material, Figures S1–S5). The polymerization degree, molecular formula, fragmentations, and the associated errors in ppm for the found phlorotannins are shown in Table 1. The ionization conditions used in the MS method caused a first fragmentation at the source that produced phlorotannin fragments in MS^1, i.e., it provided information similar to an MS^2 analysis. In particular, fragmentation PGU + H_2O + $2H^+$ was observed in several of the identified phlorotannins (Table 1). This fragmentation pattern, similar to that previously observed in a study on *Fucus vesiculosus*, indicated a possible fragmentation of the aryl–ether–aryl (C–O–C) bond [63]. Indeed, several identified phlorotannins exhibited comparable daughter ion fragmentation patterns linked to PGU + H_2O + $2H^+$ with distinct molecular weights (Table 1), suggesting the presence of fucophloroethol isomers in the purification streams, which was confirmed due to the losses of PGU + H^+ corresponding to aryl–aryl (C–C) bonds.

Based on these aspects, a structural hypothesis is formulated for certain phlorotannins identified in brown seaweeds such as *Ascophyllum nodosum*, *Fucus vesiculosus*, and *Durvillea incurvata*, postulating breaks in the C–O–C and C–C bonds as the basic units of phloroglucinol. These tentative structures are categorized based on their type (Figure 3). It should be mentioned that the phlorotannins found in these extracts (carmalols, fucols, and eckols) are among those that have shown potent effects on the inflammatory response in previous studies [12,14,15].

Table 1. Tentative identification of phlorotannins in the CE from *Durvillaea incurvata* frond using UHPLC-QToF-MS.

Polymerization Degree	Molecular Formula	Theoretical [M − H]$^-$ (*m/z*)	Observed [M − H]$^-$ (*m/z*)	Error (ppm)	MS^1 (*m/z*)	MS^2 (*m/z*)
Trimer	$C_{18}H_{11}O_{12}$ (Carmalol)	419.0251	419.0241	2.3	389 (−30), 243 (−PGU + H_2O + 2H$^+$), 116 (PGU + H$^+$)	N.D.
Pentamer	$C_{30}H_{15}O_{17}$ (Carmalol)	647.0309	647.0368	−9.1	535 (111), 389 (PGU + H2O + 2H$^+$), 243 (PGU + H2O + 2H$^+$), 116 (PGU + H$^+$)	116, 100 (–O)
Hexamer	$C_{36}H_{25}O_{17}$ (Fucol)	729.1092	729.1088	0.5	681 (47), 535 (−PGU + H_2O + 2H$^+$), 389 (−PGU + H_2O + 2H$^+$), 243 (−PGU + H_2O + 2H$^+$), 116 (PGU + H$^+$)	A) 116, 100 (–O)
Heptamer	$C_{42}H_{25}O_{27}$ (Eckol)	961.0583	961.0578	0.6	815 (145), 681 (134), 535 (−PGU + H_2O + 2H$^+$), 389 (−PGU + H_2O + 2H$^+$), 243 (−PGU + H_2O + 2H$^+$), 116 (PGU + H$^+$)	N.D.
Octamer	$C_{48}H_{31}O_{21}$ (Eckol)	943.1358	943.1305	5.6	791 (151), 501 (−2PGU + H_2O + 2H$^+$), 389 (112), 243 (−PGU + H_2O + 2H$^+$), 116 (PGU + H$^+$)	A) 791, 225 (566), 167 (60), 81 (84)

N.D., non-detected; PGU, phloroglucinol units.

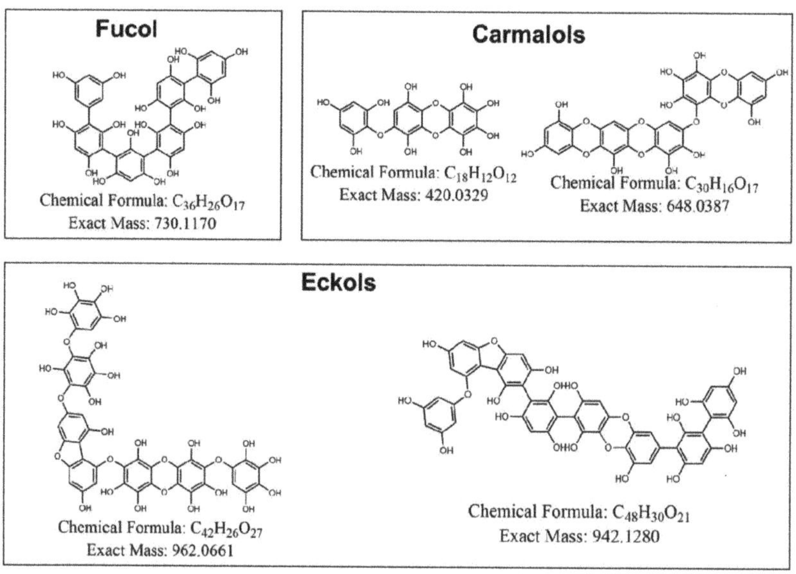

Figure 3. Proposed structures of the five identified phlorotannins in the extracts of *Durvillaea incurvata* frond.

The five proposed phlorotannins identified in CE were also present in ES. The trimer and the hexamer were found in CS, while the pentamer was present in WS. Although the five CE phlorotannins were detected in ES, according to the relative areas, two compounds ($C_{18}H_{12}O_{12}$ and $C_{30}H_{16}O_{17}$) were more abundant in the intermediate RP streams (CS and WS, respectively) (Figure 4). The premature elution of these phlorotannins ($C_{18}H_{12}O_{12}$ and $C_{30}H_{16}O_{17}$) may be attributed to their smaller size, which reduces their residence time in the column relative to larger ones [27,64]. In this case, molecular size seems to play a more significant role than the high polarity of larger phlorotannins, which contain numerous hydroxyl groups that enhance their polarity and water affinity. It is important to consider these aspects when seeking extracts enriched in specific phlorotannins. Although ES typically exhibits the highest TPhC of the RP streams, its levels of certain target polyphenols can be low. Notably, each RP stream contained a phlorotannin with a relative abundance exceeding 70% (Figure 4), which is particularly relevant for targeted separation and has not been previously reported for RP purification of phlorotannins.

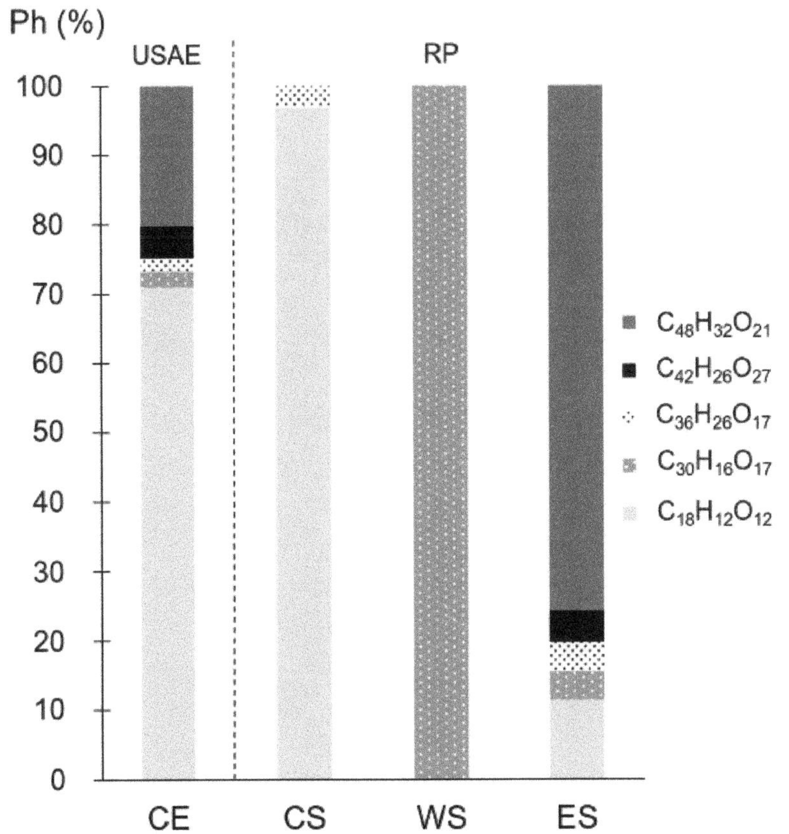

Figure 4. Distribution of the five proposed phlorotannins identified in the process streams (CE, CS, WS, and ES) from *Durvillaea incurvata* frond.

3.5. Process Evaluation: Quality Parameters and RP Mass Balance

This complementary analysis provided a more comprehensive understanding of the separation process and allowed for a better assessment of its potential. The proportion of ExS generated by USAE was small, with the majority of solids remaining in the extraction cake (Figure 5). This was reflected in the $yield_{ExS}$, which was 15%. Typically, the extraction yield from brown seaweed ranged between 10% and 37% when employing various extraction technologies, such as HPLE-EtOH, USAE-EtOH, and ASLE-EtOH [17,35,65,66]. Interestingly, the recoveries of polyphenols (Rec_{TP}) and phlorotannins (Rec_{TPh}) were significantly higher than the extraction yield ($yield_{ExS}$), at 45% and 33%, respectively. This disparity highlights the potential for specific compounds to be extracted more efficiently than the overall yield might suggest [38]. Hence, both the extraction yield of solids and the recovery of specific bioactive compounds should be considered when evaluating extraction processes. While the first parameter offers insight into the overall extraction efficiency, the recoveries of polyphenols and phlorotannins may provide guidance on how to optimize the process for targeted extraction.

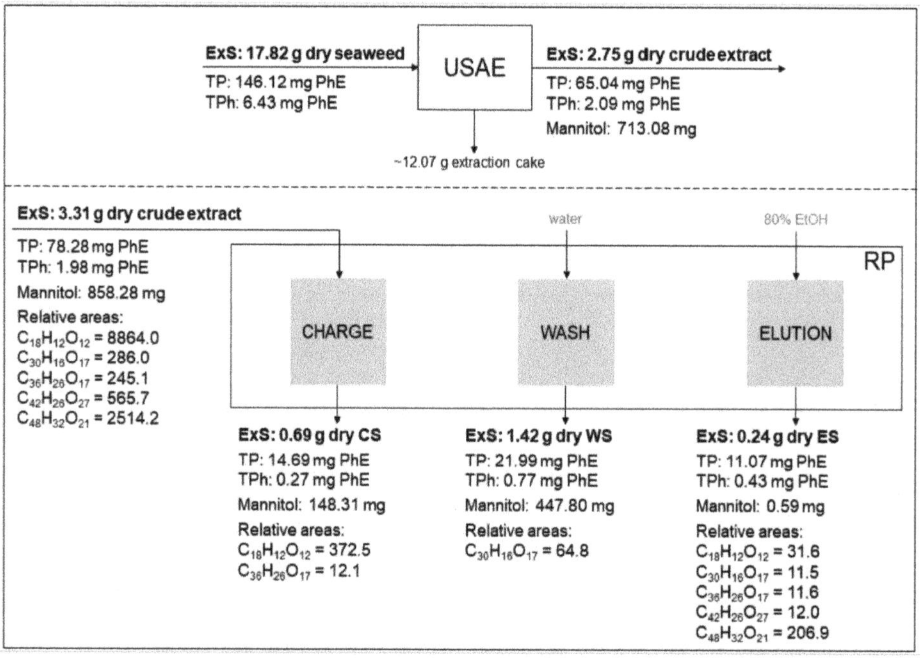

Figure 5. Relative mass balance of the integrated process of USAR-RP from *Durvillaea incurvata* frond. Relative abundances correspond to the ratio of the relative area of phlorotannin and ExS of the stream.

The mannitol recovered in the CE stream was predominantly found in WS (52%, Figure 6), primarily containing polar compounds removed from the resin by water [38]. In contrast, ES contained a negligible amount of the mannitol originally present in CE, thereby achieving one of the main objectives of the RP process: the reduction or elimination of mannitol, a known undesirable compound in brown seaweed bioactive extracts [10,67]. The efficiency of mannitol removal by RP (Eff_{Ma}) was ~100%, corroborating the findings mentioned above and indicating a good performance of the RP process. Due to mannitol removal, the TPC and TPhC in ES increased with respect to CE. However, the TP and TPh amounts in ES were significantly lower than in CE. This decrease is particularly marked in the relative areas of the five tentatively identified phlorotannins in the RP streams (Figure 5).

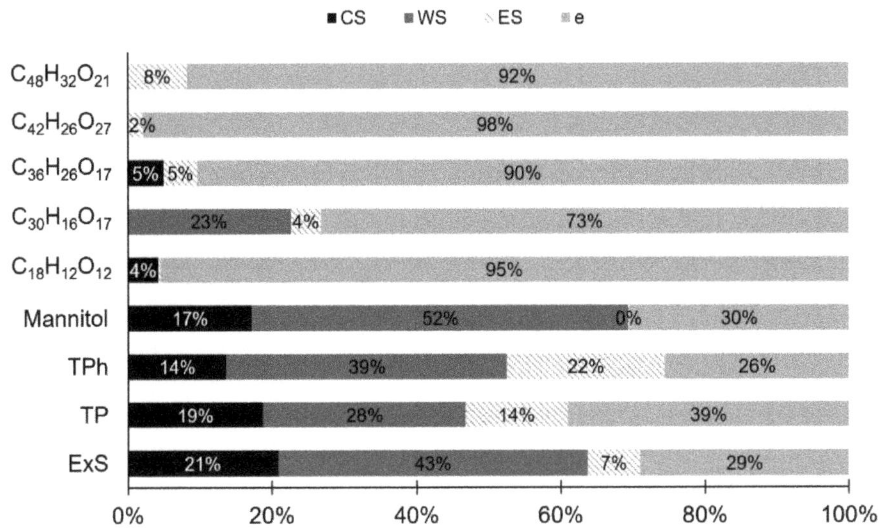

Figure 6. Percentage distribution of ExS, TP, TPh, Mannitol, and specific phlorotannins of CE in the RP streams.

This behavior could be attributed to the RP, which distributed the TP and TPh present in CE among the three generated streams. Ideally, the proportions of TP and TPh in CS and WS should have been negligible. The presence of all four components in CS as compared to CE: ExS: 21%, TP: 19%, TPh: 14%, and Ma: 17% (Figure 6) suggests a possible overestimation of the CE loading rate, which may result in an insufficient residence time for complete adsorption of the target compounds. Alternatively, it could indicate early saturation of the resin in the RP system [38]. Readjustment of operational conditions could provide a viable solution to address these issues effectively. Although the operating conditions of the RP were not optimal, as the desired components (TP and TPh) were not exclusively found in ES, the RP still achieved a purification factor ($F_{Purification}$) of ~2, indicating that the RP was effective in concentrating the target components. The polyphenol purity (P_{TP}) in CE (~2%) was lower than that of RE (~5%). Nevertheless, through the RP, a P_{TP} of ~5% was obtained in ES. Concurrently, both CS and WS exhibited a purity of around ~2%, likely due to the premature elution of phlorotannins.

The variation in AC reflected the trend observed in the P_{TP} of the samples. Notably, both RE and ES, with a P_{TP} of 5%, showed statistically equivalent AC_{ORAC} values. Likewise, CE, CS, and WS, with 2% purity, showed AC_{ORAC} values without significant differences.

The relative mass balance confirmed that the intermediate streams (CS and WS), which are typically discarded, contain significant proportions of ExS, TP, and TPh. Additionally, for these four components, the difference in relative mass between the input and output (*e*) ranged from 26% to 39%, suggesting that a significant proportion of phlorotannins may have been strongly adsorbed onto the resin, with their release likely occurring during resin regeneration. This loss is consistent with the characteristics of the HP-20 resin, which, despite having the highest adsorption capacity for phlorotannins compared to other macroporous resins, also exhibited the lowest desorption capacity [37]. When mass balance is applied to the relative areas of the five tentative identified compounds, *e* values showed a considerable increase ranging from 73% to 98%. These results would indicate the critical need to consider a deeper and more comprehensive chemical characterization to develop a precise process performance evaluation.

4. Conclusions

This study advances the development of sustainable technologies by demonstrating the effectiveness of combining USAE with RP to obtain mannitol-free purified extracts of phlorotannins from *Durvillaea incurvata* frond. The application of RP after USAE resulted in a significant enrichment of the crude extract, as evidenced by increased purities of both polyphenols and phlorotannins, resulting in a purification factor of ~2. The FTIR assay demonstrated the presence of phlorotannins and mannitol in the crude extract (CE) and intermediate RP streams (CS and WS), while no bands corresponding to this sugar alcohol were observed in the elution stream (ES). This finding confirms the phlorotannin selectivity of the USAE-RP process, which was also reflected in a 1.5-fold and 3-fold increase in their ORAC and DPPH antioxidant capacity, respectively, in the ES.

Furthermore, UHPLC-QToF allowed us to identify five new phlorotannins in *Durvillaea incurvata* frond USAE and RP extracts. The distribution of these phlorotannins in the different RP streams suggested that molecular size is the main factor that influences their separation. Future research should focus on optimizing RP's operating conditions to enhance phlorotannin's desorption and minimize target compound losses. The findings of this study lay a robust foundation for the scalable production of bioactive phlorotannin extracts, paving the way for more efficient and sustainable utilization of marine biomass in natural medical applications. To further optimize the integrated process in terms of biological activity, future studies should include cell-based assays. These will provide a more realistic evaluation of the bioactivity of the extracts, helping to refine their potential for therapeutic use.

Supplementary Materials: The following supporting information can be downloaded at: https://www.mdpi.com/article/10.3390/antiox14030250/s1. Figure S1: Mass Spectra (MS^1) of $C_{18}H_{11}O_{12}$ in crude extract using UHPLC-QToF, Figure S2: Mass Spectra of $C_{30}H_{15}O_{17}$ in crude extract using UHPLC-QToF: A) MS^1 and B) MS^2, Figure S3: Mass Spectra of $C_{36}H_{25}O_{17}$ in crude extract using UHPLC-QToF: A) MS^1 and B) MS^2, Figure S4: Mass Spectra (MS^1) of $C_{42}H_{25}O_{27}$ in crude extract using UHPLC-QToF, Figure S5: Mass Spectra of $C_{48}H_{31}O_{21}$ in crude extract using UHPLC-QToF: (A) MS^1 and (B) MS^2.

Author Contributions: P.R.R.-T.: Methodology, Investigation, Formal analysis, Data curation, Writing—original draft; G.C.-C.: Investigation, Data curation; P.I.R.-R.: Methodology, Investigation; J.P.-J.: Writing—review and editing; M.M.-C.: Data curation, Writing—review and editing; J.R.P.-C.: Conceptualization, Supervision, Writing—review and editing, Resources; M.S.M.-C.: Conceptualization, Supervision, Writing—review and editing, Resources. All authors have read and agreed to the published version of the manuscript.

Funding: Financial support was received from FONDECYT Regular projects number 1220097 and 1230115, and FOVI project number 240263.

Data Availability Statement: The data was provided in Supplementary Materials.

Acknowledgments: We are grateful to Luis Omar Mamani for their support in carrying out the spectrophotometric determinations. During the preparation of this work, the authors used Grammarly, DeepL, and ChatGPT to enhance the readability of this text. After using these tools, the authors reviewed and edited the content as necessary; they take full responsibility for the content of the published article.

Conflicts of Interest: The authors declare that they have no known competing financial interests or personal relationships that could have appeared to influence the work reported in this paper.

References

1. Soquetta, M.B.; Terra, L.M.; Bastos, C.P. Green technologies for the extraction of bioactive compounds in fruits and vegetables. *CyTA—J. Food* **2018**, *16*, 400–412. [CrossRef]
2. Li, Y.; Lee, S.H.; Le, Q.T.; Kim, M.M.; Kim, S.K. Anti-allergic effects of phlorotannins on histamine release via binding inhibition between IgE and FcεRI. *J. Agric. Food Chem.* **2008**, *56*, 12073–12080. [CrossRef] [PubMed]
3. Machu, L.; Misurcova, L.; Ambrozova, J.V.; Orsavova, J.; Mlcek, J.; Sochor, J.; Jurikova, T. Phenolic content and antioxidant capacity in algal food products. *Molecules* **2015**, *20*, 1118–1133. [CrossRef]
4. Huaman-Castilla, N.L.; Mariotti-Celis, M.S.; Perez-Correa, J.R. Polyphenols of Carménère Grapes. *Mini Rev. Org. Chem.* **2017**, *14*, 176–186. [CrossRef]
5. Alara, O.R.; Abdurahman, N.H.; Ukaegbu, C.I. Extraction of phenolic compounds: A review. *Curr. Res. Food Sci.* **2021**, *4*, 200–214. [CrossRef]
6. Zujko, M.E.; Witkowska, A.M. Antioxidant potential and polyphenol content of selected food. *Int. J. Food Prop.* **2011**, *14*, 300–308. [CrossRef]
7. Mateos, R.; Pérez-Correa, J.R.; Domínguez, H. Bioactive Properties of Marine Phenolics. *Mar. Drugs* **2020**, *18*, 501. [CrossRef] [PubMed]
8. Phang, S.J.; Teh, H.X.; Looi, M.L.; Arumugam, B.; Fauzi, M.B.; Kuppusamy, U.R. Phlorotannins from brown algae: A review on their antioxidant mechanisms and applications in oxidative stress-mediated diseases. *J. Appl. Phycol.* **2023**, *35*, 867–892. [CrossRef]
9. Simón, L.; Arazo-Rusindo, M.; Quest, A.F.G.; Mariotti-Celis, M.S. Phlorotannins: Novel Orally Administrated Bioactive Compounds That Induce Mitochondrial Dysfunction and Oxidative Stress in Cancer. *Antioxidants* **2023**, *12*, 1734. [CrossRef] [PubMed]
10. Erpel, F.; Mariotti-Celis, M.S.; Parada, J.; Pedreschi, F.; Pérez-Correa, J.R. Pressurized Hot Liquid Extraction with 15% v/v Glycerol-Water as An Effective Environment-Friendly Process to Obtain *Durvillaea incurvata* and *Lessonia spicata* Phlorotannin Extracts with Antioxidant and Antihyperglycemic Potential. *Antioxidants* **2021**, *10*, 1105. [CrossRef] [PubMed]
11. Kwon, Y.J.; Kwon, O.I.; Hwang, H.J.; Shin, H.-C.; Yang, S. Therapeutic effects of phlorotannins in the treatment of neurodegenerative disorders. *Front. Mol. Neurosci.* **2023**, *16*, 1193590. [CrossRef] [PubMed]
12. Corona, G.; Ji, Y.; Anegboonlap, P.; Hotchkiss, S.; Gill, C.; Yaqoob, P.; Spencer, J.P.E.; Rowland, I. Gastrointestinal modifications and bioavailability of brown seaweed phlorotannins and effects on inflammatory markers. *Br. J. Nutr.* **2016**, *115*, 1240–1253. [CrossRef] [PubMed]
13. Baldrick, F.R.; McFadden, K.; Ibars, M.; Sung, C.; Moffatt, T.; Megarry, K.; Thomas, K.; Mitchell, P.; Wallace, J.M.W.; Pourshahidi, L.K.; et al. Impact of a (poly)phenol-rich extract from the brown algae *Ascophyllum nodosum* on DNA damage and antioxidant activity in an overweight or obese population: A randomized controlled trial. *Am. J. Clin. Nutr.* **2018**, *108*, 688–700. [CrossRef]
14. Bahar, B.; O'Doherty, J.V.; Smyth, T.J.; Ahmed, A.M.; Sweeney, T. A cold water extract of *Fucus vesiculosus* inhibits lipopolysaccharide (LPS) induced pro-inflammatory responses in the porcine colon ex-vivo model. *Innov. Food Sci. Emerg. Technol.* **2016**, *37*, 229–236. [CrossRef]
15. Pasca, S.; Jurj, A.; Petrushev, B.; Tomuleasa, C.; Matei, D. *MicroRNA*-155 Implication in M1 Polarization and the Impact in Inflammatory Diseases. *Front. Immunol.* **2020**, *11*, 625. [CrossRef] [PubMed]
16. Parada, J.; Pérez-Correa, J.R.; Pérez-Jiménez, J. Design of low glycemic response foods using functional ingredients from seaweed. *J. Funct. Foods.* **2019**, *56*, 33–39. [CrossRef]
17. Ummat, V.; Tiwari, B.K.; Jaiswal, A.K.; Condon, K.; Garcia-Vaquero, M.; O'Doherty, J.; O'Donnell, C.; Rajauria, G. Optimisation of Ultrasound Frequency, Extraction Time and Solvent for the Recovery of Polyphenols, Phlorotannins and Associated Antioxidant Activity from Brown Seaweeds. *Mar. Drugs* **2020**, *18*, 250. [CrossRef]
18. Pacheco, L.V.; Parada, J.; Pérez-Correa, J.R.; Mariotti-Celis, M.S.; Erpel, F.; Zambrano, A.; Palacios, M. Bioactive Polyphenols from Southern Chile Seaweed as Inhibitors of Enzymes for Starch Digestion. *Mar. Drugs* **2020**, *18*, 353. [CrossRef] [PubMed]
19. Noticias Ambientales. Chile Faces the Disappearance of 80 Beaches: A Concerning Call to Action. Available online: https://noticiasambientales.com/environment-en/chile-faces-the-disappearance-of-80-beaches-a-concerning-call-to-action/ (accessed on 2 February 2025).
20. Food and Agriculture Organization (FAO). Living Climate Change on the Coastline of Chile. Available online: https://www.fao.org/newsroom/story/Living-climate-change-on-the-coastline-of-Chile/en (accessed on 2 February 2025).
21. López, B.A.; Macaya, E.C.; Jeldres, R.; Valdivia, N.; Bonta, C.C.; Tala, F.; Thiel, M. Spatio-temporal variability of strandings of the southern bull kelp *Durvillaea antarctica* (Fucales, Phaeophyceae) on beaches along the coast of Chile—Linked to local storms. *J. Appl. Phycol.* **2019**, *31*, 2159–2173. [CrossRef]
22. Plaza, M.; Turner, C. Pressurized hot water extraction of bioactives. *TrAC—Trends Anal. Chem.* **2015**, *71*, 39–54. [CrossRef]
23. Gullón, B.; Gagaoua, M.; Barba, F.J.; Gullón, P.; Zhang, W.; Lorenzo, J.M. Seaweeds as promising resource of bioactive compounds: Overview of novel extraction strategies and design of tailored meat products. *Trends Food Sci. Technol.* **2020**, *100*, 1–18. [CrossRef]
24. Ferreira-Anta, T.; Torres, M.D.; Vilarino, J.M.; Dominguez, H.; Flórez-Fernández, N. Green Extraction of Antioxidant Fractions from *Humulus lupulus* Varieties and Microparticle Production via Spray-Drying. *Foods* **2023**, *12*, 3881. [CrossRef] [PubMed]

25. Zhang, Y.; Wang, H.; Yang, R.; Zhang, Y.; Chen, Y.; Jiang, C.; Li, X. Synergistic Therapeutic Effects of D-Mannitol–Cerium–Quercetin (Rutin) Coordination Polymer Nanoparticles on Acute Lung Injury. *Molecules* **2024**, *29*, 2819. [CrossRef] [PubMed]
26. André, P.; Villain, F. Free radical scavenging properties of mannitol and its role as a constituent of hyaluronic acid fillers: A literature review. *Int. J. Cosmet. Sci.* **2017**, *39*, 355–360. [CrossRef] [PubMed]
27. Erpel, F.; Camilo, C.; Mateos, R.; Pérez-Correa, J.R. A macroporous resin purification process to obtain food-grade phlorotannin-rich extracts with α-glucosidase inhibitory activity from Chilean brown seaweeds: An UHPLC-MSn profiling. *Food Chem.* **2023**, *402*, 134472. [CrossRef] [PubMed]
28. Ford, L.; Theodoridou, K.; Sheldrake, G.N.; Walsh, P.J. A critical review of analytical methods used for the chemical characterisation and quantification of phlorotannin compounds in brown seaweeds. *Phytochem. Anal.* **2019**, *30*, 587–599. [CrossRef]
29. Ojeda, J.J.; Dittrich, M. Fourier Transform Infrared Spectroscopy for Molecular Analysis of Microbial Cells. In *Microbial Systems Biology. Methods in Molecular Biology 881*; Navid, A., Ed.; Humana Press: Totowa, NJ, USA, 2012; pp. 187–211. [CrossRef]
30. Gheda, S.; Hamouda, R.A.; Naby, M.A.; Mohamed, T.M.; Al-Shaikh, T.M.; Khamis, A. Potent Effect of Phlorotannins Derived from *Sargassum linifolium* as Antioxidant and Antidiabetic in a Streptozotocin-Induced Diabetic Rats Model. *Appl. Sci.* **2023**, *13*, 4711. [CrossRef]
31. Olate-Gallegos, C.; Barriga, A.; Vergara, C.; Fredes, C.; García, P.; Giménez, B.; Robert, P. Identification of Polyphenols from Chilean Brown Seaweeds Extracts by LC-DAD-ESI-MS/MS. *J. Aquat. Food. Prod. Technol.* **2019**, *28*, 375–391. [CrossRef]
32. Liu, X.; Robinson, A.L.; Jarratt, G.; Haritos, V.S. Analysis of the fate of valuable bioactive polyphenols during commercial winemaking and their partitioning into wastes for valorization. *Clean Waste Syst.* **2023**, *5*, 100104. [CrossRef]
33. Huaman-Castilla, N.L.; Martínez-Cifuentes, M.; Camilo, C.; Pedreschi, F.; Mariotti-Celis, M.; Pérez-Correa, J.R. The impact of temperature and ethanol concentration on the global recovery of specific polyphenols in an integrated HPLE/RP process on Carménère pomace extracts. *Molecules* **2019**, *24*, 3145. [CrossRef]
34. AOAC International. *Official Methods of Analysis of AOAC International*, 20th ed.; AOAC International: Rockville, MD, USA, 2020.
35. Wang, T.; Jónsdóttir, R.; Liu, H.; Gu, L.; Kristinsson, H.G.; Raghavan, S.; Ólafsdóttir, G. Antioxidant capacities of phlorotannins extracted from the brown algae *Fucus vesiculosus*. *J. Agric. Food Chem.* **2012**, *60*, 5874–5883. [CrossRef]
36. Kim, J.; Yoon, M.; Yang, H.; Jo, J.; Han, D.; Jeon, Y.-J.; Cho, S. Enrichment and purification of marine polyphenol phlorotannins using macroporous adsorption resins. *Food Chem.* **2014**, *162*, 135–142. [CrossRef] [PubMed]
37. Leyton, A.; Vergara-Salinas, J.R.; Pérez-Correa, J.R.; Lienqueo, M.E. Purification of phlorotannins from *Macrocystis pyrifera* using macroporous resins. *Food Chem.* **2017**, *237*, 312–319. [CrossRef] [PubMed]
38. Mariotti-Celis, M.S.; Rivera-Tovar, P.R.; Huamán-Castilla, N.L.; Pérez-Correa, J.R. Purification and fractionation of crude seaweed extracts by adsorption-desorption processes. In *Marine Phenolic Compounds: Science and Engineering*; Pérez-Correa, J.R., Mateos, R., Domínguez, H., Eds.; Elsevier: Amsterdam, The Netherlands, 2023; pp. 187–215.
39. Singleton, V.L.; Orthofer, R.; Lamuela-Raventós, R.M. Analysis of total phenols and other oxidation substrates and antioxidants by means of Folin-Ciocalteu reagent. *Methods Enzymol.* **1999**, *299*, 152–178. [CrossRef]
40. Stern, J.L.; Hagerman, A.E.; Steinberg, P.D.; Winter, F.C.; Estes, J.A. A new assay for quantifying brown algal phlorotannins and comparisons to previous methods. *J. Chem. Ecol.* **1996**, *22*, 1273–1293. [CrossRef] [PubMed]
41. Lopes, G.; Sousa, C.; Silva, L.R.; Pinto, E.; Andrade, P.B.; Bernardo, J.; Mouga, T.; Valentão, P. Can Phlorotannins Purified Extracts Constitute a Novel Pharmacological Alternative for Microbial Infections with Associated Inflammatory Conditions? *PLoS ONE* **2012**, *7*, e31145. [CrossRef]
42. Huang, D.; Ou, B.; Hampsch-Woodill, M.; Flanagan, J.A.; Prior, R.L. High-throughput assay of oxygen radical absorbance capacity (ORAC) using a multichannel liquid handling system coupled with a microplate fluorescence reader in 96-well format. *J. Agric. Food Chem.* **2002**, *50*, 4437–4444. [CrossRef] [PubMed]
43. Brand-Williams, W.; Cuvelier, M.E.; Berset, C. Use of a free radical method to evaluate antioxidant activity. *LWT—Food Sci. Technol.* **1995**, *28*, 25–30. [CrossRef]
44. Ismail, M.M.; El Zokm, G.M.; El Sikaily, A.M.; Selim, A.I.; Ismail, G.A. Chemodiversity and bioactivity assessment of phlorotannins from some Phaeophyta species from the Red Sea. *J. Appl. Phycol.* **2023**, *35*, 1769–1788. [CrossRef]
45. Cho, H.M.; Doan, T.P.; Ha, T.K.Q.; Kim, H.W.; Lee, B.W.; Pham, H.T.T.; Cho, T.O.; Oh, W.K. Dereplication by High-Performance Liquid Chromatography (HPLC) with Quadrupole-Time-of-Flight Mass Spectroscopy (qTOF-MS) and Antiviral Activities of Phlorotannins from *Ecklonia cava*. *Mar. Drugs* **2019**, *17*, 149. [CrossRef] [PubMed]
46. Gu, C.; Howell, K.; Dunshea, F.R.; Suleria, H.A.R. LC-ESI-QTOF/MS characterisation of phenolic acids and flavonoids in polyphenol-rich fruits and vegetables and their potential antioxidant activities. *Antioxidants* **2019**, *8*, 405. [CrossRef]
47. Meng, W.; Mu, T.; Sun, H.; Garcia-Vaquero, M. Phlorotannins: A review of extraction methods, structural characteristics, bioactivities, bioavailability, and future trends. *Algal. Res.* **2021**, *60*, 102484. [CrossRef]
48. Rivera-Tovar, P.R.; Mariotti-Celis, M.S.; Pérez-Correa, J.R. Maqui (*Aristotelia chilensis* (Mol.) Stuntz) and murta (Ugni molinae Turcz): Native Chilean sources of polyphenol compounds. *Mini Rev. Org. Chem.* **2019**, *16*, 261–276. [CrossRef]

49. Muñoz-Molina, N.; Parada, J.; Simirgiotis, M.; Montecinos-González, R. Potential of Cochayuyo (*Durvillaea incurvata*) Extract Obtained by Ultrasound Assisted Extraction against Aging Related Diseases. *Foods* **2024**, *13*, 269. [CrossRef] [PubMed]
50. Rivera-Tovar, P.R.; Torres, M.D.; Camilo, C.; Mariotti-Celis, M.S.; Domínguez, H.; Pérez-Correa, J.R. Multi-response optimal hot pressurized liquid recovery of extractable polyphenols from leaves of maqui (*Aristotelia chilensis* [Mol.] Stuntz). *Food Chem.* **2021**, *357*, 129729. [CrossRef]
51. Zhong, B.; Robinson, N.A.; Warner, R.D.; Barrow, C.J.; Dunshea, F.R.; Suleria, H.A.R. LC-ESI-QTOF-MS/MS characterization of seaweed phenolics and their antioxidant potential. *Mar. Drugs* **2020**, *18*, 331. [CrossRef] [PubMed]
52. Schaich, K.M.; Tian, X.; Xie, J. Reprint of "Hurdles and pitfalls in measuring antioxidant efficacy: A critical evaluation of ABTS, DPPH, and ORAC assays". *J. Funct. Foods* **2015**, *18*, 782–796. [CrossRef]
53. Lee, H.A.; Lee, J.H.; Han, J.S. A phlorotannin constituent of Ecklonia cava alleviates postprandial hyperglycemia in diabetic mice. *Pharm. Biol.* **2017**, *55*, 1149–1154. [CrossRef]
54. Han, E.; Kim, H.-S.; Sanjeewa, K.K.A.; Herath, K.H.I.N.M.; Jeon, Y.-J.; Jee, Y.; Lee, J.; Kim, T.; Shim, S.-Y.; Ahn, G. Eckol from Ecklonia cava Suppresses Immunoglobulin E-mediated Mast Cell Activation and Passive Cutaneous Anaphylaxis in Mice. *Nutrients* **2020**, *12*, 1361. [CrossRef] [PubMed]
55. Obluchinskaya, E.; Pozharitskaya, O.; Shevyrin, V.; Kovaleva, E.; Flisyuk, E.; Shikov, A. Optimization of Extraction of Phlorotannins from the Arctic *Fucus vesiculosus* Using Natural Deep Eutectic Solvents and Their HPLC Profiling with Tandem High-Resolution Mass Spectrometry. *Mar. Drugs* **2023**, *21*, 263. [CrossRef]
56. Yeo, M.; Jung, W.-K.; Kim, G. Fabrication, characterisation and biological activity of phlorotannin-conjugated PCL/β-TCP composite scaffolds for bone tissue regeneration. *J. Mater. Chem.* **2012**, *22*, 3568. [CrossRef]
57. Chades, T.; Scully, S.M.; Ingvadottir, E.M.; Orlygsson, J. Fermentation of mannitol extracts from brown macro algae by thermophilic Clostridia. *Front. Microbiol.* **2018**, *9*, 1931. [CrossRef] [PubMed]
58. Rajauria, G.; Ravindran, R.; Garcia-Vaquero, M.; Rai, D.K.; Sweeney, T.; O'Doherty, J. Molecular characteristics and antioxidant activity of laminarin extracted from the seaweed species Laminaria hyperborea, using hydrothermal-assisted extraction and a multi-step purification procedure. *Food Hydrocoll.* **2021**, *112*, 106332. [CrossRef]
59. Zubia, M.; Payri, C.; Deslandes, E. Alginate, mannitol, phenolic compounds and biological activities of two range-extending brown algae, *Sargassum mangarevense* and *Turbinaria ornata* (Phaeophyta: Fucales), from Tahiti (French Polynesia). *J. Appl. Phycol.* **2008**, *20*, 1033–1043. [CrossRef]
60. Msomi, N.Z.; Erukainure, O.L.; Islam, M.S. Suitability of Sugar Alcohols as Antidiabetic Supplements: A Review. *J. Food Drug Anal.* **2021**, *29*, 1–14. [CrossRef] [PubMed]
61. Milman, B.L. General principles of identification by mass spectrometry. *TrAC—Trends Anal. Chem.* **2015**, *69*, 24–33. [CrossRef]
62. Li, Y.; Fu, X.; Duan, D.; Liu, X.; Xu, J.; Gao, X. Extraction and Identification of Phlorotannins from the Brown Alga, *Sargassum fusiforme* (Harvey) Setchell. *Mar. Drugs* **2017**, *15*, 49. [CrossRef]
63. Hermund, D.B.; Plaza, M.; Turner, C.; Jónsdóttir, R.; Kristinsson, H.G.; Jacobsen, C.; Nielsen, K.F. Structure dependent antioxidant capacity of phlorotannins from Icelandic *Fucus vesiculosus* by UHPLC-DAD-ECD-QTOFMS. *Food Chem.* **2018**, *240*, 904–909. [CrossRef] [PubMed]
64. Silva, M.; Castellanos, L.; Ottens, M. Capture and Purification of Polyphenols Using Functionalized Hydrophobic Resins. *Ind. Eng. Chem. Res.* **2018**, *57*, 5359–5369. [CrossRef] [PubMed]
65. Agregán, R.; Munekata, P.; Franco, D.; Carballo, J.; Barba, F.; Lorenzo, J. Antioxidant Potential of Extracts Obtained from Macro- (*Ascophyllum nodosum, Fucus vesiculosus* and *Bifurcaria bifurcata*) and Micro-Algae (*Chlorella vulgaris* and *Spirulina platensis*) Assisted by Ultrasound. *Medicines* **2018**, *5*, 33. [CrossRef]
66. Montero, L.; Sánchez-Camargo, A.P.; García-Cañas, V.; Tanniou, A.; Stiger-Pouvreau, V.; Russo, M.; Rastrelli, L.; Cifuentes, A.; Herrero, M.; Ibáñez, E. Anti-proliferative activity and chemical characterization by comprehensive two-dimensional liquid chromatography coupled to mass spectrometry of phlorotannins from the brown macroalga *Sargassum muticum* collected on North-Atlantic coasts. *J. Chromatogr. A* **2016**, *1428*, 115–125. [CrossRef]
67. Hamid, S.S.; Wakayama, M.; Ichihara, K.; Sakurai, K.; Ashino, Y.; Kadowaki, R.; Soga, T.; Tomita, M. Metabolome profiling of various seaweed species discriminates between brown, red, and green algae. *Planta* **2019**, *249*, 1921–1947. [CrossRef] [PubMed]

antioxidants

MDPI

Article

Lipidomic Characterization of Marine By-Product Oils: Impact of Species and Extraction Methods on Lipid Profile and Antioxidant Potential

Ioannis C. Martakos [1], Paraskeui Tzika [2], Marilena E. Dasenaki [3], Eleni P. Kalogianni [2] and Nikolaos S. Thomaidis [1,*]

[1] Laboratory of Analytical Chemistry, Department of Chemistry, National and Kapodistrian University of Athens, Panepistimiopolis Zographou, 15772 Athens, Greece; johnmrtk@chem.uoa.gr
[2] Department of Food Science and Technology, International Hellenic University, 57400 Thessaloniki, Greece; tzikap@food.ihu.gr (P.T.); elekalo@ihu.gr (E.P.K.)
[3] Laboratory of Food Chemistry, Department of Chemistry, National and Kapodistrian University of Athens, Panepistimiopolis Zographou, 15772 Athens, Greece; mdasenaki@chem.uoa.gr
* Correspondence: ntho@chem.uoa.gr; Tel.: +30-2107274317

Abstract

Marine by-products represent an important source of bioactive lipids with potential applications in nutraceuticals and functional foods. This study provides a biochemical and lipidomic characterization of oils derived from sardine, monkfish, grey mullet roe, squid, and anchovy by-products, assessing how the extraction method influences their lipid and antioxidant profiles. Fatty acids were quantified by GC-FID, antioxidant compounds by HPLC-DAD, and untargeted lipidomics by TIMS-HRMS. A total of 228 lipid species were identified, predominantly triglycerides (TGs) and diglycerides (DGs), accounting for approximately 69% of the annotated lipidome. Grey mullet roe oils exhibited the highest levels of long-chain PUFAs (EPA, DHA) and antioxidants (α-tocopherol 205–469 mg/Kg, lutein 10–125 mg/Kg, and squalene 1004–6049 mg/Kg), whereas squid oils showed high n-3/n-6 proportions. The extraction method strongly affected lipid integrity. Supercritical CO_2 extraction with ethanol (SFE–SE) preserved the greatest proportion of PUFA-rich TGs, yielding ~27–28 g EPA + DHA per 100 g oil, while wet reduction and mechanical pressing produced lower PUFA levels (~22 g/100 g) and increased hydrolysis/oxidation-associated lipids. PCA and PLS-DA revealed clear clustering driven by species and extraction class, with PUFA-containing TGs and DGs identified as major discriminating lipids. These results highlight the critical role of extraction conditions in determining the nutritional and functional value of marine oils and support the valorization of marine by-products in high-value applications.

Keywords: marine oils; marine by-products; lipidomics; tocopherols; carotenoids; extraction methods

1. Introduction

Marine by-products such as heads, viscera, skin, and roes represent a sustainable and under-utilized source of bioactive lipids and antioxidants [1,2]. Marine-derived oils are rich in omega-3 fatty acids and represent an important opportunity for valorization due to their high content of bioactive lipids, including polyunsaturated fatty acids (PUFAs) such as eicosapentaenoic acid (EPA), docosahexaenoic acid (DHA), and monounsaturated fatty acids (MUFAs), which contribute to improved metabolic health and enhanced cognitive function, while they have strong antioxidant, anti-inflammatory, and cardiovascular-protective properties, making them valuable for nutraceutical, pharmaceutical, and cosmetic applications [3–6]. In addition, marine-derived fish oils contain high concentrations of a wide range of common lipid classes and subclasses, including phospholipids, triglycerides, diglycerides, and sterols, compounds that have demonstrated superior bioavailability [7].

Additionally, marine organisms are a great source of important antioxidant compounds, which play a crucial role against oxidative stress caused by environmental factors, enhance the immune function, improve survival rates, and protect the cells by scavenging free radicals and reducing lipid peroxidation [8–11]. These compounds include carotenoids, tocopherols, and squalene, and they also contribute by enhancing the nutritional value of these organisms [9]. Beyond their role in enhancing the nutritional value of marine organisms, these antioxidants are key to the preservation and stability of seafood products [10]. Specifically, carotenoids and tocopherols have been linked to scavenging free radicals, reducing lipid peroxidation while also ensuring nutritional quality, and squalene is linked to reactive oxygen species (ROS) neutralization, with several studies in the literature focusing on the antioxidant and cell-protective nature of these compounds [10,12,13]. Consequently, the substantial bioactivity of these compounds establishes their significant value across pharmaceutical, nutraceutical, and skincare applications.

Beyond traditional dietary supplementation, marine bioactive lipids have gained prominence in the nutraceutical and cosmetic industries. The nutraceutical sector incorporates omega-3-rich oils into functional foods, supplements, and pharmaceutical formulations aimed at preventing chronic diseases [14]. Marine-derived lipids are increasingly used in cosmeceuticals for their ability to reinforce skin barriers, enhance hydration, and reduce oxidative stress-induced skin aging [15,16]. Omega-3 and omega-6 fatty acids help regulate inflammatory skin conditions such as eczema, psoriasis, and acne. Studies have shown that incorporating marine phospholipids into topical skincare formulations can lead to improved skin elasticity, moisture retention, and UV-damage protection [17].

The efficacy of these bioactive lipids depends largely on their stability and preservation during extraction. As marine lipids are highly susceptible to oxidation and degradation, selecting an appropriate extraction method is crucial for maintaining their bioactivity and functional properties. Furthermore, different methods and solvents result in the extraction of lipids of different polarities and, therefore, affect the bioactive content of the oils [18–20]. Industrial-scale oil from fish by-products is extracted mainly via "wet reduction", where the by-products undergo hydrothermal treatment, followed by separation methods to obtain the oil (pressing and/or centrifugation and/or decantation and finally drying) [21]. However, other methods, including supercritical CO_2 (SFE-CO_2) extraction, enzymatically assisted hydrolysis, and solvent extraction methods, have been studied to optimize yield while preserving the structural integrity of omega-3 fatty acids, phospholipids, and antioxidants. Furthermore, in recent years, methods considered more sustainable, such as SFE-CO_2, ultrasound-assisted extraction, microwave-assisted extraction, and enzymatically assisted hydrolysis, have been highlighted within the literature [19,22,23]. Studies comparing SFE-CO_2 extraction (SFE) with conventional solvent-based methods indicate that SFE better preserves lipid oxidation stability, significantly reducing peroxide values and maintaining the integrity of bioactive compounds such as tocopherols, phospholipids, and omega-3 fatty acids. Research on fish oil extraction from Arctic charr and fish roe highlights that supercritical CO_2 prevents oxidative degradation due to the absence of oxygen and the low-temperature conditions, resulting in superior retention of polyunsaturated fatty acids (PUFAs) and antioxidants compared to solvent-based methods [24,25]. Furthermore, recent

https://doi.org/10.3390/antiox15010095

research highlights the low-temperature continuous phase-transition extraction (LCPE) method as a promising approach for maintaining oxidative stability in Antarctic krill oil, reducing lipid degradation while maximizing bioactive compound retention [26]. These findings emphasize the need to refine extraction protocols to ensure that marine-derived lipids retain their bioactivity, making them viable for long-term nutraceutical formulations and cosmetic applications.

The aim of this study is to comprehensively evaluate how different extraction methods affect the lipid composition and antioxidant content of oils obtained from various by-products of Greek fisheries. Five distinct methods were grouped and compared based on their mechanisms of action: three conventional methods (wet reduction, enzymatic hydrolysis, and mechanical pressing) are compared with newer and more sustainable ones, such as SFE and green solvent extraction (with the use of ethanol). Particular focus is placed on omega-3 and omega-6 fatty acids, along with key antioxidant compounds such as α-tocopherol, lutein, and squalene. By integrating targeted biochemical assays with untargeted lipidomic profiling, this work seeks to clarify how extraction conditions influence the nutritional and functional potential of marine oils. The inclusion of diverse by-products from commercially important marine species reflects an effort to promote full-resource utilization and contribute to more sustainable, circular approaches in seafood processing—particularly in the context of nutraceutical and cosmetic applications.

2. Materials and Methods

2.1. *Chemicals and Standards*

Methanol (MeOH), ethanol (EtOH), isopropyl alcohol (IPA), and acetonitrile (ACN), HPLC grade, as well as methyl-tert-butyl ether (MTBE) and hexane, analytical grade, were purchased from Fisher Scientific (Geel, Belgium). ACN and IPA of LC-MS grade were obtained by Merck (Darmstadt, Germany). Ammonium formate, formic acid, potassium hydroxide, and sodium sulfate were acquired by Sigma-Aldrich (Steinheim, Germany). A Milli-Q water purification system (Direct–Q, UV, Millipore, Bedford, MA, USA) was used to provide ultra-pure water (18.2 MΩ/cm). Standard compounds, including lutein, β-carotene, α-tocopherol, γ-tocopherol, δ-tocopherol, and squalene, were obtained from Sigma Aldrich. FAME mix containing a 37-component mix was supplied by Supelco. Chromafil 0.22 μm regenerated cellulose (RC) filters were acquired by Macherey-Nagel (Düren, Germany).

2.2. *Samples and Extraction Techniques*

In total, 19 samples of marine oils were analyzed in the Laboratory of Analytical Chemistry of NKUA. The marine oil samples originated from 6 different marine species by-products, namely sardine heads (*Sardina pilchardus*), monkfish bellies (*Lophius piscatorius*), grey mullet roes (*Mugil cephalus*), squid bellies (*Nototodarus sloanii*), and anchovy heads (*Engraulis encrasicolus*). All fish were caught in the FAO 37.3.1 region, whereas the squid was caught in the FAO 81 region and processed by Atlantida S.A. A total of 5 extraction methods were used and assessed: wet reduction (WR), mechanical pressing (MP), solvent extraction with the use of ethanol (EtOH) (SE), enzymatically assisted hydrolysis extraction (EH), and supercritical fluid extraction with CO_2 and ethanol as a cosolvent (SFE-SE). All samples were extracted on the premises of the International Hellenic University (IHU).

Extraction methods (WR, MP, SE, and SFE-SE) were applied as described previously [18–20]. Briefly, WR was applied in a mixture of finely cut raw mullet roe by-product (300 g) and water (75 g). The mixture was heated under stirring (water bath, 90 °C, 10 min) and then separated by passing twice in an expeller oil press (OW 500 s-inox oil press, Oelwerk, Cottbus, Germany) and subsequently centrifuged (RC3, Sorvall) at 4132× *g* for 15 min.

https://doi.org/10.3390/antiox15010095

MP was the same as WR without the hydrothermal treatment. For SE, ethanol was used as a solvent, and a ratio of 1/10 (weight of freeze-dried by-product/volume of solvent) was applied at two temperatures (25 and 50 °C) for 2 h. The oil was obtained by centrifugation (1878× *g* for 15 min, Rotofix 32A, Hettich, Tuttlingen, Germany) and vacuum evaporation (R-210 Rotavapor, Buchi) at 37 °C. Finally, EH was applied as in Kalogianni et al. [27]. Briefly, first the endogenous enzymes were inactivated (90 °C for 10 min), then the sample was cooled to 55 °C, and 1.0% (*w*/*w*) of Alcalase ® or 0.5% (*w*/*w*) Protease ® was added and incubated at the same temperature for 1 h. Then, the added enzymes were inactivated as above, and the oil was recovered by centrifugation, as above. The mixture was incubated in a water bath at 55 °C with continuous stirring, and samples were taken at 35 min, 1 h, 2 h, and 3 h. For all samples, the added enzymes were inactivated at 90 °C for 10 min, and then the system was cooled at 4 °C. Finally, the oil was recovered using centrifugation with an RC3 Sorvall (Thermo Scientific, Waltham, MA, USA) centrifuge, operating at 4 °C at a force of 4132× *g* for 15 min [19]. Conventional extraction techniques were considered: wet reduction, enzymatically assisted hydrolysis, and mechanical pressing. These techniques offer a sustainable alternative, minimizing chemical exposure while preserving the natural structure of marine bioactive lipids.

By categorizing the samples into these three groups, this study enables a systematic comparison of lipid and antioxidant compound retention across different extraction methodologies. The classification allows for a better understanding of the effects of each technique on the final lipid profile, contributing to the optimization of extraction processes for nutraceutical and cosmetic applications. In Table S1 of the Electronic Supplementary Material (ESM), the samples alongside their detailed information are presented.

2.3. *Determination of FAME and Antioxidant Profile*

2.3.1. FAME Analysis

For the determination of FAMEs, the method 996.06 of AOAC was used. Briefly, 0.050 g of fish oil was accurately weighed in an Eppendorf tube, followed by the addition of 750 μL of hexane. The sample was vortexed to ensure proper mixing, and 100 μL of 2 M KOH/MeOH was added. After vortexing for 1 min, 0.2 g of Na_2SO_4 was added, and the mixture was mixed vigorously. The sample was subsequently frozen at −20 °C for 30 min. After freezing, 100 μL of the supernatant was collected and diluted with 900 μL of hexane. Finally, 1 μL of the prepared sample was injected into the GC-FID system for further analysis. The GC system comprised a Varian 450 GC with a Flame Ionization Detector (FID) with the settings presented in Table 1.

Table 1. GC-FID system parameters and settings.

Analytical Column	Agilent J&W DB-23 GC (60 m, 0.25 mm)
Carrier Gas	Helium
Injection System Temperature	250 °C
Detector Temperature	280 °C
Oven Temperature Program	Initial 50 °C for 1 min, ramping at 25 °C/min to 175 °C, then ramping at 4 °C/min to 230 °C, holding for 10 min
Hydrogen Flow Rate	40 mL/min
Synthetic Air Flow Rate	450 mL/min
Helium Flow Rate	30 mL/min
Split Ratio	1:50
Injection Volume	1 μL

https://doi.org/10.3390/antiox15010095

2.3.2. Tocopherols, Carotenoids, and Squalene Analysis

For the determination of tocopherols, carotenoids, and squalene, a method previously published [28] was used. Briefly, 100 μL of oil sample was weighted in an eppendorf tube, followed by the addition of 900 μL of IPA. The sample was mixed vigorously with the use of a vortex mixer and was filtered by a 0.22 μm RC filter. Finally, 20 μL were injected in the HPLC system. For this analysis, a Shimadzu 2030AD HPLC system was used. The system consisted of an autosampler, a column thermostat, and a photo-diode array detector (PDA). Samples were measured in duplicates. The identification and detailed characterization of the target compounds, including retention time and chemical structure, were performed as described in Martakos et al. (2019) [28].

2.4. *Lipidomics Analysis*

For the lipidomic analysis, 100 mg of the sample was weighed into an Eppendorf tube, followed by the addition of 600 μL of an MTBE:MeOH (3:1) extraction system. The sample underwent vigorous vortex mixing and was then placed in an ultrasonic bath for 30 min at 30 °C. After sonication, 500 μL of water was added, and the sample was centrifuged at 4 °C to separate the phases. The upper MTBE layer, containing the extracted lipids, was collected, and a second extraction was performed to ensure complete lipid recovery. The two MTBE layers were then combined, and the solvent was evaporated under a gentle nitrogen stream to dryness, at room temperature. The dried lipid extract was subsequently reconstituted in 200 μL of ACN:IPA (1:1) before injection into the LC-MS system.

For the comprehensive analysis of the lipid profile of fish oils, ultra-performance liquid chromatography coupled with trapped ion mobility spectrometry and high-resolution quadrupole time-of-flight mass spectrometry (UPLC-TIMS-QTOF-MS) was utilized. The chromatographic separation was performed using a Bruker Elute UPLC system equipped with a Thermo Acclaim RSLC 120 C18 analytical column (2.2 μm, 2.1 × 100 mm) and an Acquity UPLC BEH C18 VanGuardTM pre-column (2.1 × 5 mm, 1.7 μm; Waters). The mobile phase consisted of ACN:H_2O (65:35) as phase A and ACN:IPA (15:85) with 10 mM ammonium formate and 0.1% formic acid as phase B. A gradient elution was applied, starting at 30% B, gradually increasing to 100% B over 30 min, at a constant flow rate of 0.25 mL/min. Detection was performed using a Bruker TIMS-TOF PRO 2 mass spectrometer, enabling high-resolution analysis of lipid species by adding the 4th dimension of ion mobility. The TIMS-TOF instrument was operated using the parameters described in Martakos et al. (2024) [29].

Due to the complexity of LC–MS datasets and the extensive processing required [30], specialized software was used. MetaboScape 2023b and DataAnalysis 5.3 (Bruker Daltonics, Bremen, Germany) were employed for spectral processing, feature extraction, and compound annotation. Metaboscape 2023b was utilized for peak-picking, lipid metabolite annotation, and lipid identity confirmation, with additional verification performed using Data Analysis 5.3. Peak-picking was conducted using the T-ReX 4D algorithm, which integrates four key parameters—retention time, signal intensity, mass-to-charge ratio (m/z), and collision cross-section (CCS) value—to enhance metabolite detection accuracy. To minimize false positives, only features detected in at least 50% of samples were considered. For lipid annotation, Metaboscape's Lipid Annotation Tool was employed alongside the Lipidblast spectral library [31], applying stringent criteria: an m/z tolerance of 5.0 ppm, an isotope fit threshold of 250 mSigma, MS/MS matching scores up to 400/1000, and a CCS value tolerance within 3.0%. Features exceeding these thresholds were excluded to reduce false positives, particularly since Lipidblast incorporates in silico-generated CCS values, requiring careful validation. Batch correction was applied in Metaboscape using quality control (QC) samples to account for systematic variations, with variance thresholds set at

https://doi.org/10.3390/antiox15010095

40% before correction and 20% post-correction. Features that exceeded these thresholds or were absent in QC samples were excluded from further analysis.

Statistical analyses were carried out using MetaboAnalyst 6.0 [32], incorporating Principal Component Analysis (PCA) and Partial Least Squares Discriminant Analysis (PLS-DA), with violin and box plots used for data visualization. Probabilistic Quotient Normalization (PQN) was applied to correct for sample-to-sample variability and ensure comparability across samples, a standard approach for metabolomic datasets lacking chemical standards [33].

3. Results and Discussion

3.1. Characterization of Marine By-Product Oils: Lipid and Antioxidant Composition

A broad array of lipid and lipid-soluble molecules was identified in the samples, including omega-3 and omega-6 fatty acids, phospholipids, triglycerides, diglycerides, tocopherols, carotenoids, and squalene.

As expected, the different marine species and extraction techniques presented differences in the fatty acid composition. In Figure 1, the average content of omega-3, omega-6, EPA + DHA, and fatty acid class (saturated, MUFAs, and PUFAs) per marine species is presented. The Figure illustrates clear differences in fatty acid profiles between species, highlighting the diversity in their nutritional quality. Notably, squid bellies and grey mullet roe clearly stand out among the species by-products for their higher average levels of total omega-3 and combined EPA + DHA, as illustrated in Figure 1. Squid, in particular, exhibits the highest omega-3 and EPA + DHA content, followed by roe, while fish display comparatively lower values. These findings suggest that oils derived from squid and roe may offer superior nutritional and health benefits, given the well-established roles of long-chain polyunsaturated fatty acids in supporting cardiovascular, cognitive, and anti-inflammatory functions. Furthermore, the fatty acid class composition highlights that squid and roe oils have a more favourable balance of polyunsaturated to saturated fatty acids compared to fish, reinforcing their potential as high-quality sources of marine lipids [5,34]. The results on mullet roe by-products are in agreement with our previous works [18,19].

Conversely, fish samples exhibit notably lower average omega-3 content than roe and squid, which may limit their value as dietary sources of these essential fatty acids. A similar pattern emerges with omega-6 levels—roe samples tend to have the highest omega-6 concentrations, which could contribute to an imbalanced omega-6 to omega-3 ratio, a factor often associated with pro-inflammatory states and adverse health outcomes [35]. When examining the overall fatty acid composition by species, fish oils appear relatively richer in saturated and monounsaturated fats, while roe and squid demonstrate higher proportions of polyunsaturated fats, particularly omega-3. However, it is important to note that proportional omega-3 or PUFA levels do not fully reflect the nutritional contribution of each species. Total lipid content varies substantially, and fish typically contain considerably higher overall fat levels, making them significant dietary sources of omega-3 fatty acids despite their lower proportional values [36,37]. Thus, the present comparison focuses on compositional differences rather than nutritional adequacy, as each species contributes to omega-3 intake in different ways.

The present study evaluated the concentrations of three key bioactive lipophilic compounds—α-tocopherol, lutein, and squalene—which are recognized for their antioxidant properties and roles in supporting cardiovascular, immune, and ocular health [28]. Although limited studies are available in the literature for direct comparison, our findings are discussed in relation to previously published data to contextualize the nutritional potential of the oils analyzed. It is important to note that the composition and antioxidant content of marine oils are closely influenced by both the biological origin of the raw material

https://doi.org/10.3390/antiox15010095

and the parameters of the extraction process, including pressure and temperature [7,38]. As such, some compounds were not detected in all samples analyzed. The results are summarized in Table 2.

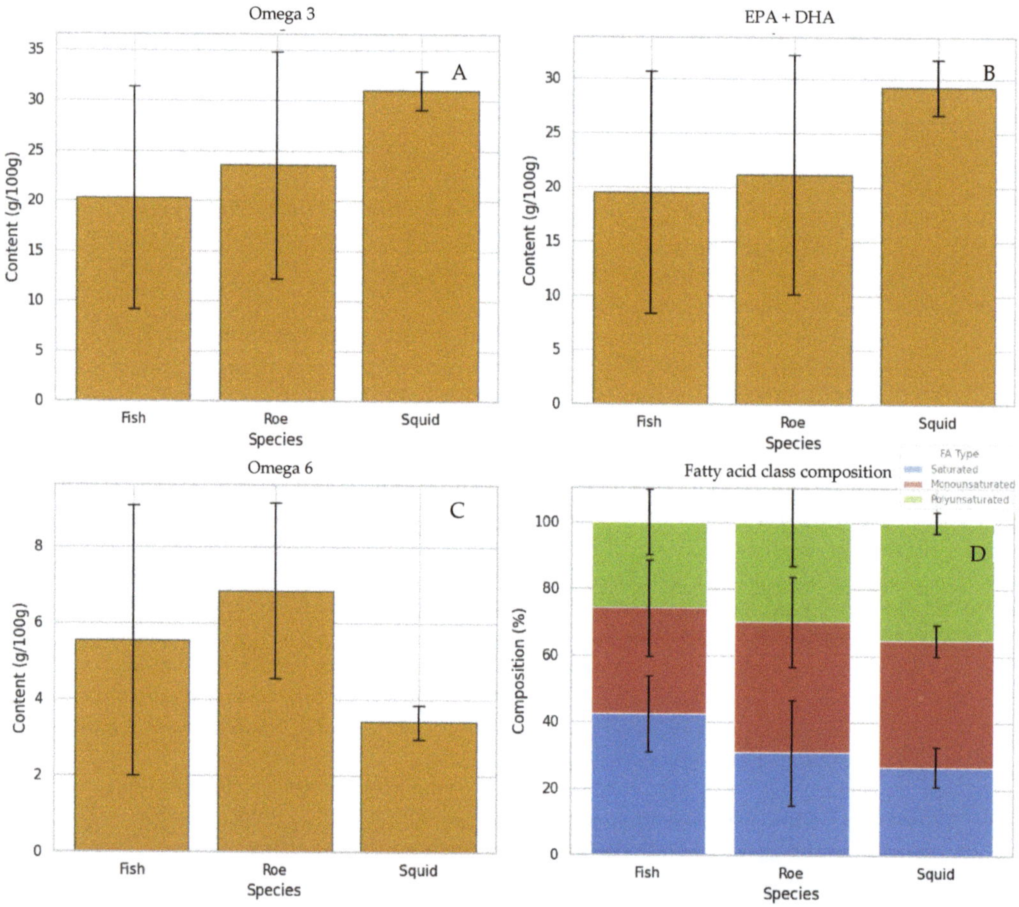

Figure 1. Average fatty acid composition across the marine species analyzed, presented as Mean ± Standard Deviation. "Fish" refers to the group of fish by-product samples listed in the ESM (Table S1), "Roe" corresponds specifically to grey mullet roe samples, and "Squid" refers to squid by-product samples (bellies). Values represent the mean of all samples within each species group. (**A**) Total omega-3 content (%), (**B**) sum of EPA and DHA (%), (**C**) total omega-6 (%), and (**D**) fatty acid class composition (%).

α-Tocopherol content ranged from 205 to 469 mg/Kg, with the highest concentration observed in sample ROE6, a sample produced by ethanol extraction at 25 °C. These levels substantially exceed those reported in unprocessed roe, such as 66.3 mg/Kg in fresh Chinook salmon roe [39] and 41 mg/Kg in avgotaracho, a traditional salted and wax-coated mullet roe product [40]. While some enriched or lipid-extracted samples, including mature roe oils, have reached up to 110 mg/100 g oil [39,41], such values pertain to the oil fraction and are not directly comparable to whole-sample measurements, as used in the present study. Lutein concentrations ranged from 10 to 125 mg/Kg, again peaking in ROE6. These values significantly exceed those previously reported for salmon roe, where concentrations between 0.24 and 0.46 mg/Kg have been documented in fresh and dried

https://doi.org/10.3390/antiox15010095

samples, respectively [42]. Although herring roe has been cited for its relatively high lutein content at 6.4 mg/g lipid, this corresponds to substantially lower levels on a whole-sample basis when typical lipid content is accounted for [41]. The elevated lutein levels observed here highlight the potential of roe oils as effective sources of dietary xanthophylls, which are highly valued for their roles in ocular function and antioxidant defence [43]. Squalene concentrations varied from 1301 to 6049 mg/Kg, with the highest value again found in ROE6. These levels greatly exceed those reported in comparable roe products such as avgotaracho (54 mg/Kg) [40] and are consistent with concentrations found in certain fish by-products, including viscera, bones, and heads, as well as in dried or salted fillets [44]. While shark liver oil remains the most concentrated known source of squalene—reportedly containing over 40% by weight [12]—its use is primarily restricted to pharmaceutical and cosmetic industries in EU when it is derived from sharks, due to sustainability and ethical concerns. The results on antioxidant species found in the present work are in alignment with the oxidative state of the oils examined previously [19]. Therefore, the appropriate choice of extraction method can provide oils rich in bioactive antioxidant species, which protect them from oxidation, as was found for mullet roe by-products [18,19], and provide valuable bioactives for humans. This is also supported by our previous findings on the in vitro biological effects of the mullet roe oils against platelet oxidation, where there seems to be a link between the concentration of antioxidants found in the present work and the in vitro antioxidant potential of the oils [20]. The high levels of squalene observed in roe oils derived from by-products in this study position these materials as nutritionally valuable and environmentally viable alternatives for dietary squalene supplementation.

Table 2. Lutein, α-tocopherol, and squalene results of the analyzed samples *. Sample codes indicate species and tissue origin.

Sample	Lutein (mg/Kg)	a-Tocopherol (mg/Kg)	Squalene (mg/Kg)
ROE1	50	345	2075
ROE2	43	205	2604
ROE3	88	354	n.d.
ROE4	103	298	n.d.
ROE5	10	440	1301
ROE6	125	469	6049
ROE7	124	307	n.d.
SAR1	87	363	n.d.
SAR2	75	312	n.d.
ANC1	n.d.	n.d.	1004
ANC2	n.d.	n.d.	4120
MON1	n.d.	n.d.	2419
SQD1	n.d.	n.d.	n.d.
SQD2	n.d.	n.d.	n.d.
SQD3	n.d.	n.d.	n.d.
SQD4	n.d.	n.d.	n.d.

* only samples where the abovementioned compounds were detected are presented. Sample codes indicate species group and extraction replicate: ROE = grey mullet roe (Mugil cephalus), SAR = sardine heads (Sardina pilchardus), ANC = anchovy heads (Engraulis encrasicolus), MON = monkfish bellies (Lophius piscatorius), and SQD = squid bellies (Nototodarus sloanii). n.d. = non-detected.

The consistently elevated levels of all three bioactive compounds—particularly in grey mullet roe-derived samples—underscore the nutritional richness and functional potential of grey mullet roe oils. These findings not only reinforce the utility of roe as a dense source of health-promoting lipids but also highlight the influence of species selection, tissue maturity, and processing conditions on nutrient retention. In comparison to published data,

https://doi.org/10.3390/antiox15010095

our results identify roe oils as among the most promising natural matrices for the delivery of α-tocopherol, lutein, and squalene. Based on the samples analyzed in this study, solvent extraction with EtOH at 50 °C (ROE6) appeared most effective for preserving α-tocopherol, lutein, and squalene, yielding the highest concentrations of all three compounds. Other techniques, including mechanical press extraction and supercritical CO_2 with EtOH as co-solvent, recovered moderate levels but did not match the antioxidant retention observed under the 50 °C solvent extraction conditions. This work focused on the composition and nutritional aspects of the examined oils. Nevertheless, in assessing an oil with respect to its potential use as a lipid source, other parameters have to be taken into account, such as the yield, the economy, and the sustainability of the process, which was beyond the aim of the present work.

Lipidomics Results

The lipidomics analysis resulted in the detection and identification of 228 distinct lipid species across the marine oil samples, with the majority of them (69%) consisting of diglycerides (DAGs) and triglycerides (TAGs) (Figure 2), an observation consistent with multiple studies reporting TAGs and DAGs as the predominant lipid classes in marine oils. Song et al. [45] showed that TAGs constituted the dominant fraction in anchovy oils, regardless of their enrichment method, while DAGs represented the second most abundant glycerolipid class. Similarly, Windarsih et al. [46] reported that TAGs and DAGs formed the core lipid composition across five different marine fish oils, supporting their recurring prominence in marine lipidomes. Several of these lipids contained essential omega-3 and omega-6 fatty acids, which is consistent with reports showing that marine by-products frequently contain nutritionally relevant long-chain PUFAs [47,48].

Ceramides and PEs were additionally detected as relevant subclasses, in agreement with lipidomics studies that have identified these lipids in various marine tissues, including roe and shellfish [49–51]. A total of 102 lipid molecules were found to contain n-3 (80 compounds) and n-6 (22 compounds) fatty acids, highlighting these fish oils as a rich source of EPA and DHA, key omega-3 fatty acids known for their anti-inflammatory and cardiovascular benefits. Samples derived from squid oil exhibited a higher abundance of diglycerides and triglycerides, while roe oil samples displayed elevated levels of triglycerides containing DHA and EPA. Research supports that omega-3 and omega-6 lipids, particularly DHA and EPA, play essential roles in reducing cardiovascular risk, mitigating inflammatory responses, and promoting cognitive health [48,52,53]. These findings further substantiate the value of fish oils as a nutritionally rich supplement, especially in forms derived from roe and other marine by-products.

Chemometric Analysis

To further examine the lipid differences between the marine species in this study, chemometric analyses involving both unsupervised and supervised techniques were used. Specifically, a PCA model (Figure 3) was constructed using the data obtained from the lipidomics analysis. Eight PCs explain almost 95% of the total variation between the samples, with the squid group clearly separating from the other two. Fish samples, represented by the red cluster, display a broad variance along both PC1 and PC2, suggesting heterogeneous lipidomic features within this group. The blue cluster for roe is elongated along the PC2 axis, indicating distinct lipid characteristics that diverge more in one specific compositional direction compared to fish and squid. Squid samples, represented by the green cluster, are tightly grouped, reflecting a more homogeneous lipid profile and distinct separation from both fish and roe, particularly along PC1.

https://doi.org/10.3390/antiox15010095

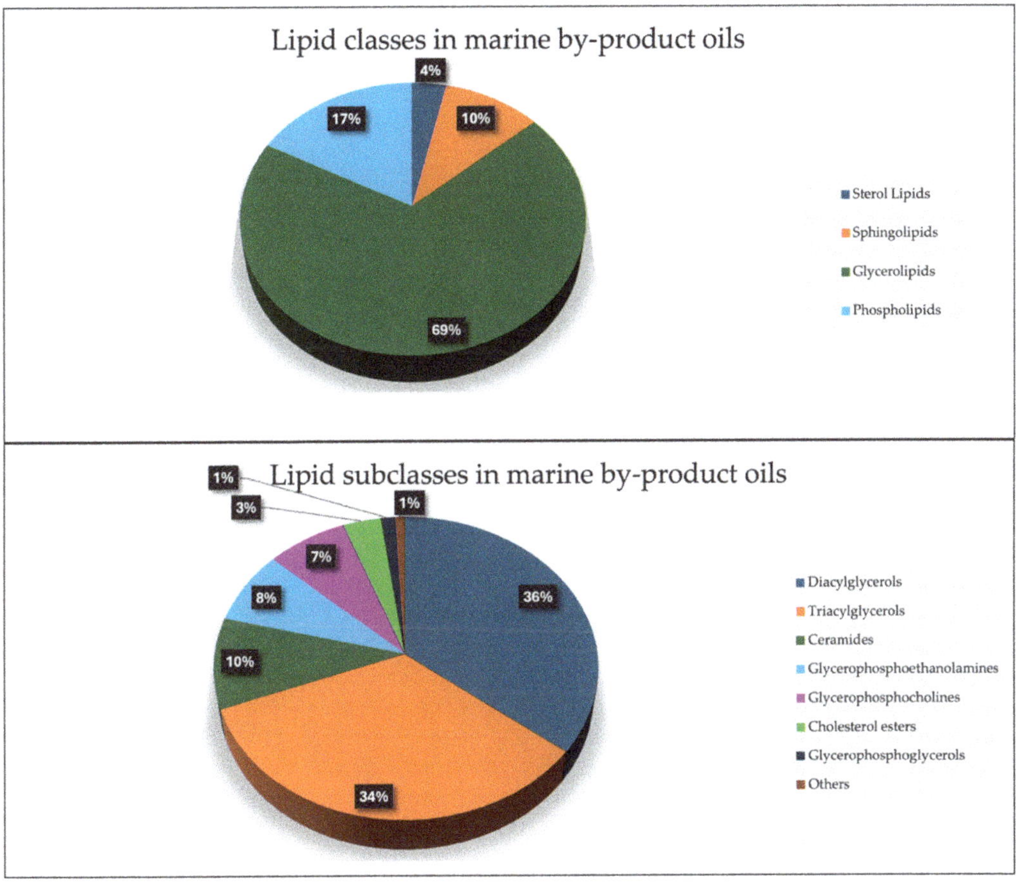

Figure 2. (**Up**): Pie chart showing the lipid classes presented in marine by-product oils. (**Bottom**): Pie chart showing the lipid subclasses presented in marine by-product oils.

These distinct clusters underscore the unique lipidomic signatures associated with each sample origin. The separation along PC1 indicates that squid lipid profiles markedly differ from fish and roe, potentially due to their unique triglyceride and fatty acid composition. Meanwhile, the positioning of squid along PC2 suggests that its lipid profile possesses specific attributes that differ from fish and roe, possibly related to structural lipids unique to its biological function.

PLS-DA was used to further study the lipid profiles of the samples. This PLS-DA scores plot (Figure 4) shows the distribution of fish, squid, roe, and QC samples, revealing distinct lipidomic patterns among these groups with the aid of supervised dimensionality reduction. Here, Component 1 explains 45.9% of the variance, while Component 3 accounts for 8.4%, collectively capturing over half of the variance in the lipid profiles within the dataset.

Fish samples, highlighted in red, occupy a broad, spread-out area along Component 1, indicating substantial lipid variability within this group. Squid samples, represented in green, form a tight cluster, suggesting a consistent lipid profile within these samples, likely reflective of their unique biochemical properties. Roe samples, in blue, are similarly distinct but form a cluster with relatively less variation along Component 3, emphasizing a unique lipid signature specific to roe tissues.

https://doi.org/10.3390/antiox15010095

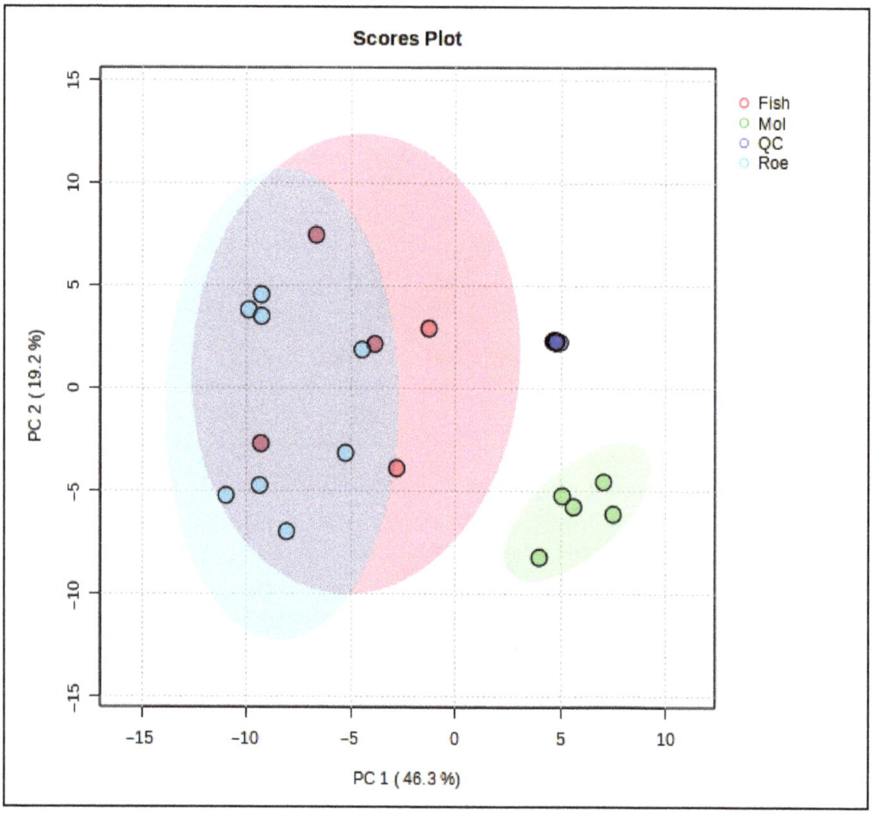

Figure 3. Principal Component Analysis (PCA) scores plot based on the untargeted lipidomic dataset of marine by-product oils. Samples are grouped according to biological origin: Fish (sardine heads, monkfish bellies, and anchovy heads) in red, Squid (squid bellies) in green, and Roe (grey mullet roe) in blue. PC1 and PC2 explain 46.3% and 19.2% of the total variance, respectively, highlight species driven differences in lipid composition.

The inclusion of QC samples (blue) provides a check for the analytical consistency and reproducibility of the data. QC samples cluster closely together, indicating stability and reliability of the lipidomic measurements across the study, affirming that observed group separations are indeed due to biological differences rather than analytical variability.

This PLS-DA plot strengthens the evidence of lipidomic differentiation among fish, squid, and roe groups, with QC samples providing confidence in the robustness of the dataset. This level of group separation could potentially facilitate the identification of lipid biomarkers specific to each sample type, contributing to applications in food authentication, quality control, and dietary profiling.

To elucidate group-specific variations in lipid profiles among fish, squid, and roe samples, we employed a heatmap visualization (Figure 5), which enables the clear representation of lipid abundance patterns across sample groups, highlighting differences in lipid composition and clustering lipid classes with similar expression trends. This heatmap provides key insights by emphasizing the relative abundance of each lipid species across groups, facilitating the identification of distinctive lipid classes that characterize each group.

https://doi.org/10.3390/antiox15010095

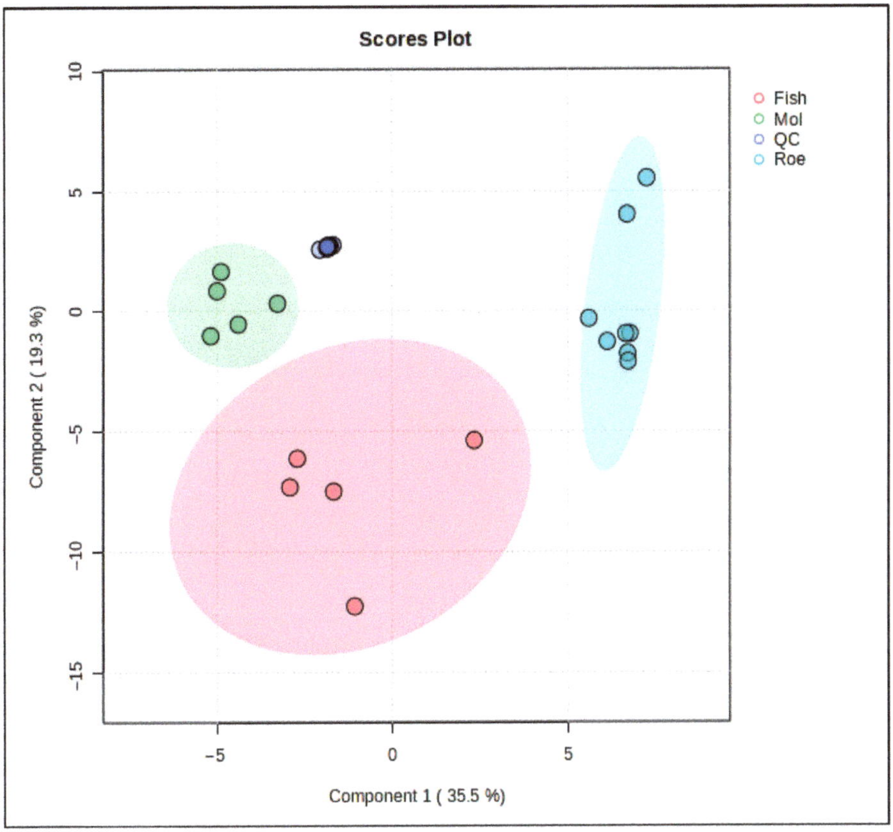

Figure 4. PLS-DA score plot for the comparison between fish-, squid-, and roe-derived oils.

The heatmap reveals several notable patterns in lipid abundance across the fish, squid, and roe sample classes. Fish samples, highlighted in red, exhibit elevated levels of specific TG and DG lipid species, as indicated by intense red coloration, suggesting a distinct lipid composition potentially tied to unique dietary or metabolic characteristics. In contrast, squid and roe samples show more moderate lipid levels, with fewer high-abundance lipids, indicating potential species-specific differences in lipid metabolism or absorption pathways. Specifically, fish oil samples exhibited a distinct lipidomic signature characterized by the significantly elevated relative abundance of multiple PUFAs containing glycerolipids, primarily within the TAG and DAG subclasses. Key differentiating species included the highly abundant TG 16:0_20:5_22:1, TG 14:0_20:5_22:6, and DG 18:1_24:0. Conversely, the roe samples demonstrated an intermediate profile, marked by an overall lower concentration of the TAG species compared to the fish group, yet were uniquely defined by the high relative intensity of DG 16:1_22:5 and the sterol ester CE 18:3. The squid oil samples profile was differentiated by uniformly low relative abundance across the majority of the major TAG species but displayed moderate enrichment in specific components like DG 20:0_22:6. These abundance patterns within the neutral lipid fraction—specifically reflecting the varying incorporation of PUFAs into TAGs and DGs—are directly linked to the species-level discrimination observed in multivariate analysis, confirming that the glycerolipid composition is a principal driver of the observed chemometric separation. The QC samples demonstrate consistent, uniform patterns, reflecting well-controlled quality assurance processes and reinforcing the reliability of the analytical results. Additionally, distinct lipid species such as CE 18:3, TG 14:0_14:0_20:5, and DG 18:1_24:0 show unique abundance profiles across the sample groups, potentially highlighting variability influenced by environmental or physiological factors. These findings underscore significant group-specific lipidomic profiles, offering insights into lipid metabolism that complement other multivariate analyses like PLS-DA.

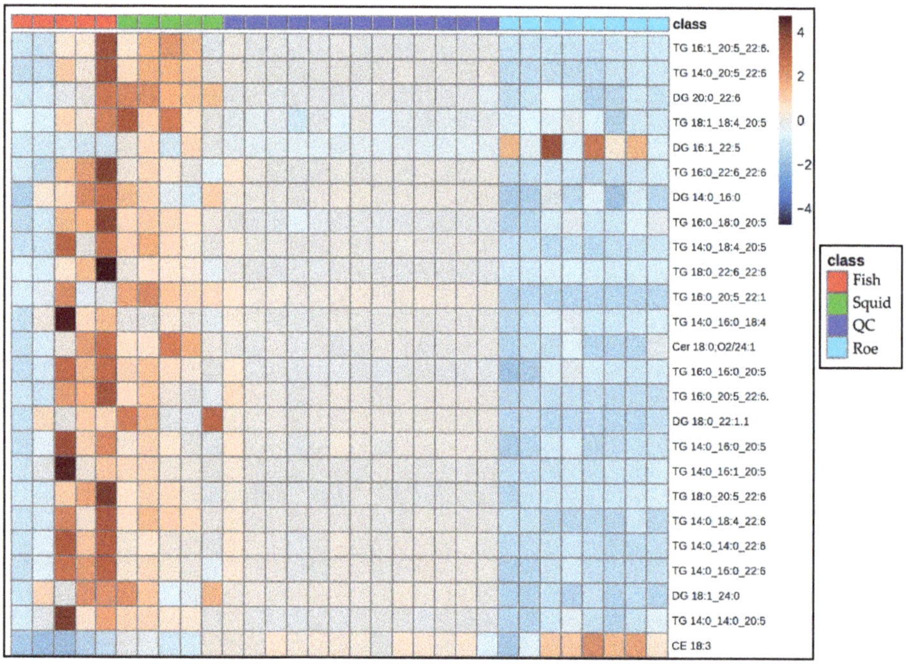

Figure 5. Heatmap showing the Top 25 VIPs from the PLS-DA, among the sample of marine by-product oil.

3.2. *Effect of Extraction Techniques on Lipid and Antioxidant Composition*

Having established the species-specific lipid and antioxidant signatures in Section 3.1, we next examined how different processing approaches modify the levels of the same compounds. The extraction technique applied during marine oil recovery exerts a critical influence on both the quantitative and qualitative profiles of lipids and lipid-soluble antioxidants. In our study, extraction methods were classified into three distinct classes, solvent extraction (SE), supercritical fluid extraction combined with solvent (SFE and SE), and conventional extraction (CE), and their effects were examined through fatty acid quantification, antioxidant profiling, and multivariate lipidomic analysis.

Comparative evaluation of total EPA + DHA content (Figure 6A) revealed that both SE and SFE and SE extractions yielded similarly high levels (~27–28 g/100 g), with SFE and SE marginally outperforming SE. Conventional methods, however, exhibited significantly lower retention (~22 g/100 g), likely reflecting oxidative or hydrolytic degradation during mechanical or enzymatic processing. It should be noted, however, that these values represent averages across all available samples within each extraction class, and the dataset is not fully balanced with respect to species and extraction technique. Therefore, the observed differences reflect overall trends rather than species-independent extraction effects. A similar trend was observed for total omega-3 content (Figure 6C), with SFE and SE demonstrating the most efficient preservation, underscoring their superior capacity for retaining long-chain polyunsaturated fatty acids (PUFAs), particularly those with nutritional significance, such as EPA and DHA.

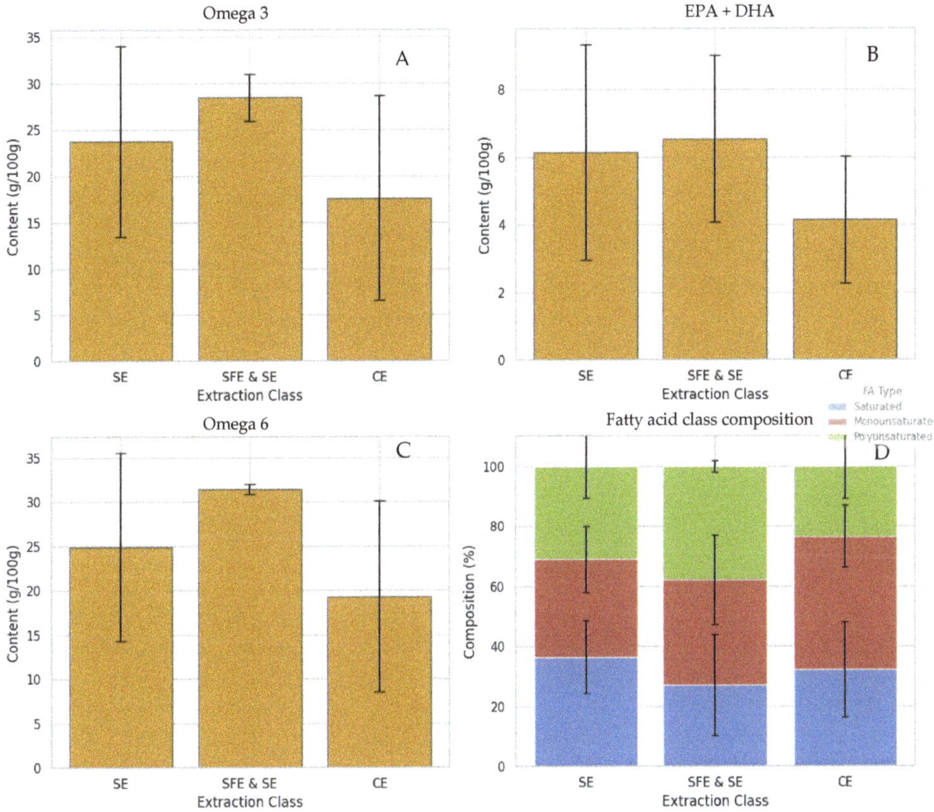

Figure 6. Fatty acid content in samples of marine oils based on the extraction technique used, presented as Mean ± Standard Deviation. Fatty acid composition was determined by GC-FID following transesterification of whole oil samples and is shown as extraction-class averages across the available samples. (**A**) total omega-3 content (%), (**B**) sum of EPA and DHA (%), (**C**) total omega-6 (%), and (**D**) fatty acid class composition (%).

Omega-6 fatty acid levels exhibited a slightly different profile. As shown in Figure 6B, SFE and SE extractions also resulted in the highest omega-6 concentrations, followed closely by SE, whereas TE consistently yielded the lowest levels. This suggests that SFE and SE preserve not only beneficial omega-3 species but also broader classes of unsaturated lipids, including omega-6, probably due to their gentle thermodynamic conditions and improved extraction efficacy across a broader polarity range [24]. Moreover, the relative abundance of different fatty acid classes—saturated, monounsaturated, and polyunsaturated—was visualized in a compositional bar plot (Figure 6D). In this plot, it is evident that SFE and SE yielded oils with the highest proportion of PUFAs. In contrast, oils extracted through conventional methods like enzymatic hydrolysis and wet reduction showed a greater composition of saturated and MUFAs. This pattern suggests that extraction processes within the CE group may contribute to partial lipid degradation or selective retention of shorter, more saturated chains, potentially due to enzyme activity or mechanical effects such as shear stress during processing. These findings are in agreement with previous studies, which have consistently reported that SFE and SE are more effective at preserving long-chain

https://doi.org/10.3390/antiox15010095

PUFAs like EPA and DHA, while enzyme-assisted methods tend to favour MUFA retention, likely due to the lipase specificity and mild processing conditions involved [35,54,55]. Formal separation of species and extraction effects would require a fully crossed experimental design, which was beyond the scope of the present study.

In addition to lipid class composition and fatty acid profiles, the antioxidant fraction of marine oils—particularly α-tocopherol, lutein, and squalene—was significantly influenced by the extraction method applied. These compounds, while minor in concentration, play a crucial role in stabilizing polyunsaturated lipids and extending the oxidative shelf-life of marine oil products. Among the three, α-tocopherol (vitamin E) was best preserved under SFE and SE conditions, where limited oxygen exposure and controlled pressure/temperature parameters minimized degradation [24]. Oils extracted by mechanical pressing or enzymatic treatment showed the lowest tocopherol concentrations, likely due to oxidative degradation during processing. This is supported by reports of substantial tocopherol loss under oxygen-rich and heat-prone conditions common to these methods [41,42,54,56].

In contrast, moderate retention was observed in selected solvent-extracted (SE) oils, particularly when extraction was performed at lower temperatures, which minimized oxidative stress and preserved a greater proportion of tocopherols [56]. A similar pattern was observed for lutein, a carotenoid with strong antioxidant and anti-inflammatory potential. Lutein was most abundant in SFE and SE and cold SE samples, while oils obtained via conventional methods showed diminished levels, consistent with the carotenoid's known sensitivity to heat, light, and oxidative stress. Lastly, squalene was significantly enriched in SFE and SE samples, reflecting the method's efficiency in extracting and preserving hydrophobic bioactive compounds. SE offered variable results depending on solvent polarity and temperature, whereas CE consistently underperformed, presumably due to mechanical degradation and air exposure. Collectively, these findings reinforce that SFE and SE extractions not only optimize lipid recovery but also protect valuable minor compounds, supporting their preferential use in the production of marine oils intended for functional or nutraceutical applications.

To complement the results of the FAME and antioxidant compound analyses, we next investigated whether these trends were also reflected in the full lipidomic profile using untargeted multivariate analysis. A PLS-DA model was constructed, based on the different extraction classes, and the results suggest that the extraction method has a clear impact on the lipidomic profile (Figure 7). Samples obtained via SFE and SE clustered tightly together, reflecting a high degree of biochemical consistency and extraction reproducibility. CE-extracted samples also formed a distinct, cohesive group, highlighting a different lipidomic composition from the other extraction techniques. In contrast, SE-extracted samples were more scattered, highlighting considerable variability within this group. This dispersion is likely due to the broad nature of the SE category, which included extractions at both 25 °C and 50 °C—conditions known to influence lipid yield and stability. Moreover, the partial overlaps between extraction groups suggest that lipid composition is not solely determined by the extraction technique but also influenced by the marine species used. Shared species-specific lipid traits likely contributed to similarities between certain samples, regardless of the extraction technique used for their production.

The compounds with the highest VIP scores—representing the most influential lipids in distinguishing between extraction classes—are listed in Table S3 of the ESM. From this group, the top 40 were used to generate the heatmap shown in Figure 8, highlighting the lipid species most responsible for the observed sample clustering. Interestingly, half of these key differentiators were TGs—complex storage lipids that often carry high-value omega-3 fatty acids such as EPA and DHA. These were most abundant in the SFE and SE extracts,

https://doi.org/10.3390/antiox15010095

reinforcing the idea that this technique is more effective at preserving intact, functional lipids. DGs accounted for another 20% of the VIPs and were more prevalent in TE and SE samples, where they likely resulted from partial hydrolysis or TG degradation [54,55,57]. Similarly, lysophosphatidylcholines (LPCs)—well-known markers of membrane damage and oxidative stress—were also more common in these conventional and solvent-based extracts. Only a few phosphatidylethanolamines (PEs) appeared among the VIPs, but those detected were DHA-rich and notably concentrated in SFE and SE samples.

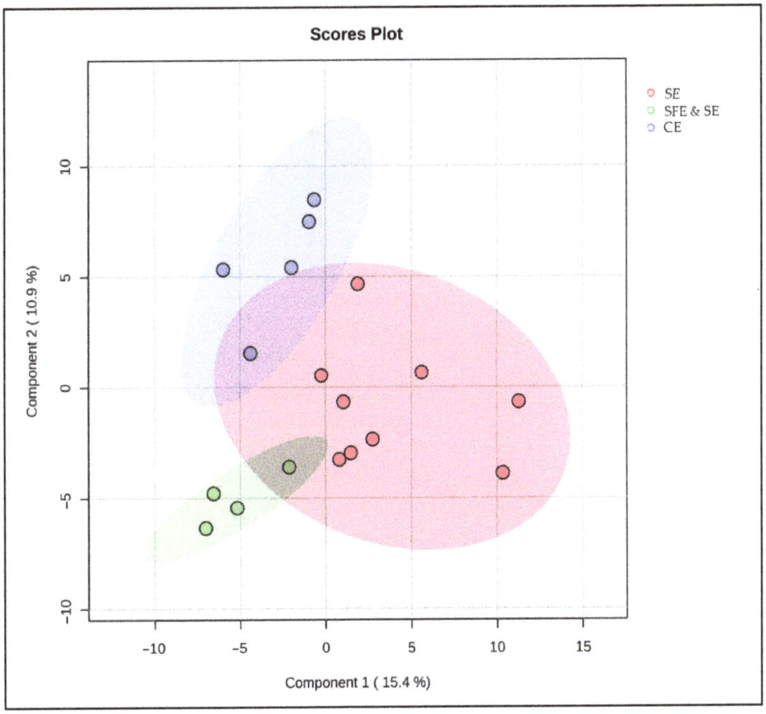

Figure 7. PLS-DA score plot for the comparison between SE, SFE and SE, and CE classes.

The heatmap visualization of these top VIP lipids added further clarity to the above-mentioned results. Samples clustered according to extraction method, reinforcing the trends seen in the PLS-DA analysis. SFE and SE samples stood out, marked by high levels of DHA- and EPA-containing triglycerides and phospholipids. In contrast, CE-extracted samples were dominated by LPCs and DGs, indicative of greater oxidative and mechanical stress during processing. The SE group again proved to be more complex: some SE samples aligned closely with SFE profiles—likely those extracted at optimized conditions—while others mirrored the degradation-prone TE group. Together, these findings paint a detailed picture of how different extraction methods influence not just the quantity, but also the quality and complexity of lipids in marine oils. SFE and SE emerge as the superior technique, offering the best preservation of long-chain, bioactive omega-3 lipids and the lowest abundance of degradation markers. In contrast, CE and conventional SE methods appear to compromise lipid integrity to varying degrees. These lipidomic insights reinforce the importance of choosing the right extraction strategy, especially when targeting high-value nutritional and functional compounds. For producers aiming to deliver potent, stable, and bioactive marine lipid formulations, SFE and SE represent the most promising and scientifically supported approach.

https://doi.org/10.3390/antiox15010095

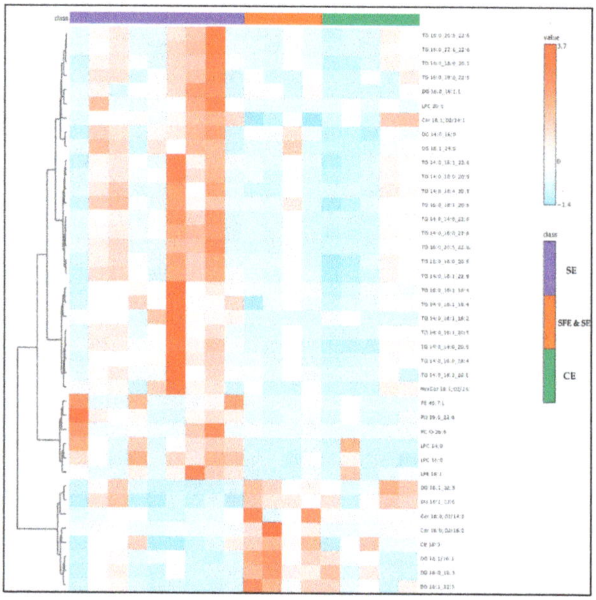

Figure 8. Heatmap showing the top 40 VIPs from the PLS-DA, among the samples of different extraction techniques.

4. Conclusions

This study comprehensively assessed how different extraction techniques influence the lipid composition and antioxidant content of oils derived from various marine by-products. The results clearly demonstrate that both the extraction method and the biological origin of the raw material significantly impact the nutritional and functional properties of the final oil product, particularly through differences in PUFA-rich triglycerides (TAGs), diglycerides (DGs), and key antioxidant molecules. Among the extraction methods examined, the supercritical extraction with CO_2 and ethanol as co-solvents (SFE-SE) emerged as the most effective in preserving bioactive and antioxidant lipid compounds. Oils obtained through this method showed consistently higher levels of essential omega-3 fatty acids—particularly EPA and DHA—as well as elevated concentrations of antioxidant molecules, such as α-tocopherol, lutein, and squalene. These compounds are critical for cardiovascular, neurological, and immune health, and their presence substantially enhances the nutritional value and oxidative stability of marine oils. In contrast, conventional extraction methods, such as wet reduction and mechanical pressing, resulted in lower retention of polyunsaturated fatty acids and antioxidants and a higher presence of hydrolysis/oxidation-associated lipids (e.g., DGs and LPCs), likely due to increased exposure to heat, oxygen, or enzymatic degradation during processing. Solvent-based extractions offered variable results, with temperature playing a decisive role in lipid and antioxidant preservation. Notably, roe-derived oils consistently demonstrated superior lipid profiles across all analyses, including higher levels of long-chain polyunsaturated fats and antioxidant compounds. This finding highlights grey mullet roe as a particularly rich and under-utilized source of health-promoting marine lipids, reinforcing its potential for use in high-value nutraceutical and cosmeceutical formulations. The application of untargeted lipidomics, coupled with advanced chemometric models (PCA and PLS-DA), further confirmed the clear biochemical separation between sample groups based on both species and extraction class. Discriminating variables included PUFA-rich TAGs and DAGs in fish oils and characteristic DG and

https://doi.org/10.3390/antiox15010095

sterol ester species in grey mullet roe and squid belly samples. This analytical approach not only provided a detailed molecular fingerprint of each sample but also revealed biomarkers that could guide future quality control, traceability, and product development efforts.

Overall, the findings presented in this work highlight how strongly the choice of extraction method shapes the nutritional quality and functional value of marine-derived oils. Optimizing conditions—particularly through supercritical CO_2–ethanol (SFE-SE)—can significantly enhance both the yield and preservation of key bioactive lipids. This work also strengthens the case for marine by-products as a viable, sustainable source of high-value compounds, supporting their integration into circular bioeconomy models and value-added applications.

Supplementary Materials: The following supporting information can be downloaded at https://www.mdpi.com/article/10.3390/antiox15010095/s1: Table S1. List of samples and sample information; Table S2. Lipid compounds detected in marine oils with their molecular formula, lipid class, and subclass; Table S3. List of VIP compounds with VIP score > 1 found in marine oils.

Author Contributions: Conceptualization, N.S.T.; methodology, I.C.M. and E.P.K.; validation, I.C.M.; investigation, I.C.M. and P.T.; resources, N.S.T.; data curation, I.C.M.; writing—original draft preparation, I.C.M.; writing—review and editing, M.E.D., E.P.K. and N.S.T.; visualization, I.C.M.; supervision, M.E.D. and N.S.T.; project administration, I.C.M., M.E.D. and N.S.T.; funding acquisition, N.S.T. and E.P.K. All authors have read and agreed to the published version of the manuscript.

Funding: This research has been co-financed by the European Regional Development Fund of the European Union and Greek national funds through the Operational Program Competitiveness, Entrepreneurship and Innovation, under the call RESEARCH—CREATE—INNOVATE (project code:T2EΔK- 02465–project title "Research and development of high nutritional value anti-inflammatory functional foods, enriched in ω-3 fatty acids sourced from Greek fisheries and farming byproducts".

Institutional Review Board Statement: Not applicable.

Data Availability Statement: The original contributions presented in this study are included in the article/supplementary material. Further inquiries can be directed to the corresponding author.

Conflicts of Interest: The authors declare no conflicts of interest.

References

1. Liu, L.; Chen, Y.; Chen, B.; Xu, M.; Liu, S.; Su, Y.; Qiao, K.; Liu, Z. Advances in Research on Marine-Derived Lipid-Lowering Active Substances and Their Molecular Mechanisms. *Nutrients* **2023**, *15*, 5118.
2. Lobine, D.; Rengasamy, K.R.R.; Mahomoodally, M.F. Functional Foods and Bioactive Ingredients Harnessed from the Ocean: Current Status and Future Perspectives. *Crit. Rev. Food Sci. Nutr.* **2021**, *62*, 5794–5823. [CrossRef]
3. Mutalipassi, M.; Esposito, R.; Ruocco, N.; Viel, T.; Costantini, M.; Zupo, V. Bioactive Compounds of Nutraceutical Value from Fishery and Aquaculture Discards. *Foods* **2021**, *10*, 1495. [CrossRef]
4. Ahmmed, M.K.; Ahmmed, F.; Stewart, I.; Carne, A.; Tian, H.; Bekhit, A.E.D.A. Omega-3 Phospholipids in Pacific Blue Mackerel (*Scomber Australasicus*) Processing by-Products. *Food Chem.* **2021**, *353*, 129451. [CrossRef]
5. Calder, P.C. The Role of Marine Omega-3 (n-3) Fatty Acids in Inflammatory Processes, Atherosclerosis and Plaque Stability. *Mol. Nutr. Food Res.* **2012**, *56*, 1073–1080. [PubMed]
6. Lordan, R.; Redfern, S.; Tsoupras, A.; Zabetakis, I. Inflammation and Cardiovascular Disease: Are Marine Phospholipids the Answer? *Food Funct.* **2020**, *11*, 2861–2885. [CrossRef]
7. Rodríguez, A.; Trigo, M.; Aubourg, S.P.; Medina, I. Optimised Extraction of Phospholipids and Omega-3 Fatty Acids from Patagonian Squid Waste with Low-Toxicity Solvents. *Int. J. Food Sci. Technol.* **2023**, *58*, 6541–6551. [CrossRef]
8. de Carvalho, C.C.C.R.; Caramujo, M.J. Carotenoids in Aquatic Ecosystems and Aquaculture: A Colorful Business with Implications for Human Health. *Front. Mar. Sci.* **2017**, *4*, 93. [CrossRef]
9. Nakano, T.; Wiegertjes, G. Properties of Carotenoids in Fish Fitness: A Review. *Mar. Drugs* **2020**, *18*, 568. [CrossRef] [PubMed]
10. Afonso, C.; Bandarra, N.M.; Nunes, L.; Cardoso, C. Tocopherols in Seafood and Aquaculture Products. *Crit. Rev. Food Sci. Nutr.* **2013**, *56*, 128–140. [CrossRef]

https://doi.org/10.3390/antiox15010095

11. Hamre, K. Metabolism, Interactions, Requirements and Functions of Vitamin E in Fish. *Aquac. Nutr.* **2010**, *17*, 98–115. [CrossRef]
12. Kim, S.K.; Karadeniz, F. Biological Importance and Applications of Squalene and Squalane. In *Advances in Food and Nutrition Research*; Academic Press Inc.: San Diego, CA, USA, 2012; Volume 65, pp. 223–233.
13. Gopakumar, K. Therapeutic Applications of Squalene—A review. *Fish. Technol.* **2012**, *49*, 1–9.
14. Saidaiah, P.; Banu, Z.; Khan, A.A.; Geetha, A.; Somraj, B. A Comprehensive Review of Omega-3 Fatty Acids: Sources, Industrial Applications, and Health Benefits. *Ann. Phytomedicine Int. J.* **2024**, *13*, 209–225. [CrossRef]
15. Fonseca, S.; Amaral, M.N.; Reis, C.P.; Custódio, L. Marine Natural Products as Innovative Cosmetic Ingredients. *Mar. Drugs* **2023**, *21*, 170. [CrossRef]
16. Alves, A.; Sousa, E.; Sousa, E.; Kijjoa, A.; Pinto, M.; Pinto, M. Marine-Derived Compounds with Potential Use as Cosmeceuticals and Nutricosmetics. *Molecules* **2020**, *25*, 2536. [CrossRef] [PubMed]
17. Giménez Martínez, R.J.; Rivas García, F.; March Cerdá, J.C.; Hernández-Ruíz, Á.; González Castro, M.I.; Valverde-Merino, M.-I.; Huertas Camarasa, F.J.; Lloris Meseguer, F.; López-Viota Gallardo, M. Bioactive Substances and Skin Health: An Integrative Review from a Pharmacy and Nutrition Perspective. *Pharmaceuticals* **2025**, *18*, 373. [CrossRef] [PubMed]
18. Georgiou, D.; Tzika, P.; Charisis, A.; Kalogianni, E.P. Effect of Extraction Method on the Yield and Quality of Mullet Roe Oil. *Eur. J. Lipid Sci. Technol.* **2023**, *125*, 2300008. [CrossRef]
19. Kalogianni, E.P.; Georgiou, D.; Charisis, A.; Exarhopoulos, S.; Tzika, P. Valorization of Mullet Roe By-Products for the Production of Polyunsaturated Fatty Acids Rich Oils. *JAOCS J. Am. Oil Chem. Soc.* **2023**, *100*, 317–327. [CrossRef]
20. Tsamesidis, I.; Tzika, P.; Georgiou, D.; Charisis, A.; Hans, S.; Lordan, R.; Zabetakis, I.; Kalogianni, E.P. Oil from Mullet Roe Byproducts: Effect of Oil Extraction Method on Human Erythrocytes and Platelets. *Foods* **2023**, *13*, 79. [CrossRef]
21. Food and Agriculture Organization of the United Nation. *The Production of Fish Meal and Oil*; Food and Agriculture Organization of the United Nation: Rome, Italy, 1986.
22. Ivanovs, K.; Blumberga, D. Extraction of Fish Oil Using Green Extraction Methods: A Short Review. *Energy Procedia* **2017**, *128*, 477–483. [CrossRef]
23. Marsol-Vall, A.; Aitta, E.; Guo, Z.; Yang, B. Green Technologies for Production of Oils Rich in N-3 Polyunsaturated Fatty Acids from Aquatic Sources. *Crit. Rev. Food Sci. Nutr.* **2021**, *62*, 2942–2962. [CrossRef]
24. Semenoglou, I.; Eliasson, L.; Uddstål, R.; Tsironi, T.; Taoukis, P.; Xanthakis, E. Supercritical CO_2 Extraction of Oil from Arctic Charr Side Streams from Filleting Processing. *Innov. Food Sci. Emerg. Technol.* **2021**, *71*, 102712. [CrossRef]
25. Luksta, I.; Mika, T.; Spalvins, K. Supercritical CO_2 Extraction of Fish Roe. *Environ. Clim. Technol.* **2024**, *28*, 12–20. [CrossRef]
26. Liu, Y.; Li, Y.; Yu, W.; Ma, Y.; Cao, Y.; Zhao, L. Extraction of High-Quality Antarctic Krill (*Euphausia Superba*) Oil Using Low-Temperature Continuous Phase-Transition Extraction Equipment. *J. Food Sci.* **2024**, *89*, 9073–9087. [CrossRef] [PubMed]
27. Kalogianni, E.P.; Tzika, P.; Georgiou, D. Enzyme Assisted Oil Extraction from Fishery by Products: Insight Gained from Interfacial Phenomena. *J. Food Eng.* **2025**, *407*, 112829. [CrossRef]
28. Martakos, I.; Kostakis, M.; Dasenaki, M.; Pentogennis, M.; Thomaidis, N. Simultaneous Determination of Pigments, Tocopherols, and Squalene in Greek Olive Oils: A Study of the Influence of Cultivation and Oil-Production Parameters. *Foods* **2019**, *9*, 31. [CrossRef]
29. Martakos, I.C.; Tzavellas, I.F.; Dasenaki, M.E.; Thomaidis, N.S. Food Lipidomics: Development and Application of a Cutting-Edge Untargeted 4D HRMS Workflow for the Lipid Profiling of Food of Animal Origin. *J. Food Compos. Anal.* **2024**, *131*, 106232. [CrossRef]
30. Jimenez-Carvelo, A.M. Data Mining/Machine Learning Methods in Foodomics. *Curr. Opin. Food Sci.* **2021**, *37*, 76–82. [CrossRef]
31. Kind, T.; Liu, K.-H.; Lee, D.Y.; DeFelice, B.; Meissen, J.K.; Fiehn, O. LipidBlast in Silico Tandem Mass Spectrometry Database for Lipid Identification. *Nat. Methods* **2013**, *10*, 755–758. [CrossRef]
32. Pang, Z.; Lu, Y.; Zhou, G.; Hui, F.; Xu, L.; Viau, C.; Spigelman, A.F.; Macdonald, P.E.; Wishart, D.S.; Li, S.; et al. MetaboAnalyst 6.0: Towards a Unified Platform for Metabolomics Data Processing, Analysis and Interpretation. *Nucleic Acids Res.* **2024**, *52*, W398–W406. [CrossRef]
33. Dieterle, F.; Ross, A.; Schlotterbeck, G.; Senn, H. Probabilistic Quotient Normalization as Robust Method to Account for Dilution of Complex Biological Mixtures. Application in 1H NMR Metabonomics. *Anal. Chem.* **2006**, *78*, 4281–4290. [CrossRef]
34. Ahmmed, M.K.; Ahmmed, F.; Tian, H.; Carne, A.; Bekhit, A.E.D. Marine Omega-3 (n-3) Phospholipids: A Comprehensive Review of Their Properties, Sources, Bioavailability, and Relation to Brain Health. *Compr. Rev. Food Sci. Food Saf.* **2019**, *19*, 64–123. [CrossRef] [PubMed]
35. Linda Mgbechidinma, C.; Zheng, G.; Buyagao Baguya, E.; Zhou, H.; Ukpong Okon, S.; Zhang, C. Fatty Acid Composition and Nutritional Analysis of Waste Crude Fish Oil Obtained by Optimized Milder Extraction Methods. *Environ. Eng. Res.* **2022**, *28*, 220034. [CrossRef]
36. Durmuş, M. Fish Oil for Human Health: Omega-3 Fatty Acid Profiles of Marine Seafood Species. *Food Sci. Technol.* **2019**, *39*, 454–461. [CrossRef]

https://doi.org/10.3390/antiox15010095

37. Richter, C.K.; Skulas-Ray, A.C.; Kris-Etherton, P.M. Recommended Intake of Fish and Fish Oils Worldwide. In *Fish and Fish Oil in Health and Disease Prevention*; Elsevier Inc.: Amsterdam, The Netherlands, 2016; pp. 27–48, ISBN 9780128028445.
38. Getachew, A.T.; Jacobsen, C.; Sørensen, A.D.M. Supercritical CO_2 for Efficient Extraction of High-Quality Starfish (*Asterias rubens*) Oil. *J. Supercrit. Fluids* **2023**, *206*, 106161. [CrossRef]
39. Bunga, S.J.; Ahmmed, M.K.; Lawley, B.; Carne, A.; Bekhit, A.E.D.A. Physicochemical, Biochemical and Microbiological Changes of Jeotgal-like Fermented Chinook Salmon (*Oncorhynchus tshawytscha*) Roe. *Food Chem.* **2022**, *398*, 133880. [CrossRef]
40. Kalogeropoulos, N.; Nomikos, T.; Chiou, A.; Fragopoulou, E.; Antonopoulou, S. Chemical Composition of Greek Avgotaracho Prepared from Mullet (*Mugil cephalus*): Nutritional and Health Benefits. *J. Agric. Food Chem.* **2008**, *56*, 5916–5925. [CrossRef]
41. Moriya, H.; Kuniminato, T.; Hosokawa, M.; Fukunaga, K.; Nishiyama, T.; Miyashita, K. Oxidative Stability of Salmon and Herring Roe Lipids and Their Dietary Effect on Plasma Cholesterol Levels of Rats. *Fish. Sci.* **2007**, *73*, 668–674. [CrossRef]
42. Bunga, S.; Ahmmed, M.K.; Carne, A.; Bekhit, A.E.D.A. Effect of Salted-Drying on Bioactive Compounds and Microbiological Changes during the Processing of Karasumi-like Chinook Salmon (*Oncorhynchus tshawytscha*) Roe Product. *Food Chem.* **2021**, *357*, 129780. [CrossRef]
43. Galasso, C.; Corinaldesi, C.; Sansone, C. Carotenoids from Marine Organisms: Biological Functions and Industrial Applications. *Antioxidants* **2017**, *6*, 96. [CrossRef] [PubMed]
44. López-Puebla, S.; Arias-Santé, M.F.; Romero, J.; Costa de Camargo, A.; Rincón-Cervera, M.Á. Analysis of Fatty Acid Profile, α-Tocopherol, Squalene and Cholesterol Content in Edible Parts and By-Products of South Pacific Wild Fishes. *Mar. Drugs* **2025**, *23*, 104. [CrossRef]
45. Song, R.; Wang, X.; Deng, S.; Tao, N. Lipidomic Analysis and Triglyceride Profiles of Fish Oil: Preparation through Silica Gel Column and Enzymatic Treatment. *Food Res. Int.* **2022**, *162*, 112100. [CrossRef] [PubMed]
46. Windarsih, A.; Irnawati; Suratno; Warmiko, H.D.; Alam, L.P.M.; Utami, I.D.; Rohman, A.; Indrianingsih, A.W. Lipidomics Analysis of Different Marine Fish Oils Using Untargeted Liquid Chromatography–Orbitrap High-Resolution Mass Spectrometry and Chemometrics. *Chromatographia* **2024**, *87*, 203–214. [CrossRef]
47. Vazquez, J.A.; Meduina, A.; Duran, A.I.; Nogueira, M.; Fernandez-Compas, A.; Perez-Martin, R.I.; Rodriguez-Amado, I. Production of Valuable Compounds and Bioactive Metabolites from By-Products of Fish Discards Using Chemical Processing, Enzymatic Hydrolysis, and Bacterial Fermentation. *Mar. Drugs* **2019**, *17*, 139. [CrossRef]
48. Farag, M.A.; Abib, B.; Tawfik, S.; Shafik, N.; Khattab, A.R. Caviar and Fish Roe Substitutes: Current Status of Their Nutritive Value, Bio-Chemical Diversity, Authenticity and Quality Control Methods with Future Perspectives. *Trends Food Sci. Technol.* **2021**, *110*, 405–417. [CrossRef]
49. Liang, P.; Li, R.; Sun, H.; Zhang, M.; Cheng, W.; Chen, L.; Cheng, X.; Akoh, C.C. Phospholipids Composition and Molecular Species of Large Yellow Croaker (*Pseudosciaena crocea*) Roe. *Food Chem.* **2018**, *245*, 806–811. [CrossRef]
50. Li, H.; Song, Y.; Zhang, H.; Wang, X.; Cong, P.; Xu, J.; Xue, C. Comparative Lipid Profile of Four Edible Shellfishes by UPLC-Triple TOF-MS/MS. *Food Chem.* **2020**, *310*, 125947. [CrossRef] [PubMed]
51. Yin, M.; Chen, M.; Matsuoka, R.; Song, X.; Xi, Y.; Zhang, L.; Wang, X. UHPLC-Q-Exactive Orbitrap MS/MS Based Untargeted Lipidomics Reveals Fatty Acids and Lipids Profiles in Different Parts of Capelin (*Mallotus villosus*). *J. Food Compos. Anal.* **2022**, *116*, 105096. [CrossRef]
52. Monteiro, M.; Marques, A.; Costa, R.S.; Salgado, M.A.; Castro, C.; Conceição, L.; Valente, L.M.P. Beyond Fish Oil: Assessing the Implications of Alternative Dietary Lipid Sources for Turbot (*Scophthalmus maximus*) on Growth, Nutrient Utilization and Muscle Quality. *Aquaculture* **2023**, *578*, 740073. [CrossRef]
53. Rosa, A.; Scano, P.; Melis, M.P.; Deiana, M.; Atzeri, A.; Dessì, M.A. Oxidative Stability of Lipid Components of Mullet (*Mugil cephalus*) Roe and Its Product "Bottarga". *Food Chem.* **2009**, *115*, 891–896. [CrossRef]
54. Hao, S.; Wei, Y.; Li, L.; Yang, X.; Cen, J.; Huang, H.; Lin, W.; Yuan, X. The Effects of Different Extraction Methods on Composition and Storage Stability of Sturgeon Oil. *Food Chem.* **2015**, *173*, 274–282. [CrossRef] [PubMed]
55. Rubio-Rodríguez, N.; De Diego, S.M.; Beltrán, S.; Jaime, I.; Sanz, M.T.; Rovira, J. Supercritical Fluid Extraction of Fish Oil from Fish By-Products: A Comparison with Other Extraction Methods. *J. Food Eng.* **2012**, *109*, 238–248. [CrossRef]
56. Park, J.S.; Roy, V.C.; Kim, S.Y.; Lee, S.C.; Chun, B.S. Extraction of Edible Oils and Amino Acids from Eel By-Products Using Clean Compressed Solvents: An Approach of Complete Valorization. *Food Chem.* **2022**, *388*, 132949. [CrossRef]
57. Taati, M.M.; Shabanpour, B.; Ojagh, M. Investigation on Fish Oil Extraction by Enzyme Extraction and Wet Reduction Methods and Quality Analysis. *AACL Bioflux* **2018**, *11*, 83–90.

https://doi.org/10.3390/antiox15010095

antioxidants

Article

Bioactive Compounds and Antioxidant Efficacy of Djulis (*Chenopodium formosanum*) Leaves: Implications for Sustainable Cosmeceutical Development

Chien-Shan Chiu [1,2,3], Yung-Jia Chan [3], Yan-Zhen Wu [4], Wen-Chien Lu [5], Po-Yuan Chiang [6] and Po-Hsien Li [4,*]

1 Department of Dermatology, Taichung Veterans General Hospital, Taichung 40705, Taiwan; cschiu@nchu.edu.tw or dermvgh@gmail.com
2 Department of Post-Baccalaureate Medicine, College of Medicine, National Chung Hsing University, Taichung 40227, Taiwan
3 College of Biotechnology and Bioresources, Da-Yeh University, Changhua 51591, Taiwan; d0867601@cloud.dyu.edu.tw
4 Department of Food and Nutrition, Providence University, Taichung 43301, Taiwan
5 Department of Food and Beverage Management, Chung-Jen Junior College of Nursing, Health Sciences and Management, Chia-Yi City 60077, Taiwan; m104046@cjc.edu.tw
6 Department of Food Science and Biotechnology, National Chung Hsing University, Taichung 40227, Taiwan
* Correspondence: pohsien0105@pu.edu.tw; Tel.: +886-4-2632-8001 (ext. 15326)

Abstract: *Chenopodium formosanum* (djulis), well known for its vivid color variation during diverse senescence stages, has attracted attention for its perceived health benefits and antioxidant capacity. Djulis leaves, often discarded as biowaste after harvesting, were evaluated for their potential as a source of antioxidant compounds. The current study analyzes the physicochemical and antioxidant activities of red, green, and yellow djulis leaf extracts across various senescence stages to probe their prospective utility in cosmeceuticals. Various plant compounds including total carbohydrates, phenolics and flavonoids, chlorophyll a and b, and betaxanthins and betacyanins were measured using spectrophotometric techniques. Antioxidant potential was assessed using DPPH, FRAP, CUPRAC, TEAC, and DMPD assays. Green djulis leaves displayed elevated total carbohydrate and chlorophyll levels, whereas red djulis leaves exhibited heightened phenolic, flavonoid, betaxanthin, and betacyanin content, indicating its suitability for cosmeceutical applications. Antioxidant evaluations revealed substantial disparities among the extracts, with red djulis leaf extract demonstrating superior antioxidant activity in most assays. These findings revealed the distinct antioxidant profiles of djulis leaf extracts influenced by color and senescence stage. These findings advance our understanding of the bioactive attributes of djulis leaves and their potential for incorporation into functional products.

Keywords: *Chenopodium formosanum*; djulis leaves; antioxidant; cosmeceuticals

1. Introduction

Chenopodium formosanum, commonly known as djulis, has a significant place in Taiwanese culture and traditional medicine because of its rich nutritional content and purported health benefits [1]. Djulis, a member of the Amaranthaceae family, is indigenous to Taiwan, where it has been cultivated and consumed for centuries [2]. Beyond its aesthetic appeal, djulis contains a treasure trove of bioactive compounds with potential health-promoting properties [3,4]. Among these, antioxidants have garnered considerable attention because of their ability to neutralize harmful free radicals and mitigate oxidative stress, a process implicated in various chronic diseases and aging [5]. The antioxidant

 https://doi.org/10.3390/antiox14020202

capacity of djulis is attributed to their phenolic compounds, vitamins, and minerals, which act synergistically to scavenge free radicals and modulate oxidative pathways within the body [6]. Previous studies have investigated the effect of bioactive compounds on vasodilation, angiotensin-converting enzyme activity (ACE), and hypertension in vitro and in vivo, demonstrating that aqueous extracts of djulis displayed an obvious inhibitory effect on ROS generation, peroxynitrite levels, and ACE activity [7].

The seeds of djulis have been extensively utilized in food manufacturing and pharmaceutical research [8]; however, their leaves have long been underappreciated. In addition to its seeds, djulis is characterized by its strikingly colorful leaves, transitioning from shades of green to red and yellow as they progress through various senescence stages [9]. In the early stages of growth, djulis leaves are predominantly green due to the high concentration of chlorophyll [10]. As the plant matures or under certain environmental conditions, the amount of chlorophyll may decrease, and other pigments like betacyanins become more dominant [11]. Betacyanins are part of the betalain pigment family and are responsible for the red-to-purple colors in djulis [12]. These pigments absorb green light and reflect red and violet light, giving the leaves a reddish hue. Betacyanins also have antioxidant properties, which help protect the plant from environmental stress, such as UV radiation and oxidative stress [13]. The yellow color in djulis leaves is due to the presence of pigments like betaxanthins [14]. Betaxanthins, another group of betalain pigments, are responsible for the yellow-to-orange colors [14]. As the plant ages, chlorophyll may break down, causing the yellow pigments to become more prominent [15]. However, the leaves of djulis have often been overlooked and considered waste materials even though djulis leaves are allowed to be used as food ingredients by the Taiwan Food and Drug Administration (TFDA). Nevertheless, the distinct features of djulis leaves have captured the attention of researchers and enthusiasts alike, prompting investigations into the underlying factors influencing their coloration and nutritional profile. Recent studies have highlighted the potential value of djulis leaves as a rich source of bioactive compounds, including antioxidants and phytochemicals [16]. There is an increasing interest in utilizing agricultural by-products and waste materials for sustainable product development [17].

In recent years, there has been growing interest in utilizing natural plant extracts as antioxidants for developing cosmeceutical products in the cosmetic and skincare industries [18]. Cosmeceuticals, a hybrid category of cosmetics and pharmaceuticals, offer functional benefits beyond traditional skincare by targeting specific skin concerns while providing nourishment and protection. Related studies using unhulled red djulis extracts to treat human fibroblasts (CCD-966SK cells) demonstrated that unhulled red djulis extracts effectively increased the repairing ability of skin cells and amplified skin-barrier-related and antioxidant-related gene expression [19]. Furthermore, unhulled red djulis extracts increased collagen-related gene expression and inhibited AGEs and melanin formation [19]. Additionally, unhulled djulis was extracted using water at a ratio of 1:10 (w/v), which decreased UV-induced intracellular ROS generation and triggered the antioxidant defense system by activating the nuclear factor erythroid 2-related factor 2 (Nrf2)/heme oxygenase-1 (HO-1) signaling pathway in human skin fibroblasts (Hs68 cells) [20]. Djulis extracts, with their potent antioxidant activity, are promising natural ingredients for cosmeceutical formulations aimed at addressing oxidative-stress-induced skin damage, premature aging, and other dermatological conditions. Nevertheless, the use of djulis leaf extract in cosmeceutical applications is still not well researched.

Despite the growing interest in djulis leaves as a potential source of antioxidants for cosmeceutical applications, comprehensive studies examining the antioxidant profiles of leaves of different colors at various senescence stages remain limited. Understanding how color variation and senescence influence the antioxidant composition of djulis leaves is essential for optimizing their efficacy and safety in cosmeceutical formulations. Elucidating the bioactive constituents responsible for the observed antioxidant activity will facilitate the development of standardized extracts of consistent quality and potency.

Based on the factors mentioned above, this study aimed to investigate the antioxidant activities of red, green, and yellow djulis leaf extracts, taking into account their distinct senescence stages. Through a comprehensive analysis of their antioxidant profiles, this study provides valuable insights into the potential applications of djulis leaves in cosmeceuticals and other functional products. By harnessing the natural antioxidant properties of djulis leaves, this study seeks to contribute to the development of innovative skincare formulations that promote skin health and vitality, reflecting the rich botanical heritage of Taiwan and its indigenous plant species. By using djulis leaves as a material, this study may contribute to the valorization of agricultural waste and the development of value-added products with potential benefits for both human health and environmental sustainability.

2. Materials and Methods

2.1. Materials and Sample Preparation

Djulis (*Chenopodium formosanum* Koidz), cultivated in Yunlin County (Taiwan), was supplied by Hearty Farm (Yunlin, Taiwan) (Figure 1A). The leaves of djulis were separated by color (Figure 1B–D) before being oven-dried at a temperature of 40 °C, powdered using a blender (model CB-5A, Chuang Pao Special Precision Industry Co., Ltd., Kaohsiung City, Taiwan), and passed through a 0.5 mm mesh. For the extraction, 10 g of the powder was mixed with 100 mL of distilled water and stirred (150 rpm) at 4 °C overnight. After 24 h, the mixture was filtered, and the filtrate was frozen and lyophilized in a freeze dryer (Uniss FDM-20 Freeze Dryer, Xinbei City, Taiwan). The chemicals used in this study, such as sulfuric acid, Folin-Ciocalteu reagent, gallic acid, ferric acid, sodium hydroxide, and dimethyl sulfoxide (DMSO), were of analytical grade (purity > 99%) and purchased from Sigma-Aldrich (St. Louis, MO, USA).

2.2. Physicochemical Properties

2.2.1. Total Carbohydrate

The freeze-dried sample powder (4.2 mg) was dissolved in distilled water to a final volume of 25 mL. After adding 1 mL of the sample solution to 1 mL of 5% phenol solution, 5 mL of 98% concentrated sulfuric acid was quickly added. After thorough mixing, the solution was left at room temperature for 15 min and the absorbance was measured at 490 nm using a spectrophotometer (SH-U830, Shishin Technologies Co., Ltd., Taipei City, Taiwan). A standard curve was created using D-glucose concentrations ranging from 0.009 to 0.100 mg glucose/mL (Figure S1A) while the result was performed in mg glucose/g to determine the total carbohydrate content (mg glucose/g DW) of the samples.

2.2.2. Total Phenolic Content (TPC)

A total of 5.6 mg of freeze-dried sample powder was dissolved in distilled water to a final volume 5 mL. Next, 1 mL of the sample solution was mixed with 1 mL of 0.5 Folin-Ciocalteu reagent and 2 mL of 20% sodium carbonate (Na_2CO_3). After mixing thoroughly, the absorbance was measured at 750 nm. The TPC (mg gallic acid/g) of the sample was determined using a standard curve prepared with gallic acid (0.001–0.04 mg gallic acid/mL) (Figure S1B). In contrast, the results in the samples will be expressed in mg gallic acid/g of the lyophilized sample.

2.2.3. Total Flavonoid Content (TFC)

Freeze-dried sample powder (8.2 mg) was dissolved in distilled water to a final volume of 5 mL. In total, 1 mL of the sample solution was mixed with 0.3 mL of 5% sodium nitrite ($NaNO_2$) and 0.3 mL of 10% aluminum chloride ($AlCl_3$). After 5 min, 2 mL of 1 M sodium hydroxide (NaOH) was added, and the volume was made up to 10 mL using distilled water. After thorough mixing, the absorbance was measured at 510 nm using a spectrophotometer (SH-U830, Shishin Technologies Co., Ltd., Taipei City, Taiwan), and the TFC (mg QE/g) of the sample was determined using a standard curve prepared with Quercetin (Sigma-Aldrich, St. Louis, MO, USA) at the concentrations ranging from 0.004 to 0.06 mg quercetin/mL (Figure S1C). In contrast, the result was performed in mg quercetin/g to determine the total flavonoid content (mg quercetin/g DW) of the samples.

Figure 1. (**A**) Djulis (*Chenopodium formosanum*). (**B**) Green djulis leaf. (**C**) Red djulis leaf. (**D**) Yellow djulis leaf.

2.2.4. Chlorophyll

Freeze-dried sample powder (0.1 g) was mixed with 1 mL of dimethyl sulfoxide (DMSO) and heated in a 65 °C water bath for 15 min. The mixture was centrifuged (model: Heraeus Megafuge 8R, ThermoFisher, Onlyscience Co., Ltd., Taichung City, Taiwan) at 2268× *g* for 10 min and the supernatant was collected. The precipitate was re-extracted by adding 1 mL of DMSO, and the supernatants collected after centrifugation were combined. Chlorophyll content (mg/g) was determined using a spectrophotometer at wavelengths of 645 nm and 663 nm using a spectrophotometer (model: SH-U830, Shishin Technologies Co., Ltd., Taipei City, Taiwan) and calculated using the following Formulas (1) and (2).

$$\text{Chlorophyll a} = [(12.7 \times A\ 663) - (2.69 \times A\ 645)] \times \text{mL DMSO/mg djulis leaves} \quad (1)$$

$$\text{Chlorophyll b} = [(22.9 \times A\ 645) - (4.68 \times A\ 663)] \times \text{mL DMSO/mg djulis leaves} \quad (2)$$

At wavelengths 645 nm and 663 nm, the absorbance is represented by A 645 and A 663, respectively.

2.2.5. Betacyanin and Betaxanthin

The analysis of betacyanin (BC) and betaxanthin (BX) in djulis leaf was conducted using high-performance liquid chromatography (HPLC) [6]. Djulis samples were first freeze-dried and finely ground into a uniform powder. The pigments were extracted by mixing 1 g of djulis powder with 20 mL of 80% methanol containing 0.1% formic acid, followed by ultrasonication for 30 min at 25 °C. The extract was centrifuged at 10,000 rpm for 15 min, and the supernatant was filtered through a 0.45 μm membrane filter before analysis.

The HPLC analysis was performed using a reverse-phase C18 column (4.6 × 250 mm, 5 μm particle size). The mobile phase consisted of solvent A (0.1% formic acid in water) and solvent B (0.1% formic acid in acetonitrile) with a gradient elution program: 0–5 min, 95% A; 5–15 min, 95–75% A; 15–20 min, 75% A. The flow rate was set at 1.0 mL/min, and the injection volume was 20 μL. Detection was carried out using a diode array detector (DAD) at 538 nm for betacyanin and 476 nm for betaxanthin. The identification and quantification of the pigments were achieved by comparing the retention times and absorption spectra with those of standard betacyanin and betaxanthin compounds. A standard curve for betacyanin was prepared using concentrations ranging from 1.0 to 6.0 μg/g, while the standard curve for betaxanthin used concentrations ranging from 1.0 to 8.0 μg/g.

2.3. *Antioxidant Capacity*

2.3.1. DPPH Free Radical Scavenging Assay

Freeze-dried sample powder (1 mg) was dissolved in methanol to prepare the sample solution. Then, 0.1 mL of the sample solution was mixed with 0.9 mL of DPPH (2,2-diphenyl-1-picrylhydrazyl) solution (0.15 mM in methanol) [17]. The reaction mixture was incubated at room temperature in the dark for 30 min. The absorbance was measured at 517 nm using a spectrophotometer.

2.3.2. Ferric-Reducing Antioxidant Power (FRAP)

FRAP solution was prepared by mixing 0.1 M acetate buffer (pH 3.6), 10 mM ferric-tripyridyltriazine (TPTZ, 0.01 M), and 20 mM ferric acid (0.02 M) at a ratio of 10:1:1 (*v/v/v*) [17]. Freeze-dried sample powder (1 mg) was dissolved in distilled water to prepare the sample solution. Then, 0.1 mL of the sample solution was mixed with 1.9 mL of FRAP solution (Figure S2A). After incubation at room temperature for 10 min, absorbance was measured at 593 nm using a spectrophotometer (SH-U830, Shishin Technologies Co., Ltd., Taipei City, Taiwan).

2.3.3. Trolox Equivalent Antioxidant Capacity (TEAC)

The 2,2′-azino-bis-(3-ethylbenzothiazoline-6-sulfonic acid) ($ABTS^{\bullet+}$) solution was prepared by dissolving 20 mg of ABTS in 2.6 mL of 0.0049 mol/L potassium persulfate ($K_2S_2O_8$) and 2.6 mL of distilled water [17]. Absorbance was measured at 734 nm. The ABTS solution was diluted with phosphate buffer (5 mmol/L, pH 7.4) to achieve an absorbance of 0.700 ± 0.020 (Figure S2B). The freeze-dried sample powder was dissolved in distilled water, and 10 μL of the sample solution was mixed with 900 μL ABTS solution. The absorbance was measured at 734 nm using a spectrophotometer (SH-U830, Shishin Technologies Co., Ltd., Taipei City, Taiwan).

2.3.4. N, N-Dimethyl-p-Phenylenediamine (DMPD)

A 100 mM DMPD solution was prepared by dissolving 209 mg of DMPD in 10 mL of distilled water [17]. The DMPD solution (1 mL) was mixed with 100 mL of 0.1 M acetate buffer (pH 5.3). The freeze-dried powder samples were dissolved in distilled water, and 200 μL of the sample solution was mixed with 2 mL of DMPD solution. After incubation at room temperature in the dark for 15 min, absorbance was measured at 505 nm using a spectrophotometer (SH-U830, Shishin Technologies Co., Ltd., Taipei City, Taiwan).

2.3.5. Cupric-Ion-Reducing Antioxidant Capacity Assay (CUPRAC)

A 10 mM copper (II) chloride solution, 7.5 mM neocuproine solution, and 1 M ammonium acetate buffer (pH 7.0) solution were prepared [17]. The freeze-dried powder samples were dissolved in distilled water. Then, 100 μL of the sample solution was mixed with 1 mL of copper (II) chloride solution, neocuproine solution, ammonium acetate buffer (pH 7.0) solution, and distilled water. After incubation at room temperature for 30 min, absorbance was measured at 450 nm using a spectrophotometer (SH-U830, Shishin Technologies Co., Ltd., Taipei City, Taiwan) (Figure S2C).

2.4. Dermatology Skin Test

Table 1 shows the compositions of the djulis leaf extract test formulations, which were prepared based on the previous method with modifications [21]. The carbopol gel was prepared using hot water. Next, djulis leaf extract, glycerin, and triethanolamine (TEA) were added and stirred vigorously. The gel sample without djulis leaf extract was designated as the control (blank, B) sample. Gel samples containing 0.0625% (*w/v*) of green, red, or yellow djulis leaf extract were labelled as G-1, R-1, and Y-1, respectively. Gel samples containing 0.125% (*w/v*) of green, red, or yellow djulis leaf extract were labelled as G-2, R-2, and Y-2, respectively. Gel samples containing 0.25% (*w/v*) of green, red, or yellow djulis leaf extract were labelled as G-3, R-3, and Y-3, respectively.

Table 1. Composition of djulis gel.

Djulis Leaves Extracts (g)	Water (mL)	Carbopol (g)	TEA (g)	Glycerin (g)	Anti-Microbial Agent (g)
0	100	0.6	0.2	5	0.3
0.0625	100	0.6	0.2	5	0.3
0.125	100	0.6	0.2	5	0.3
0.25	100	0.6	0.2	5	0.3

The color of the formulated djulis leaf gel samples was evaluated using a Color Meter ZE-2000 (Nippon Denshku Industries Co., Ltd., Tokyo, Japan). The parameters L* (lightness), a* (green-to-red spectrum, where negative values indicate a shift towards green and positive values indicate a shift towards red), and b* (blue-to-yellow spectrum, where negative values indicate a shift towards blue and positive values indicate a shift towards yellow) were recorded to assess variations in color quality. Before measurement, the device was calibrated with a standard black-and-white ceramic tile. All color assessments were conducted at room temperature, with each sample measured in triplicate.

The test samples were freshly prepared prior to testing. Approximately 35 students from the Department of Food and Nutrition, Providence University (Taichung City, Taiwan), were interviewed, and 30 were selected to participate in the single-application closed-patch epicutaneous test under semi-occlusion conditions. Before participating in the study, the participants were required to provide written informed consent. The participants completed the personal data form. Skin properties such as the moisture, oil, texture, complexion, and 3D levels were analyzed using a skin analyzer (deViso skin analyzer, Prismatique, FL,

USA) at the beginning of the test. Next, 1 mL of the test sample was dipped into the cotton pad and placed on the inner wrist. The control sample contained sterilized water on the wrist instead of djulis leaf extracts. After 20 min, the skin properties were analyzed using a skin analyzer (Multi Skin Test Center MC 1000, Courage-Khazaka Electronic GmbH, Cologne, Germany).

This study protocol followed Good Clinical Practices and the Declaration of Helsinki and agreed with the appropriate institutional review board (IRB) regulations. The experimental protocol was registered at the Taichung Jen Ai Hospital. This protocol was approved by the IRB of Taichung Jen Ai Hospital, Taichung, Taiwan (clinical trial approval certificate no. 110-59). The participants gave informed consent before beginning any study-related procedures or medications.

2.5. Statistical Analysis

All analyses of physicochemical parameters and antioxidant capacities were conducted in triplicate (n = 3), and the data are expressed as mean values ± standard deviations (X ± SD). Statistical analyses were carried out using Microsoft Office 365 for data calculations, with one-way ANOVA performed and Duncan's multiple range test applied to assess significant differences at $p < 0.05$ using SPSS software (version 23.0; SPSS Inc., Chicago, IL, USA). Additionally, correlation analysis was performed using XLSTAT software (version 2020; Addinsoft, New York, NY, USA).

3. Results and Discussion

3.1. Physicochemical Properties of Djulis Leaves

The amount of carbohydrates present in the djulis leaf extracts was determined using the phenol-sulfuric acid method. This method involves the hydrolysis of carbohydrates and subsequent reaction with concentrated sulfuric acid, resulting in a yellow coloration that is directly proportional to the carbohydrate content. The total carbohydrate content of the three djulis leaf extracts was assessed, as shown in Figure 2A. Among the djulis leaf extracts, the highest total carbohydrate content was observed in green djulis leaves (0.09 mg/g), followed by yellow (0.06 mg/g) and red djulis leaves (0.04 mg/g).

Phenolic compounds are prevalent functional substances in plants, recognized for their antioxidative properties and health benefits, including their role in reducing stress-induced ailments [22]. The total phenolic content was determined using a reduction assay, in which phenolic compounds in the sample were oxidized to form dark-colored compounds, with absorbance directly proportional to TPC. Figure 2B illustrates the TPC of djulis leaf extracts, with red djulis leaves exhibiting the highest content (13.88 mg GA/g), followed by yellow (10.73 mg GA/g), and green djulis leaves had the lowest (6.75 mg GA/g). Variations in phenolic content among djulis leaf extracts may arise from differences in pigment synthesis pathways or external environmental factors such as light exposure, temperature, moisture, and nutrients during growth [17].

Flavonoids, another subclass of polyphenolic compounds widely present in plants, exhibit numerous health benefits including anti-inflammatory and antioxidant properties [6]. The TFC of djulis leaf extracts was determined, and the results are presented in Figure 2C. Among the extracts, red djulis leaves showed the highest flavonoid content (4.46 mg QE/g), followed by yellow (4.52 mg QE/g), and green djulis leaves had the lowest content (1.75 mg QE/g). The correlation between flavonoid and phenolic content suggests similar biosynthetic pathways or regulatory mechanisms governing their accumulation [23].

Chlorophylls serve as indicators of plant coloration and are vital pigments in photosynthesis, aiding in the conversion of carbon dioxide to organic compounds and oxygen release [24]. The chlorophyll content (a and b) of the djulis leaf extracts was assessed and is presented in Figure 2D,E. Green djulis leaves exhibited the highest chlorophyll content (10.53 mg/g; 5.62 mg/g), characterized by deep-green coloration. Conversely, yellow and red djulis leaves displayed lower chlorophyll content, corresponding to their respective coloration. The chlorophyll content in the current study was higher than the previous study that investigated the effect of calcium on the antioxidant enzyme activity of djulis sprouts [25].

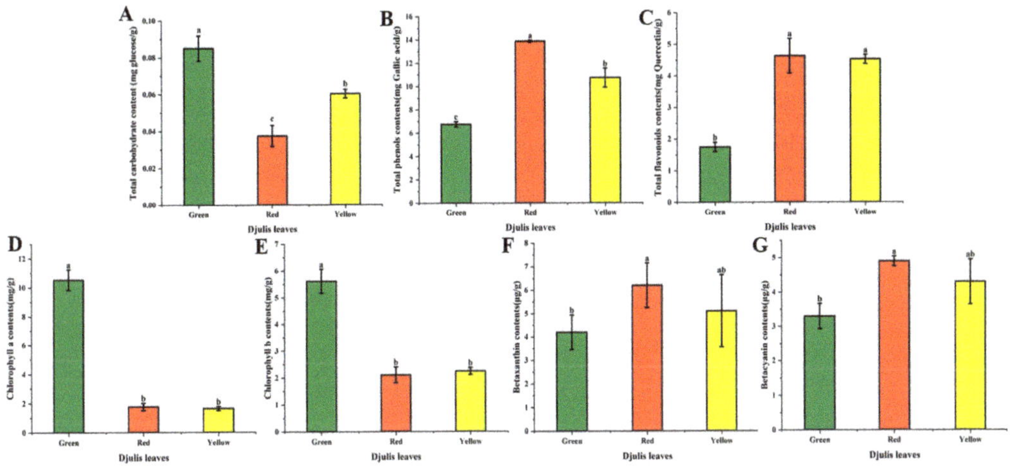

Figure 2. Bioactive compound content of djulis leaves at different senescence stages. (**A**) Total carbohydrate content (mg glucose/g DW), where 'mg/g DW' represents milligrams of carbohydrate per gram of dry weight. (**B**) Total phenolic content (mg gallic acid/g DW), quantified as milligrams of gallic acid equivalents per gram of dry weight. (**C**) Total flavonoid content (mg quercetin/g DW), expressed as milligrams of quercetin equivalents per gram of dry weight. (**D**) Chlorophyll a content (mg/g DW). (**E**) Chlorophyll b content (mg/g DW). (**F**) Betaxanthin content (µg/g DW). (**G**) Betacyanin content (µg/g DW). The results highlight significant differences in the bioactive profiles of green, red, and yellow djulis leaves. Error bars represent standard deviations (n = 3). Means with different letters indicate statistically significant differences at $p < 0.05$ based on Duncan's multiple range test.

Betacyanins and betaxanthins are natural pigments commonly found in plants, imparting purple–red and orange–yellow hues, respectively [26]. The betaxanthin (BX) and betacyanin (BC) content of djulis leaf extracts was determined and is presented in Figure 2F,G (Figure S3). The green djulis leaf extract exhibited the lowest content of both BX and BC (4.2 µg/g and 3.3 µg/g, respectively), while yellow and red djulis leaf extracts showed slightly higher levels.

The diverse phytochemical profiles observed in red, green, and yellow djulis leaf extracts offer valuable insights into the dynamic biochemical changes that occur during leaf senescence. The variation in total carbohydrate content among the extracts reflects the shifting metabolic priorities and energy storage strategies across different stages of leaf development [27]. Green djulis leaves, representing an early senescence stage, exhibited the highest total carbohydrate content, which is indicative of active photosynthesis and carbohydrate accumulation to support growth and development. In contrast, the lower total carbohydrate content observed in red djulis leaves, representing a later senescence stage, suggests a shift toward carbohydrate utilization or allocation to other metabolic pathways [28]. Additionally, seasonal and environmental factors, such as temperature, soil nutrient availability, and water stress, may influence the antioxidant profiles of djulis leaves. For instance, higher temperatures or water scarcity during specific growing seasons can induce oxidative stress in plants, leading to the increased production of secondary metabolites like phenolics and flavonoids as part of the plant's defense mechanism [29]. Similarly, variations in soil nutrient composition may impact the biosynthesis of pigments like betaxanthins and betacyanins, contributing to the differences observed in the phytochemical and antioxidant profiles of leaves harvested under different conditions. Future studies should consider these environmental variables to better understand the factors influencing the bioactive composition of djulis leaves.

Significant differences in TPC among the three djulis leaf extracts highlight their varying antioxidant capacities. Red djulis leaves, characterized by advanced senescence, displayed the highest TPC, potentially as a response to increased oxidative stress and environmental challenges. Conversely, green djulis leaves exhibited relatively low TPC, reflecting their earlier developmental stage and reduced demand for antioxidant defense. These findings highlight the dynamic modulation of phenolic metabolism during leaf senescence and its implications for plant physiology and stress response [30].

Furthermore, the variation in TFC among the extracts underscores the dynamic regulation of secondary metabolites during leaf development. The higher flavonoid content observed in red and yellow djulis leaves may serve as a protective mechanism against environmental stressors and pathogens, particularly during the later stages of senescence when plants are more vulnerable [31]. In contrast, green djulis leaves showed relatively low flavonoid content, reflecting their early developmental stage and potentially reduced need for flavonoid-mediated defense mechanisms.

Chlorophyll content analysis revealed distinct pigment composition patterns across the different senescence stages of djulis leaves. Green djulis leaves exhibited the highest chlorophyll content, consistent with their active photosynthetic machinery and primary role in light absorption and energy capture [32]. As leaves mature, their chlorophyll content decreases, causing changes in pigment composition and color [32]. The observed variations in chlorophyll a and chlorophyll b content among djulis leaf extracts highlight the dynamic nature of chlorophyll metabolism during leaf development and its impact on photosynthetic efficiency and pigment composition.

The determination of betaxanthin and betacyanin content provides insights into the pigment profile of djulis leaves at different senescence stages. The higher betaxanthin and betacyanin content observed in red djulis leaves may be attributed to enhanced pigment accumulation during senescence stages, contributing to their distinctive red coloration [6]. Conversely, green djulis leaves displayed relatively lower betaxanthin and betacyanin content, consistent with their early developmental stage and reduced pigment synthesis. These findings underscore the dynamic regulation of pigment biosynthesis pathways during leaf senescence and their role in determining leaf coloration and visual appeal.

3.2. *Antioxidant Capacity of Djulis Leaves*

DPPH (2, 2-diphenyl-1-picrylhydrazyl) is commonly used to assess the antioxidant capacity of free radicals. The antioxidant ability of the samples was evaluated by measuring their ability to scavenge DPPH radicals. Antioxidants in the sample donate electrons to DPPH radicals, reducing them. This reduction process changes the deep purple color of the DPPH radicals to a light yellow or colorless hue, accompanied by a decrease in absorbance [17]. In this experiment, the DPPH free radical scavenging ability of the three djulis extracts was determined, as shown in Figure 3A, with gallic acid used as a control. Green djulis leaves exhibited the highest DPPH radical scavenging ability (56.91%, *w/v*), followed by yellow djulis leaves (49.45%, *w/v*), while red djulis leaves displayed the lowest activity (18.58%, *w/v*).

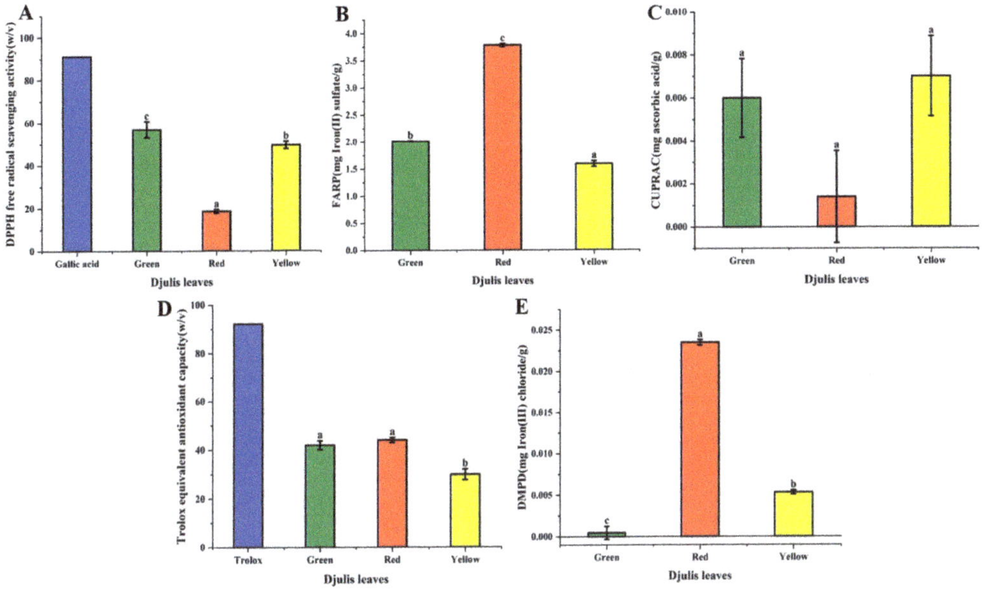

Figure 3. Antioxidant activities of djulis leaf extracts at different senescence stages measured by various in vitro assays: (**A**) DPPH radical scavenging activity (%, *w/v*), expressed as the percentage of free radical scavenging relative to the standard (gallic acid). (**B**) Ferric-reducing antioxidant power (FRAP), expressed in mg $FeSO_4$ equivalents per gram of dry weight (mg $FeSO_4$/g DW). (**C**) Cupric-reducing antioxidant power (CUPRAC), expressed in mg ascorbic acid equivalents per gram of dry weight (mg ascorbic acid/g DW). (**D**) Trolox equivalent antioxidant capacity (TEAC), expressed as a percentage of antioxidant capacity compared to the Trolox standard. (**E**) DMPD antioxidant capacity, expressed in mg ferric chloride equivalents per gram of dry weight (mg $FeCl_3$/g DW). These units provide a standardized measure of antioxidant activity, enabling comparisons across different assays and extracts. Error bars represent standard deviations (n = 3). Means with different letters indicate statistically significant differences at $p < 0.05$ based on Duncan's multiple range test.

The FRAP assay is a simple, rapid, and reproducible method for assessing the antioxidant capacity. It evaluates the ability of the sample to reduce ferric ions (Fe^{3+}) to ferrous ions (Fe^{2+}), reflecting its antioxidant potential [33]. The FRAP values of the three extracts were determined, as shown in Figure 3B. The red djulis leaves had the highest FRAP value (3.78 mg $FeSO_4$/g), followed by the green djulis leaves and yellow djulis leaves, with values of 2.01 mg $FeSO_4$/g and 1.59 mg $FeSO_4$/g, respectively.

The CUPRAC assay is based on the reduction in copper ions (Cu^{2+}) to copper ions (Cu^{+}), primarily used to assess the sample's reducing power. In this experiment, the copper ion reduction ability of three djulis extracts was determined, as shown in Figure 3C. The CUPRAC values of green, yellow, and red djulis leaf extracts were 0.006 mg ascorbic acid/g, 0.007 mg ascorbic acid/g, and 0.0014 mg ascorbic acid/g, respectively, while yellow djulis leaves exhibited the highest activity, followed by green djulis leaves, and the lowest was shown by red djulis leaves.

The Trolox equivalent antioxidant capacity (TEAC) is an analytical method used to measure the antioxidant capacity of the samples. It assesses the ability of the sample to scavenge a synthetic antioxidant (Trolox), and reflects its overall antioxidant potential. Using 2,2′-Azino-bis(3-ethylbenzothiazoline-6-sulfonic acid) diammonium salt (ABTS) reagent, the assay generates $ABTS^{\bullet+}$ radicals, and the ability of the sample to quench $ABTS^{\bullet+}$ radicals was measured [33]. In this experiment, the TEAC of three djulis leaf extracts was determined, and Figure 3D illustrates the results, with Trolox serving as a control sample (92.1%, *w/v*). Red djulis leaves had the highest antioxidant capacity (43.9%, *w/v*), followed by green djulis leaves (41.9%, *w/v*), while yellow djulis leaves exhibited the lowest antioxidant capacity (29.8%, *w/v*).

The DMPD assay was used to determine the antioxidant capacity of the samples based on the electron transfer reaction between the reagent and the antioxidants in the sample. This assay assesses the ability of the samples to transfer electrons, changing the color of the DMPD radical cation from blue to colorless or pale. In this experiment, the DMPD antioxidant capacity of the three djulis extracts was determined, as shown in Figure 3E; the red djulis leaves exhibited the highest DMPD value (0.024 mg $FeCl_3$/g), followed by the yellow djulis leaves (0.005 mg $FeCl_3$/g), while the green djulis leaves had the lowest value of only 0.0005 mg $FeCl_3$/g. Compared to other natural ingredients commonly used in cosmeceutical products, such as green tea, which is rich in catechins—particularly epigallocatechin gallate (EGCG), known for its strong radical-scavenging ability as demonstrated by DPPH and FRAP assays—red djulis leaf extract shows noteworthy phenolic content and antioxidant capacity [34]. This suggests that it may play a competitive or complementary role in formulations. Furthermore, the betaxanthin and betacyanin content of red djulis leaves provides additional antioxidative benefits, distinguishing it from these conventional ingredients. These unique properties position djulis leaves as a promising and sustainable ingredient for cosmeceutical development, with the potential to cater to growing demands for natural and eco-friendly products.

The observed variations in antioxidant activities among red, green, and yellow djulis leaf extracts provided valuable insights into the dynamic antioxidant capacity at different senescence stages. Antioxidant assays, including DPPH, FRAP, CUPRAC, TEAC, and DMPD, reflect the diverse phytochemical profiles and oxidative stress responses exhibited by djulis leaves during senescence. The DPPH assay, commonly used to assess free radical scavenging activity, revealed significant differences in antioxidant capacity among red, green, and yellow djulis leaf extracts. The highest DPPH scavenging activity was observed in the green djulis leaf extract, indicating its potent antioxidant properties. This may be attributed to the presence of high levels of chlorophyll compounds, which are known to contribute to antioxidant activity. Conversely, red and yellow djulis leaf extracts exhibited relatively lower DPPH scavenging activity, suggesting variations in antioxidant composition and efficacy across different senescence stages [35].

Similarly, the FRAP and CUPRAC assays, which evaluate the ferric-reducing antioxidant power, provided insights into the reducing capacity of red, green, and yellow djulis leaf extracts. The red djulis leaf extract displayed the highest FRAP values, whereas yellow djulis leaves exhibited the highest CUPRAC value, indicating superior ferric-reducing ability. In contrast, yellow and red djulis leaf extracts exhibited lower FRAP and CUPRAC values, suggesting differences in their capacity to donate electrons and reduce oxidized species. These findings underscore the influence of senescence stage on antioxidant activity and the potential implications for oxidative stress mitigation and cellular protection [35].

Moreover, the TEAC assay, which measures the ability of antioxidants to scavenge ABTS radicals, revealed distinct antioxidant capacities among red, green, and yellow djulis leaf extracts. While all extracts exhibited TEAC values indicative of antioxidant activity, the red djulis leaf extract demonstrated the highest TEAC value, followed by the green and yellow djulis leaf extracts. This variation in TEAC values may be attributed to differences in the phytochemical composition, particularly flavonoids and phenolic acids, which contribute to the scavenging of ABTS radicals and other reactive oxygen species [33].

Additionally, the DMPD assay evaluates the antioxidant capacity based on electron transfer reactions, providing further insights into the antioxidant properties of the three djulis leaf extracts. The red djulis leaf extract exhibited the highest DMPD value, indicating its superior electron transfer ability and potential for oxidative stress mitigation. Conversely, yellow and green djulis leaf extracts displayed lower DMPD values, suggesting differences in their capacity to donate electrons and neutralize free radicals. The comprehensive evaluation of antioxidant activities in djulis leaf extracts highlights the dynamic antioxidant capacity at different senescence stages. The observed variations in DPPH, FRAP, CUPRAC, TEAC, and DMPD values underscore the complex interplay between the phytochemical composition, senescence stage, and antioxidant efficacy.

3.3. Color and Appearance of Djulis Leaf Gel

The color changes in the gels with djulis extract are illustrated in Figure 4A–C. The gel matrix was transparent, but with the addition of djulis extract, the gel turned yellow. As the concentration of the djulis extract increased (from 0.0625% to 0.25%), color changes were observed. The L^* values of the gels with red, green, and yellow djulis leaves decreased, and higher concentrations of the djulis extract resulted in gels that were yellow and opaque. Variations in the djulis leaf extracts led to different gel colors. The addition of red djulis leaf extract resulted in gels with colors ranging from tan to dark brown as the concentration increased. The gel color changed from light yellow to golden yellow with increasing concentrations of green djulis leaves. Meanwhile, the addition of yellow djulis leaves results in a color change from yellow to deep yellow. A higher a^* value indicates a more yellow color and, in gels containing 0.25% djulis extract, the a^* values for red, green, and yellow djulis leaves (-0.11 ± 0.25, -0.55 ± 0.15, and -0.11 ± 0.09, respectively) showed higher values for red and yellow djulis leaf gels. In contrast, a higher b^* value indicates a more yellow color. In gels containing red, green, and yellow djulis extracts (-0.11 ± 0.25, -0.55 ± 0.15, and -0.11 ± 0.09, respectively), the b^* values are higher for red and yellow djulis extract gels, which is consistent with the leaf color results.

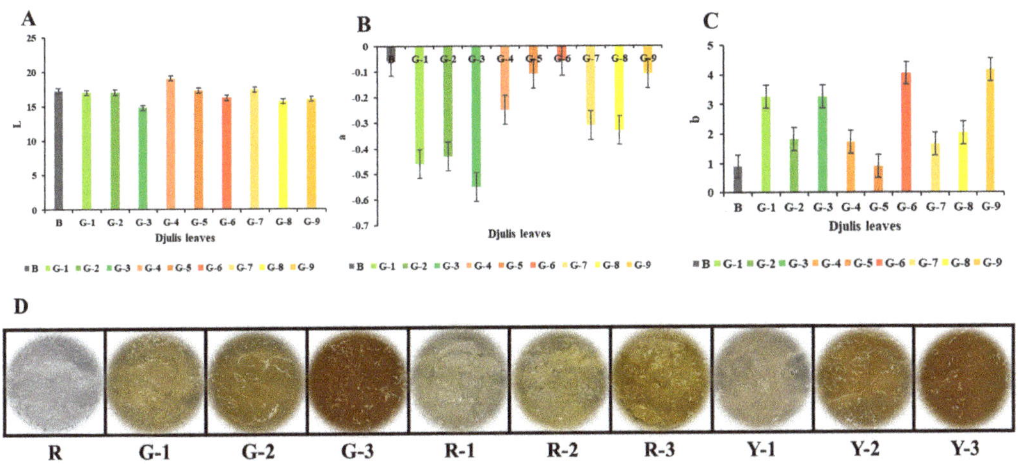

Figure 4. (**A**) L^*, (**B**) a^*, and (**C**) b^* values representing the color parameters of djulis leaf gels, and (**D**) visual appearance of the gels prepared from different djulis leaf extracts. Abbreviations: B refers to the blank control gel (no djulis extract added); G-1, G-2, and G-3 refer to gels prepared from green djulis leaf extracts at three different concentrations; R-1, R-2, and R-3 refer to gels prepared from red djulis leaf extracts at three different concentrations; Y-1, Y-2, and Y-3 refer to gels prepared from yellow djulis leaf extracts at three different concentrations. Error bars represent standard deviations (n = 3).

The observed variations in the color of gels supplemented with djulis leaf extract can be attributed to multiple factors. First, the inherent differences in pigment composition and chemical properties of the djulis leaf extracts contributed significantly to the color variations observed. The presence of different pigments in djulis leaves results in varying color responses when used in gel matrices [36]. Additionally, the antioxidant activity of these pigments may influence gel formation mechanisms and color stability during gelation processes [37]. Furthermore, the gel color changes observed, such as the transition from transparent to yellow hues with increasing concentrations of djulis extract, align with the expected changes in L^*, a^*, and b^* values indicative of color shifts towards higher yellowness. These findings underscore the intricate interplay between pigment composition, antioxidant activity, gel formation dynamics, and resultant color changes, providing valuable insights into the development of functional food products enriched with djulis extracts.

3.4. Dermatological Test of Djulis Leaf Gel

The experiment aimed to evaluate skin properties after applying gel formulations containing djulis leaf extracts for 20 min. The results depicted in Figure 5 illustrate the changes in the skin characteristics. Figure 5A (moisture level) and E (3D level) indicate an upward trend in moisture and 3D level, while Figure 5B (oil level), C (texture level), and D (complexion level) display a downward trend in oil secretion, skin texture, and complexion, respectively. Upon the application of djulis leaf extract gels for 20 min, the gel containing yellow djulis leaf extract decreased skin oiliness, texture irregularities, and discoloration, thereby improving skin wrinkles, inhibiting oil secretion, and reducing skin dullness. Red djulis leaf gel led to an increase in skin 3D level, while green djulis leaf gel enhanced skin hydration, augmenting its moisture content. After applying the djulis leaf extract gel formulations to the inner forearm, testers noticed significant improvements in skin hydration, elasticity, radiance, and firmness, and reduced wrinkle scores. These positive effects on skin quality and functionality suggest the potential application of beauty gel formulations containing djulis leaf extracts, highlighting their biocompatibility and biodegradability with the skin.

The observed improvements in skin properties following the application of gel formulations containing djulis leaf extracts underscore the potential of these formulations in skincare applications. The significant increase in moisture levels and 3D skin texture, coupled with the reduction in oil secretion, texture irregularities, and complexion dullness, highlights the multifaceted benefits of djulis leaf extract on skin health. The distinct effects observed with different colored djulis leaf extracts suggest the presence of unique bioactive compounds in each extract, which may contribute to specific skin-enhancing properties [19]. The ability of yellow djulis leaf extract to mitigate oiliness and discoloration indicates its potential to address concerns related to acne-prone and uneven skin tones [38]. Conversely, the enhancement of 3D skin texture by red djulis leaf extract suggests its role in promoting skin firmness and elasticity [21]. Moreover, the hydration-improving effects of green djulis leaf extract emphasize its suitability for moisturizing skincare formulations [21]. These findings support the notion that djulis-leaf-extract-based gel formulations are promising natural and effective ingredients in skincare products, offering diverse benefits for improving skin health and appearance. Further research is necessary to fully understand the potential of these formulations for skincare, including their long-term effects and underlying mechanisms.

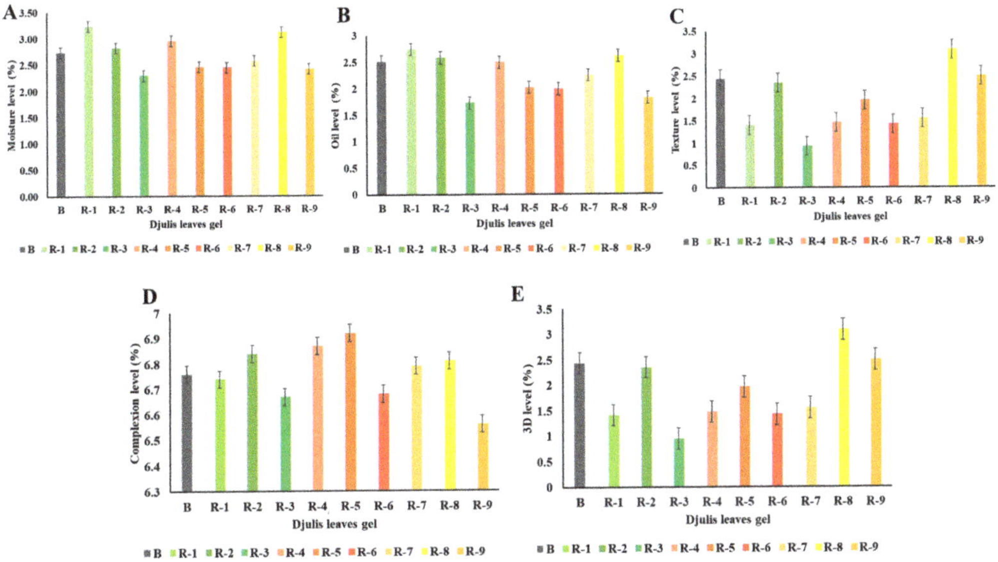

Figure 5. Dermatological analysis of djulis leaf gel formulations over 20 min, showing changes in (**A**) moisture level, (**B**) oil level, (**C**) texture level, (**D**) complexion level, and (**E**) 3D level. Abbreviations: B refers to the blank control gel (no djulis extract added); G-1, G-2, and G-3 represent gels prepared from green djulis leaf extracts at three different concentrations; R-1, R-2, and R-3 represent gels prepared from red djulis leaf extracts at three different concentrations; Y-1, Y-2, and Y-3 represent gels prepared from yellow djulis leaf extracts at three different concentrations. Error bars represent standard deviations (n = 3).

3.5. Correlation Analysis

Figures 6 and S4 show the correlation analysis between the physicochemical properties and antioxidant capacity of three different colors of djulis leaves. The total carbohydrate content exhibited intriguing correlations, showing negative associations with TPC, TF, betaxanthin, betacyanin, and DMPD antioxidant activity, and was positively correlated with chlorophyll content and DPPH free radical scavenging activities. Conversely, TPC was negatively correlated with total carbohydrate, chlorophyll content, and DPPH antioxidant activity, while it was positively correlated with betaxanthin, betacyanin, and DMPD antioxidant activity. These correlations align with previous studies, which demonstrate the distinct roles of chlorophyll and phenolic compounds in antioxidant activity. Chlorophyll content has been positively correlated with DPPH free radical scavenging activity in other plants, as chlorophyll and its derivatives effectively quench reactive oxygen species (ROS) and reduce oxidative stress [29]. For example, Ferruzzi and Blakeslee (2007) highlighted the radical scavenging capacity of chlorophyll-derived compounds, such as pheophytins, which contribute to antioxidant activity in chlorophyll-rich plant extracts [39]. On the other hand, phenolic compounds, including TPC, have been shown to correlate with betacyanin and betaxanthin levels, as they share biosynthetic pathways under stress conditions that induce secondary metabolite production [40]. A related study also observed these associations in leafy vegetables, where chlorophyll levels and phenolic compounds exhibit inverse correlations due to different metabolic priorities during senescence [41]. Together, these comparisons emphasize that the antioxidant capacity of djulis leaves is influenced by the interplay of carbohydrate, chlorophyll, and phenolic compound levels across different senescence stages.

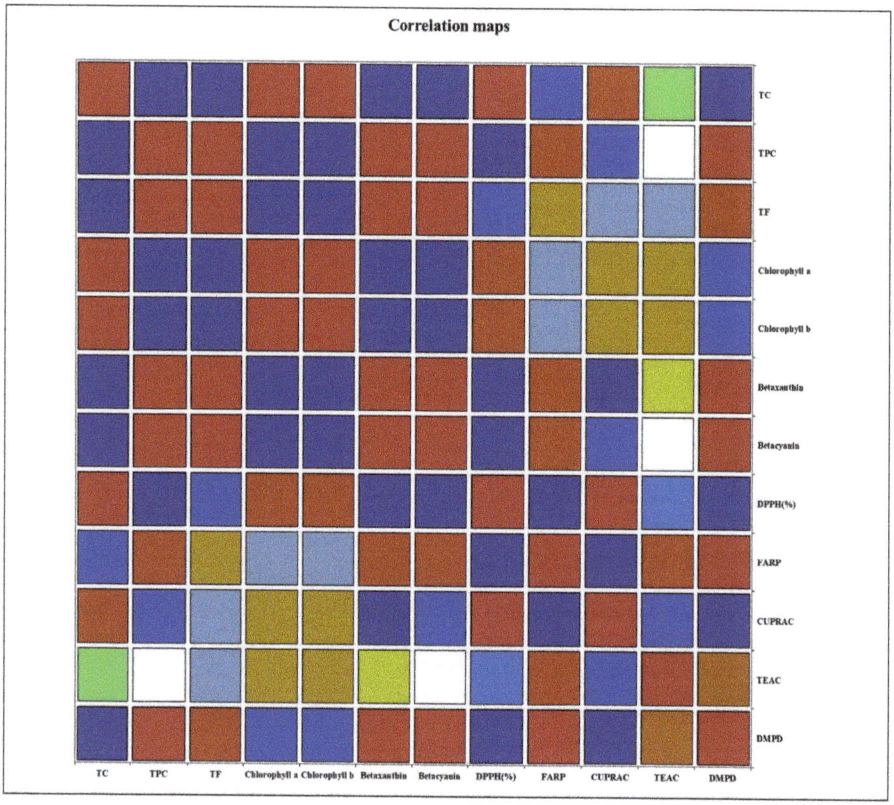

Figure 6. Correlation map showing the relationships between total carbohydrate (TC), total phenolic content (TPC), total flavonoid content (TF), chlorophyll a, chlorophyll b, betaxanthin, betacyanin, and antioxidant activities (DPPH, FRAP, CUPRAC, TEAC, and DMPD). Box colors represent the correlation coefficients: blue shades indicate negative correlations (darker blue = stronger negative correlation), red shades indicate positive correlations (darker red = stronger positive correlation), and white indicates no significant correlation.

The significantly elevated DPPH free radical scavenging activities observed in green djulis leaf extracts align with increased total carbohydrate levels, leading to elevated chlorophyll a and chlorophyll b content. Additionally, the FRAP value was negatively correlated with total carbohydrates, whereas it was moderately positively correlated with TPC, betaxanthin, and betacyanin. Similarly, the DMPD assay exhibited negative correlations with total carbohydrates, but moderate positive correlations with TPC, betaxanthin, and betacyanin. These findings suggest that changes in TPC, betaxanthin, and betacyanin levels moderately influence the FRAP and DMPD activities of red djulis leaf extract. In contrast, the CUPRAC assay demonstrated negative correlations with TPC, betaxanthin, and betacyanin, but moderately positive correlations with total carbohydrate and chlorophyll content. This underscores the impact of total carbohydrate and chlorophyll content on CUPRAC activity, particularly in the extract of green djulis leaves. These findings highlight the complex interplay between physicochemical composition and antioxidant capacity in djulis leaves, highlighting their potential as valuable sources of natural antioxidants with diverse health-promoting properties.

4. Conclusions

The color variation observed in djulis leaves is attributed to several factors, including differences in chlorophyll content, betacyanin and betaxanthin accumulation, and overall plant metabolism during different growth stages. Chlorophyll, the green pigment responsible for photosynthesis, undergoes degradation as the leaves mature, leading to changes in leaf coloration. Simultaneously, the accumulation of betacyanins and betaxanthins, which are water-soluble pigments responsible for red hues in plants, contributes to the vibrant red color observed in mature djulis leaves.

The findings from this study elucidate the intricate relationship between the physicochemical composition and antioxidant capacity of red, yellow, and green djulis leaf extracts. The negative correlations observed between the total carbohydrate content and TPC, betaxanthin, and betacyanin underscore the dynamic balance between carbohydrate metabolism and the accumulation of antioxidant compounds. The notably high DPPH free radical scavenging activity observed in green djulis leaf extracts suggests a potential association with increased total carbohydrate levels, leading to elevated chlorophyll content. Moreover, the negative correlations of FRAP and DMPD assays with total carbohydrates, along with their positive correlations with TPC, betaxanthin, and betacyanin, highlight the influence of these antioxidant compounds on the overall antioxidant capacity of red djulis leaf extract. Conversely, the positive correlations between CUPRAC activities and total carbohydrate and chlorophyll content emphasize the impact of these components on the antioxidant potential, particularly in green djulis leaf extracts. In short, these findings enhance our understanding of the intricate relationship between the physicochemical composition and antioxidant capacity of djulis leaves. This suggests their potential as valuable sources of natural antioxidants for cosmeceutical products with diverse health-promoting properties. Red djulis leaves, rich in phenolic compounds and flavonoids, exhibit strong antioxidant activity, making them ideal for anti-aging creams aimed at reducing oxidative stress and preventing skin damage. Additionally, the betacyanin and betaxanthin content in red djulis leaves, known for their light-absorbing properties, could be utilized in UV protection lotions to enhance sun protection. Green djulis leaves, characterized by their high carbohydrate content, offer moisturizing and skin barrier repair properties, making them suitable for incorporation into hydrating gels or creams. These applications demonstrate the multifunctional potential of djulis extracts, providing a natural and sustainable alternative for cosmeceutical formulations.

While the current study provides valuable insights into the bioactive compounds and antioxidant efficacy of djulis leaves, several limitations should be acknowledged. First, the variability in bioactive compound content due to environmental factors such as climate, soil conditions, and cultivation practices can affect the reproducibility of results. Future studies should explore standardized agricultural practices to minimize such variability. Second, the scalability of the extraction process poses challenges, including potential cost and efficiency concerns. Optimizing extraction methods for industrial-scale production is critical for commercial applications. Finally, while the antioxidant potential of djulis leaves was demonstrated through in vitro assays, additional testing is required to evaluate their safety and efficacy in real-world cosmeceutical formulations. This includes cytotoxicity testing, skin irritation studies, and product stability assessments under various conditions. Addressing these limitations in future research will ensure the successful translation of djulis leaves into sustainable cosmeceutical products.

Supplementary Materials: The following supporting information can be downloaded at: https://www.mdpi.com/article/10.3390/antiox14020202/s1, Figure S1: (A) Standard curve of total carbohydrate content of glucose, (B) Standard curve of total phenols content of Gallic acid, and (C) Standard curve of total flavonoids content of quercetin of djulis leaves; Figure S2: (A) Standard curve of ferric reducing antioxidant power (FRAP) of ferrous sulfate, (B) Standard curve of cupric reducing antioxidant power (CUPRAC) of ascorbic acid, (C) Standard curve of trolox equivalent antioxidant capacity (TEAC) of trolox, and (D) Standard curve of DMPD antioxidant capacity of iron(III) chloride of djulis leaves; Figure S3: (A) Red djulis leaves, (B) Green djulis leaves, and (C) Yellow djulis leaves of HPLC analysis; Figure S4: Scatter plots of djulis leaves.

Author Contributions: Conceptualization, Y.-J.C. and P.-H.L.; Methodology, C.-S.C., Y.-J.C., Y.-Z.W., W.-C.L. and P.-H.L.; Software, C.-S.C., Y.-J.C. and P.-H.L.; Validation, Y.-J.C., Y.-Z.W., W.-C.L. and P.-H.L.; Formal analysis, P.-Y.C. and P.-H.L.; Investigation, Y.-J.C., W.-C.L., P.-Y.C. and P.-H.L.; Resources, C.-S.C. and P.-H.L.; Data curation, Y.-Z.W., W.-C.L., P.-Y.C. and P.-H.L.; Writing—original draft, P.-H.L.; Writing—review & editing, P.-H.L.; Visualization, P.-H.L.; Supervision, P.-H.L.; Project administration, C.-S.C. and P.-Y.C.; Funding acquisition, C.-S.C. All authors have read and agreed to the published version of the manuscript.

Funding: This research was financially supported by Taichung Veterans General Hospital/Providence University Joint Research Program, (TCVGH-PU-1138102), Taiwan.

Institutional Review Board Statement: The study was conducted in accordance with the Declaration of Helsinki, and approved by the Taichung Jen Ai Hospital (protocol code 110-59 and date of approval was 21 October 2021) for studies involving humans.

Informed Consent Statement: Informed consent was obtained from all subjects involved in the study.

Data Availability Statement: Data is contained within the article or Supplementary Materials.

Conflicts of Interest: The authors declare that they have no known competing financial interests or personal relationships that could have influenced the work reported in this paper.

References

1. Cheng, Y.-T.; Lu, W.-C.; Chan, Y.-J.; Huang, P.-H.; Chiang, P.-Y.; Chen, R.-S.; Li, P.-H. Effect of extruded djulis (*Chenopodium formosanum*) snacks on the ameliorative potential against diabetic cardiomyopathy. *J. Funct. Foods* **2024**, *116*, 106154. [CrossRef]
2. Lu, W.-C.; Chan, Y.-J.; Tseng, F.-Y.; Chiang, P.-Y.; Li, P.-H. Production and physicochemical properties of starch isolated from djulis (*Chenopodium formosanum*). *Foods* **2019**, *8*, 551. [CrossRef] [PubMed]
3. Chio, E.H.; Yang, E.-C.; Huang, H.-T.; Hsu, E.-L.; Chen, C.-R.; Huang, C.-G.; Huang, R.-N. Toxicity and repellence of Taiwanese indigenous djulis, *Chenopodium formosaneum*, against *Aedes albopictus* (*Diptera: Culicidae*) and *Forcipomyia taiwana* (*Diptera: Ceratopogonidae*). *J. Pest Sci.* **2013**, *86*, 705–712. [CrossRef]
4. Ker, Y.-B.; Wu, H.-L.; Chen, K.-C.; Peng, R.Y. Nutrient composition of *Chenopodium formosanum* Koidz. bran: Fractionation and bioactivity of its soluble active polysaccharides. *PeerJ* **2022**, *10*, e13459. [CrossRef]
5. Chu, C.-C.; Chen, S.-Y.; Chyau, C.-C.; Wang, S.-C.; Chu, H.-L.; Duh, P.-D. Djulis (*Chenopodium formosanum*) and its bioactive compounds protect human lung epithelial A549 cells from oxidative injury induced by particulate matter via Nrf2 signaling pathway. *Molecules* **2021**, *27*, 253. [CrossRef]
6. Hsu, B.; Lin, S.; Inbaraj, B.S.; Chen, B. Simultaneous determination of phenolic acids and flavonoids in *Chenopodium formosanum* Koidz.(djulis) by HPLC-DAD-ESI–MS/MS. *J. Pharm. Biomed. Anal.* **2017**, *132*, 109–116. [CrossRef]
7. Chen, S.-Y.; Chu, C.-C.; Chyau, C.-C.; Yang, J.-W.; Duh, P.-D. Djulis (*Chenopodium formosanum*) and its bioactive compounds affect vasodilation, angiotensin converting enzyme activity, and hypertension. *Food Biosci.* **2019**, *32*, 100469. [CrossRef]
8. Li, P.-H.; Chan, Y.-J.; Hou, Y.-W.; Lu, W.-C.; Chen, W.-H.; Tseng, J.-Y.; Mulio, A.T. Functionality of djulis (*Chenopodium formosanum*) by-products and in vivo anti-diabetes effect in type 2 diabetes mellitus patients. *Biology* **2021**, *10*, 160. [CrossRef]
9. Ye, J.-W.; Ong, W.A.; Chao, Y.-Y. Analysis of antioxidant capacity of different colour strain of djulis (*Chenopodium formosanum Koidz.*). *Int. J. Agric. Innov. Technol. Glob.* **2021**, *2*, 157–172. [CrossRef]
10. Chao, Y.-Y.; Hsueh, I.E. Insights into Physiological Mechanisms of Salt Stress Tolerance in Djulis (*Chenopodium formosanum* Koidz.) Sprouts. *J. Plant Biol.* **2019**, *62*, 273. [CrossRef]
11. Baranski, R.; Goldman, I.; Nothnagel, T.; Budahn, H.; Scott, J.W. Chapter 22—Improving color sources by plant breeding and cultivation. In *Handbook on Natural Pigments in Food and Beverages (Second Edition)*; Schweiggert, R., Ed.; Woodhead Publishing: Cambridge, UK, 2024; pp. 507–553.
12. Xie, G.-R.; Chen, H.-J. Comprehensive Betalain Profiling of Djulis (*Chenopodium formosanum*) Cultivars Using HPLC-Q-Orbitrap High-Resolution Mass Spectrometry. *J. Agric. Food Chem.* **2021**, *69*, 15699–15715. [CrossRef] [PubMed]

13. Sepúlveda-Jiménez, G.; Rueda-Benítez, P.; Porta, H.; Rocha-Sosa, M. Betacyanin synthesis in red beet (*Beta vulgaris*) leaves induced by wounding and bacterial infiltration is preceded by an oxidative burst. *Physiol. Mol. Plant Pathol.* **2004**, *64*, 125–133. [CrossRef]
14. Miguel, M.G. Betalains in Some Species of the Amaranthaceae Family: A Review. *Antioxidants* **2018**, *7*, 53. [CrossRef] [PubMed]
15. Kuai, B.; Chen, J.; Hörtensteiner, S. The biochemistry and molecular biology of chlorophyll breakdown. *J. Exp. Bot.* **2017**, *69*, 751–767. [CrossRef]
16. Lin, Y.-M.; Chung, Y.-C.; Chen, P.-Y.; Chang, Y.-C.; Chen, W.-L. Fermentation of *Chenopodium formosanum* Leaf Extract with *Aspergillus oryzae* Significantly Enhanced Its Physiological Activities. *Appl. Sci.* **2023**, *13*, 2917. [CrossRef]
17. Li, P.-H.; Chan, Y.-J.; Lu, W.-C.; Huang, D.-W.; Chang, T.-C.; Chang, W.-H.; Nie, X.-B.; Jiang, C.-X.; Zhang, X.-L. Bioresource utilization of djulis (*Chenopodium formosanum*) biomass as natural antioxidants. *Sustainability* **2020**, *12*, 5926. [CrossRef]
18. del Carpio-Jiménez, C.; Molleda-Gutierrez, R.S.; Tapia-Delgado, P. Evaluation of Properties of *Chenopodium pallidicaule* (*Cañihua*) Oil for Possible Use in Cosmetic Formulations. *J. Oleo Sci.* **2023**, *72*, 501–509. [CrossRef]
19. Lin, Y.-K.; Chung, Y.-M.; Lin, Y.-H.; Lin, Y.-H.; Hu, W.-C.; Chiang, C.-F. Health functional properties of unhulled red djulis (*Chenopodium formosanum*) in anti-aging. *Int. J. Food Prop.* **2021**, *24*, 833–844. [CrossRef]
20. Lyu, J.-L.; Liu, Y.-J.; Wen, K.-C.; Chiu, C.-Y.; Lin, Y.-H.; Chiang, H.-M. Protective Effect of Djulis (*Chenopodium formosanum*) Extract against UV- and AGEs-Induced Skin Aging via Alleviating Oxidative Stress and Collagen Degradation. *Molecules* **2022**, *27*, 2332. [CrossRef]
21. Lu, W.-C.; Chiu, C.-S.; Chan, Y.-J.; Guo, T.-P.; Lin, C.-C.; Wang, P.-C.; Lin, P.-Y.; Mulio, A.T.; Li, P.-H. An In Vivo Study to Evaluate the Efficacy of Blue Shark (*Prionace glauca*) Cartilage Collagen as a Cosmetic. *Mar. Drugs* **2022**, *20*, 633. [CrossRef]
22. Dzah, C.S.; Kpodo, F.M.K.; Asante-Donyinah, D.; Boateng, N.A.S. The influence of Morinda citrifolia fruit maturity level, parts and storage length on total phenols, ascorbic acid, antioxidant activity and ethylene gas emission. *Food Chem. Adv.* **2024**, *4*, 100599. [CrossRef]
23. Hernanz, D.; Palomar, M.Á.; Moujanni, A.; Essamadi, A.; Heredia, F.J.; Terrab, A. Phenolic compounds and color of labeled resin spurge honey and their correlations with pollen content. *LWT* **2022**, *169*, 113987. [CrossRef]
24. Sato, T.; Shimoda, Y.; Matsuda, K.; Tanaka, A.; Ito, H. Mg-dechelation of chlorophyll a by Stay-Green activates chlorophyll b degradation through expressing Non-Yellow Coloring 1 in *Arabidopsis thaliana*. *J. Plant Physiol.* **2018**, *222*, 94–102. [CrossRef]
25. Chao, Y.-Y.; Wang, W.-J.; Liu, Y.-T. Effect of Calcium on the Growth of Djulis (*Chenopodium formosanum* Koidz.) Sprouts. *Agronomy* **2021**, *11*, 82. [CrossRef]
26. López, M.A.H.; Luna-Suárez, S.; Macuil, R.J.D.; Cárdenas, F.d.F.R. Simple and efficient protocol for amaranth betalains extraction and stability analysis by ATR-FTIR spectroscopy. *J. Cereal Sci.* **2023**, *113*, 103745. [CrossRef]
27. Lin, P.-H.; Chao, Y.-Y. Different drought-tolerant mechanisms in quinoa (*Chenopodium quinoa* Willd.) and djulis (*Chenopodium formosanum* Koidz.) based on physiological analysis. *Plants* **2021**, *10*, 2279. [CrossRef]
28. Zhang, M.; Shen, J.; Wu, Y.; Zhang, X.; Zhao, Z.; Wang, J.; Cheng, T.; Zhang, Q.; Pan, H. Comparative transcriptome analysis identified ChlH and POLGAMMA2 in regulating yellow-leaf coloration in Forsythia. *Front. Plant Sci.* **2022**, *13*, 1009575. [CrossRef]
29. Akula, R.; Ravishankar, G.A. Influence of abiotic stress signals on secondary metabolites in plants. *Plant Signal. Behav.* **2011**, *6*, 1720–1731. [CrossRef]
30. Singh, A.K.; Dhanapal, S.; Yadav, B.S. The dynamic responses of plant physiology and metabolism during environmental stress progression. *Mol. Biol. Rep.* **2020**, *47*, 1459–1470. [CrossRef]
31. Chowdhary, V.; Alooparampil, S.; Pandya, R.V.; Tank, J.G. Physiological function of phenolic compounds in plant defense system. In *Phenolic Compounds—Chemistry, Synthesis, Diversity, Non-Conventional Industrial, Pharmaceutical and Therapeutic Applications*; IntechOpen Limited: London, UK, 2021.
32. Ustin, S.L.; Jacquemoud, S. How the optical properties of leaves modify the absorption and scattering of energy and enhance leaf functionality. In *Remote Sensing of Plant Biodiversity*; Springer: Berlin/Heidelberg, Germany, 2020; pp. 349–384.
33. Rumpf, J.; Burger, R.; Schulze, M. Statistical evaluation of DPPH, ABTS, FRAP, and Folin-Ciocalteu assays to assess the antioxidant capacity of lignins. *Int. J. Biol. Macromol.* **2023**, *233*, 123470. [CrossRef]
34. Paiva, L.; Rego, C.; Lima, E.; Marcone, M.; Baptista, J. Comparative Analysis of the Polyphenols, Caffeine, and Antioxidant Activities of Green Tea, White Tea, and Flowers from Azorean *Camellia sinensis* Varieties Affected by Different Harvested and Processing Conditions. *Antioxidants* **2021**, *10*, 183. [CrossRef] [PubMed]
35. Maxiselly, Y.; Anusornwanit, P.; Rugkong, A.; Chiarawipa, R.; Chanjula, P. Morpho-physiological traits, phytochemical composition, and antioxidant activity of canephora coffee leaves at various stages. *Int. J. Plant Biol.* **2022**, *13*, 106–114. [CrossRef]
36. Anju, T.R.; Parvathy, S.; Valiya Veettil, M.; Rosemary, J.; Ansalna, T.H.; Shahzabanu, M.M.; Devika, S. Green synthesis of silver nanoparticles from Aloe vera leaf extract and its antimicrobial activity. *Mater. Today Proc.* **2021**, *43*, 3956–3960. [CrossRef]
37. Ghosh, S.; Sarkar, T.; Das, A.; Chakraborty, R. Natural colorants from plant pigments and their encapsulation: An emerging window for the food industry. *Lwt* **2022**, *153*, 112527. [CrossRef]

38. Majeed, M.; Majeed, S.; Jain, R.; Mundkur, L.; Rajalakshmi, H.; Lad, P.S.; Neupane, P. An open-label single-arm, monocentric study assessing the efficacy and safety of natural pterostilbene (*Pterocarpus marsupium*) for skin brightening and antiaging effects. *Clin. Cosmet. Investig. Dermatol.* **2020**, *13*, 105–116. [CrossRef]
39. Ferruzzi, M.G.; Blakeslee, J. Digestion, absorption, and cancer preventative activity of dietary chlorophyll derivatives. *Nutr. Res.* **2007**, *27*, 1–12. [CrossRef]
40. Tossi, V.E.; Martínez Tosar, L.; Pitta-Álvarez, S.I.; Causin, H.F. Casting light on the pathway to betalain biosynthesis: A review. *Environ. Exp. Bot.* **2021**, *186*, 104464. [CrossRef]
41. Stajčić, S.; Ćetković, G.; Tumbas Šaponjac, V.; Travičić, V.; Ilić, P.; Brunet, S.; Tomić, A. Bioactive Compounds and the Antioxidant Activity of Selected Vegetable Microgreens: A Correlation Study. *Processes* **2024**, *12*, 1743. [CrossRef]

antioxidants

MDPI

Article

Targeting Aging Skin with GABALAGEN®: A Synergistic Marine Nutricosmetic Ingredient Validated Through Human Randomized Trials

Jimin Hyun [1], Kyoung-Min Rheu [2], Bae-Jin Lee [2] and Bomi Ryu [1,*]

[1] Department of Food Science & Nutrition, Pukyong National University, Busan 48513, Republic of Korea
[2] Marine Bioprocess Co., Ltd., Busan 47281, Republic of Korea; hansola82@hanmail.net (B.-J.L.)
* Correspondence: bmryu@pknu.ac.kr; Tel.: +82-51-629-5858

Abstract: This study introduces GABALAGEN® (GBL), a marine-derived ingredient combining low-molecular-weight fish collagen and gamma-aminobutyric acid (GABA) produced via lactobacillus fermentation. GBL contains approximately 10% GABA, making up 39% of its free amino acid profile. A 12-week, randomized, double-blind, placebo-controlled trial with 100 adults (aged 35–60) assessed its effects on aging skin. Participants consumed 1500 mg/day of GBL in jelly form, with 94% completing the study. By Week 12, the GBL group showed a 20% increase in skin hydration and a 15% reduction in wrinkle depth. Improvements in skin density and elasticity were also observed, with no adverse effects reported. In vitro tests demonstrated strong antioxidant and anti-inflammatory effects, including enhanced superoxide dismutase activity and reduced pro-inflammatory cytokine expression in UVB-irradiated keratinocytes. GBL exemplifies sustainable innovation by upcycling fishery byproducts into high-value materials while addressing stability issues common to seafood-derived products. The fermentation process ensures safety and enhances GABA's antioxidant activity and bioavailability. This scalable method aligns with circular economic principles and global sustainability goals, extending GBL's potential to other functional materials which were proved their safety. GBL represents a breakthrough in nutricosmetics, combining efficacy, environmental sustainability, and industrial innovation.

Keywords: GABARAGEN®; GABA; collagen; fermentation; randomized trial; skin aging; nutricosmetics

1. Introduction

The overconsumption of resources and the mismanagement of waste have led to severe environmental pollution, posing one of the most critical threats to human health [1]. Among its many consequences, climate change and pollution have profoundly impacted skin health, the body's largest organ and primary barrier against external aggressors [2]. Rising ultraviolet (UV) radiation, air pollution, and extreme temperature fluctuations accelerate skin aging and damage, manifesting as wrinkles, pigmentation, and reduced elasticity [3]. These environmental stressors not only compromise skin integrity but also contribute to a growing global burden of skin-related diseases, creating significant psychological distress for many individuals [4]. Recognizing this, the World Health Organization has identified climate change as a major global health threat [5].

Amid these challenges, sustainable consumption has emerged as a global priority, driving interest in eco-friendly solutions that mitigate environmental harm while deliv-

 https://doi.org/10.3390/antiox14030245

ering high-value benefits [6]. Marine bioresources have gained particular attention for their potential to address these dual objectives. By upcycling fish byproducts into functional materials such as low-molecular-weight collagen, these resources align with circular economy principles while offering superior bioavailability compared to land-based collagen sources [7]. Furthermore, combining marine collagen with bioactive compounds like gamma-aminobutyric acid (GABA), produced through microbial fermentation, presents an innovative approach to enhancing skin health [8].

Skin health research is particularly critical in this context due to the skin's dual role as both a protective barrier and a mirror of internal health [9]. Chronic exposure to UV radiation and oxidative stress accelerates skin aging, which not only affects physical appearance but also impacts mental well-being [10,11]. In recent years, there has been a significant shift toward well-aging: a proactive approach to aging that emphasizes long-term health and vitality rather than merely addressing superficial signs of aging [12]. This shift has fueled the rapid growth of the nutricosmetics market, which bridges nutrition and cosmetics by offering ingestible products designed to enhance skin health from within. Projected to grow at an annual rate of over 8% by 2032, this market reflects increasing demand for scientifically validated, natural ingredients that align with sustainability and wellness trends [13].

In response to these trends, we tested the well-aging potential of GABALAGEN® (GBL)—an innovative ingredient that combines low-molecular-weight fish collagen (FC) derived from fish scales with GABA produced through probiotic fermentation—at the pre-clinical level [8,14–16]. Furthermore, unlike previous studies that focused on single components such as collagen or GABA alone, this research investigates the synergistic effects of these two components on key indicators of skin aging such as hydration loss, reduced elasticity, and wrinkles. Through randomized trials, this study aims to establish GBL as a novel nutricosmetic solution that not only addresses consumer needs for well-aging but also aligns with global efforts toward sustainability by utilizing marine byproducts [17]. This research contributes to advancing both environmental responsibility and human wellness while setting a new standard in the nutricosmetics field.

2. Materials and Methods

2.1. Production of GBL

GBL was produced by Marine Bioprocess Co., Ltd. (Busan, Republic of Korea), a company affiliated with the authors who participated in the current research. Before fermentation, a fish-scale-byproduct-derived collagen from GELTECH Co., Ltd. (Busan, Republic of Korea) was enzymatically hydrolyzed at 55 ± 2 °C for 12 h with Prozyme 2000P (BISION-Biochem Inc., Gyeonggi-do, Republic of Korea). The FC production required two consecutive fermentations by *Lactobacillus brevis* BJ20 (accession No. KCTC 11377BP) and *Lactobacillus brevis* BJ21 (accession No. KCTC 18911P). A seed medium composed of these raw materials—1% yeast extract (Choheung, Ansan, Republic of Korea), 0.5% hydrated glucose (Choheung, Ansan, Republic of Korea), 1% L-glutamic acid (Samin chemical Co., Ltd., Siheung, Republic of Korea), and 98.5% water—was sterilized for 15 min at 121 ± 5 °C before being inoculated with 0.002% BJ20 and 0.002% *L. brevis* BJ21. These microorganisms were then cultured for 24 h at 37 °C separately. For the first fermentation, 10% (v/v) of the *L. brevis* BJ20 cultured seed medium was fermented in a fermentation medium (yeast extract 2%, glucose 0.28%, enzymatically hydrolyzed fish-scale-byproduct-derived collagen 29% (GELTECH Co., Ltd.), L-glutamic acid 5.5% (Samin chemical Co., Ltd.), water 63.22%) at 37 °C for 24 h. Then, 10% (v/v) of BJ21 cultured seed medium was added and fermented at 37 °C for another 24 h. The fermentation medium was sterilized and spray-dried to prepare GBL powder product (Lot. MBP-GABALAGEN-20230516). The ingredients including GBL, formulated in test and control supplements for randomized trials, are presented in the Supplementary Results (Table S2).

2.2. Free Amino Acid Analysis in GBL

The GBL was hydrolyzed and analyzed with an amino acid analyzer (L-8900, Hitachi Inc., Tokyo, Japan). Specifically, 10 mg of the GBL was hydrolyzed in 10 mL of 6.0 N HCl within a sealed vacuum ampoule at 110 °C for 24 h to determine its amino acid composition. After hydrolysis, the HCl was removed using a rotary evaporator, and the GBL sample was adjusted to a final volume of 10 mL. Amino acids were then quantified using the L-8900 amino acid analyzer (Hitachi Inc.). For free amino acid (FAA) analysis, 3.0 g of the GBL was mixed with an equal volume of 16% trichloroacetate solution and homogenized using a vortex mixer for 2 min. The homogenized sample was centrifuged at 3000 rpm for 15 min. Supernatants from two extractions were pooled and filtered through Whatman No. 41 filter paper (Whatman Inc., Florham Park, NJ, USA). The filtrate was acidified to pH 2.2 with a 10 M HCl solution and diluted to a final volume of 50 mL using distilled water. The resulting sample was analyzed using the same amino acid analyzer.

2.3. Cell Study

Human HaCaT keratinocytes were acquired from American Type Culture Collection (ATCC, Manassas, VA, USA). The cells were cultured in growth media consisting of Dulbecco's Modified Eagle Medium (DMEM; Gibco, Thermo Fisher, Waltham, MA, USA), 10% fetal bovine serum (FBS, Gibco), and 1% penicillin–streptomycin (100 units/mL; Gibco) at a temperature of 37 °C with 5% CO_2. To evaluate the cytotoxicity of GBL, HaCaT cells were cultured in 96-well plates at a density of 1×10^5 cells/mL. After reaching full confluence, the cells were exposed to FCs at concentrations ranging from 1 to 100,000 μg/mL for 24 h. DMEM was replaced after treatment, and cells were gently washed with phosphate-buffered saline (PBS; Gibco). Subsequently, 10% of CCK-8 reagent (Dojindo Molecular Technologies Inc., Kumamoto, Japan) along with 0.1 mL of cultured medium was supplemented to each well. The plates were incubated at 37 °C for 1 h. The optical density of cell supernatants was measured at 450 nm using a microplate reader (Synergy HTX; Agilent Technologies Co., Ltd., Santa Clara, CA, USA). All of the data were investigated in triplicate to ensure consistent and reliable analysis.

2.3.1. Enzyme-Linked Immunosorbent Assay

To evaluate the protective effect of GL against UVB irradiation, we determined the production levels of two representative biomarkers related to oxidative stress damage—8-hydroxy-2′-deoxyguanosine (8-OHdG) and superoxide dismutase (SOD)—using enzyme-linked immunosorbent assay (ELISA). HaCaT keratinocytes were grown to over 70% confluency. DMEM was then replaced with PBS (Gibco). Following that, the cells were exposed to 30 s UV (50 mJ/cm^2) and treated with GBL (ranging from 50 to 350 μg/mL) for 48 h, and the cell cultured supernatants were collected for further analysis. Microplates were initially incubated overnight at 4 °C with 100 nM carbonate and bicarbonate mixed buffer (pH 9.6; Sigma-Aldrich, St. Louis, MO, USA) and then washed three times with PBS containing 0.1% Tween 20 (LPS solution, Daejeon, Republic of Korea) (TPBS) to eliminate unbound substances. To prevent nonspecific protein binding, the plates were blocked overnight at 4 °C with 5% skim milk (LPS solution) dissolved in 0.1% TPBS. After washing three more times with 0.1% TPBS, each well received 30 μg of cell supernatant sample and was incubated overnight at 4 °C. Following another wash with 0.1% TPBS, primary antibodies diluted in PBS (8-OHdG; sc-393871 and SOD; sc-515404, Santacruz Biotechnology Co., Ltd., Santa Cruz, CA, USA) were added to the wells and incubated overnight at 4 °C. After a final PBS wash, horseradish peroxidase-conjugated secondary antibodies (1:1000;

Vector Laboratories, Burlingame, CA, USA) were introduced, and the plates were kept in the dark at room temperature for 4 h. Protein expression was visualized by adding tetramethylbenzidine solution (Sigma-Aldrich Co., Ltd., St. Louis, MA, USA) to each well and incubating at room temperature in the dark for 30 min. The reaction was terminated by adding sulfuric acid (Sigma-Aldrich), and the absorbance was measured at 450 nm with a microplate reader (Synergy HTX; Agilent Technologies Co., Ltd., Santa Clara, CA, USA). Absorbance values were normalized against a control group, and each experiment was conducted in triplicate for reliability and precision.

2.3.2. Extraction of RNA and Quantitative Real-Time Polymerase Chain Reaction

RNA extraction and quantitative real-time polymerase chain reaction (qRT-PCR) were performed as described previously [18]. Homogenized muscle samples were treated with 500 μL of TRIzol (Takara Inc., Shiga, Japan), followed by the addition of chloroform (Sigma-Aldrich) to separate the clear supernatant (centrifuged at 12,000× g, 4 °C, for 15 min). RNA was precipitated by mixing the supernatant with 2-propanol (Sigma-Aldrich) and centrifuging at 12,000× g, 4 °C, for 10 min. The resulting RNA pellets were washed with 70% ethanol, air-dried at room temperature, and dissolved in diethyl pyrocarbonate (DEPC)-treated water (Sigma-Aldrich). The isolated RNA was reverse-transcribed into cDNA using M-MLV Reverse Transcriptase (Thermo Fisher). qRT-PCR was performed on a CFX 384 Touch™ Real-Time PCR Detection System (Bio-Rad, Irvine, CA, USA), and threshold cycle values were analyzed using CFX Manager™ software version 3.1 (Bio-Rad). Acbt served as internal control. The primer sequences for the target genes are provided in Table 1.

Table 1. List of primers for qRT-PCR.

Gene		Primers
Actb (Gene ID: 60)	*Forward* *Reverse*	*5′-CTGGAACGGTGAAGGTGACA-3′* *5′-AAGGGACTTCCTGTAACAATGCA-3′*
Tnf-a (Gene ID: 7124)	*Forward* *Reverse*	*5′-CCTCTCTCTAATCAGCCCTCTG-3′* *5′-GAGGACCTGGGAGTAGATGAG-3′*
Il-1a (Gene ID: 3552)	*Forward* *Reverse*	*5′-GGCATCACTGTTGCTTCGG-3′* *5′-GCCAGTCTAATTCTCCTGGTCA-3′*
Il-6 (Gene ID: 3569)	*Forward* *Reverse*	*5′-ACTCACCTCTTCAGAACGAATTG-3′* *5′-CCATCTTTGGAAGGTTCAGGTTG-3′*
IL10 (Gene ID: 3586)	*Forward* *Reverse*	*5′-GCCTAACATGCTTCGAGATC-3′* *5′-TGATGTCTGGGTCTTGGTTC-3′*
AP-1 (Gene ID: 3725)	*Forward* *Reverse*	*5′-ACTCATACACAGCTACGGGATACG-3′* *5′-GGGTCGGCCAGGTTGAC-3′*
MMP1 (Gene ID: 4312)	*Forward* *Reverse*	*5′-ACATTGCAGGATGTGCAGGCTCTT-3′* *5′-CTTGGGTACTGGTGACCGGTGTCA-3′*
MMP3 (Gene ID: 4314)	*Forward* *Reverse*	*5′-AGC TGA GTA CCG AGA GAT CGA C-3′* *5′-TCA GCC ACA TCA AGT ATT GGT C-3′*
MMP9 (Gene ID: 4318)	*Forward* *Reverse*	*5′-GAC ATG CCC AAG ACT CAG AAG T-3′* *5′-ACT TCC TTT CCT TCT CCT TTG C-3′*

2.4. *HPLC for the Indication Component Assessment in GBL*

2.4.1. Chemicals and Reagents

GABA and sodium acetate (50 mM, pH 6.5) were sourced from Sigma-Aldrich (USA). HPLC-grade acetonitrile, methanol, and distilled water (DW) were obtained from Samchun Pure Chemical Co., Ltd. (Pyeongtaek, Republic of Korea). Hydrochloric acid solution was purchased from Biosesang Inc. (Seongnam, Republic of Korea). Borate buffer (0.4 N in water, pH 10.2; Agilent P/N 5061-3339) and o-phthaldialdehyde reagent (10 mg/mL, Agilent P/N 5061-3335) were acquired from Agilent Technologies Inc. (Santa Clara, CA, USA). Acetic acid was supplied by Junsei Chemical Co., Ltd. (Tokyo, Japan).

2.4.2. Standard Solution and Sample Preparation

To prepare a standard solution, 0.1 g of the standard sample was dissolved in 100 mL of DW using a volumetric flask. The resulting solution was filtered through a polytetrafluoroethylene syringe filter (25 mm/0.2 μm, Whatman Inc., Kent, UK) and stored at −80 °C. For the preparation of a 5% aqueous sample solution, 5 g of GBL was dissolved in DW in a 100 mL volumetric flask and similarly filtered through a polytetrafluoroethylene syringe filter (Whatman Inc., Kent, UK).

2.4.3. HPLC Analysis Method

The analysis was conducted using a Dionex U3000 series HPLC system (Thermo Fisher Scientific) equipped with an ultraviolet (UV) detector operating at a flow rate of 1 mL/min. The samples were analyzed via UV–vis spectrophotometry at a wavelength of 338 nm. The GABA content in GBL was determined using the following formula:

$$\text{Substance}(\text{mg}/\text{g}) = \frac{\text{measurement}(\text{mg}/\text{mL}) \times \text{dilution factor}}{\text{amount}(\text{g})} \times 100(\text{mL})$$

2.5. *Randomized Trials*

2.5.1. Study Design

This randomized trial (Approval No. M202301; 4 May 2023~23 November 2023) was a double-blind, placebo-controlled, single-center study designed to evaluate the effects of GBL on skin hydration and protection against UV-induced skin damage. The study adhered to Good Clinical Practice (GCP) guidelines and was approved by the Institutional Review Board (IRB). Participants were randomly assigned to either the experimental group (receiving GBL) or the control group (receiving a placebo) in a 1:1 ratio using block randomization following the randomization method. Specifically, the study employs a block randomization method to assign selected participants to the test and control groups in a 1:1 ratio. To maintain the integrity of the randomization process, a statistician arbitrarily or randomly determines the block size, which remains undisclosed to prevent prediction of group assignments. The randomization table is generated for approximately 120% of the target recruitment number. Participants are assigned unique three-digit identification codes (randomized numbers) in order of recruitment, according to the randomization schedule. The assignment of randomization numbers to subjects was performed without the involvement of guardians. They received either the test or control product packaged corresponding to their assigned code. Once a randomization number is assigned to a participant, it cannot be reused, even if that participant withdraws from the study. The statistician generates the randomization table and provides it to the sponsor. The sponsor then seals each participant's randomization details in separate opaque envelopes and supplies them to the principal investigator for secure storage and management. This process ensures the double-blind nature of the study is maintained throughout its duration.

2.5.2. Participants for Randomized Trial

The study recruited 100 healthy adults (men and women) aged 35–60 years with visible wrinkles in the crow's feet area (Grade ≥ 3 based on expert visual assessment) and low skin hydration levels (Corneometer® value ≤ 49). Participants were excluded if they had uncontrolled hypertension, diabetes, liver or kidney dysfunction, chronic diseases, or were using medications or supplements that could impact skin health within the specified washout period. Pregnant or lactating individuals were also excluded. More detailed exclusion criteria are presented in the Supplementary Material (Table S1).

2.5.3. Intervention and Dermatologic Outcomes Measurement

Participants in the experimental group consumed GBL in stick jelly form (1500 mg/day) once daily for 12 weeks. The control group received an identical placebo jelly containing dextrin (1500 mg/day) (Table S2). Both products were indistinguishable in taste, appearance, and packaging to maintain blinding. The primary outcome was skin hydration, measured using Corneometer CM 825 (Courage + Khazaka Electronic GmbH Inc., Köln, Germany) at baseline, 6 weeks, and 12 weeks. The secondary outcomes were as follows: wrinkle severity, assessed via expert visual grading and 3D imaging using PRIMOSCR (Canfield Scientific Inc., Parsippany, NJ, USA); skin elasticity, measured with Cutometer MPA 580 (Courage + Khazaka Electronic GmbH Inc., Köln, Germany); transepidermal water loss (TEWL), evaluated using Vapometer (Delfin Technologies Co., Ltd., Kuopio, Finland); skin density, measured with DUB Skin Scanner (Courage + Khazaka Electronic GmbH Inc., Köln, Germany); skin roughness, quantified via PRIMOSCR imaging (Courage + Khazaka Electronic GmbH Inc., Köln, Germany); skin gloss, measured with Skin Gloss Meter (Delfin Technologies Co., Ltd., Kuopio, Finland); skin desquamation, analyzed using Visioscan VC 98 (Courage + Khazaka Electronic GmbH Inc., Köln, Germany).

Skin Wrinkles Assessment

Skin wrinkles were evaluated using a three-dimensional (3D) optical skin imaging device (Canfield Scientific Inc., Parsippany, NJ, USA). Parallel projection stripes were projected onto the skin in 3D; variations in stripe displacement, caused by differences in skin surface height, were quantified by the device's software. Images captured before and after each measurement were matched for an identical reference area, enabling comparative analysis of skin wrinkles. Subjects were instructed to cleanse their face and rest under constant temperature (22–24 °C) and humidity (45–55%) conditions for 30 min prior to measurement. Additionally, they were asked to refrain from consuming water for 1 h before the test. The crow's feet area was scanned, and the resulting images were analyzed to collect the following parameter values: Ra, Rmax, Rp, Rv, and Rz.

Measurement of Subsurface Skin Hydration

Subsurface skin hydration was assessed using a Moisturemeter D Compact (Delfin Technologies Co., Ltd., Kuopio, Finland). This instrument transmits a 265 MHz high-frequency electromagnetic field to a depth of approximately 2–2.5 mm beneath the skin surface and measures the reflected signal. The dielectric constant of the targeted tissue is converted into a percentage value of partial water content (PWC, %) for quantitative evaluation. Prior to the assessment, participants cleansed their faces, rested for 30 min under controlled temperature (22–24 °C) and humidity (45–55%) conditions, and refrained from fluid intake for at least 1 h. Three measurements were taken at the perpendicular intersection between the lateral canthus and the nasal bridge, and the mean of these three values was used for analysis.

Skin Elasticity Measurement

Skin elasticity was evaluated using a Cutometer MPA 580 (Courage + Khazaka Electronic GmbH Inc.). A negative pressure of 450 mbar was applied with an on/off time of 2.0 s, and measurements were taken three times. During each measurement, an infrared sensor determined the length of the skin drawn into a 2 mm diameter probe, allowing calculation of the elasticity parameters. Prior to measurement, subjects washed their faces and then rested for 30 min under controlled temperature (22–24 °C) and humidity (45–55%). In addition, water intake was restricted for 1 h before measurement. The measurement site was defined as the perpendicular intersection of the lateral canthus and the tip of the nose. Among the elasticity parameters (R0–R9) derived from the device, R2 (Ua/Uf), R5 (Ur/Ue), and R7 (Ur/Uf) values were used for analysis.

Measurement of TEWL

TEWL was measured using a Vapometer (Delfin Technologies Co., Ltd.). The device features a cylindrical measurement chamber equipped with sensitive humidity sensors, which monitor the increase in relative humidity (RH) within the chamber during the measurement process. The system automatically calculates the rate of water evaporation. Participants were instructed to wash their faces and then rest under controlled conditions of temperature (22–24 °C) and humidity (45–55%) for 30 min before the measurement. To minimize variability, participants were restricted from consuming water for at least 1 h prior to the assessment. TEWL was measured at the intersection of the lateral canthus and the tip of the nose. Each measurement was performed three times, and the average value was used for analysis.

Skin Density Measurement Method

Skin density was assessed using the DUB Skin Scanner (Courage + Khazaka Electronic GmbH Inc.), which operates with a 22 MHz ultrasound transducer. This device emits short electrical pulses to evaluate the density of the epidermis and dermis layers. Participants underwent the measurement under controlled environmental conditions of 22–24 °C temperature and 45–55% humidity. Prior to the assessment, participants washed their faces and acclimated to the controlled conditions for 30 min. To ensure accuracy, participants were instructed to refrain from water intake for at least 1 h before the measurement. The measurement focused on a point located 3 cm from the outer corner of the eye. Images captured during this process were analyzed to collect skin density values, providing a quantitative assessment of skin structure integrity.

Skin Roughness Measurement

Skin roughness was assessed using an optical 3D skin imaging device (Canfield Scientific Inc.). This device projects parallel projection stripes onto the skin surface, and the deformation of these stripes due to variations in skin height is quantitatively analyzed by a computer. Pre- and post-measurement images were matched to ensure consistent analysis of the same skin area. The measurement procedure was conducted under controlled temperature and humidity conditions (22–24 °C, 45–55%). Participants were instructed to refrain from water intake for at least 1 h prior to the measurement. After cleansing their faces, they acclimated to the controlled environment for 30 min before testing. Images of the cheek area were captured and analyzed, focusing on the Ra parameter value as the primary metric for skin roughness.

Skin Gloss Measurement

Skin gloss was measured using a Skin Gloss Meter (Delfin Technologies Co., Ltd.). This device is equipped with a 635 nm red semiconductor diode laser that projects a laser beam onto the skin surface and analyzes the light reflected at the same angle. Subjects washed their faces and then acclimated for 30 min in a temperature- and humidity-controlled environment (22–24 °C, 45–55% relative humidity) before measurements were taken. Water intake was restricted to 1 h prior to measurement. The measurement site was at the perpendicular intersection of the eye corner and the tip of the nose. Three measurements were taken at this site, and the average value was used for analysis.

Skin Desquamation Measurement

Skin desquamation was assessed using the Visioscan VC 98 (Courage + Khazaka Electronic GmbH Inc.). The measurement was performed by collecting skin scales using a special film (D-squame disk). The collected sample was then imaged and analyzed with the Visioscan VC 98. The Desquamation Index (D.I., %) obtained from this analysis was used as the evaluation parameter for skin desquamation.

2.5.4. Data Recruitment Procedures for Study

Participants attended four visits: Screening Visit (1st visit, −21 to 0 days): informed consent, demographic data collection, physical examination, and baseline assessments; Baseline Visit (2nd visit, Start, Week 0): randomization, initial product distribution, and baseline measurements; Interim Visit (3rd visit, Week 6 ± 7 days): compliance check, adverse event monitoring, and interim assessments; Final Visit (4th visit, Terminal, Week 12 ± 7 days): final measurements of all outcomes and safety evaluations. The significance of skin health improvement following GBL intake was presented by comparing the start (2st visit) and terminal (4th visit) points.

2.5.5. Interruption of Participants

Participants could be excluded from the study without their consent under the following circumstances: A. violation of inclusion/exclusion criteria; B. participant's request to discontinue the trial; C. occurrence of adverse events (including serious adverse events); D. participant non-compliance (below 80%) or inability to observe; E. violation of the human trial protocol; F. consumption or need to consume medications or health functional foods that may affect the human trial; G. failure to take the test food for more than 5 consecutive days; H. other reasons based on the researcher's judgment. Once a participant's involvement in the study was terminated prematurely, they were supposed to be tested to ensure their safety. This precautionary measure was designed to protect the participant's well-being after withdrawal from the study.

2.5.6. Safety Assessments

Adverse events were monitored throughout the study via participant interviews and laboratory tests conducted at baseline and Week 12. Blood samples were analyzed for glucose, liver enzymes (SGOT/SGPT), creatinine, cholesterol levels, and other safety markers (Table S3).

2.5.7. Statistical Analysis

The statistical analysis plan adopts the results of the Per Protocol (PP) analysis as the primary method for evaluating efficacy, while also conducting an analysis of the Full Analysis Set (FAS) group for reference. For the definition of analysis sets, subjects included in the efficacy evaluation for PP analysis must meet specific criteria. These include being part of the FAS group and completing the randomized trial up to Visit 4 without any significant protocol violations. Additionally, subjects included in the FAS analysis must have consumed test food at least once and have data available for primary efficacy variables after consumption. All statistical significance tests were conducted at a 5% significance level. The primary analysis of data obtained from study participants was performed on the PP set, with additional analyses conducted on the FAS set. For demographic data analysis, demographic variables measured at Visit 1 were summarized for each group and compared between groups. Continuous variables were tested using *t*-tests or Wilcoxon's rank-sum tests, while categorical variables were analyzed using Chi-square tests or Fisher's exact tests.

Efficacy analyses evaluated baseline characteristics and functional assessment data using both FAS and PP. The FAS includes participants who underwent at least one functional assessment after consuming the test or control food, while PP includes participants who adhered to the randomized trial protocol to a certain extent and completed the study. Safety evaluations will target the Safety Set (SS), which includes all participants who consumed the test or control food at least once after randomization. The normality of data was assessed using the Shapiro–Wilk test. For continuous variables, intergroup comparisons were analyzed using methods such as ANOVA, ANCOVA, or Kruskal–Wallis tests, with post hoc tests conducted if significant differences were found. Intragroup comparisons before and after intervention used paired *t*-tests or Wilcoxon's signed rank tests. Outliers identified before unblinding might be considered in analyses. Adjustments for confounding factors, stratified analyses, and multiple comparison analyses were also performed as necessary. Categorical variables were compared between groups using Chi-square tests or Fisher's exact tests. 'p-value < 0.05' was considered statistically significant. If there were differences in alcohol consumption history that could affect primary efficacy variables, covariate analyses controlling for this variable were additionally conducted.

3. Results

3.1. Production Process and Analytical Characterization of GBL

Environmental pollution and climate change pose significant threats to human health, particularly affecting skin integrity. These stressors accelerate skin aging and contribute to various skin-related diseases. In response, sustainable consumption has become a global priority, with marine bioresources gaining attention for their eco-friendly potential. In our research, GBL, combining marine collagen derived from fish scale byproducts with bioactive compounds such as GABA produced through microbial fermentation, presents an innovative approach to enhancing skin health while addressing environmental concerns. Primarily, FC was enzymatically hydrolyzed by Prozyme 2000P peptidase (1% enzyme-add, 10% substrate, pH 7.0, 50 °C, and 12 h reaction) (Figure 1A). The low-molecularized FC, with added raw materials (1% L-glutamic acid, 1% yeast extract, 0.5% hydrated glucose, and water), was sequentially fermented twice by two types of lactobacillus to produce GBL extract. Finally, the product labeled as 'MBP-GABALAGEN-20230516', was powdered and used in pre-clinical and clinical randomized studies in the form of a stick jelly formula to investigate its biological potential for skin health (Figure 1A).

To recognize a key indicating component in GBL, HPLC analysis was applied (Figure 1B). These stacked comparison chromatograms showed that a distinct level of GABA substance is present in GBL compared to the reference material and blanks (Figure 1B). Therefore, as GBL was clearly shown to contain GABA, a concentration-dependent reference material standard calibration curve was prepared under the same analytical conditions to accurately quantify the GABA content ($y = 21.2720x + 8.0540$) (Figure 1C). Based on the standard calibration curve, we evaluated the GABA content produced during the sequential lactic acid bacteria fermentation process of low-molecular-weight fish collagen by assessing the amount of GABA generated over different fermentation time periods (Figure 1D). As a result, the highest GABA production rate (10.08%) was observed at 48 h fermentation (24 h and 24 h) after primary inoculation during the two-stage sequential lactobacillus culture process (fermentation 1st and 2nd) (Figure 1A,D). In conclusion, GBL produced under optimal microbial fermentation conditions for 48 h from low-molecular-weight-fish-scale-derived FC was verified to contain approximately 9.955 ± 0.08–9.958 ± 0.10% of GABA, regardless of the sample amount used for HPLC assessment (Figure 1E).

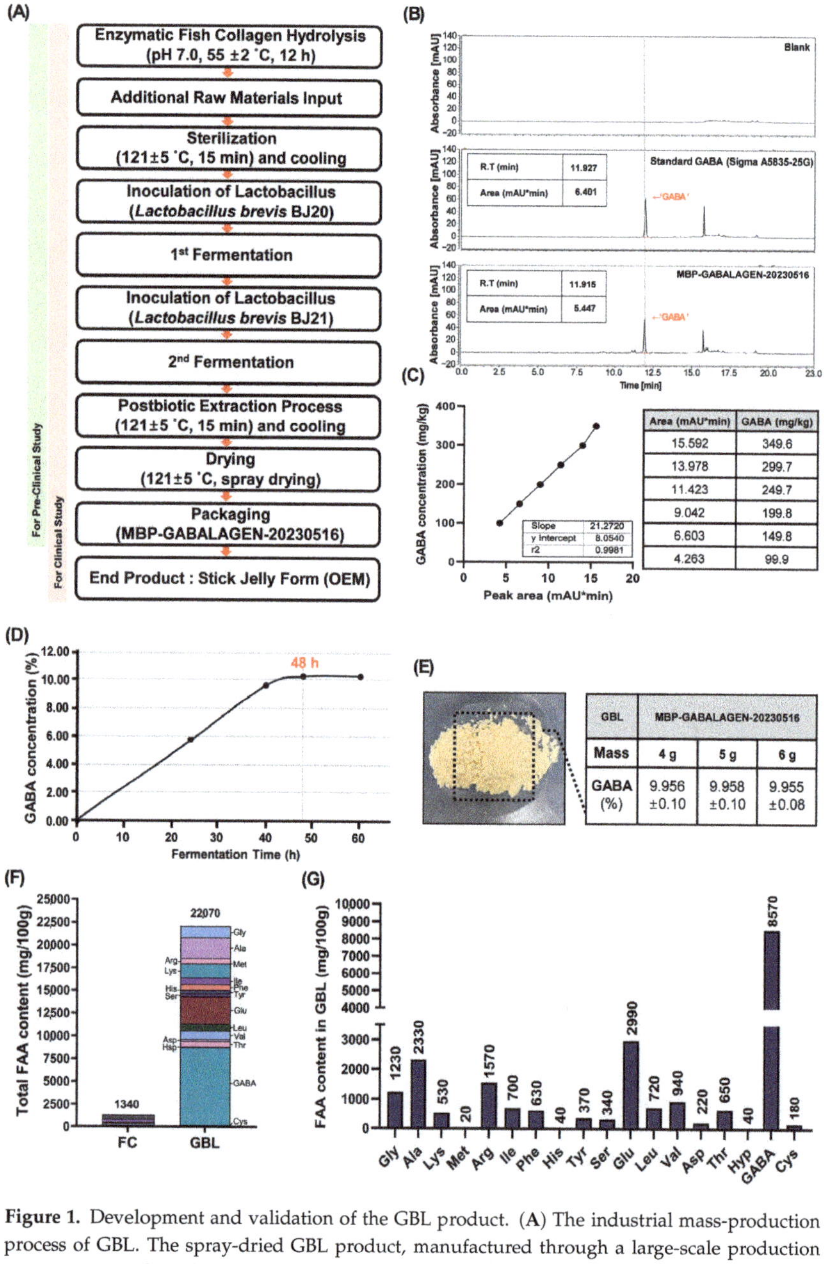

Figure 1. Development and validation of the GBL product. (**A**) The industrial mass-production process of GBL. The spray-dried GBL product, manufactured through a large-scale production process, was evaluated for its anti-skin-aging effects at the in vitro level. At the clinical level, GBL was provided in the form of stick jelly to examine its protective role against photo-induced skin aging. (**B**) The chromatograms of the marker component in GBL. GABA: Gamma-aminobutyric acid; R.T: Retention time. (**C**) Calibration curve of GABA standard material. (**D**) Time-dependent GABA production during fermentation process of GBL production. (**E**) The verified GABA content in GBL product. (**F**) Comparison of FFA content in FC and GBL. (**G**) The FFA profile in GBL. Gly: Glycine; Ala: Alanine; Lys: Lysine; Met: Methione; Arg: Arginine; Ile: Iso-leucine; Phe: Phenylalanine; His: Histidine; Tyr: Tyrosine; Ser: Serine; Glu: Glutamine; Leu: Leucine; Val: Valine; Asp: Aspartic acid; Thr: Threonine; Hyp: Hydroxyproline; Cys: Cystine.

Next, to elucidate the nutritional value of GBL, the content of free amino acids (FAAs) in the material was evaluated. GBL exhibited an FAA content of 22,070 mg/100 g, which was approximately 16.47-fold higher than that of FC, the raw material used in GBL production (Figure 1F). Among the 18 identified FAAs in GBL, GABA accounted for the highest proportion at 38.83% of the total FAA content, followed by glutamine (Glu) at 13.55%, which was the second highest in GBL (Figure 1G). In summary, this research successfully transformed marine collagen derived from fishery byproducts into a high-value biomaterial containing high concentrations of GABA through microbial fermentation processes. Our study emphasized the nutritional value of GBL, which demonstrates superior bioavailability compared to conventional collagen materials, based on its rich FAA content.

3.2. GBL Supplementation Safely Mitigates Aging Dermal Health in Human Randomized Trial

This randomized, double-blind, placebo-controlled trial aimed to evaluate the efficacy and safety of GBL supplementation in improving age-related skin damage in adults aged 35 to 60 years old. Those who did not meet the inclusion criteria (n = 15) were excluded from the assessed eligibility (n = 115) (Figure 2). A total of 100 participants with significant skin aging concerns, such as crow's feet wrinkles graded at level 3 or higher and skin moisture levels below 49%, were finally enrolled (Figure 2). A total of 6 participants dropped out during the study, leaving 94 participants who completed the trial, resulting in a compliance rate of 94% (Figures 2 and 3A). The FAS included all 100 participants (50 per group), while the PP analysis included the 94 participants who completed the trial (47 per group), to assess GBL's impact on skin health (Figure 3A).

Baseline characteristics, including age, height, weight, cardiovascular conditions, skin condition, gender, smoking status, and drinking status, showed no significant differences between the test and control groups in either the FAS or PP populations (Table 2). Nutritional intake parameters such as total calorie consumption, protein intake, and water intake were monitored throughout the trial from Week 0 ('Start') to Week 12 ('Terminal') (Table 3). No notable changes were observed in these parameters between the two groups during the study period (Table 3). Safety assessments revealed that while γ-GTP levels in the test group were slightly lower at baseline compared to the control group, no abnormal findings were detected among individual subjects (Table S3). All safety-related factors remained within normal ranges throughout the trial.

Table 2. Characteristics of the participants in the current randomized trial.

Variable		FAS Population Control (n = 50)	FAS Population Test (n = 50)	p-Value **	PP Population Control (n = 47)	PP Population Test (n = 47)	p-Value **
Age (years)		47.76 ± 6.10	47.82 ± 6.18	0.961 [1)]	47.83 ± 6.11	48.00 ± 6.26	0.894 [1)]
Height (cm)		164.01 ± 8.47	163.35 ± 8.31	0.692 [1)]	163.94 ± 8.67	163.16 ± 8.22	0.653 [1)]
Weight (kg)		63.11 ± 10.26	59.59 ± 10.56	0.094 [1)]	62.89 ± 10.45	59.19 ± 9.82	0.080 [1)]
Diastolic BP (mmHg)		71.74 ± 14.67	69.04 ± 11.06	0.301 [1)]	71.81 ± 14.83	69.17 ± 11.07	0.331 [1)]
Systolic BP (mmHg)		119.52 ± 18.62	118.14 ± 14.39	0.679 [1)]	119.45 ± 18.84	118.26 ± 14.52	0.732 [1)]
Pulse rate		76.32 ± 10.20	77.00 ± 8.63	0.720 [1)]	76.15 ± 10.49	77.23 ± 8.62	0.585 [1)]
Skin Hydration		43.55 ± 6.35	44.50 ± 3.79	0.367 [1)]	43.25 ± 6.43	44.39 ± 3.88	0.302 [1)]
Sex	Male	12 (24.0%)	13 (26.0%)	0.817 [2)]	11 (23.4%)	12 (25.5%)	0.810 [2)]
	Female	38 (76.0%)	37 (74.0%)		36 (76.6%)	35 (74.5%)	
Smoker	None	46 (92.0%)	46 (92.0%)	1.000 [3)]	44 (93.6%)	44 (93.6%)	1.000 [3)]
	Ex	1 (2.0%)	1 (2.0%)		1 (2.1%)	1 (2.1%)	
	Current	3 (6.0%)	3 (6.0%)		2 (4.3%)	2 (4.3%)	
Drinker	None	17 (34.0%)	24 (48.0%)	0.155 [2)]	17 (36.2%)	23 (48.9%)	0.211 [2)]
	Moderate	33 (66.0%)	26 (52.0%)		30 (63.8%)	24 (51.1%)	
	Heavy	0 (0.0%)	0 (0.0%)		50(0.0%)	0 (0.0%)	

** p-values were compared between groups; [1)] independent t-test; [2)] Chi-square test; [3)] Fisher's exact test.

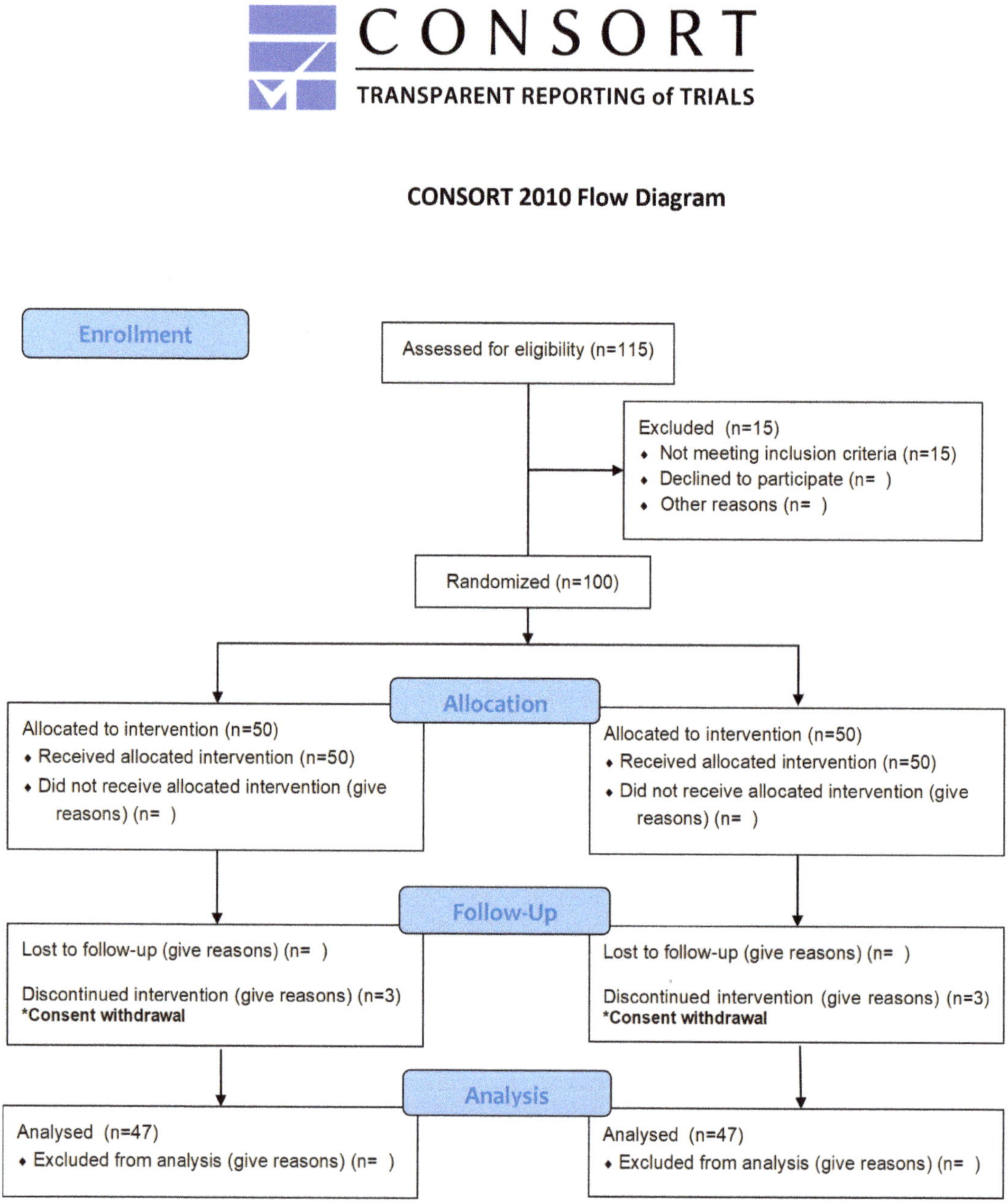

Figure 2. The regular Consolidated standards of reporting trials (CONSORT) 2010 flow diagram for patient selection.

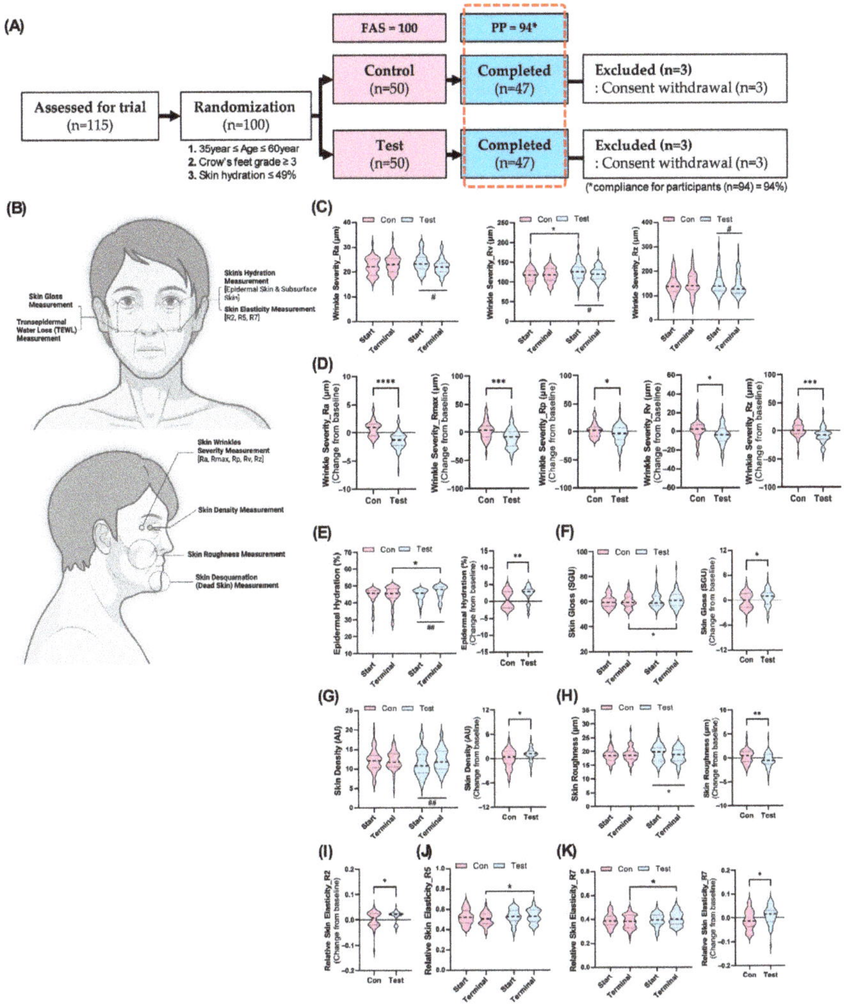

Figure 3. The regular GBL treatment protects skin aging in human randomized trials. (**A**) The randomized study scheme designed for control and test groups, and the number of participants excluded in each group for reasons indicated. (**B**) Assessment diagram of facial areas for skin health evaluations by analytical techniques. (**C**) The comparison measurement of average wrinkle severity (Ra, Rv, and Rz) between the groups before and after sample intake. (**D**) The changes from baseline of wrinkle severity in GBL-supplemented test and control groups. (**E**) The comparison measurement of average epidermal hydration rates between the groups before and after sample intake, as well as the changes in epidermal hydration rates from baseline for each group. (**F**) The comparison measurement of average skin gloss rates between the groups before and after sample intake, as well as the changes in skin gloss rates from baseline for each group. (**G**) The comparison measurement of average skin density rates between the groups before and after sample intake, as well as the changes in skin density rates from baseline for each group. (**H**) The comparison measurement of average skin roughness rates between the groups before and after sample intake, as well as the changes in skin roughness rates from baseline for each group. (**I**) The changes from baseline of relative skin elasticity (R2) in GBL-supplemented test and control groups. (**J**) The comparison measurement of relative skin elasticity (R5) between the groups before and after sample intake. (**K**) The comparison measurement of average relative skin elasticity (R7) rates between the groups before and after sample intake, as well as the changes in relative skin elasticity (R7) rates from baseline for each group. Data are shown to mean ± S.D. (n = 47). $^{*,\#}$ $p < 0.05$, $^{**,\#\#}$ $p < 0.01$, *** $p < 0.001$, and **** $p < 0.0001$.

Table 3. The caloric and nutritional consumption by the study groups.

Variable	Observed Value (PP)			Change from Baseline				
	Control (n = 47)	Test (n = 47)	p-Value **	Control	p-Value *	Test	p-Value *	p-Value **
Total calorie consumption (Kcal)								
Start	1331 ± 395	1252 ± 342	0.306 [1)]					
Terminal	1323 ± 337	1303 ± 333	0.771 [1)]	−8. ± 369	0.884 [2)]	60 ± 303	0.259 [2)]	0.404 [1)]
Protein intake (g)								
Start	53.56 ± 15.62	50.20 ± 12.38	0.250 [1)]					
Terminal	53.12 ± 14.04	51.70 ± 12.27	0.603 [1)]	−0.44 ± 18.84	0.873 [2)]	1.50 ± 15.70	0.515 [2)]	0.588 [1)]
Water intake (mL)								
Start	612.0 ± 174.3	562.4 ± 158.6	0.152 [1)]					
Terminal	590.0 ± 161.8	598.3 ± 145.3	0.794 [1)]	−22.0 ± 191.2	0.434 [2)]	35.9 ± 218.2	0.265 [2)]	0.174 [1)]

* p-values were compared within each group; ** p-values were compared between groups; [1)] Independent t-test; [2)] paired t-test.

Efficacy evaluations focused on various skin health parameters, including hydration, wrinkle severity, elasticity, density, roughness, glossiness, and desquamation. These indicators were measured across specific facial areas using advanced imaging and analytical techniques (Figure 3B). These dermal health-related indicators were then compared to determine whether GBL supplementation had a statistically significant impact on skin health compared to the control group. The influence of GBL supplementation on wrinkle severity was assessed using a 3D optical imaging device based on fringe projection technology. Key wrinkle parameters analyzed included Ra (average roughness), Rmax (maximum peak-to-valley distance), Rp (highest peak height), Rv (wrinkle depth), and Rz (average peak-to-valley distance) (Figure 3C,H). Regular GBL consumption demonstrated significant improvements in wrinkle severity, particularly in the crow's feet area. While no significant differences were observed between groups for most indices except Rv at baseline, within-group comparisons revealed that GBL intake led to a marked reduction in wrinkle depth (Rv) and average roughness (Ra) (Figure 3C). The Rz index also showed significant improvement in the test group (Figure 3C). However, no changes were observed for Rmax or Rp (Figure S1A,B). Further analysis showed significant improvements within the test group across all five wrinkle indicators when comparing changes from baseline to post-GBL intervention (Figure 3D). Additionally, GBL supplementation significantly improved other skin health indicators such as epidermal hydration, density, roughness, and glossiness (Figure 3E–H). Notably, except for skin glossiness, these improvements were statistically significant at the end of the intake period within the test group (Figure 3E,G,H).

Skin elasticity was also assessed through gross elasticity (R2), net elasticity (R5), and biological elasticity (R7) (Figure 3I–K). Regular GBL intake resulted in significant increases in R2 and R7 from baseline (Figure 3I,J). At the end of the trial, between-group comparisons indicated that R5 and R7 values were significantly higher in the test group compared to controls (Figure 3J,K). These findings suggest that GBL supplementation may enhance skin elasticity. However, other measured indicators such as subsurface hydration, desquamation levels, and TEWL did not show significant improvements with GBL consumption (Figure S1C–E). In conclusion, cumulative intake of GBL demonstrated its potential to alleviate various signs of age-related skin deterioration by improving parameters such as wrinkle severity, hydration levels, density, roughness, and elasticity.

These findings indicate that GBL supplementation may serve as an effective intervention for enhancing overall skin health in aging adults.

3.3. GBL Treatment Restore UVB-Irradiated Oxidative Stress in Human Skin Cells

Upon trial, GBL application significantly improved aging-linked dermal health phenotypes. GBL was evaluated to determine whether it protects against UVB-driven oxidative stress in HaCaT human keratinocytes. To investigate this, GBL was supplemented post-UVB irradiation and tested after 48 h of incubation (Figure 4A). A range of GBL treatment concentrations (350–50 µg/mL) did not exhibit cytotoxicity in human keratinocytes; however, the group treated with 50 µg/mL of GBL displayed relatively lower cell viability compared to others. Therefore, the 150–350 µg/mL range was selected for further experiments (Figure 4B). UVB stimulation notably elevated the levels of 8-OHdG, a marker of oxidative-stress-induced DNA damage, but this increase was statistically mitigated by GBL treatment in a dose-dependent manner (Figure 4C). In addition, UVB exposure distinctly downregulated SOD, an enzyme with an antioxidant role under oxidative stress conditions (Figure 4D). However, this oxidative stress-relieving mechanism was significantly restored by dose-dependent GBL treatment (Figure 4C). Taken together, these results suggest that oxidative stress and damage to skin cells caused by UVB occur through various mechanisms. However, these effects are inhibited by GBL supplementation, which enhances antioxidant enzyme activity and promotes DNA repair.

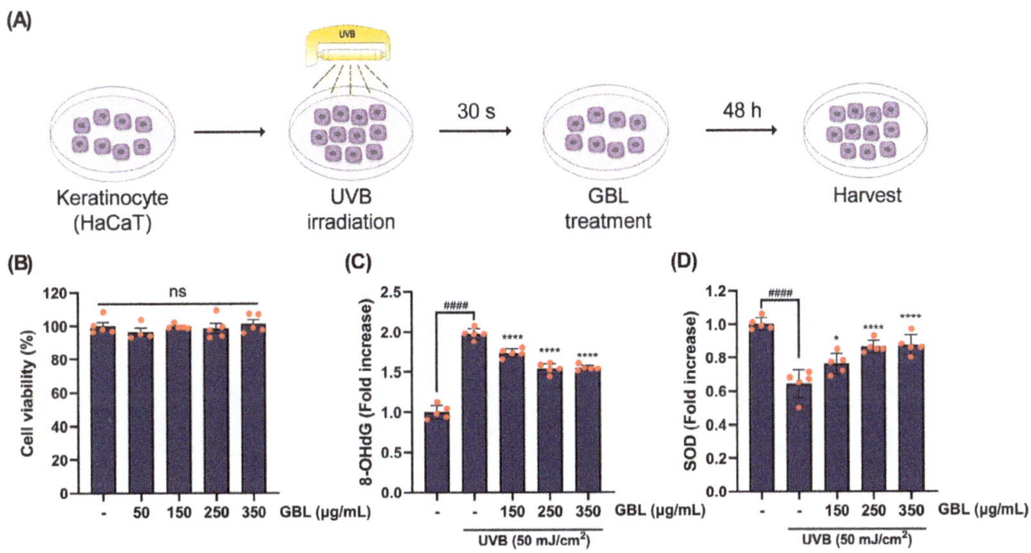

Figure 4. The protective effect of GBL against UVB-stimulated photo-irradiation on human keratinocytes. (**A**) The diagram of experimental procedure for the potential role of GBL in UVB-irradiated HaCaT cells. (**B**) The cytotoxicity of GBL treatment with the indicated range of concentrations. (**C**) Evaluation of GBL's protective effects on UVB-stimulated HaCaT cells through measurement of 8-OHdG secretion. (**D**) Evaluation of GBL's protective effects on UVB-stimulated HaCaT cells through measurement of SOD secretion. Data are displayed to mean ± S.D. ($n = 5$). ns; not significant. * $p < 0.05$ and **** $p < 0.0001$ vs. UVB-only-stimulated blank group. #### $p < 0.0001$ vs. control group.

3.4. GBL Treatment Improves Gene Expressions Related to Dermal Damage Caused by UVB Exposure

The restorative effect of GBL application on skin cells was also demonstrated at the gene expression level (Figure 5). Significant UVB irradiation (50 mJ/cm^2) on HaCaT cells markedly increased the expression of the UV-induced transcription factor AP-1 gene by 170.45%. Additionally, the expression of target genes related to dermal collagen degradation increased by 235.26%, 195.84%, and 194.87% for metalloproteinase (MMP)-1, MMP3, and MMP9, respectively, compared to the control group (Figure 5A). However, when GBL was applied to UVB-stimulated HaCaT cells in a concentration-dependent manner (150–350 μg/mL), the expression of these UVB-induced AP-1 downstream genes was significantly reduced compared to the control group. In the group treated with the highest concentration of GBL, the reduction rates were 25.39%, 24.70%, and 19.08%, respectively (Figure 5A).

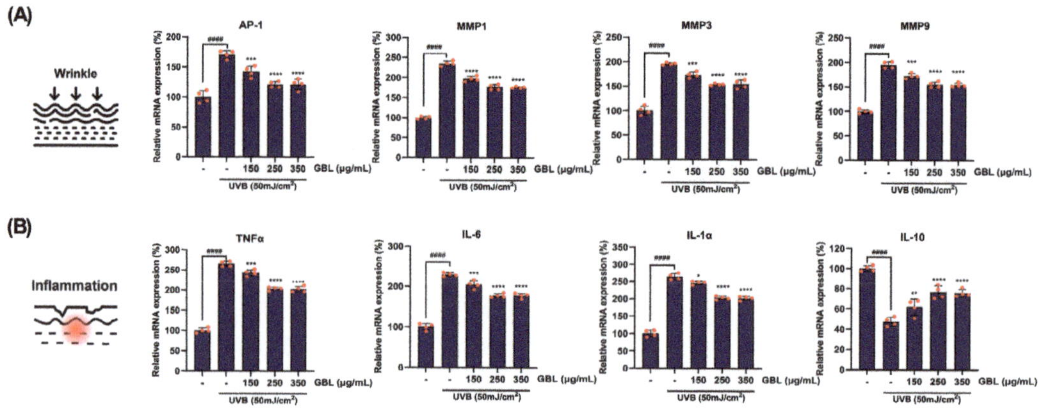

Figure 5. Quantitative evaluations of the impacts of GBL treatment on the gene expressions associated with UVB stress in human keratinocytes. (**A**) Relative mRNA gene expressions regulated by the dose-dependent GBL exposure under UVB stimulation, linked to wrinkle formation. (**B**) Relative mRNA gene expressions regulated by the dose-dependent GBL application under UVB stimulation, linked to skin inflammation. Data are depicted to mean ± S.D. (n = 4). * $p < 0.05$, ** $p < 0.01$, *** $p < 0.001$, and **** $p < 0.0001$ vs. only-UVB-stimulated blank group. #### $p < 0.0001$ vs. control group.

Among the target genes of AP-1 induced by UVB exposure, the major pro-inflammatory cytokines TNF-α, IL-6, and IL-1α were regulated in a dose-dependent manner by GBL treatment in human keratinocytes (Figure 5B). At the highest concentration, the maximum inhibitory effects compared to the control group were observed as 21.22%, 22.49%, and 19.67%, respectively, suggesting that GBL treatment alleviates inflammatory responses (Figure 5B). Furthermore, the suppression of IL-10 (an immunosuppressive regulatory cytokine) gene expression, which is activated as a negative feedback mechanism of UVB radiation, was observed (Figure 5B). This immunosuppressive mechanism was significantly improved by high-dose GBL treatment, resulting in a 37.90% increase, and this activation was found to occur in a dose-dependent manner (Figure 5B).

Exposure to GBL was found to significantly alleviate oxidative-stress-induced cell death (including collagenase activation and inflammatory responses) caused by UVB stimulation, a major environmental factor contributing to aging, in human skin cells. These findings suggest that the activation of oxidative stress defense mechanisms played a key role in the significant improvements observed in aging-related skin health indicators at the human level following GBL consumption.

4. Discussion

GBL, developed by combining low-molecular-weight collagen derived from fish scales with GABA through lactobacillus fermentation, represents a significant advancement in nutricosmetics [8]. Compared to traditional collagen-based products, GBL offers enhanced benefits for skin health through its synergistic composition and innovative production process in our previous study [14]. Traditional marine collagen-based nutricosmetics, such as hydrolyzed FC peptides, have been shown to improve skin hydration, elasticity, and wrinkle reduction [19]. Randomized trials have demonstrated that fish-derived low-molecular-weight collagen (LMWC), due to its smaller molecular size, is absorbed more efficiently into the bloodstream and dermis, enhancing its bioavailability and efficacy. Several studies have reported significant improvements in skin hydration and elasticity after 12 weeks of LMWC supplementation, with reductions in wrinkle depth and skin roughness [20,21]. Also, marine-derived collagen peptides have been particularly praised for their superior bioavailability compared to land-based sources [22].

Despite its proven benefits for skin health, there are several hurdles that hinder the industrial application of LMWC. One significant limitation is its thermal stability. FC has a lower denaturation temperature compared to mammalian collagen, which makes it less suitable for applications that require high thermal stability, such as medical-grade products [23,24]. Another challenge lies in its oxidative stability. Fish collagen is susceptible to oxidative degradation during processing and storage, which can negatively affect the quality of the final product [25]. Therefore, in the industrial setting, these limitations that shorten the shelf life of FC are ultimately cited as the primary factor driving up overall product costs, and there is growing demand for improvements. Interestingly, several studies indicate that lactobacillus fermentation can extend the shelf life of protein-based products [26,27]. Lactobacillus fermentation achieves this by creating an acidic environment, producing antimicrobial compounds, and modifying protein structures, which collectively inhibit spoilage and pathogenic microorganisms, and fermentation further reduces the molecular weight of collagen peptides, enhancing their absorption by the body [15,28]. Therefore, to apply the benefits conferred by fermentation, we attempted GBL production via bioconversion through LMWC fermentation (Figure 1A).

Furthermore, we were able to anticipate additional improvements in stability and the introduction of functional properties through the generation of GABA during the lactobacillus fermentation process. Studies have shown that GABA exhibits significant antioxidant properties. For instance, during fresh cheese fermentation, GABA production was associated with increased ABTS antioxidant and metal-chelating activities, indicating its potential to reduce oxidative stress in food products [29]. As shown in Figure 1, GBL was found to produce a significant amount of GABA during the production process, with GABA being the most abundant FAA contained within GBL (Figure 1D,G). These facts indicate that GABA's inclusion in GBL introduces unique benefits that traditional collagen products lack. According to a variety of previous reports, GABA upregulates type I collagen expression in dermal fibroblasts, promoting skin elasticity and reducing wrinkles [30]. And GABA treatment not only significantly inhibited UVB-induced matrix MMP-1, which degrades collagen, but also improved skin hydration by expression of aquaporin-3 in keratinocytes [31].

The efficacy of GBL was validated through a randomized, double-blind, placebo-controlled trial involving 100 participants (aged 35–60 years) with visible signs of aging (Figures 2 and 3A). While traditional collagen supplements primarily target hydration and elasticity, GBL's unique combination of LMWC and GABA widely offers significant improvements in multiple skin health parameters, including hydration, elasticity, wrinkle reduction, and skin density, without adverse effects during the study period, confirming its safety for long-term use (Figure 3 and Table S3). Supplementation of GBL significantly showed a skin improvement in hydration (increase in 20%) and wrinkle depth (decrease in 15%) after 12 weeks (Figure 3C–E). In addition, GBL consumption notably enhanced skin density and reduced roughness and significantly upregulated the elasticity-indicating parameters (R2, R5, and R7), outperforming traditional collagen formulation (Figure 3G,H).

The in vitro experiments using UVB-irradiated HaCaT keratinocytes further elucidated the molecular mechanisms underlying these clinical effects (Figures 4 and 5). GBL treatment restored oxidative stress markers such as 8-OHdG and SOD levels in a dose-dependent manner (Figure 4). This suggests that GBL mitigates UVB-induced oxidative stress through its potent antioxidant properties. Additionally, GBL reduced the gene expression of pro-inflammatory cytokines and MMPs, which are known to degrade collagen and exacerbate skin aging (Figure 5). The clinical improvements observed in skin hydration, elasticity, and wrinkle depth can be attributed to GBL's dual action at the cellular level: the UVB-linked oxidative stress and antioxidant pathway mechanism in human skin. These effects are attributed to GBL's ability to combat UVB-induced oxidative stress in human keratinocytes, reduce pro-inflammatory cytokine expression, and enhance antioxidant enzyme activity (Figures 3–5).

GBL maximizes functionality and not only increases the added value of FC, which uses fish scale byproducts from the fishery industry as its main raw material but also proposes a new solution for improving stability—a weakness of seafood—thereby contributing to the revitalization of related industries and aligning with global sustainability trends. Additionally, the results of this study, which produce GABA through the curated lactobacillus fermentation, can overcome the limitations of traditionally recognized functional materials with established safety and provide novelty in bioavailability [32]. This can be applied to enhance the value of various materials. Such an upcycling approach not only reduces cost inefficiencies and environmental impacts associated with the development of new materials but also supports the principles of a circular economy.

While the randomized trial demonstrated GBL's efficacy in improving skin health, limitations include the lack of subgroup analyses based on age and gender. These factors may influence treatment responses due to physiological differences, such as collagen metabolism and hormonal variations. Future studies should stratify participants by these demographics to assess differential effects, ensuring a comprehensive understanding of GBL's efficacy across diverse populations. For instance, populations should be studied in the variable types of skin hydration, wrinkle depth, and elasticity separately for men and women to identify potential sex-specific responses. Similarly, different skin types may need different doses of GBL to see the desired effects.

5. Conclusions

This study highlights the development and evaluation of GBL, a novel nutricosmetic ingredient combining low-molecular-weight fish collagen and GABA, produced through sustainable microbial fermentation. GBL demonstrated significant improvements in skin health, including enhanced hydration, elasticity, wrinkle reduction, and skin density, as validated through a randomized trial. These effects were achieved without adverse events, confirming the safety of GBL consumption for long-term use. Additionally, GBL exhibited protective properties against UVB-induced oxidative stress and inflammation in human keratinocytes, underscoring its multifunctional benefits. By upcycling marine byproducts into high-value biomaterials, GBL aligns with global sustainability goals and offers an innovative solution for addressing age-related skin concerns while promoting environmental responsibility. Future research should explore its efficacy across diverse populations to optimize its broader application.

Supplementary Materials: The following supporting information can be downloaded at: https://www.mdpi.com/article/10.3390/antiox14030245/s1, Figure S1: Measurement of additional skin health indicators following GL administration in human clinical trial; Table S1: Detailed exclusion criteria for clinical study; Table S2: The test and placebo ingredient's formular for the clinical trial; Table S3: The clinical safety test results in participants' serum in between test and control groups.

Author Contributions: Conceptualization, K.-M.R. and B.-J.L.; methodology, J.H. and K.-M.R.; validation, J.H. and K.-M.R.; formal analysis, K.-M.R.; investigation, K.-M.R.; resources, B.-J.L.; data curation, J.H.; writing—original draft preparation, J.H.; visualization, B.R.; supervision and project administration, B.-J.L.; funding acquisition. All authors have read and agreed to the published version of the manuscript.

Funding: This research was supported by Korea Institute of Marine Science & Technology Promotion (KIMST) funded by the Ministry of Oceans and Fisheries (RS-2023-00253195). And this work was supported by the National Research Foundation of Korea (NRF) grant funded by the Korea government (MSIT) (RS-2024-00348214) and this work was supported by the National Research Foundation of Korea (NRF) grant funded by the Korea government (RS-2023-00208025). Also, this research was supported by the Pukyong National University Industry-university Cooperation Foundation's 2024 Post-Doc.

Institutional Review Board Statement: The human randomized trial was conducted in accordance with the Declaration of Helsinki, and approved by the Institutional Review Board of Global Medical Research Center Inc. (protocol code GIRB-23420-OT and 3 May 2023) for studies involving humans.

Informed Consent Statement: Informed consent was obtained from all subjects involved in the study.

Data Availability Statement: Data are contained within this article.

Conflicts of Interest: Authors Kyoung-Min Rheu and Bae-Jin Lee are from Marine Bioprocess Co., Ltd. company and GABALAGEN® product is from Marine Bioprocess Co., Ltd. Company. The Marine Bioprocess Co., Ltd. company had no role in the design of the study; in the collection, analyses, or interpretation of data; in the writing of the manuscript, or in the decision to publish the results.

References

1. Wackernagel, M.; Hanscom, L.; Jayasinghe, P.; Lin, D.; Murthy, A.; Neill, E.; Raven, P. The importance of resource security for poverty eradication. *Nat. Sustain.* **2021**, *4*, 731–738. [CrossRef]
2. Belzer, A.; Parker, E.R. Climate Change, Skin Health, and Dermatologic Disease: A Guide for the Dermatologist. *Am. J. Clin. Dermatol.* **2023**, *24*, 577–593. [CrossRef] [PubMed]
3. Martic, I.; Jansen-Dürr, P.; Cavinato, M. Effects of Air Pollution on Cellular Senescence and Skin Aging. *Cells* **2022**, *11*, 2220. [CrossRef]
4. Garg, A.; Chren, M.M.; Sands, L.P.; Matsui, M.S.; Marenus, K.D.; Feingold, K.R.; Elias, P.M. Psychological stress perturbs epidermal permeability barrier homeostasis: Implications for the pathogenesis of stress-associated skin disorders. *Arch. Dermatol.* **2001**, *137*, 53–59. [CrossRef]
5. Johnson, S.; Williams, K.D.; Clark, B.; Stewart, E., Jr.; Peyton, C.; Johnson, C. *Perspective Chapter: Climate Change and Health Inequities*; IntechOpen: London, UK, 2024.
6. Mont, O.; Lehner, M.; Dalhammar, C. Sustainable consumption through policy intervention—A review of research themes. *Front. Sustain.* **2022**, *3*, 921477. [CrossRef]
7. Rajabimashhadi, Z.; Gallo, N.; Salvatore, L.; Lionetto, F. Collagen Derived from Fish Industry Waste: Progresses and Challenges. *Polymers* **2023**, *15*, 544. [CrossRef] [PubMed]
8. Byun, K.-A.; Lee, S.Y.; Oh, S.; Batsukh, S.; Jang, J.-W.; Lee, B.-J.; Rheu, K.-m.; Li, S.; Jeong, M.-S.; Son, K.H.; et al. Fermented Fish Collagen Attenuates Melanogenesis via Decreasing UV-Induced Oxidative Stress. *Mar. Drugs* **2024**, *22*, 421. [CrossRef] [PubMed]
9. Baker, P.; Huang, C.; Radi, R.; Moll, S.B.; Jules, E.; Arbiser, J.L. Skin Barrier Function: The Interplay of Physical, Chemical, and Immunologic Properties. *Cells* **2023**, *12*, 2745. [CrossRef] [PubMed]
10. Cao, L.; Qian, X.; Min, J.; Zhang, Z.; Yu, M.; Yuan, D. Cutting-edge developments in the application of hydrogels for treating skin photoaging. *Front. Mater.* **2024**, *11*, 1443514. [CrossRef]
11. Poljšak, B.; Dahmane, R. Free Radicals and Extrinsic Skin Aging. *Dermatol. Res. Pr.* **2012**, *2012*, 135206. [CrossRef]

12. Hautekiet, P.; Saenen, N.D.; Martens, D.S.; Debay, M.; Van der Heyden, J.; Nawrot, T.S.; De Clercq, E.M. A healthy lifestyle is positively associated with mental health and well-being and core markers in ageing. *BMC Med.* **2022**, *20*, 328. [CrossRef] [PubMed]
13. Dini, I.; Laneri, S. Nutricosmetics: A brief overview. *Phytother. Res. PTR* **2019**, *33*, 3054–3063. [CrossRef] [PubMed]
14. Oh, S.; Lee, S.Y.; Jang, J.-W.; Son, K.H.; Byun, K. Fermented Fish Collagen Diminished Photoaging-Related Collagen Decrease by Attenuating AGE–RAGE Binding Activity. *Curr. Issues Mol. Biol.* **2024**, *46*, 14351–14365. [CrossRef] [PubMed]
15. Park, H.-J.; Rhie, S.J.; Jeong, W.; Kim, K.-R.; Rheu, K.-M.; Lee, B.-J.; Shim, I.J.B. GABALAGEN Alleviates Stress-Induced Sleep Disorders in Rats. *Biomedicines* **2024**, *12*, 2905. [CrossRef] [PubMed]
16. Ye, M.; Rheu, K.-M.; Lee, B.-J.; Shim, I. GABALAGEN Facilitates Pentobarbital-Induced Sleep by Modulating the Serotonergic System in Rats. *Curr. Issues Mol. Biol.* **2024**, *46*, 11176–11189. [CrossRef] [PubMed]
17. Hyun, J.; Kang, S.-I.; Lee, S.-W.; Amarasiri, R.P.G.S.K.; Nagahawatta, D.P.; Roh, Y.; Wang, L.; Ryu, B.; Jeon, Y.-J. Exploring the Potential of Olive Flounder Processing By-Products as a Source of Functional Ingredients for Muscle Enhancement. *Antioxidants* **2023**, *12*, 1755. [CrossRef]
18. Hyun, J.; Ryu, B.; Oh, S.; Chung, D.-M.; Seo, M.; Park, S.J.; Byun, K.; Jeon, Y.-J. Reversibility of sarcopenia by Ishige okamurae and its active derivative diphloroethohydroxycarmalol in female aging mice. *Biomed. Pharmacother.* **2022**, *152*, 113210. [CrossRef] [PubMed]
19. Kim, S.K. Marine cosmeceuticals. *J. Cosmet. Dermatol.* **2014**, *13*, 56–67. [CrossRef]
20. Lee, M.; Kim, E.; Ahn, H.; Son, S.; Lee, H. Oral intake of collagen peptide NS improves hydration, elasticity, desquamation, and wrinkling in human skin: A randomized, double-blinded, placebo-controlled study. *Food Funct.* **2023**, *14*, 3196–3207. [CrossRef] [PubMed]
21. Kim, D.U.; Chung, H.C.; Choi, J.; Sakai, Y.; Lee, B.Y. Oral Intake of Low-Molecular-Weight Collagen Peptide Improves Hydration, Elasticity, and Wrinkling in Human Skin: A Randomized, Double-Blind, Placebo-Controlled Study. *Nutrients* **2018**, *10*, 826. [CrossRef]
22. Seong, S.H.; Lee, Y.I.; Lee, J.; Choi, S.; Kim, I.A.; Suk, J.; Jung, I.; Baeg, C.; Kim, J.; Oh, D.; et al. Low-molecular-weight collagen peptides supplement promotes a healthy skin: A randomized, double-blinded, placebo-controlled study. *J. Cosmet. Dermatol.* **2024**, *23*, 554–562. [CrossRef]
23. Jafari, H.; Lista, A.; Siekapen, M.M.; Ghaffari-Bohlouli, P.; Nie, L.; Alimoradi, H.; Shavandi, A. Fish Collagen: Extraction, Characterization, and Applications for Biomaterials Engineering. *Polymers* **2020**, *12*, 2230. [CrossRef]
24. Chen, J.; Wang, G.; Li, Y. Preparation and Characterization of Thermally Stable Collagens from the Scales of Lizardfish (Synodus macrops). *Mar. Drugs* **2021**, *19*, 597. [CrossRef] [PubMed]
25. Wallenwein, C.M.; Weigel, V.; Hofhaus, G.; Dhakal, N.; Schatton, W.; Gelperina, S.; Groeber-Becker, F.K.; Dressman, J.; Wacker, M.G. Pharmaceutical Development of Nanostructured Vesicular Hydrogel Formulations of Rifampicin for Wound Healing. *Int. J. Mol. Sci.* **2022**, *23*, 16207. [CrossRef] [PubMed]
26. Borremans, A.; Smets, R.; Van Campenhout, L. Fermentation Versus Meat Preservatives to Extend the Shelf Life of Mealworm (*Tenebrio molitor*) Paste for Feed and Food Applications. *Front. Microbiol.* **2020**, *11*, 1510. [CrossRef] [PubMed]
27. Wang, D.; Cheng, F.; Wang, Y.; Han, J.; Gao, F.; Tian, J.; Zhang, K.; Jin, Y. The Changes Occurring in Proteins during Processing and Storage of Fermented Meat Products and Their Regulation by Lactic Acid Bacteria. *Foods* **2022**, *11*, 2427. [CrossRef] [PubMed]
28. Stadnik, J.; Kęska, P.; Gazda, P.; Siłka, Ł.; Kołożyn-Krajewska, D. Influence of LAB Fermentation on the Color Stability and Oxidative Changes in Dry-Cured Meat. *Appl. Sci.* **2022**, *12*, 11736. [CrossRef]
29. Woraratphoka, J.; Innok, S.; Soisungnoen, P.; Tanamool, V.; Soemphol, W. γ-Aminobutyric acid production and antioxidant activities in fresh cheese by Lactobacillus plantarum L10-11. *Food Sci. Technol.* **2021**, *42*, e03121. [CrossRef]
30. Uehara, E.; Hokazono, H.; Sasaki, T.; Yoshioka, H.; Matsuo, N.J.B. Effects of GABA on the expression of type I collagen gene in normal human dermal fibroblasts. *Biosci. Biotechnol. Biochem.* **2017**, *81*, 376–379. [CrossRef]
31. Zhao, H.; Park, B.; Kim, M.J.; Hwang, S.H.; Kim, T.J.; Kim, S.U.; Kwon, I.; Hwang, J.S. The Effect of γ-Aminobutyric Acid Intake on UVB- Induced Skin Damage in Hairless Mice. *Biomol. Ther.* **2023**, *31*, 640–647. [CrossRef] [PubMed]
32. Cui, Y.; Miao, K.; Niyaphorn, S.; Qu, X. Production of Gamma-Aminobutyric Acid from Lactic Acid Bacteria: A Systematic Review. *Int. J. Mol. Sci.* **2020**, *21*, 995. [CrossRef] [PubMed]

antioxidants MDPI

Review

Betaine Dietary Supplementation: Healthy Aspects in Human and Animal Nutrition

Giovanni Buonaiuto [1], Alessia Federiconi [1], Carla Giuditta Vecchiato [1], Elisa Benini [2] and Attilio Luigi Mordenti [1,*]

1 Dipartimento di Scienze Mediche Veterinarie (DIMEVET), Alma Mater Studiorum-Università di Bologna, Via Tolara di Sopra 50, 40064 Ozzano Emilia, BO, Italy; giovanni.buonaiuto@unibo.it (G.B.); alessia.federiconi3@unibo.it (A.F.); carla.vecchiato2@unibo.it (C.G.V.)

2 Dipartimento di Scienze Mediche Veterinarie (DIMEVET), Alma Mater Studiorum-Università di Bologna-Unità Operativa di Sede (Cesenatico), Viale Vespucci 2, 47042 Cesenatico, FC, Italy; elisa.benini7@unibo.it

* Correspondence: attilio.mordenti@unibo.it; Tel.: +39-05-1209-7381; Fax: +39-05-1209-7373

Abstract

Betaine, a naturally occurring compound primarily derived from sugar beet by-products, has attracted increasing attention for its multifaceted roles in human and animal nutrition. Acting as both an osmolyte and a methyl group donor, betaine contributes to cellular hydration, methylation balance, antioxidant defense, and metabolic regulation. This review provides a comprehensive overview of betaine's biological functions and its health-promoting effects across species. In humans, betaine supports hepatic function, cardiovascular health, renal protection, and physical performance, mainly by modulating homocysteine metabolism, lipid profiles, and oxidative stress. In animal production systems, it enhances growth, feed efficiency, reproductive performance, and resilience to heat stress, with species-specific applications in monogastrics, ruminants, aquaculture species, and companion animals. The review also explores the molecular mechanisms underlying betaine's effects, including epigenetic regulation and mitochondrial function, and presents updated evidence on its biosynthesis, bioavailability, and nutrient interactions. Furthermore, the use of betaine derived from agro-industrial by-products aligns with the principles of the circular economy, promoting the sustainable reuse of valuable compounds within the agri-food chain. Despite promising findings, further research is needed to standardize effective dosages and clarify species-specific responses under different physiological and environmental conditions. Overall, betaine emerges as a promising and sustainable functional ingredient with wide-ranging applications in nutrition and health.

Keywords: betaine dietary supplementation; human nutrition; animal nutrition; antioxidant activity; nutraceutical activity

1. Introduction

Betaine, or N-trimethylglycine $(CH_3)_3N^+CH_2COO^-$, is a naturally occurring zwitterionic compound derived from the amino acid glycine. First identified in the 19th century in sugar beet (*Beta vulgaris*) juice, it is ubiquitously distributed across plants, animals, and microorganisms. Beyond its widespread biological presence, betaine is also used industrially as a solubilizing agent and humectant [1]. Endogenously, betaine is synthesized through the irreversible oxidation of choline and is abundant in several dietary sources. It is most concentrated in plant-derived foods, particularly sugar beet and cereal grains [2]. However, certain seafoods—especially bivalve mollusks such as mussels, oysters, clams, and scallops—also contain appreciable amounts of betaine [2]. Functionally, betaine performs two major physiological roles: as an organic osmolyte, it contributes to cellular hydration and protein stabilization under hyperosmotic stress [3]; and as a methyl donor in one-carbon metabolism, it participates in the remethylation of homocysteine to methionine (Met), catalyzed by betaine-homocysteine methyltransferase (BHMT) [4]. Emerging evidence highlights its metabolic importance in sustaining methylation capacity, modulating lipid and amino acid metabolism, and regulating gene expression via epigenetic pathways. These functions are closely linked to redox homeostasis, as both methylation potential and osmotic regulation influence mitochondrial activity and antioxidant defenses. Moreover, betaine has been implicated in modulating redox-sensitive transcription factors, notably Nrf2, with downstream effects on glutathione synthesis, lipid peroxidation, and inflammatory signaling. In human nutrition, betaine supplementation has been investigated for its efficacy in lowering plasma homocysteine levels, improving hepatic lipid metabolism, and enhancing physical performance. In animal nutrition, dietary inclusion of betaine has been associated with improved growth rates, feed efficiency, carcass composition, thermotolerance, and oxidative status—particularly under environmental or physiological stress conditions.

This review provides an in-depth examination of betaine's chemical structure, metabolic functions, and health-related effects, with an emphasis on its role in oxidative balance and nutritional physiology. Specific attention is given to its applications in human health and animal production systems, including ruminants, monogastric species, companion animals, and aquaculture. The discussion highlights species-specific responses to dietary supplementation, implications for production sustainability, and recent advancements in betaine sourcing and formulation strategies.

2. Industrial Extraction of Betaine from Sugar Beet

The industrial extraction of betaine is intrinsically linked to the evolution of sugar production from *Beta vulgaris*, a process that began approximately two hundred years ago and has since undergone significant technological advancements. Historically, calco-carbonic purification has been the cornerstone of sugar beet processing, involving sequential liming and carbonation steps to purify raw juice. This multi-stage process—comprising lime addition, CO_2 carbonation, solid-liquid filtration, and ion-exchange resin treatment—effectively removes suspended solids, color bodies, and non-sugar impurities, thereby enhancing the purity of the final sugar product. However, this traditional method presents notable drawbacks, including excessive lime consumption, the generation of calcium-rich sludge, high energy demands, and environmental concerns related to CO_2 emissions and waste disposal [5]. Furthermore, the harsh calcareous conditions and elevated processing temperatures can degrade or partially destroy thermolabile compounds such as betaine, limiting its recovery from the molasses stream. In response, both industry and research have pursued eco-friendly, lime-free alternatives aimed at preserving valuable co-products while reducing resource consumption and environmental impact [6]. One such innovation involves the direct concentration and crystallization of raw juice, bypassing lime treatment altogether. This approach not only reduces color formation and sugar degradation but also improves crystallization kinetics and product yields.

Simultaneously, membrane filtration techniques—particularly microfiltration—have been adopted to remove suspended solids and colloidal impurities, thereby simplifying the juice matrix prior to further refining. Chromatographic separation, either in batch mode or via simulated moving bed (SMB) systems, has emerged as a powerful addition to the processing toolkit. These methods enable selective partitioning of betaine-containing fractions. SMB systems, in particular, facilitate continuous or semi-continuous processing where eluent streams are recycled within the system, reducing fresh solvent consumption while increasing throughput and betaine purity [7–9]. The combination of chromatographic separation with low-temperature crystallization further enhances sugar quality: slower crystal growth at reduced thermal stress produces high-purity sugar while preserving thermolabile co-products. Quantitatively, chromatographic systems have demonstrated betaine removal efficiencies exceeding 96%, representing a significant improvement over traditional purification methods [7,10,11]. These technologies offer precise selectivity, operate under mild conditions, and avoid the use of strong alkaline or inorganic reagents. Energy savings result not only from milder operating conditions and lower thermal requirements for cooling crystallization but also from reduced chemical consumption and minimized waste streams.

However, the adoption of membrane-based and chromatographic technologies requires higher capital investment and specialized operational expertise, which may limit their implementation in legacy plants. Nevertheless, economies of scale, regulatory incentives promoting environmental sustainability, and potential revenues from betaine valorisation increase the attractiveness of these innovative approaches [7–9]. By recovering high-purity betaine alongside sugar, sugar beet processing plants can evolve toward integrated biorefineries, aligning with circular economy principles and diversifying their product portfolios.

3. Betaine: Chemical Characteristics and Nutritional Properties/Applications

Betaine is considered an important compound due to its role as an organic osmolyte, its nutritional properties, and its ability to donate methyl groups. Betaine can be obtained through dietary sources or synthesized de novo within tissues [11]. In animals, betaine is primarily synthesized in the liver and kidneys via the oxidation of choline. This process involves a two-step enzymatic reaction: choline is first converted to betaine aldehyde by choline oxidase in the mitochondria, followed by oxidation to betaine by betaine aldehyde dehydrogenase [1,3,12].

In plants, betaine is synthesized through two pathways—one involving choline and another involving glycine—occurring in both the chloroplasts and cytosol. One of betaine's key functions is its role as an osmolyte, which contributes to heat stress tolerance. By stabilizing cellular structures under stress, betaine activates calcium-dependent protein kinases and mitogen-activated protein kinases, which in turn activate stress-responsive transcription factors [13]. This signaling cascade enhances antioxidant enzyme activity, helping to mitigate the damaging effects of reactive oxygen species (ROS) during heat stress. These adaptations enable plants to survive under varying climatic conditions and improve water-use efficiency [14].

In microorganisms, betaine is also synthesized via a two-step choline oxidation process; however, some species additionally employ a three-step methylation of glycine, a pathway exclusive to certain microorganisms [15]. Beyond osmolyte function, betaine serves as a methyl donor in the remethylation of homocysteine to methionine within the betaine-homocysteine methyltransferase pathway. This process influences the concentration of S-adenosylmethionine, thereby affecting DNA methylation of cytosine bases, gene transcription, genomic imprinting, and stability. As a result, betaine supplementation can impact gene silencing and alter gene expression. Notably, betaine is more efficient than phosphatidylcholine in homocysteine remethylation, as phosphatidylcholine must first be metabolized to betaine before contributing to this process. As an osmolyte, betaine regulates cell volume, stabilizes proteins, and mitigates the denaturing effects of urea [2,16]. While betaine is not considered an essential nutrient—since it can be synthesized endogenously—it is necessary to meet daily requirements due to insufficient endogenous production. Betaine's functional applications extend beyond nutrition. It improves osmotic tolerance in cells by modulating osmolarity, protects microbial strains during storage and recovery, promotes cell growth, and safeguards intracellular components, including enzymes, from stress [15].

4. Betaine in Human Nutrition

Betaine is a nutritional component of the human diet, naturally present in a variety of foods and also available as a dietary supplement in cases of deficiency. It is considered a derivative of choline [17–19]. As previously described, betaine is utilized by higher organisms due to its unique chemical structure, which confers two primary physiological functions. First, it acts as an osmolyte, protecting cells under stress by promoting water retention, replacing inorganic salts and shielding intracellular enzymes from osmotic or thermal damage [1,3,4]. Second, it serves as a methyl group donor, playing a central role in restoring methionine homeostasis through trans-methylation reactions that support multiple biochemical pathways [1,3,4]. These two functions make betaine essential for maintaining the health of several organs, including the heart, liver and kidneys.

The daily intake of betaine in Western populations depends on both dietary sources and endogenous synthesis. Although the commonly referenced value is the adequate intake of choline (550 mg/d for men [20]), reported values for actual betaine intake vary widely in the literature. For example, Arumugam et al. [1] and Hoffman et al. [21] report a range of 1000–2500 mg/day, whereas Craig [3] and Ashtary-Larky et al. [22] mention much higher intakes of 9000–15,000 mg/day (with an average of approximately 12,000 mg/day). In contrast, other studies by Willingham et al. [4], Moro et al. [23] and Cho et al. [24] report substantially lower intakes, ranging from 100 to 400 mg/day (mean: 208 ± 90 mg/day). However, based on hematological studies in rats, even an intake of 12,000 mg/day may be considered safe in humans [25,26].

Betaine is efficiently absorbed in the small intestine. Initially, it accumulates in enterocytes via active transport mechanisms involving sodium or chloride-coupled amino acid transport systems, and it may also be absorbed through passive, sodium-independent pathways [3–5]. Due to its methylated structure, betaine transport is further mediated by the membrane-bound betaine/γ-aminobutyric acid transporter 1 (BGT-1) and amino acid transport system A, particularly under hypertonic stress [1,3,4,27–29]. After ingestion, betaine is rapidly absorbed in the duodenum [30,31]. Within the body, it is catabolized to dimethylglycine, which is further degraded to sarcosine and ultimately converted to glycine as part of the methionine cycle. These transformations mainly occur in the mitochondria of liver and kidney cells [1,3,4,31,32].

4.1. Betaine and Hepatic Function

Betaine plays a critical role in hepatic physiology, primarily as a methyl donor in one-carbon metabolism and due to its lipotropic properties. It is increasingly recognized for its hepatoprotective effects in both alcoholic liver disease (ALD) and metabolism-associated fatty liver disease (MAFLD) [1,3,33]. Despite differing etiologies—chronic alcohol intake and metabolic dysfunction, respectively—ALD and MAFLD share pathological features including hepatic steatosis, inflammation, fibrosis, and progression to cirrhosis or hepatocellular carcinoma. Hepatic steatosis, the initial stage of both ALD and MAFLD, is characterized by excessive lipid accumulation in hepatocytes [33–35]. In both conditions, hepatic steatosis marks the earliest stage and is driven by increased free fatty acid (FFA) influx from adipose tissue, elevated de novo lipogenesis, impaired mitochondrial β-oxidation, and reduced very-low-density lipoprotein (VLDL) secretion [33–35]. As a methyl donor, betaine supports methionine regeneration via betaine-homocysteine methyltransferase (BHMT), thereby restoring the hepatic S-adenosylmethionine (SAM): S-adenosylhomocysteine (SAH) ratio—crucial for methylation capacity and epigenetic regulation.

Chronic ethanol exposure disrupts this balance, impairing methylation reactions and phosphatidylcholine synthesis via phosphatidylethanolamine N-methyltransferase (PEMT), thereby compromising VLDL secretion [11]. Betaine supplementation restores PEMT activity, enhances VLDL export, and reduces hepatic triglyceride accumulation. Moreover, betaine attenuates oxidative stress and inflammation by enhancing antioxidant capacity and downregulating cytokines such as TNF-α and COX-2 [36]. In animal models of alcohol-induced liver injury, betaine supplementation reduced endoplasmic reticulum stress, improved mitochondrial ultrastructure, and reversed ethanol-induced hypomethylation of genes involved in nitric oxide synthesis [11]. Betaine has also been shown to prevent the methyl-folate trap caused by chronic alcohol intake, reduce vitamin A depletion, and preserve erythrocyte membrane integrity. In the context of MAFLD, betaine offers additional benefits beyond hepatic lipid metabolism. It strengthens the intestinal barrier by upregulating tight junction proteins and modulates gut microbiota composition, thereby reducing endotoxemia and hepatic inflammation [11]. Betaine also improves adipose tissue function by restoring normal adipokine profiles and enhancing mitochondrial activity.

The liver is the primary site of ethanol metabolism, which occurs via three main enzymatic pathways: (i) alcohol dehydrogenase (ADH), converting ethanol to acetaldehyde using NAD+; (ii) the microsomal ethanol-oxidizing system (MEOS), primarily involving cytochrome P450 2E1; and (iii) catalase-mediated oxidation in peroxisomes [1,37]. Acetaldehyde, a toxic metabolite, is further oxidized to acetate by mitochondrial aldehyde dehydrogenase (ALDH) [1,38,39]. Ethanol metabolism impairs the methionine cycle and increases homocysteine levels, leading to reduced methylation potential. Betaine restores the SAM:SAH ratio and supports hepatic methylation reactions, thereby reducing triglyceride accumulation and liver injury [40–42].

Moreover, betaine counters oxidative stress and inflammation induced by chronic alcohol intake by boosting antioxidant defenses and repressing proinflammatory cytokines. It also reverses epigenetic changes such as the hypomethylation of genes like nitric oxide synthase (NOS), thus preserving normal gene expression [1,43,44]. MAFLD, formerly known as non-alcoholic fatty liver disease (NAFLD), shares several pathophysiological mechanisms with ALD but is mainly driven by metabolic imbalances, including excess caloric intake, insulin resistance, and altered adipose tissue function. The disease progresses from simple steatosis to steatohepatitis, fibrosis, and hepatocellular carcinoma [45–47]. Betaine supplementation exerts protective effects by improving intestinal barrier integrity, modulating gut microbiota, reducing systemic inflammation, and enhancing mitochondrial and adipose tissue function [1,48,49].

4.2. Cardiovascular and Metabolic Benefits of Betaine Supplementation

The combination of betaine and guanidinoacetate (glycocyamine) has been shown to alleviate symptoms in individuals with chronic diseases, including cardiovascular disorders, without evidence of toxicity. Clinical observations report improvements in general well-being, reductions in fatigue, enhanced muscular strength and endurance, and improvements in both physical and cognitive performance [1,50,51]. In patients with cardiovascular conditions such as cardiac decompensation and congestive heart failure, betaine supplementation has been associated with improved cardiac function [52]. Additional reported benefits include weight gain in undernourished individuals, attenuation of symptoms in arthritis and asthma, and enhanced libido [3,51,53,54].

Betaine has also demonstrated transient antihypertensive effects and improved glucose tolerance in both diabetic and non-diabetic individuals [55]. Moreover, several studies suggest that betaine positively modulates lipid metabolism and may contribute to the prevention of atherosclerosis [9,56–58]. Notably, higher dietary intake of betaine has been associated with reduced cardiovascular-related mortality in individuals with atherosclerotic disease [9,56–59].

Hyperhomocysteinemia, caused by disruptions in the methionine cycle due to genetic polymorphisms or nutritional deficiencies, is strongly associated with an increased risk of cardiovascular disease, stroke, Alzheimer's disease, and other metabolic disorders [3,33,58,60–62]. Betaine acts as a methyl donor in the remethylation of homocysteine to methionine via BHMT, thereby reducing plasma homocysteine levels and mitigating related health risks. Clinical studies have confirmed that betaine supplementation effectively lowers homocysteine concentrations in individuals with pre-existing cardiovascular disease, while epidemiological data show an inverse relationship between dietary intake of betaine-rich foods and cardiovascular risk in healthy populations [3,33,63–65].

Mild hyperhomocysteinemia—a common metabolic condition characterized by moderately elevated homocysteine levels—represents a significant contributor to cardiovascular risk. It is commonly managed using a combination of betaine, folic acid, and vitamin B6 [10,33,66,67]. However, betaine alone has also been shown to prevent increases in homocysteine concentrations and support cardiovascular health, particularly in individuals with impaired methylation capacity [68]. Furthermore, genetic polymorphisms affecting homocysteine metabolism may influence the individual response to betaine supplementation and modulate susceptibility to cardiovascular disease. In such cases, betaine supplementation may serve as an effective therapeutic intervention to restore metabolic balance [69].

Importantly, the dosage of betaine plays a critical role in ensuring cardiovascular safety. According to Ashtary-Larky et al. [70], daily doses of 4 g or more may adversely affect the lipid profile, particularly in individuals with underlying metabolic disorders. In contrast, supplementation with lower doses (<4 g/day) appears to offer comparable reductions in plasma homocysteine without negatively impacting blood lipid levels [70]. These findings suggest that a cardiovascular-safe range for betaine intake is likely below 4 g/day. Therefore, the prescribed dosage of betaine should be carefully tailored to the individual's health status, balancing therapeutic efficacy with metabolic safety.

4.3. Kidney Function and Osmoregulation

BHMT is abundantly expressed in the kidneys of primates and pigs, whereas it is generally absent in the organs of other monogastric species. Within the kidney, BHMT expression is restricted to the proximal tubules of the renal cortex and is not detected in the medulla. The enzyme is localized in the cytosol of both the liver and kidney, and its activity is regulated by extracellular tonicity [3,71]. Under hypertonic conditions, BHMT activity is downregulated to preserve intracellular betaine for its role as an osmoprotectant. Conversely, in hypotonic environments, BHMT expression is upregulated to support the regulation of cellular turgor pressure. In addition to its osmoregulatory functions, betaine has demonstrated nephroprotective effects. In patients with renal disease receiving folic acid supplementation, betaine has been shown to reduce plasma homocysteine levels following methionine loading, although it does not significantly affect fasting homocysteine concentrations [72,73].

4.4. Performance, Body Mass Composition, Heat Tolerance and General Well-Being

Animal studies have demonstrated that betaine supplementation can positively influence body composition by reducing fat mass. However, findings from human studies remain inconclusive and, in some cases, contradictory [22,74]. Betaine plays a key role in creatine biosynthesis and may contribute to improved muscle recovery, endurance, and strength. Dietary supplementation with a combination of betaine and guanidinoacetate has shown clinical benefits in patients with poliomyelitis, although limited efficacy has been observed in motor neuron diseases; nonetheless, psychological improvements have been reported [74,75]. Moreover, betaine has been proposed as an ergogenic aid capable of enhancing athletic performance by improving energy metabolism and delaying fatigue, particularly when administered in combination with electrolyte solutions. In addition to its potential ergogenic properties, betaine supports the physiological function of the liver, heart, and kidneys, modulates plasma homocysteine concentrations, and may regulate gene expression through epigenetic mechanisms [75–77].

5. Betaine in Monogastrics Nutrition

5.1. *Swine*

Betaine is widely used in pig nutrition due to its dual role as a methyl group donor and osmolyte [3,78,79]. It plays a key role in enhancing energy metabolism, growth performance, and carcass traits. As a methyl donor, betaine contributes to essential biochemical processes such as the remethylation of homocysteine to methionine, thereby sparing methionine and choline for other physiological functions [80,81]. As an osmolyte, it helps maintain cellular water balance under osmotic stress, improving intestinal integrity, nutrient absorption, and overall metabolic resilience [78,82].

Numerous studies have demonstrated that dietary betaine supplementation improves average daily gain (ADG) and feed conversion ratio (FCR), particularly under conditions of heat stress or nutrient restriction [3,78,83,84]. However, responses may vary depending on factors such as diet composition, environmental conditions, and genetic merit of the animals. Notably, there is no evidence that betaine supplementation enhances fat-free mass in humans [22,74]. Betaine has been shown to reduce fat deposition and promote lean tissue accretion by modulating hepatic lipid metabolism and stimulating fatty acid oxidation. Improvements in carcass characteristics—such as reduced backfat thickness and increased lean mass—are particularly evident when betaine is administered during specific growth phases (e.g., between 10 and 16 weeks of age in sows) [79,83–88]. Meta-analyses further confirm betaine's beneficial effects on meat quality, including improved colour, enhanced water-holding capacity, and reduced cooking and drip losses [79]. Improvements have also been reported in blood biochemical parameters, including reductions in plasma urea and cholesterol and increases in albumin and total protein levels—indicators of enhanced protein metabolism [84]. Through its osmolytic function, betaine helps maintain cellular hydration and volume under osmotic stress, particularly in renal and intestinal tissues. This protective effect preserves cellular function by replacing inorganic ions, stabilising proteins and membranes, and reducing energy expenditure associated with ion transport [89,90]. These mechanisms are especially important during gastrointestinal challenges such as coccidiosis, where betaine supplementation has been shown to improve gut morphology, reduce lesion severity, and enhance nutrient absorption [91,92]. Its ability to reduce diarrhoea and strengthen intestinal integrity has been documented in several studies [93,94]. Given the anticipated restrictions on in-feed antibiotics, betaine is increasingly regarded as a promising bioactive compound for promoting gut health [78].

From a practical standpoint, recommendations for betaine use vary across different stages of pig production. Identifying the most critical phases for supplementation is essential for optimizing its functional benefits while reducing the duration of inclusion and, consequently, production costs [95]. Reported effective inclusion rates are approximately 0.1% in diets for piglets and grower pigs, and up to 0.2% in breeder diets. The effectiveness and cost-efficiency of betaine depend largely on the timing and duration of supplementation. The critical phases for betaine supplementation are the first week after weaning for weaner pigs (4–10 weeks), especially for light weight pigs (it is not beneficial when other methyl donors are sufficient), the growth phase for fattening/finisher pigs (10–16 weeks) (it is recommended in female grower pigs for backfat reduction), and in sows during summer lactation and gestation to manage heat stress [96,97].

5.2. Poultry

Dietary betaine supplementation has demonstrated beneficial effects across various poultry species (Table 1), improving performance [98–102], nutrient utilization [103,104], and physiological resilience, particularly under environmental stress conditions [105–107]. In broiler chickens, betaine enhanced growth performance indicators such as feed intake, body weight gain, and FCR [108,109], including under heat stress [110]. Supplementation has also been associated with increased pectoral muscle yield [106,108,111] and improved performance under disease challenges, including coccidiosis [112]. These effects may be attributed to betaine's role as an osmolyte, preserving intestinal epithelial integrity, enhancing nutrient digestibility, and promoting the expansion of intestinal mucosa [113]. Furthermore, betaine acts as a methyl donor, potentially improving energy efficiency by conserving ATP normally expended for Na^+/K^+ pump function, especially under thermal stress [114,115]. Enhanced availability of sulfur amino acids has also been noted [116]. Betaine's positive influence extends to gut microbiota modulation. Supplementation increased beneficial microbial populations such as Enterococci, while reducing total bacterial counts in the crop [113]. It also elevated concentrations of short-chain fatty acids (SCFAs), including acetic and propionic acids, essential for beneficial bacteria such as Lactobacillus and Bifidobacterium [105,117]. However, not all studies report consistent benefits; some have shown no significant effects on performance at certain dietary inclusion levels [118–120]. In laying hens, betaine has shown promising effects on egg production, FCR, and eggshell quality, particularly under heat stress [102,104]. It improved laying performance, hormonal profiles, and bone strength [113,121]. The compound also mitigated heat stress–induced reductions in vitellogenin and very-low-density lipoprotein levels in egg yolk precursors [113]. When combined with other nutrients (e.g., folic acid, choline, vitamins E and C), betaine further improved performance in heat-challenged layers [104]. These effects appear to be mediated through endocrine modulation, particularly via increased luteinizing hormone, estradiol, melatonin, and follicle-stimulating hormone levels [113].

In ducks, particularly under heat stress conditions, dietary or water-based betaine supplementation has been effective in improving live weight gain (LWG), feed efficiency, and physiological biomarkers [117]. Combination with antioxidants, such as Vitamin C, further enhanced these effects. Betaine also improved laying, hatchability, semen quality, nutrient digestibility, and economic performance in ducks under summer conditions [122]. Interestingly, when compared with DL-methionine, betaine proved more effective in enhancing feed efficiency, carcass traits, and growth in starter ducks [98]. In quails, results are less consistent. Some studies have reported increased egg weight and body weight gain with dietary betaine supplementation, particularly when combined with specific energy and protein levels [123]. However, other studies observed no significant improvements in egg production during early or peak laying periods [100]. Research on turkeys is limited. However, available evidence suggests that oral administration of anhydrous betaine (2.5 g/L in drinking water) over two days led to a favorable response in 96% of infected birds over 70 days of age [113], indicating potential immunomodulatory and stress-alleviating properties worthy of further investigation.

Table 1. Effect of betaine supplementation in poultry.

Source	Dosage	Species	Main Effects [1]	Reference
Feed-grade betaine	0.5 g/kg	Ducks, White Pekin, 1–41 days	BW, BMY	[98]
Water-added betaine	50, 100 g/kg	Broilers, Cobb, 1–42 days	RT, H/Lr	[124]
Feed-grade betaine	1 g/kg	Broilers, Cobb 500, 1–56 days	FCR, BMY	[108]
Feed-grade betaine	0.75, 1.5, 2.25 g/kg	Broilers, Ross 308, 1–42 days	BWG, FCR, Nd	[103]
Feed-grade betaine	0.06, 1.2 g/kg	Broilers, Ross 308, 1–42 days	RtC	[112]
Feed-grade betaine	0.8 g/kg	Broilers, Cobb 400, 1–42 days	BWG, Sbp	[125]
Feed-grade betaine	0.75, 1.5, 2.25 g/kg	Turkey, Bronze, 12–20 weeks	BWG, Sbp, CW, %D	[99]
Feed-grade betaine	1 g/kg	Laying hens, Mandarah, 32–48 weeks	BWG, EP, FI, FCR, Nd	[104]
Feed-grade betaine	0.7, 1, 1.3 g/kg	Ducks, NR, 1–42 days	BWG, FI, FCR, Bh, SCFA	[105]
Feed-grade betaine	1.2 g/kg	Ducks, Cherry Valley, 1–42 days	BW, FI, FE, Bh, SCFA	[117]
Feed-grade betaine	0.6, 12 g/kg	Japanese quails, 42–98 days	Nd, FI, EP, EW, FCR, EQ	[123]
Betaine hydrochloride	0.25 g/kg BW	Broilers, Ross, 1–42 days	Bh, ADG, FCR, FI, WI	[109]
Feed-grade betaine	1 g/kg	Broilers, Ross 308, 1–42 days	ADG, FCR, BMY, RR	[106]
Anhydrous betaine	0.5, 1, 2 g/kg	Broilers, Huaixiang, 1–28 days	BWG, FI, IH	[126]
Anhydrous betaine	1.2 g/kg	Laying quails, 23–27 weeks	Nd, FI, EP, EW, FCR, EQ	[100]
Feed-grade betaine	0.5, 1, 2 g/kg	Broilers, Huaixiang, 1–35 days	BWG, FI, NR	[127]
Feed-grade betaine	0.8 g/kg	Ducks, NR, 1–42 days	BW, FI, FCR, Bh, Im, SCFA	[122]
Anhydrous betaine	1, 2, 3 g/kg	Broilers, Ross 308, 1–35 days	M, BMY, BMfl, BMpH, a*, Dl, CL, SF, OS	[111]
Feed-grade betaine	1 g/kg	Broilers, Huaixiang, 5–15 weeks	IBF	[107]
Active natural betaine	1 g/kg	Broilers, Ross 308, 1–35 days	BWG, FI, FCR, IH	[101]
Water-added betaine	1 g/L	Broilers, Ross 308, 21–35 days	%D, CC, WHC, CL, Mt	[128]
Feed-grade betaine	2 g/kg	Broilers, Ross 308, 1–35 days	BWG, FI, Sbp, Mmp, Ir	[110]
Feed-grade betaine	8 g/kg	Laying hens, Hy-Line Brown, 21–27 weeks	Lr, EQ	[102]
Feed-grade betaine	1.5, 3 g/kg	Broilers, NR, 1–21 days	FCR, Sbp, Lbp	[129]
Feed-grade betaine	0.5, 1, 1.5 g/kg	Broilers, Caribro-vishal, 0–42 days	BWG, FI, FCR, CMIR, Gm; Bh, BMW	[130]
Feed-grade betaine	1, 3 g/kg	Laying hens, Hy-Line Brown, 71–77 weeks	Im, Lr; Mc	[121]

[1] %D: dressing percentage; a*: meat redness; ADG: average daily weight gain; Bh: blood hematology; BMfl: Breast muscle fibre length; BMpH: fresh breast muscle pH; BMW: breast muscle weight; BMY: breast muscle yield; BWG: body weight gain; CL: cooking loss; CMIR: cell mediated immune response; CW: carcass weight; Dl: drip loss; EP: Egg production; EQ: egg quality; EW: egg weight; FI: Feed intake; Gm: Gut morphology; H/Lr: Heterophil/lymphocyte ratio; IBF: intestinal barrier function; IH: Intestinal histomorphology; Im: Intestinal morphology; Ir: Immune response; Lbp: Liver biochemistry parameters; Lr: Laying rate; M: Mortality; Mc: Microbiota composition; Mmp: muscle mineral profile; Mt: meat texture; Nd: Nutrient digestibility; NR: Nutrient Retention; OS: oxidative status; RR: respiration rate; RT: rectal temperature; RtC: Resistance to coccidiosis infection; Sbp: Serum biochemical parameters; SCFA: Short Chain Fatty Acid Profile of Ceca; SF: Shear force; WHC: water-holding capacity; WI: water intake.

5.3. Rabbits

Heat stress leads to major physiological and reproductive issues in rabbits: they are particularly vulnerable due to their poor sweating capacity and difficulty in dissipating heat. A temperature-humidity index (THI) $\geq$ 30 is classified as severe heat stress, resulting in reduced performance and health problems [131]. As regards the use of betaine in rabbit diet formulation, in recent decades, numerous experiments have been performed on rabbits, in particular under heat stress. The major goal, with a feed inclusion rate that varied from 0.5 to 2 g/kg/day, was to evaluate the positive effects of dietary betaine alone, or in association with other ingredients, to enhance resistance to heat stress by improving rabbit performance, nutrient digestibility, intestinal health, immune response, and antioxidant status [131–138]. As already mentioned in the previous sections, the criticism highlighted in these studies is precisely that of the use of betaine under specific stressed environmental conditions and not under normal conditions.

5.4. Horses

While nutritional supplements and specific feeding strategies are widely used in equine sports, few have demonstrated clear, reliable performance benefits in horses. In the past, some research groups [139,140] have investigated the use of betaine, as an antioxidant, in cryopreservation of stallion semen, with controversial results. The use of betaine supplementation alone, or in association with other metil donors, in trained horses has been studied with the aim of enhancing performance; while some earlier studies [141–143] showed no clear effect of DMG (Dimethylglycine) on equine performance, the study of de Oliveira et al. [144] supports the hypothesis that longer supplementation periods (20–30 days) may improve lactate metabolism and extend exercise duration, possibly through enhanced activation of the pyruvate dehydrogenase enzyme, aiding glucose oxidation.

5.5. Dogs and Cats

The potential benefits of betaine in companion animal nutrition arise from its involvement in three key metabolic domains: (i) its central role in one-carbon metabolism, where it donates methyl groups for essential biochemical processes; (ii) its influence on fatty acid metabolism, affecting lipid utilization and distribution; and (iii) its function as an osmolyte, supporting cellular homeostasis under osmotic stress [145]. These functions are particularly relevant in the cat, given its unique metabolic profile as an obligate carnivore with elevated protein requirements and limited flexibility in amino acid metabolism [146].

As a methyl donor, betaine supports the remethylation of homocysteine to methionine, a pathway critical for sustaining methylation-dependent processes such as DNA synthesis, gene expression, and antioxidant defence. This may also help spare dietary methionine, an essential and often limiting amino acid in feline diets. Furthermore, cats display distinct features within their one-carbon metabolism that may increase their reliance on dietary methyl donors. The synthesis of felininine, a sulfur-containing amino acid used in urinary pheromone signalling, along with a limited capacity for taurine biosynthesis and elevated methionine adenosyltransferase (MAT) activity, all contribute to an increased demand for methylation [146].

In most studied species, betaine supplementation is known to support lipid metabolism by enhancing fatty acid uptake and oxidation and improving intramuscular fat composition. However, in cats, dietary polyunsaturated fatty acid (PUFA) intake is associated with a reduction in circulating betaine, suggesting a distinct metabolic demand for methyl donors during lipid metabolism and highlighting species-specific differences in fatty acid handling [147].

Recent studies have highlighted the interactions between dietary betaine and PUFAs in modulating lipid and one-carbon metabolism in dogs and cats [148,149]. In dogs, 60 days of supplementation with betaine (ranging from 0.46% to 0.58% diet as fed), alone or in combination with flaxseed (a source of alpha-linolenic acid, ALA, 0.83% diet) or fish oil (providing eicosapentaenoic acid [EPA, 0.15% diet] and docosahexaenoic acid [DHA, 0.10% diet]), significantly altered circulating metabolites. Betaine increased plasma concentrations of key one-carbon metabolites including dimethylglycine, methionine, and N-methylalanine, while also enhancing EPA synthesis from ALA and reducing circulating carnitine and acylcarnitines, suggesting a shift toward enhanced fatty acid utilization and mitochondrial efficiency [148].

In cats, similar dietary interventions revealed species-specific metabolic responses. Betaine supplementation for 60 days (0.62% diet as fed), particularly in conjunction with fish oil (EPA 0.15%, DHA 0.11%), elevated plasma levels of betaine, dimethylglycine, and several long-chain fatty acids, including DHA and ALA [149]. These findings are supported by an earlier study, which demonstrated that dietary enrichment with either fish oil (as a source of EPA/DHA) or arachidonic acid in cats resulted in significant reductions in circulating betaine, dimethylglycine, and sarcosine, suggesting increased utilization of one-carbon donors during PUFA metabolism [147]. These results highlight a potential nutritional demand for enhanced one-carbon metabolism support, such as betaine, when high levels of PUFA are included in feline diets.

Interestingly, the effect of betaine on circulating fatty acid profiles appears to diverge between dogs and cats. In dogs, the combination of betaine and ALA promoted an intermediate increase in EPA levels, even without direct fish oil supplementation, pointing to a possible role for betaine in enhancing fatty acid desaturation and elongation pathways [148]. Conversely, in cats, no such increase in EPA was observed with added ALA, even in the presence of betaine [149]. While betaine influenced circulating PUFA concentrations in both species, the metabolic response to ALA was more pronounced in dogs. These findings reflect underlying species differences, likely associated with the cat's limited delta-6 desaturase activity and reduced capacity to convert precursor fatty acids into longer-chain forms. Furthermore, reductions in acylcarnitines and dicarboxylates suggested improved fatty acid oxidation and mitochondrial function [149].

As an osmolyte, betaine plays a crucial role in maintaining cellular homeostasis during osmotic stress. This is particularly important for kidney health and function, as betaine helps mitigate the damage caused by hypertonicity and the accumulation of homocysteine. Research has shown that betaine can reduce hypertonicity-induced cell death in renal cell lines derived from dogs, offering significant insights into its potential for managing kidney disease [150].

Recent studies investigated the effects of diets enriched with betaine (0.5% of diet) and different soluble fibre supplementations (short-chain fructooligosaccharides, apple pomace, and oat beta-glucan) on dogs and cats with spontaneous chronic kidney disease (CKD) in different stages [151,152]. In CKD dogs, the dietary treatment with betaine and increased soluble fibre was shown to decrease creatinine levels and increase circulating methionine [151]. In the one-carbon metabolism pathway, betaine donates a methyl group to convert the uremic toxin homocysteine into methionine, during which it is transformed into dimethylglycine. This metabolite was shown to increase in CKD cats when fed a diet enriched with betaine and apple pomace [152]. These findings suggest the potential of dietary interventions with betaine, offering a promising approach to managing renal disease in dogs and cats.

Collectively, these findings highlight betaine's capacity to modulate both one-carbon and lipid metabolism in dogs and cats.

5.6. Fishes

Betaine plays a crucial role in aquaculture, bringing a variety of positive effects, among others, by enhancing feed intake, growth performance, immunity, osmoregulation, and metabolic functions [153–159] (Table 2). Betaine is typically provided to aquatic species through feed supplementation, ensuring uniform distribution of the nutrient or feeding stimulant to all animals during feeding. However, the study by Lu et al. [160] showed how intraperitoneal betaine injection can also be successfully used as a method to provide betaine to ayu (*Plecoglossus altivelis*). Despite this, the industry-standard method for betaine delivery remains direct feed supplementation in commercial aquafeed.

The primary benefit of betaine in aquaculture is its role as a feed attractant, improving the palatability and acceptance of aquafeeds, particularly when alternative plant-based protein sources are used [161]. Betaine stimulates the cephalic reflex, enhancing feed intake and nutrient absorption. Studies have shown that *Salmo trutta* fed with 1.4% betaine in plant-based diets exhibited improved feed acceptance, growth performance, immune response, antioxidant defense, and digestive activity [156,161]. Similarly, *Oncorhynchus mykiss* demonstrated increased feed intake and utilization when supplemented with 1.5% betaine [162]. The positive effects of dietary betaine supplementation on growth have been observed in various other species, including Nile trout *Oreochromis niloticus* [158,163], common carp *Cyprinus carpio* fingerlings [164], rainbow trout *Oncorhynchus mykiss* [162,165,166], and *Lithobates catesbeianus* [167]. Furthermore, betaine has also been used for Sea Bream larvae (*Sparus aurata*) as a potent chemical feeding stimulant, significantly enhancing microdiet ingestion rates when included in the rearing environment, either alone or synergistically with other attractants like arginine, alanine, and glycine [168,169].

Beyond its role as a feed attractant, betaine functions as an osmoprotectant, helping aquatic species maintain cellular balance under stressful environmental conditions, such as fluctuating salinity levels. Research has demonstrated that dietary supplementation with betaine improves survival rates when stressed by water salinity variation. For instance, *Labeo bata* fed with 0.025% betaine exhibited enhanced survival in brackish water conditions [170], while *Oreochromis aureus* showed increased survival when reared in seawater with 1.0% betaine supplementation [171].

In addition to its role in osmoregulation, betaine significantly influences metabolic functions, particularly in protein and lipid metabolism. As a methyl donor, betaine modulates fat accumulation and lipid metabolism, reducing hepatic fat deposition. Studies have shown that betaine supplementation downregulates the expression of fatty acid synthase (FASS) and acetyl-CoA carboxylase (ACC) genes, mitigating hepatic steatosis in species like *Carassius auratus gibelio* [172]. It also promotes fatty acid oxidation by upregulating carnitine palmitoyltransferase 1 (CPT1) and peroxisome proliferator-activated receptor alpha (PPARα) genes. In black seabream *Acanthopagrus schlegelii*, betaine supplementation in high-fat diets increased Sirt1 expression, a gene responsible for enhancing lipid breakdown and reducing fat accumulation in the liver [173].

Betaine also plays a critical role in enhancing immune function and resistance to disease. It has been shown to boost antioxidant enzyme activity, such as superoxide dismutase (SOD) and catalase (CAT), thereby protecting cells from oxidative stress in Caspian trout *Salmo trutta* [161]. Research on Ayu *Plecoglossus altivelis* demonstrated that betaine injection stimulated the immune response, enhancing monocyte/macrophage activity against *Vibrio anguillarum* [160]. Similarly, in *Macrobrachium rosenbergii*, dietary supplementation with 0.5% betaine significantly improved innate immunity against *Vibrio cholera*, leading to better survival rates [157]. Betaine also improves the phagocytic activity and bacterial-killing ability of immune cells, providing an additional layer of defense against infections in Nile tilapia (*Oreochromis niloticus*) [174].

Another significant benefit of betaine supplementation in aquaculture is its impact on meat quality. By influencing muscle composition, betaine increases the ratio of beneficial fatty acids while reducing saturated fat content in blunt snout bream fed a high-fat diet [154,157]. This not only improves the nutritional value of aquaculture products but also enhances their commercial appeal.

Overall, betaine is a multifunctional feed additive in aquaculture, contributing to improved growth, feed utilization, immune function, osmoregulation, metabolism, and meat quality. The optimal dietary levels of betaine, typically ranging between 0.5% and 1.5%, provide the best results, while excessive supplementation may negatively affect survival rates. Its versatility makes it a valuable component in aquafeeds, helping to optimize production efficiency and the overall health of farmed aquatic species. In light of the increasing environmental stressors driven by climate change—such as rising water temperatures, fluctuating salinity levels, ocean acidification, and increased pathogen pressure—future research on betaine in aquaculture should focus on its potential as a strategic additive for enhancing resilience in farmed aquatic species. Given its proven roles in osmoregulation, immune modulation, and stress mitigation, betaine represents a promising tool in developing adaptive feeding strategies to support fish health and performance under climate-induced challenges. Notably, the Mediterranean Sea, home to economically important species such as European sea bass (*Dicentrarchus labrax*) and gilthead sea bream (*Sparus aurata*), is considered a climate change hotspot, where marine ecosystems are already experiencing rapid shifts in temperature and salinity. Despite this, studies examining the effects of betaine supplementation in Mediterranean aquaculture species remain scarce. This represents a critical knowledge gap, especially given the region's vulnerability and its reliance on aquaculture as a major food and economic resource. Future investigations should not only evaluate species-specific responses to betaine under stress conditions but also explore the molecular mechanisms involved in its protective effects. Understanding how betaine interacts with physiological stress responses in Mediterranean species could significantly contribute to the development of climate-resilient aquaculture practices in this increasingly fragile marine environment.

Table 2. Overview of the recommended doses of betaine in different water animals and observed positive effects.

Species	Betaine Recommended	Observed Effects	Source
Hybrid Grouper *Epinephelus lanceolatus* ♂× *Epinephelus fuscoguttatus* ♀	0.5%	Improved feed efficiency, protein and methionine retention, growth performance (SGR), and reduced ammonia excretion.	[159,175]
Rainbow Trout (*Onchorincus mykiss*)	0.5–1%	In combination with 25% soybean meal, positively influenced growth, feed utilization, and fatty acid profiles.	[156,166,176]
Nile Tilapia (*Oreochromis niloticus*)	0.05%	Improved weight gain, body protein content, reduced FCR and body fat.	[158,163]
Common Carp (*Cyprinus carpio*)	0.3%	Growth performance (BWG, DWG, FCR, FCE, RGR%, SGR%) improved and enhanced digestive enzyme activity.	[177–179]
Caspian Trout (*Salmo trutta*)	1.4%	Higher weight gain, SGR, and PER compared to the baseline (plant-protein diet). Digestive enzyme activity (lipase, amylase) increased.	[161]
Giant Freshwater Prawn (*Macrobrachium rosenbergii*)	0.5%	Improved weight gain, FCR, daily growth, and feed intake vs. control. Digestive enzyme activity was higher. Survival rate under bacterial stress also increased.	[157]
Black Tiger Prawn (*Penaeus monodon*)	2%	Effects on intestinal histology, lipid metabolism, and immune response.	[180,181]
Pikeperch (*Sander luciperca*)	2%	Increased palatability and acceptability of food for fingerlings.	[182,183]
Ayu (*Plecoglossus altivelis*)	100–200 mg/kg (injection)	Protection against *Vibrio anguillarum*.	[160]
Bullfrog Bass (*Lithobates catesbeianus*)	0.4%	Increased whole-body protein deposition.	[167]

6. Betaine in Ruminant Nutrition

Betaine, has gained increasing attention in ruminant nutrition due to its multifaceted roles in enhancing metabolic efficiency, maintaining osmotic balance, and supporting overall animal performance. Naturally derived from sugar beet processing or synthesized endogenously from choline oxidation, betaine is widely used as a feed additive in the form of anhydrous betaine, betaine monophosphate, or betaine hydrochloride [184]. It functions primarily through three key mechanisms: acting as an osmolyte, serving as a methyl donor, and functioning as a molecular chaperone [185,186]. Additionally, its role as a methyl donor in one-carbon metabolism supports critical physiological functions, including growth, lactation, and liver function, by donating methyl groups to methionine via S-adenosylmethionine [187]. These metabolic contributions allow betaine to spare choline for other essential processes, enhancing overall metabolic efficiency [78]. Due to these properties, betaine supplementation has been investigated for its potential to improve milk yield and composition, growth performance, and carcass quality in ruminants. Its benefits are particularly evident under both thermoneutral and heat stress conditions, where it aids in reducing energy expenditure, protecting gut tissues, and maintaining osmotic balance [188]. This section will provide an overview of betaine's physiological roles in ruminants, its dietary sources, and the latest findings on its impact across different ruminant species, with a focus on dairy and beef cattle, goats, and sheep.

6.1. Betaine and Rumen Microorganisms

Betaine can influence the growth and activity of the microbes in the gastrointestinal tract of cattle due to its ability to maintain native protein structure and prevent cell lysis. Under stress conditions, microbial cells increase the uptake of exogenous betaine, which reduces its availability to the rumen epithelium. However, some betaine can escape the rumen and reach the duodenum [189]. Betaine acts as a promoter of bacterial growth under osmotic stress conditions and can be metabolized and converted to acetate in the rumen, contributing to milk fat synthesis [190], which suggested an interaction between betaine and fatty acid production in dairy cows.

Mahmood et al. [191] demonstrated that betaine has numerous applications in in vitro rumen fermentation studies. It stabilizes the bacterial community and enhances the ruminal fermentation process. Their study explains that betaine, due to its chemical structure with methyl groups, can act as a methyl donor for methanogenesis and directly affect methylotrophic archaea. Archaea are more stress-tolerant microorganisms than bacteria, but they are sensitive to several external factors, including dietary components. These factors lead to cellular dehydration, cell death, dysbiosis, and reduction of animal production and health. Archaea represent only 3% of microbial biomass in the rumen, and they occupy an important ecological position in the rumen microbiome and rumen fermentation [191,192]. A group called "rumen cluster C" Thermoplasmata has been discovered as a new order of methanogens. They were classified as methylotrophic methanogens and occupied a different niche separate from the other methanogens, as suggested by Poulsen et al. [193]. Thermoplasmata can obtain energy and carbon from methylamine derived from the degradation of betaine and choline. The use of methylamine by Thermoplasmata impacts the metabolic process of rumen bacteria producing ammonia nitrogen [193]. This production leads to a positive change in the protease activity of rumen bacteria such as Butyrivibrio fibrisolvens, indicating a stimulatory effect on crude protein degradation [194]. The ammonia nitrogen produced does not alter its concentration in the rumen, suggesting that it could be used for microbial protein synthesis. Betaine also influences the activities of protozoa and fibrolytic bacteria such as Ruminococcus albus and Fibrobacter succinogenes, increasing fiber digestibility [195].

6.2. Betaine Supplementation in Cattle

In cattle, betaine supplementation has been explored mainly in dairy and beef production systems. Studies have investigated its impact on DMI, milk yield, and milk composition in dairy cows, particularly under thermoneutral and heat-stress conditions [185,196]. Betaine functions as an osmoprotectant, helping maintain cellular hydration and reducing energy expenditure under environmental stressors [78]. Additionally, it acts as a methyl donor in one-carbon metabolism, potentially sparing methionine for protein synthesis and improving lactation performance [195]. While several studies have demonstrated a positive impact of betaine on milk yield and fat content, findings remain inconsistent, with some reports indicating no significant effects on performance parameters [197,198]. Furthermore, research has highlighted the importance of betaine availability in the rumen and its influence on microbial fermentation, suggesting the need for targeted formulations to enhance its efficacy [199]. A summary of key studies on betaine supplementation in cattle is provided in Table 3.

Table 3. Effects of betaine supplementation in cattle.

Source	Dosage	Breed and Category	Diet Type	Main Effects [1]	Reference
Feed-grade betaine	10.5, 21 g/d	Not reported, steers	Steam-flaked and dry-rolled corn	FI, D%, CC	[200]
Feed-grade betaine	10.5, 21 g/d	Various breed, steers	Steam-flaked and dry-rolled corn	DMI, D%, CC	[184]
Feed-grade betaine	20 g/d	Aberdeen Angus, steers	50:50 rolled corn:rolled milo	ADG, FT	[201]
Feed-grade betaine	0.25 g/kg of BW	Zebu, steers	Not reported	TrS	[202]
Rumen-protected betaine	45 g/d	Holstein, lactating cows	Corn silage-based TMR [2]	STC; LDL	[203]
Anhydrous betaine	50, 100, 150 g/d	Holstein, lactating cows	Corn silage-based TMR	MY, FCM, %F, RpH, TrVFA, A:Pr; Nd; NEFA, BHB	[204]
Rumen-unprotected betaine	25, 50, 100 g/d	Holstein, lactating cows	Triticale silage-based TMR	MY; %P	[205]
Feed-grade betaine	10, 15, 20 g/d	Holstein, lactating cows	Corn silage-based TMR	FI, MY; %L; %P, T-AOC, MDA, SOD, CPK	[206]
Feed-grade betaine	57 and 114 mg/kg BW	Holstein, lactating cows	Hay-based TMR	DMI, MY, BGl	[189]
Betaine-containing molasses	1.1 and 1.4 kg DM/d	Holstein, dry-off cows	Corn silage-based TMR	MY, %F, FCM, ECM	[207]
Natural betaine	2 g/kg DM	Holstein, lactating cows	Pasture with concentrate	MY, MPY, MFY, MF, CI, Ru	[185]
Feed-grade betaine	50 g/d	Karan Fries, lactating cows	Concentrate mixture and roughages	NEFA, Cortisol, FCy	[208]
Feed-grade betaine	20 g/d	Holstein, dry-off cows and newborn calves	Corn silage-based TMR	BTP, Gc	[199]
Feed-grade betaine	4.0 g/kg DM	Holstein, lactating cows	Corn silage-based TMR	MY, FCM, %F, Fe, Nd, TrVFA, NH_3-N, Rmea, BGl,	[190]
Feed-grade betaine	15 and 30 g/d	Holstein, lactating cows	Corn silage-based TMR	MY, SCC, TrVFA, MCP; NH_3-N, Nd, T-AOC, MDA, GSH-Px, SOD	[209]
Rumen-protected betaine	20 g/d	Holstein, lactating cows	Corn silage-based TMR	DMI, MY, %P, %F, %L, SCC, NEFA, PT, BGl, BUN,	[210]
Betaine hydrochloride	30 g/d	Aberdeen Angus cows	Concentrate mixture and wheat straw	HGB, RBC, RT, PR	[211]
Feed-grade betaine	80 g/d	Holstein, lactating cows	Corn silage-based TMR	Urea, ALT, SOD, BGl, MY, FCM, %F, DMI	[196]

[1] %F: Milk fat; %L: Milk lactose; %P: Milk protein; A:Pr: acetate to propionate rate; ADG: average daily body weight gain; ALT: alanine aminotransferase; BGl: Blood glucose; BHB: b-hydroxybutyrate; BTP: Blood total protein; BUN: blood urea nitrogen; CC: carcass characteristics; CI: Concentrate intake; CPK: creatine phosphate kinase; D%: Dressing %; DMI: Dry Matter Intake; ECM: Energy Corrected Milk; FCM: fat corrected milk; FCy: follicular cyst; Fe: Feed efficiency; FI: Feed intake; FT: Fat thickness; Gc: Globulin concentrations; GSH-Px: glutathione peroxidase; HGB: hemoglobin concentration; LDL: Low-density lipoproteins; MCP: microbial protein; MDA: malondialdehyde; MF: Milking frequency; MFY: Milk fat yield; MPY: Milk protein yield; MY: Milk yield; Nd: nutrient digestibility; NEFA: non-esterified fatty acids; NH_3-N: ammonia; PR: Pulse rate; PT: Plasma triglyceride; RBC: red blood cell; Rmea: ruminal microbial enzyme activity; RpH: ruminal pH; RT: Rectal temperature; Ru: Rumination; SCC: somatic cell count; SOD: superoxide dismutase activity; STC: Serum total cholesterol; T-AOC: total antioxidant capacity; TrS: transport stress; TrVFA: Total ruminal volatile fatty acid. [2] TMR: Total mixed ration.

Betaine supplementation has been shown to enhance DMI in both heat-stressed and non-heat-stressed dairy cows. This finding is consistent with previous studies in which dietary betaine was provided [185,209]. Notably, no significant differences were observed in the effects of betaine on DMI between heat-stressed and non-heat-stressed cows. The precise mechanism by which betaine improves DMI remains unclear; however, it is hypothesized that betaine may influence ruminal fermentation by serving as a source of either rumen-available nitrogen [204] or methyl groups [212]. The degradation of betaine by ruminal microbes has been previously reported, supporting its potential role in modifying ruminal fermentation dynamics [213]. Another proposed mechanism is the enhancement of ruminal nitrogen availability, which could promote the proliferation of beneficial microbes and, consequently, improve DMI and the total tract digestibility of NDF and ADF [204]. Betaine supplementation in heat-stressed dairy cows has been associated with improved total tract digestibility of dry matter (DM), organic matter (OM), crude protein (CP), NDF, and ADF. These findings are in agreement with those of Wang et al. [204], who reported similar improvements in nutrient digestibility upon betaine inclusion in the diet. The beneficial effects of betaine may be attributed to its dual role as an osmolyte and methyl donor, which helps mitigate heat stress, maintain optimal rumen pH, and enhance the microbial composition in the rumen [78]. Several studies [204,206,210] have demonstrated that betaine supplementation during the periparturient period and early lactation increases milk yield in dairy cows, regardless of ambient temperature conditions. A recent meta-analysis by Abhijith et al. [186] further corroborated this effect, reporting an average increase of 1 kg of milk per day in betaine-supplemented cows compared to controls. This positive impact on milk yield has also been observed in cows under heat stress conditions [189,196] and in grazing cows during the summer. Notably, the increase in milk production is evident in both heat-stressed cows (+1.67 kg/day) and cows housed under thermoneutral conditions (+1.03 kg/day) [198]. Additionally, energy-corrected milk (ECM) and fat-corrected milk (FCM) production were enhanced in betaine-supplemented cows [190,196,207]. This effect is likely due to betaine's role in the remethylation of homocysteine to Met, sparing Met from acting as a methyl donor. Wang et al. [210] suggested that betaine donates a methyl group to facilitate the conversion of homocysteine to Met, particularly in cases of dietary Met insufficiency. However, the effects of betaine supplementation on milk production may vary depending on dietary composition and lactation stage. For example, Davidson et al. [203] found no significant changes in milk yield or milk solids when betaine was supplemented at 45 g/day during early lactation, possibly because the control diet was not deficient in Met. Such discrepancies highlight the importance of dietary composition in determining betaine's efficacy. Betaine supplementation has also been associated with improvements in milk quality, particularly an increase in milk fat content [189,190,196,204,207]. This enhancement may be linked to betaine's ability to stimulate ruminal acetate production [189,206] and improve the digestibility of OM, CP, NDF, and ADF [209]. Nevertheless, some studies have reported no significant effects of betaine on milk fat percentage (e.g., Peterson et al., [197], with 100 g/day of rumen-unprotected betaine). Such inconsistencies may be attributed to differences in basal diet composition rather than lactation stage per se. Moreover, Wang et al. [210] observed that supplementing 20 g/day of rumen-protected betaine reduced milk somatic cell count (SCC), suggesting a potential role in maintaining mammary gland health. Although the precise mechanisms remain unclear, possible explanations include betaine's ability to support mammary cell proliferation and maintain cellular osmolarity [207]. Additionally, studies have suggested that betaine supplementation can reduce oxidative stress in mammary epithelial cells under heat stress conditions [214] and decrease pro-inflammatory cytokine production, indicating its potential as a feed additive for mastitis management [215]. Another potential mechanism underlying betaine's beneficial effects on nutrient digestibility and animal performance is its influence on mitochondrial function. Betaine has been shown to enhance mitochondrial fission and fusion dynamics, thereby promoting cell homeostasis and survival [216]. Furthermore, betaine has been reported to stabilize cellular and subcellular membranes by restoring both enzymatic and non-enzymatic antioxidants, positively regulating mitochondrial function [217,218]. Given that part of betaine can be metabolized by rumen microbes [219], it is plausible that betaine improves rumen microbial function, thereby contributing to enhanced feed efficiency and overall performance. Beyond its effects on nutrient digestibility and milk production, betaine supplementation influences plasma metabolic parameters. Wang et al. [210] reported a significant increase in plasma glucose concentrations following supplementation with 20 g/day of rumen-protected betaine. Similarly, Hall et al. [189] observed elevated plasma glucose and insulin levels in heat-stressed cows supplemented with betaine. The increased plasma glucose concentration may result from a reduction in peripheral glucose uptake [210]. Additionally, several studies have reported a decrease in non-esterified fatty acids (NEFA) and plasma triglyceride concentrations in betaine-supplemented cows compared to controls [208,210]. Given that NEFA levels reflect energy metabolism balance, their reduction suggests improved dietary energy utilization. Wang et al. [204] observed a linear decrease in plasma NEFA concentrations with increasing levels of betaine supplementation (50, 100, and 150 g/cow/day), potentially due to increased NEFA uptake by the mammary gland for milk fat synthesis, as supported by reports of increased milk fat content [185,207]. Furthermore, the observed reduction in plasma triglycerides [210] may be attributed to their transport to the liver for metabolism into NEFA, facilitated by enhanced fatty acid absorption and methyl group availability for chylomicron synthesis and secretion.

6.3. Betaine Supplementation in Small Ruminants

The effects of betaine supplementation in small ruminants, including sheep and goats, have been examined primarily in the context of heat stress mitigation and carcass quality improvement. As an osmolyte, betaine contributes to thermoregulation by reducing endogenous heat production and preserving cellular integrity. Studies in sheep and goats suggest that betaine supplementation can enhance nutrient utilization efficiency, improve growth performance, and support lactation under challenging environmental conditions. However, similar to cattle, responses to betaine supplementation in small ruminants appear to be dose-dependent and variable across different production systems. Research gaps remain in understanding its long-term effects on metabolic pathways and optimal dietary inclusion levels. Table 4 presents a compilation of studies examining betaine supplementation in small ruminants.

Several authors [219–221] have reported an increase in milk yield and milk solids, consistent with findings in dairy cows. Specifically, Fernández et al. [219] observed an increase of 0.37 kg/day in milk production in goats fed a betaine-supplemented diet. Supporting this result, Fernández et al. [220] reported a comparable increase in milk yield (0.37 kg/day) after three months of betaine supplementation at a rate of 2 g/kg of diet. Regarding milk quality, the literature presents conflicting data. Ghoneem and El-Tanany [221] observed a 10% increase in milk production when using dehydrated condensed molasses fermentation solubles and a 7% increase with anhydrous betaine. These authors also reported an increase in fat-corrected milk (FCM) yield (1.46 vs. 1.29 for betaine supplementation and control, respectively). This increase aligns with the observed rise in milk fat percentage (4.47% and 4.55% vs. 4.26% in the control) and total solids, the latter of which was statistically significant. These findings are consistent with an increase in ruminal acetate concentration (28.11 and 29.05 mmol/L, respectively) compared to the control group (25.96 mmol/L). It has been suggested that betaine supplementation can alter rumen fermentation patterns, favoring a higher acetate concentration, which may in turn promote fat synthesis in the mammary gland [209]. Similarly, Fernández et al. [220] observed an increase in milk fat percentage (4.69% vs. 4.40% for betaine vs. control, respectively) in lactating Murciano-Granadina goats fed betaine-supplemented diets. However, others [219,222] found no significant differences in the chemical composition of milk. An interesting aspect of milk composition is its fatty acid profile. Fernández et al. [220] reported significantly higher ($p < 0.05$) levels of caprylic (C8:0) and capric (C10:0) fatty acids in the betaine-supplemented group compared to the control (3.03% vs. 2.8% for caprylic acid; 11.14% vs. 10.05% for capric acid). Additionally, betaine supplementation significantly increased ($p < 0.05$) the concentration of short-chain fatty acids (mainly C8:0, C10:0, C10:1, and C12:0), as well as C18:3. Conversely, the levels of other C18 fatty acids (except for C18:3) decreased, with a significant reduction observed for C18:1 ($p < 0.05$). Several studies on betaine supplementation in small ruminants have focused on growth performance and birth outcomes [223–226]. For instance, Sahraei et al. [227] reported that betaine supplementation in pregnant ewes tended to increase lamb birth weight (4.41 vs. 3.95 kg). Additionally, lamb body weight and average daily gain (ADG) at 56 days of age were significantly different between groups (ADG: 221 $\pm$ 18 g in the control group vs. 224 $\pm$ 23 g in the betaine group). Similar findings were reported by Brougham et al. [226], who observed that lambs born to betaine-supplemented ewes tended to have higher body weights at weaning. In another study, Brougham et al. [226] investigated whether maternal betaine supplementation during pregnancy could improve twin lamb body weight, thermoregulatory capacity, time to stand and suckle, colostrum intake, and survival from birth to weaning. Supplementing ewes with 4 g/day of betaine during the second half of pregnancy increased twin lamb survival up to day 7 and reduced the time required to initiate suckling. However, feeding ewes 2 g/day of betaine throughout pregnancy improved postnatal growth at weaning but reduced twin lamb survival rates and increased the time needed to reach key behavioral milestones. Overall, betaine supplementation during late gestation appears to have a positive effect on early postnatal twin lamb survival.

Regarding the effect of betaine supplementation in lambs and meat goats [219,223–225,228,229], a review of the literature does not show a clear effect on weight gain and final body weight following betaine supplementation, except for Dong et al. [225], who reported an improvement in lamb growth performance. These authors also observed a positive trend in average daily feed intake. Although growth performance was not consistently enhanced, several studies have shown improvements in meat quality, particularly a reduction in subcutaneous fat and intramuscular lipids, with limited effects on fatty acid composition [223]. Banskalieva et al. [224] reported that 3 g/day of rumen-protected betaine supplementation could improve the growth of goats, possibly by increasing peripheral tissue lipolysis or decreasing lipogenesis. Conversely, Fernández et al. [219,223] found that 2 g/kg dietary rumen-protected betaine supplementation had no effect on growth performance and carcass fatness in both male and female lambs. However, betaine supplementation (1.1 g/day unprotected betaine or 1.1, 2.2, or 3.3 g/day rumen-protected betaine) increased ADG and ADFI in lambs. The observed discrepancies among studies may be due to differences in experimental subjects and feeding doses. There is considerable interest in the effects of betaine on lipid metabolism in muscles. Some studies suggest that betaine accelerates fatty acid oxidation and promotes their transfer to meat, leading to a redistribution of body fat [223]. In contrast, other studies indicate that betaine has a limited effect or may reduce intermuscular fat content [219]. However, few studies have explored betaine's impact on the fatty acid composition of lamb meat. Liu et al. [230] found that 8 g/day of betaine supplementation had no effect on fatty acid content in the longissimus muscle of finishing sheep. However, Dong et al. [225] reported that betaine supplementation increased polyunsaturated fatty acids (PUFA) in the gluteus muscle. Additionally, rumen-protected betaine supplementation decreased the saturated fatty acid (SFA) content in the longissimus dorsi while increasing unsaturated fatty acids (UFA), monounsaturated fatty acids (MUFA), and the UFA/SFA ratio compared to unprotected betaine treatment.

Table 4. Effects of betaine supplementation in sheep and goats.

Source	Dosage	Breed and Category	Diet Type	Main Effects [1]	Reference
Feed-grade betaine	2 g/kg BW	Manchego, growing lambs	Concentrate mixture and barley straw	FT, NL	[223]
Rumen escape betaine	2 g/kg BW	Manchego, growing lambs	Concentrate mixture and barley straw	ADG, FCR, FT, LMA, CW	[219]
Feed-grade betaine	4 g/kg BW	Murciano–Granadina, lactating goats	Basal pelleted diet and alfalfa hay	MY, MFA,	[220]
Rumen-protected betaine	3 g/d	Boer × Spanish, meat goats	Concentrate mixture and pasture	PTG, NEFA	[224]
Glycine betaine	0.2 g/kg BW	Barbari, non-pregnant goats	Not reported	Heat stress amelioration	[231]
Feed-grade betaine	2, 4 g/d	Merino, ewes	Oat chaff and alfalfa hay	WI, ADG,	[232]
Rumen escape betaine	2, 4 g/d	Merino, pregnant ewes	Basal pelleted diet and oaten hay	LWW, LADG, LBL, BUN, LSur	[233]
Rumen-protected betaine	1.1, 2.2, 3.3 g/d	Hu, growing lambs	Concentrate mixture	FI, ADG, MFA,	[225]
Betaine hydrochloride	5 g/d	Sanjab, pregnant ewes	Concentrate mixture and forages	BHB, MDA,	[227]
Rumen-protected betaine	2, 4, 6 g/d	Hu, growing lambs	Concentrate mixture	Ng, ADG, F/G, GSH-Px, MDA, SOD, CAT,	[228]
Betaine hydrochloride	1, 3 g/d	Hu, growing lambs	Hay-based TMR [2]	HDL-C,	[229]
Feed-grade betaine	2, 4 g/d	Merino, ewes	Oat chaff and alfalfa hay	I, NEFA,	[234]
Feed-grade betaine	4 g/kg DM	Damascus, lactating goats	Concentrate mixture and Egyptian clover hays	MY, FCM, Nd, NH_3-N, TrVFA,	[221]
Feed-grade betaine	2, 4 g/d	Merino, ewes and neonatal lambs	Basal pelleted diet and oaten hay	LWW	[235]
Feed-grade betaine	4 g/d	Merino, ewes and neonatal lambs	Pasture and barley grains	RT, LBL; BCS, GL	[226]

[1] ADG: average daily body weight gain; BCS: Body condition score; BHB: b-hydroxybutyrate; BUN: blood urea nitrogen; CAT: catalase activity; CW: carcass weight; F/G: feed to gain ratio; FCR: Feed conversion rate; FI: Feed intake; FT: Fat thickness; GL: gestation length; GSH-Px: glutathione peroxidase; HDL-C: high-density lipoprotein cholesterol; I: Insulin; LADG: Lamb average daily gain; LBL: Lamb behavioural latencies; LMA: *longissimus* muscle area; LSur: Lamb cumulative survival; LWW: Lamb weight at weaning; MDA: malondialdehyde; MFA: meat fatty acid composition; MY: Milk yield; Nd: nutrient digestibility; NEFA: non-esterified fatty acids; Ng: Net gain; NH_3-N: ammonia; NL: neutral lipids; PTG: Plasma triacylglycerols; RT: Rectal temperature; SOD: superoxide dismutase activity; TrVFA: Total ruminal volatile fatty acid; WI: Water intake. [2] TMR: Total mixed ration.

7. Conclusions and Future Perspectives

Betaine has emerged as a biologically active compound with wide-ranging applications in both human and animal nutrition. As a naturally occurring methyl donor and organic osmolyte, betaine exerts crucial physiological functions that support cellular protection, metabolic homeostasis, and organ function. Its ability to mitigate oxidative damage, regulate one-carbon metabolism, and maintain cellular hydration has positioned it as a valuable nutritional supplement in various contexts, particularly under stress-related conditions. In human health, the therapeutic potential of betaine has been demonstrated in a range of disorders, including metabolic-associated fatty liver disease (MAFLD), hyperhomocysteinemia, cardiovascular dysfunction, and renal impairment. The evidence also points to its role in improving body composition, performance, and thermoregulation, although further studies are warranted to clarify optimal dosages and long-term outcomes. Importantly, the interactions of betaine with other nutrients, such as choline, folate, and B-vitamins, underline the need for integrated nutritional strategies rather than isolated supplementation. In the field of animal science, betaine has gained attention as a sustainable and functional feed additive capable of enhancing productive performance and welfare across multiple species. In monogastrics, betaine contributes to improved feed efficiency, growth rate, meat quality, and gastrointestinal integrity—especially under heat stress or pathogenic challenge. In ruminants, it enhances ruminal fermentation, milk yield, nutrient digestibility, and energy balance, particularly in thermally stressed animals. Betaine's role in sparing methionine and choline in one-carbon metabolism also has implications for feed formulation and cost efficiency. In aquaculture, betaine not only improves feed palatability and growth but also plays a role in osmoregulation, immunity, and stress resilience—traits of increasing relevance under changing climate and water conditions.

Nevertheless, despite the promising results across species and physiological stages, the scientific community must address several unresolved questions. First, a more precise characterization of betaine's dose-dependent effects is needed, considering factors such as animal age, sex, health status, environmental temperature, and basal diet composition. Second, the variability in response to supplementation observed among studies suggests complex metabolic and microbiota-mediated mechanisms that require deeper molecular understanding. Integration of omics technologies (metabolomics, transcriptomics, proteomics, and epigenomics) may provide critical insights into tissue-specific effects and gene–nutrient interactions modulated by betaine. In addition, the regulatory pathways through which betaine modulates cellular stress responses, mitochondrial bioenergetics, and epigenetic modifications remain partially unexplored. These represent fertile grounds for future investigation, particularly in the context of chronic inflammation, metabolic dysregulation, and reproductive efficiency. Moreover, there is increasing interest in how betaine may interact with the gut–organ axis, influencing systemic health via microbiota-mediated pathways. From a sustainability perspective, the use of betaine derived from agro-industrial by-products—such as sugar beet molasses—exemplifies circular economy principles and represents an opportunity to reduce food system waste while enhancing feed and food quality. This aligns with global objectives for more sustainable and resilient agri-food chains. Given its multifunctionality, betaine could be considered a model compound for the development of next-generation functional additives capable of responding to both nutritional and environmental challenges. Looking forward, targeted research should focus on (i) the standardization of supplementation protocols for different species and production goals, (ii) the synergistic effects of betaine with other dietary interventions, (iii) its role in precision nutrition strategies tailored to individual metabolic profiles, and (iv) its integration in climate-adaptive livestock and aquaculture systems. Particularly in regions such as the Mediterranean, where environmental stress is expected to intensify, the strategic inclusion of betaine in nutritional programs may play a key role in preserving animal health and productivity.

In conclusion, betaine is not merely a feed additive but a metabolic modulator with wide applicability in health promotion and sustainable production. Its future use will benefit from multidisciplinary approaches combining animal nutrition, molecular biology, food technology, and environmental science to fully exploit its potential for the benefit of both human and animal populations.

Author Contributions: A.L.M. and G.B.—Conceptualization, supervision, and visualization of the study. G.B., A.F., C.G.V., E.B. and A.L.M.—Data collection and curation. G.B., A.L.M., A.F., C.G.V. and E.B.—Writing. All authors have read and agreed to the published version of the manuscript.

Funding: The author Elisa Benini was supported by the Project code CN_00000033, Concession Decree No. 1034 of 17 June 2022 adopted by the Italian Ministry of University and Research, CUP J33C22001190001, Project title "National Biodiversity Future Centre—NBFC" under the National Recovery and Resilience Plan (NRRP), Mission 4 Component 2 Investment 1.4—Call for tender No. 3138 of 16 December 2021, rectified by Decree No. 3175 of 18 December 2021 of the Italian Ministry of University and Research, funded by the European Union—NextGenerationEU.

Data Availability Statement: The supplementary data are available from the corresponding author upon reasonable request.

Conflicts of Interest: The authors declare no conflict of interest.

Abbreviations

SMB	Simulated moving bed
ALD	Alcoholic fatty liver disease
MAFLD	Metabolism-associated fatty liver disease
ADH	Alcohol dehydrogenase
MEOS	Microsomal ethanol-oxidizing system
Met	Methionine
SAM	S-adenosylmethionine
SAH	S-adenosylhomocysteine
FFAs	Free fatty acids
VLDL	Very low-density lipoproteins
NOS	Nitric oxide synthase
BHMT	Homocysteine methyl transferase
ADG	Average daily gain
FCR	Feed conversion ratio
MAT	Methionine adenosyltransferase
PUFAs	Polyunsaturated fatty acid
ALA	Alpha-linolenic acid
EPA	Eicosapentaenoic acid
DHA	Docosahexaenoic acid
CKD	Chronic kidney disease
FASS	Fatty acid synthase
ACC	Acetyl-CoA carboxylase

References

1. Arumugam, M.K.; Paal, M.C.; Donohue, T.M.; Ganesan, M.; Osna, N.A.; Kharbanda, K.K. Beneficial Effects of Betaine: A Comprehensive Review. *Biology* **2021**, *10*, 456. [CrossRef] [PubMed]
2. Dobrijević, D.; Pastor, K.; Nastić, N.; Özogul, F.; Krulj, J.; Kokić, B.; Bartkiene, E.; Rocha, J.M.; Kojić, J. Betaine as a Functional Ingredient: Metabolism, Health-Promoting Attributes, Food Sources, Applications and Analysis Methods. *Molecules* **2023**, *28*, 4824. [CrossRef]
3. Craig, S.A. Betaine in human nutrition. *Am. J. Clin. Nutr.* **2004**, *80*, 539–549. [CrossRef]
4. Willingham, B.D.; Ragland, T.J.; Ormsbee, M.J. Betaine Supplementation May Improve Heat Tolerance: Potential Mechanisms in Humans. *Nutrients* **2020**, *12*, 2939. [CrossRef] [PubMed]
5. Muir, B.M.; Anderson, A.R. Development and Diversification of Sugar Beet in Europe. *Sugar Tech* **2022**, *24*, 992–1009. [CrossRef]
6. Vaccari, G. Nuove tecnologie in zuccherificio. *RICHMAC Mag.* **2002**, *84*, 47–51.
7. Vaccari, G.; Sgualdino, G.; Tamburini, E.; Pezzi, G.; Citterio, P.; Verardi, R.; Urbainec, K. New eco-friendly proposal for the crystallization of beet raw juice. *J. Clean. Prod.* **2005**, *13*, 1447–1460. [CrossRef]
8. Paananen, H.; Kuisma, J. Chromatographic separation of molasses components. *Zuckerindustrie* **2000**, *125*, 978–981.
9. Altinisik, S.; Zeidan, H.; Deniz Yilmaz, M.D.; Marti, M.E. Reactive Extraction of Betaine from Sugarbeet Processing Byproducts. *ACS Omega* **2023**, *8*, 11029–11038. [CrossRef]
10. Ehrenhuber, B. Molasses desugarizing with betaine production. *Zuckerindustrie* **2005**, *130*, 463–469.
11. Ueland, P.M. Choline and betaine in health and disease. *J. Inherit. Metab. Dis.* **2011**, *34*, 3–15. [CrossRef] [PubMed]
12. Olthof, M.R.; Brink, E.J.; Katan, M.B.; Verhoef, P. Choline supplemented as phosphatidylcholine decreases fasting and postmethionine-loading plasma homocysteine concentrations in healthy men. *Am. J. Clin. Nutr.* **2005**, *82*, 111–117. [CrossRef] [PubMed]
13. Zulfiqar, F.; Ashraf, M.; Siddique, K.H.M. Role of Glycine Betaine in the Thermotolerance of Plants. *Agronomy* **2022**, *12*, 276. [CrossRef]
14. Haghighi, P.; Habibi, D.; Mozafari, H.; Sani, B.; Sadeghishoae, M. Impact of Methanol and Glycine Betaine on Yield and Quality of Fodder Beet Genotypes (*Beta vulgaris* subsp. vulgaris). *Agronomy* **2021**, *11*, 2122. [CrossRef]

15. Zou, H.; Chen, N.; Shi, M.; Xian, M.; Song, Y.; Liu, J. The metabolism and biotechnological application of betaine in microorganisms. *Appl. Microbiol. Biotechnol.* **2016**, *100*, 3865–3876. [CrossRef]
16. Kelleppan, V.T.; King, J.P.; Butler, C.S.G.; Williams, A.P.; Tuck, K.L.; Tabor, R.F. Heads or tails? The synthesis, self-assembly, properties and uses of betaine and betaine-like surfactants. *Adv. Colloid Interface Sci.* **2021**, *297*, 102528. [CrossRef]
17. Waggle, D.H.; Lambert, M.A.; Miller, G.D.; Farrel, E.P.; Deyoe, C.W. Extensive analyses of flours and millfeeds made from nine different wheat mixes. II. Amino acids, minerals, vitamins, and gross energy. *Cereal Chem.* **1967**, *44*, 48–60.
18. Sakamoto, A.; Nishimura, Y.; Ono, H.; Sakura, N. Betaine and homocysteine concentrations in foods. *Pediatr. Int.* **2002**, *44*, 409–413. [CrossRef]
19. Zeisel, S.H.; Mar, M.-H.; Howe, J.C.; Holden, J.M. Concentrations of choline-containing compounds and betaine in common foods. *J. Nutr.* **2003**, *133*, 1302–1307. [CrossRef]
20. U.S. Institute of Medicine. *Dietary Reference Intakes for Folate, Thiamin, Riboflavin, Niacin, Vitamin B12, Panthothenic Acid, Biotin, and Choline*; National Academy Press: Washington, DC, USA, 1998; Volume 1.
21. Hoffman, J.R.; Ratamess, N.A.; Kang, J.; Rashti, S.L.; Faigenbaum, A.D. Effect of betaine supplementation on power performance and fatigue. *J. Int. Soc. Sports Nutr.* **2009**, *6*, 7–10. [CrossRef]
22. Ashtary-Larky, D.; Bagheri, R.; Tinsley, G.M.; Asbaghi, O.; Salehpour, S.; Kashkooli, S.; Kooti, W.; Wong, A. Betaine supplementation fails to improve body composition: A systematic review and meta-analysis. *Br. J. Nutr.* **2022**, *128*, 975–988. [CrossRef] [PubMed]
23. Moro, T.; Badiali, F.; Fabbri, I.; Paoli, A. Betaine Supplementation Does not Improve Muscle Hypertrophy or Strength Following 6Weeks of Cross-Fit Training. *Nutrients* **2020**, *12*, 1688. [CrossRef] [PubMed]
24. Cho, E.; Zeisel, S.H.; Jacques, P.; Selhub, J.; Dougherty, L.; Colditz, G.A.; Willett, W.C. Dietary choline and betaine assessed by food-frequency questionnaire in relation to plasma total homocysteine concentration in the Framingham Offspring Study. *Am. J. Clin. Nutr.* **2006**, *83*, 905–911. [CrossRef]
25. Hayes, K.C.; Pronczuk, A.; Cook, M.W.; Robbins, M.C. Betaine in sub-acute and sub-chronic rat studies. *Food Chem. Toxicol.* **2003**, *41*, 1685–1700. [CrossRef]
26. Cholewa, J.M.; Guimarães-Ferreira, L.; Zanchi, N.E. Effects of betaine on performance and body composition: A review of recent findings and potential mechanisms. *Amino Acids* **2014**, *46*, 1785–1793. [CrossRef]
27. Wettstein, M.; Weik, C.; Holneicher, C.; Häussinger, D. Betaine as an osmolyte in rat liver: Metabolism and cell-to-cell interactions. *Hepatology* **1998**, *27*, 787–793. [CrossRef] [PubMed]
28. Peters-Regehr, T.; Bode, J.G.; Kubitz, R.; Haussinger, D. Organic osmolyte transport in quiescent and activated rat hepatic stellate cells (Ito cells). *Hepatology* **1999**, *29*, 173–180. [CrossRef] [PubMed]
29. Kim, S.K.; Choi, K.H.; Kim, Y.C. Effect of acute betaine administration on hepatic metabolism of S-amino acids in rats and mice. *Biochem. Pharmacol.* **2003**, *65*, 1565–1574. [CrossRef]
30. Kettunen, H.; Tiihonen, K.; Peuranen, S.; Saarinen, M.T.; Remus, J.C. Dietary betaine accumulates in the liver and intestinal tissue and stabilizes the intestinal epithelial structure in healthy and coccidia-infected broiler chicks. *Comp. Biochem. Physiol. Part A: Mol. Integr. Physiol.* **2001**, *130*, 759–769. [CrossRef]
31. Kettunen, H.; Peuranen, S.; Tiihonen, K.; Saarinen, M. Intestinal uptake of betaine in vitro and the distribution of methyl groups from betaine, choline, and methionine in the body of broiler chicks. *Comp. Biochem. Physiol. Part A: Mol. Integr. Physiol.* **2001**, *128*, 269–278. [CrossRef]
32. Anioła, M.; Dega-Szafran, Z.; Katrusiak, A.; Szafran, M. NH· · ·O and OH· · ·O interactions of glycine derivatives with squaric acid. *New J. Chem.* **2014**, *38*, 3556–3568. [CrossRef]
33. Ripa, R.; Ripa, S. Homocysteine: A neglected cardiovascular risk factor. *Prog. Nutr.* **2003**, *5*, 248–261.
34. Singh, S.; Osna, N.A.; Kharbanda, K.K. Treatment options for alcoholic and non-alcoholic fatty liver disease: A review. *World J. Gastroenterol.* **2017**, *23*, 6549–6570. [CrossRef] [PubMed]
35. Matteoni, C.A.; Younossi, Z.M.; Gramlich, T.; Boparai, N.; Liu, Y.C.; McCullough, A.J. Nonalcoholic fatty liver disease: A spectrum of clinical and pathological severity. *Gastroenterology* **1999**, *116*, 1413–1419. [CrossRef] [PubMed]
36. Guangfu, Z.; Fang, H.; Chenlu, W.; Pan, L.; Nengzhang, L.; Jinping, D.; Guoqiang, Z.; Wenkai, R.; Yuanyi, P. Betaine in Inflammation: Mechanistic Aspects and Applications. *Front. Immunol. Sec. Nutr. Immunol.* **2018**, *9*, 1070. [CrossRef]
37. Oshino, N.; Oshino, R.; Chance, B. The characteristics of the 'peroxidatic' reaction of catalase in ethanol oxidation. *Biochem. J.* **1973**, *131*, 555–563. [CrossRef]
38. Cederbaum, A.I. Alcohol Metabolism. *Clin. Liver Dis.* **2012**, *156*, 667–685. [CrossRef]
39. Zakhari, S. Overview: How Is Alcohol Metabolized by the Body? *Alcohol Res. Health* **2006**, *29*, 245–254. [PubMed] [PubMed Central]
40. Barak, A.J.; Beckenhauer, H.C.; Kharbanda, K.K.; Tuma, D.J. Chronic ethanol consumption increases homocysteine accumulation in hepatocytes. *Alcohol* **2001**, *25*, 77–81. [CrossRef]

41. Kharbanda, K.K. Role of transmethylation reactions in alcoholic liver disease. *World J. Gastroenterol.* **2007**, *13*, 4947–4954. [CrossRef]
42. Kharbanda, K.K. Methionine metabolic pathway in alcoholic liver injury. *Curr. Opin. Clin. Nutr. Metab. Care* **2013**, *16*, 89–95. [CrossRef] [PubMed]
43. Barak, A.J.; Beckenhauer, H.C.; Mailliard, M.E.; Kharbanda, K.K.; Tuma, D.J. Betaine Lowers Elevated S-Adenosylhomocysteine Levels in Hepatocytes from Ethanol-Fed Rats. *J. Nutr.* **2003**, *133*, 2845–2848. [CrossRef]
44. Kharbanda, K.K.; Mailliard, M.E.; Baldwin, C.R.; Beckenhauer, H.C.; Sorrell, M.F.; Tuma, D.J. Betaine attenuates alcoholic steatosis by restoring phosphatidylcholine generation via the phosphatidylethanolamine methyltransferase pathway. *J. Hepatol.* **2007**, *46*, 314–321. [CrossRef]
45. Da Silva, R.P.; Eudy, B.J.; Deminice, R. One-Carbon Metabolism in Fatty Liver Disease and Fibrosis: One-Carbon to Rule Them All. *J. Nutr.* **2020**, *150*, 994–1003. [CrossRef]
46. Lau, J.K.C.; Zhang, X.; Yu, J. Animal models of non-alcoholic fatty liver disease: Current perspectives and recent advances. *J. Pathol.* **2017**, *241*, 36–44. [CrossRef] [PubMed]
47. Kathirvel, E.; Morgan, K.; Nandgiri, G.; Sandoval, B.C.; Caudill, M.A.; Bottiglieri, T.; French, S.W.; Morgan, T.R. Betaine improves nonalcoholic fatty liver and associated hepatic insulin resistance: A potential mechanism for hepatoprotection by betaine. *Am. J. Physiol. Liver Physiol.* **2010**, *299*, G1068–G1077. [CrossRef] [PubMed]
48. He, Z.; Li, M.; Zheng, D.; Chen, Q.; Liu, W.; Feng, L. Adipose tissue hypoxia and low-grade inflammation: A possible mechanism for ethanol-related glucose intolerance? *Br. J. Nutr.* **2015**, *113*, 1355–1364. [CrossRef]
49. Kaelin, B.R.; McKenzie, C.M.; Hempel, K.W.; Lang, A.L.; Arteel, G.E.; Beier, J.I. Adipose tissue-liver crosstalk during pathologic changes caused by vinyl chloride metabolites in mice. *Toxicol. Appl. Pharmacol.* **2020**, *399*, 115068. [CrossRef]
50. Borsook, H.; Borsook, M.E. The biochemical basis of betaine glycocyamine therapy. *Ann. West. Med. Surg.* **1951**, *5*, 825–829.
51. Borsook, M.E.; Borsook, H. Treatment of cardiac decompensation with betaine and glycocyamine. *Ann. West. Med. Surg.* **1951**, *5*, 830–835.
52. Olthof, M.; Verhoef, P.; van Vliet, T.; Boelsma, E. Low Dose Betaine Supplementation Leads to Immediate and Long Term Lowering of Plasma Homocysteine in Healthy Men and Women. *J. Nutr.* **2003**, *133*, 4135–4138. [CrossRef]
53. Van Zandt, V.; Borsook, H. New biochemical approach to the treatment of congestive heart failure. *Ann. West. Med. Surg.* **1951**, *5*, 856–862. [PubMed]
54. Graybiel, A.; Patterson, C.A. Use of betaine and glycocyamine in the treatment of patients with heart disease: Preliminary report. *Ann. West. Med. Surg.* **1951**, *5*, 863–865.
55. Lever, M.; George, P.M.; Slow, S.; Bellamy, D.; Young, J.M.; Ho, M.; McEntyre, C.J.; Elmslie, J.L.; Atkinson, W.; Molyneux, S.L.; et al. Betaine and Trimethylamine-N-Oxide as Predictors of Cardiovascular Outcomes Show Different Patterns in Diabetes Mellitus: An Observational Study. *PLoS ONE* **2014**, *9*, e114969. [CrossRef] [PubMed]
56. Morrison, L.M. Results of betaine treatment of atherosclerosis. *Am. J. Dig. Dis.* **1952**, *19*, 381–384. [CrossRef] [PubMed]
57. Morrison, L.M. Use of betaine-lipotropic combinations in clinical practice. *Geriatrics* **1953**, *8*, 649–655.
58. Selhub, J. Homocysteine metabolism. *Annu. Rev. Nutr.* **1999**, *19*, 217–246. [CrossRef]
59. Morrison, L.M. Arteriosclerosis—Recent advances in the dietary and medicinal treatment. *J. Am. Med. Assoc.* **1951**, *145*, 1232–1236. [CrossRef]
60. Finkelstein, J.D. Homocysteine: A history in progress. *Nutr. Rev.* **2000**, *58*, 193–204. [CrossRef]
61. Seshadri, S.; Beiser, A.; Selhub, J.; Jacques, P.F.; Rosenberg, I.H.; D'Agostino, R.B.; Wilson, P.W.F.; Wolf, P.A. Plasma homocysteine as a risk factor for dementia and Alzheimer's disease. *N. Engl. J. Med.* **2002**, *346*, 476–483. [CrossRef]
62. Boers, G.H. Hyperhomocysteinaemia: A newly recognized risk factor for vascular disease. *Neth. J. Med.* **1994**, *45*, 34–41. [PubMed]
63. Steenge, G.R.; Verhoef, P.; Katan, M.B. Betaine supplementation lowers plasma homocysteine in healthy men and women. *J. Nutr.* **2003**, *133*, 1291–1295. [CrossRef]
64. Matthews, A.; Johnson, T.N.; Rostami-Hodjegan, A.; Chakrapani, A.; Wraith, J.E.; Moat, S.J.; Bonham, J.R.; Tucker, G.T. An indirect response model of homocysteine suppression by betaine: Optimizing the dosage regimen of betaine in homocystinuria. *Br. J. Clin. Pharmacol.* **2002**, *54*, 140–146. [CrossRef]
65. Brouwer, I.A.; Verhoef, P.; Urgert, R. Betaine supplementation and plasma homocysteine in healthy volunteers. *Arch. Intern. Med.* **2000**, *160*, 2546–2547. [CrossRef] [PubMed]
66. Franken, D.G.; Boers, G.H.; Blom, H.J.; Trijbels, F.J.; Kloppenborg, P.W. Treatment of mild hyperhomocysteinemia in vascular disease patients. *Arterioscler. Thromb.* **1994**, *14*, 465–470. [CrossRef]
67. van den Berg, M.; Boers, G.H. Homocystinuria: What about mild hyperhomocysteinaemia? *Postgrad. Med. J.* **1996**, *72*, 513–518. [CrossRef]
68. Schwahn, B.C.; Wendel, U.; Lussier-Cacan, S.; Mar, M.-H.; Zeisel, S.H.; Leclerc, D.; Castro, C.; Garrow, T.A.; Rozen, R. Effects of betaine in a murine model of mild cystathionine-β-synthase deficiency. *Metabolism* **2004**, *53*, 594–599. [CrossRef] [PubMed]

69. Xu, G.; Pan, H.; Fan, L.; Zhang, L.; Li, J.; Cheng, S.; Meng, L.; Shen, N.; Liu, Y.; Li, Y.; et al. Dietary Betaine Improves Glucose Metabolism in Obese Mice. *J. Nutr.* **2024**, *154*, 1309–1320. [CrossRef]
70. Ashtary-Larky, D.; Bagheri, R.; Ghanavati, M.; Asbaghi, O.; Tinsley, G.M.; Mombaini, D.; Kooti, W.; Kashkooli, S.; Wong, A. Effects of betaine supplementation on cardiovascular markers: A systematic review and Meta-analysis. *Crit. Rev. Food Sci. Nutr.* **2022**, *62*, 6516–6533. [CrossRef]
71. Delgado-Reyes, C.V.; Wallig, M.A.; Garrow, T.A. Immunohistochemical detection of betaine-homocysteine S-methyltransferase in human, pig, and rat liver and kidney. *Arch. Biochem. Biophys.* **2001**, *393*, 184–186. [CrossRef]
72. van Guldener, C.; Janssen, M.J.; de Meer, K.; Donker, A.J.; Stehouwer, C.D. Effect of folic acid and betaine on fasting and postmethionine-loading plasma homocysteine and methionine levels in chronic haemodialysis patients. *J. Intern. Med.* **1999**, *245*, 175–183. [CrossRef]
73. Bostom, A.G.; Shemin, D.; Nadeau, M.R.; Shih, V.; Stabler, S.P.; Allen, R.H.; Selhub, J. Short term betaine therapy fails to lower elevated fasting total plasma homocysteine concentrations in hemodialysis patients maintained on chronic folic acid supplementation. *Atherosclerosis* **1995**, *113*, 129–132. [CrossRef]
74. Gao, X.; Zhang, H.; Guo, X.-F.; Li, K.; Li, S.; Li, D. Effect of Betaine on Reducing Body Fat—A Systematic Review and Meta-Analysis of Randomized Controlled Trials. *Nutrients* **2019**, *11*, 2480. [CrossRef] [PubMed]
75. Roti, M.W.; Hatch, H.L.; Sutherland, J.W.; Mahood, N.V.; Clements, J.M.; Seen, A.D.; Fiala, K.A.; Craig, S.A.S.; Casa, D.J.; Armstrong, L.E. Homocysteine, lipid and glucose responses to betaine supplementation during running in the heat. *Med. Sci. Sports Exerc.* **2003**, *35*, S271. [CrossRef]
76. Liversedge, L.A. Glycocyamine and betaine in motor-neurone disease. *Lancet* **1956**, *2*, 1136–1138. [CrossRef]
77. Armstrong, L.; Roti, M.; Hatch, H.; Sutherland, J.; Mahood, N.; Clements, J.; Seen, A.; Fiala, K.; Craig, S.; Casa, D.; et al. Rehydration with fluids containing betaine: Running performance and metabolism in a 31 C environment. *Med. Sci. Sports Exerc.* **2003**, *35*, S311. [CrossRef]
78. Eklund, M.; Bauer, E.; Wamatu, J.; Mosenthin, R. Potential nutritional and physiological functions of betaine in livestock. *Nutr. Res. Rev.* **2005**, *18*, 31–48. [CrossRef]
79. Sales, J. A meta-analysis of the effects of dietary betaine supplementation on finishing performance and carcass characteristics of pigs. *Anim. Feed Sci. Technol.* **2011**, *165*, 68–78. [CrossRef]
80. Kidd, M.T.; Ferket, P.R.; Garlich, J.D. Nutritional and osmoregulatory functions of betaine. *World's Poult. Sci. J.* **1997**, *53*, 125–139. [CrossRef]
81. Fernandez-Figares, I.; Wray-Cahen, D.; Steele, N.C.; Campbell, R.G.; Hall, D.D.; Virtanen, E.; Caperna, T.J. Effect of dietary betaine on nutrient utilization and partitioning in the young growing feed-restricted pig. *J. Anim. Sci.* **2002**, *80*, 421–428. [CrossRef]
82. Ratriyanto, A.; Mosenthin, R.; Bauer, E.; Eklund, M. Metabolic, Osmoregulatory and Nutritional Functions of Betaine in Monogastric Animals. *Asian-Aust. J. Anim. Sci.* **2009**, *22*, 1461–1476. [CrossRef]
83. Matthews, J.O.; Southern, L.L.; Bidner, T.D. Estimation of the total sulfur amino acid requirement and the effect of betaine in diets deficient in total sulfur amino acids for the weanling pig. *J. Anim. Sci.* **2001**, *79*, 1557–1565. [CrossRef]
84. Matthews, J.O.; Southern, L.L.; Pontif, J.E.; Higbie, A.D.; Bidner, T.D. Interactive effects of betaine, crude protein, and net energy in finishing pigs. *J. Anim. Sci.* **1998**, *76*, 2444–2455. [CrossRef]
85. Lawrence, B.V.; Schinckel, A.P.; Adeola, O.; Cera, K. Impact of betaine on pig finishing performance and carcass composition. *J. Anim. Sci.* **2002**, *80*, 475–482. [CrossRef]
86. Schrama, J.W.; Heetkamp, M.J.; Simmins, P.H.; Gerrits, W.J. Dietary betaine supplementation affects energy metabolism of pigs. *J. Anim. Sci.* **2003**, *81*, 1202–1209. [CrossRef]
87. Lipiński, K.; Szramko, E.; Jeroch, H.; Matusevičius, P. Effects of betaine on energy utilization in growing pigs–A review. *Ann. Anim. Sci.* **2012**, *12*, 291–300. [CrossRef]
88. Fu, R.; Zhang, H.; Chen, D.; Tian, G.; Zheng, P.; He, J.; Yu, J.; Mao, X.; Huang, Z.; Pu, J.; et al. Long-Term Dietary Supplementation with Betaine Improves Growth Performance, Meat Quality and Intramuscular Fat Deposition in Growing-Finishing Pigs. *Foods* **2023**, *12*, 494. [CrossRef] [PubMed]
89. Petronini, P.G.; de Angelis, E.M.; Borghetti, P.; Borghetti, A.F. Modulation by betaine of cellular response to osmotic stress. *J. Biochem.* **1992**, *282*, 69–73. [CrossRef]
90. Alfieri, R.R.; Cavazzoni, A.; Petronini, P.G.; Bonelli, M.A.; Caccamo, A.E.; Borghetti, A.F.; Wheeler, K.P. Compatible osmolytes modulate the responses of porcine endothelial cells to hypertonicity and protect them from apoptosis. *J. Physiol.* **2002**, *540*, 499–508. [CrossRef]
91. Augustine, P.C.; McNaughton, J.L.; Virtanen, E.; Rosi, L. Effects of betaine on the growth performance of chicks inoculated with mixed cultures of avian Eimeria species and on invasion and development of Eimeria tenella and Eimeria acervulina in vitro and in vivo. *Poult. Sci.* **1997**, *76*, 802–809. [CrossRef]
92. Klasing, K.C.; Adler, K.L.; Remus, J.C.; Calvert, C.C. Dietary betaine increases intraepithelial lymphocytes in the duodenum of coccidian-infected chicks and increases functional properties of phagocytes. *J. Nutr.* **2002**, *132*, 2274–2282. [CrossRef] [PubMed]

93. Teeter, R.G.; Remus, J.C.; Belay, T.; Mooney, M.; Virtanen, E.; Augustine, P. The Effects of Betaine on Water Balance and Performance in Broilers Reared under Differing Environmental Conditions. In Proceedings of the Australian Poultry Science Symposium, Balnave, Sydney, Australia, 9–10 February 1999; University of Sydney Printing Service: Sydney, Australia, 1999; Volume 165.
94. Yu, D.Y.; Feng, J.; Xu, Z.R. Effects of betaine on fat and protein metabolism in different stages of swine. *Chin. J. Vet. Sci.* **2001**, *21*, 200–203.
95. Australian Government–Department of Agriculture. Optimising the Duration of Betaine Supplementation in Pig Production. Final Report 2021. APL Project 2018-0023. Available online: https://australianpork.com.au/sites/default/files/2021-06/2018-0023.pdf (accessed on 12 April 2025).
96. Cai, D.; Jia, Y.; Lu, J.; Yuan, M.; Sui, S.; Song, H.; Zhao, R. Maternal dietary betaine supplementation modifies hepatic expression of cholesterol metabolic genes via epigenetic mechanisms in newborn piglets. *Br. J. Nutr.* **2014**, *112*, 1459–1468. [CrossRef]
97. Cabezón, F.A.; Schinckel, A.P.; Richert, B.T.; Stewart, K.R.; Gandarillas, M.; Pasache, M.; Peralta, W.A. Effect of betaine supplementation during summer on sow lactation and subsequent farrowing performance. *Prof. Anim. Sci.* **2016**, *32*, 695–703. [CrossRef]
98. Wang, Y.Z.; Xu, Z.R.; Feng, J. The Effect of Betaine and Dl-Methionine on Growth Performance and Carcass Characteristics in Meat Ducks. *Anim. Feed Sci. Technol.* **2004**, *116*, 151–159. [CrossRef]
99. Dorra, T.I.; Ibrahim, S.E.; Zayed, S.M. Effect of Dietary Betaine Supplementation on Growth Performance and Carcass Traits of Growing Turkey. *J. Anim. Poult. Prod.* **2012**, *3*, 365–377. [CrossRef]
100. Ratriyanto, A.; Prastowo, S. Floor Space and Betaine Supplementation Alter the Nutrient Digestibility and Performance of Japanese Quail in a Tropical Environment. *J. Therm. Biol.* **2019**, *83*, 80–86. [CrossRef]
101. Awad, W.A.; Ruhnau, D.; Gavrău, A.; Dublecz, K.; Hess, M. Comparing Effects of Natural Betaine and Betaine Hydrochloride on Gut Physiology in Broiler Chickens. *Poult. Sci.* **2022**, *101*, 102173. [CrossRef]
102. Wang, C.; Liu, X.; Sun, X.; Li, Y.; Yang, X.; Liu, Y. Dietary Betaine Supplementation Improved Egg Quality and Gut Microbes of Laying Hens under Dexamethasone-Induced Oxidative Stress. *Poult. Sci.* **2024**, *103*, 104178. [CrossRef]
103. Honarbakhsh, S.; Mahmood, S. Interactive Effects of Dietary Betaine and Saline Water on Carcass Traits of Broiler Chicks. *J. Biol. Sci.* **2007**, *7*, 1208–1214. [CrossRef]
104. Attia, Y.A.; Abd El-Hamid, A.E.-H.E.; Abedalla, A.A.; Berika, M.A.; Al-Harthi, M.A.; Kucuk, O.; Sahin, K.; Abou-Shehema, B.M. Laying Performance, Digestibility and Plasma Hormones in Laying Hens Exposed to Chronic Heat Stress as Affected by Betaine, Vitamin C, and/or Vitamin E Supplementation. *SpringerPlus* **2016**, *5*, 1619. [CrossRef] [PubMed]
105. Park, S.O.; Kim, W.K. Effects of Betaine on Biological Functions in Meat-Type Ducks Exposed to Heat Stress. *Poult. Sci.* **2017**, *96*, 1515. [CrossRef] [PubMed]
106. Shakeri, M.; Cottrell, J.J.; Wilkinson, S.; Ringuet, M.; Furness, J.B.; Dunshea, F.R. Betaine and Antioxidants Improve Growth Performance, Breast Muscle Development and Ameliorate Thermoregulatory Responses to Cyclic Heat Exposure in Broiler Chickens. *Animals* **2018**, *8*, 162. [CrossRef] [PubMed]
107. Liu, W.-C.; Guo, Y.; An, L.-L.; Zhao, Z.-H. Protective Effects of Dietary Betaine on Intestinal Barrier Function and Cecal Microbial Community in Indigenous Broiler Chickens Exposed to High Temperature Environment. *Environ. Sci. Pollut. Res.* **2021**, *28*, 10860–10871. [CrossRef]
108. Waldroup, P.W.; Motl, M.A.; Yan, F.; Fritts, C.A. Effects of Betaine and Choline on Response to Methionine Supplementation to Broiler Diets Formulated to Industry Standards. *J. Appl. Poult. Res.* **2006**, *15*, 58–71. [CrossRef]
109. Egbuniwe, I.C.; Ayo, J.O.; Ocheja, O.B. Betaine and Ascorbic Acid Modulate Indoor Behavior and Some Performance Indicators of Broiler Chickens in Response to Hot-Dry Season. *J. Therm. Biol.* **2018**, *76*, 38–44. [CrossRef]
110. Saleh, A.A.; El-Tahan, H.M.; Shaban, M.; Morsy, W.A.; Genedy, S.; Alzawqari, M.H.; El-Tahan, H.M.; Shukry, M.; Ebeid, T.A.; El-Keredy, A.; et al. Effect of Dietary Supplementation of Betaine and Organic Minerals on Growth Performance, Serum Biochemical Parameters, Nutrients Digestibility, and Growth-Related Genes in Broilers under Heat Stress. *Poult. Sci.* **2023**, *102*, 103051. [CrossRef]
111. Nutautaitė, M.; Alijošius, S.; Bliznikas, S.; Šašytė, V.; Vilienė, V.; Pockevičius, A.; Racevičiūtė-Stupelienė, A. Effect of Betaine, a Methyl Group Donor, on Broiler Chicken Growth Performance, Breast Muscle Quality Characteristics, Oxidative Status and Amino Acid Content. *Ital. J. Anim. Sci.* **2020**, *19*, 621–629. [CrossRef]
112. Hamidi, H.; Pourreza, J.; Rahimi, H. Dietary Betaine Affect Duodenal Histology of Broilers Challenged with a Mixed Coccidial Infection. *Pak. J. Biol. Sci.* **2009**, *12*, 291–295. [CrossRef]
113. Abd El-Ghany, W.A.; Babazadeh, D. Betaine: A Potential Nutritional Metabolite in the Poultry Industry. *Animals* **2022**, *12*, 2624. [CrossRef]
114. Hassan, R.A.; Attia, Y.A.; El-Ganzory, E.H. Growth, Carcass Quality and Serum Constituents of Slow Growing Chicks as Affected by Betaine Addition to Diets Containing 1. Different Levels of Choline. *Int. J. Poult. Sci.* **2005**, *4*, 840–850. [CrossRef]

115. Sayed, M.A.M.; Downing, J. The Effects of Water Replacement by Oral Rehydration Fluids with or without Betaine Supplementation on Performance, Acid-Base Balance, and Water Retention of Heat-Stressed Broiler Chickens. *Poult. Sci.* **2011**, *90*, 157–167. [CrossRef] [PubMed]
116. Garcia Neto, M.; Pesti, G.M.; Bakalli, R.I. Influence of Dietary Protein Level on the Broiler Chicken's Response to Methionine and Betaine Supplements. *Poult. Sci.* **2000**, *79*, 1478–1484. [CrossRef]
117. Park, B.S.; Park, S.O. Effects of Feeding Time with Betaine Diet on Growth Performance, Blood Markers, and Short Chain Fatty Acids in Meat Ducks Exposed to Heat Stress. *Livest. Sci.* **2017**, *199*, 31–36. [CrossRef]
118. Esteve-Garcia, E.; Mack, S. The Effect of Dl-Methionine and Betaine on Growth Performance and Carcass Characteristics in Broilers. *Anim. Feed Sci. Technol.* **2000**, *87*, 85–93. [CrossRef]
119. Pillai, P.B.; Fanatico, A.C.; Beers, K.W.; Blair, M.E.; Emmert, J.L. Homocysteine Remethylation in Young Broilers Fed Varying Levels of Methionine, Choline, and Betaine. *Poult. Sci.* **2006**, *85*, 90–95. [CrossRef] [PubMed]
120. Al-Tamimi, H.; Mahmoud, K.; Al-Dawood, A.; Nusairat, B.; Bani Khalaf, H. Thermotolerance of Broiler Chicks Ingesting Dietary Betaine and/or Creatine. *Animals* **2019**, *9*, 742. [CrossRef]
121. Du, X.; Pian, H.; Zhao, D.; Zhang, Y.; Wu, X.; He, J.; Chen, L.; Liu, F.; Yu, D. Enhancing Gut-Ovary Health in Aged Laying Hens: The Impact of Dietary Betaine Supplementation. *Poult. Sci.* **2025**, *104*, 104894. [CrossRef]
122. Park, S.O.; Zammit, V.A. Effect of Feed Restriction with Betaine and Ascorbic Acid Supplementation on Caecal Bacteria, Short Chain Fatty Acid, Blood Biomarker, Duodenal Morphology and Growth Performance of Meat Ducks under Heat Stress. *Eur. Poult. Sci.* **2019**, *83*, 1–13. [CrossRef]
123. Ratriyanto, A.; Indreswari, R.; Nuhriawangsa, A.M.P. Effects of Dietary Protein Level and Betaine Supplementation on Nutrient Digestibility and Performance of Japanese Quails. *Braz. J. Poult. Sci.* **2017**, *19*, 445–454. [CrossRef]
124. Zulkifli, I.; Mysahra, S.A.; Jin, L.Z. Dietary Supplementation of Betaine (Betafin®) and Response to High Temperature Stress in Male Broiler Chickens. *Asian-Australas. J. Anim. Sci.* **2004**, *17*, 244–249. [CrossRef]
125. Rao, S.V.R.; Raju, M.V.L.N.; Panda, A.K.; Saharia, P.; Sunder, G.S. Effect of Supplementing Betaine on Performance, Carcass Traits and Immune Responses in Broiler Chicken Fed Diets Containing Different Concentrations of Methionine. *Asian-Australas. J. Anim. Sci.* **2011**, *24*, 662–669. [CrossRef]
126. Sun, C.-Y.; Liu, W.-C.; Xiao, M.; Zhao, Z.-H.; An, L.-L. Effects of Graded Levels of Betaine Supplementation on Growth Performance and Intestinal Morphology in Indigenous Young Yellow Feather Broilers. *Pak. J. Zool.* **2019**, *51*, 2323. [CrossRef]
127. Liu, W.; Yuan, Y.; Sun, C.; Balasubramanian, B.; Zhao, Z.; An, L. Effects of Dietary Betaine on Growth Performance, Digestive Function, Carcass Traits, and Meat Quality in Indigenous Yellow-Feathered Broilers under Long-Term Heat Stress. *Animals* **2019**, *9*, 506. [CrossRef] [PubMed]
128. Suliman, G.M.; Hussein, E.O.S.; Al-Owaimer, A.N.; Alhotan, R.A.; Al-Garadi, M.A.; Mahdi, J.M.H.; Ba-Awadh, H.A.; Qaid, M.M.; Swelum, A.A.-A. Betaine and Nano-Emulsified Vegetable Oil Supplementation for Improving Carcass and Meat Quality Characteristics of Broiler Chickens under Heat Stress Conditions. *Front. Vet. Sci.* **2023**, *10*, 1147020. [CrossRef] [PubMed]
129. Wang, H.; Liu, L.; He, X.; Bian, G. Effect of Betaine on Growth Performance, Methionine Metabolism, and Methyl Transfer in Broilers Aged 1 to 21 Days and Fed a Low-Methionine Diet. *J. Poult. Sci.* **2025**, *62*, 2025010. [CrossRef]
130. Monika, M.; Rokade, J.J.; Gopi, M.; Vispute, M.; Sonale, N.; Prasad, W.; Bhanja, S.K. Mitigation of Heat Stress in Broiler Chickens during Hot Summer Seasons Through Betaine Hydrochloride Osmoprotectant Intervention. *Indian J. Anim. Res.* **2025**, *59*, 490–496. [CrossRef]
131. Marai, I.F.M.; Habeeb, A.A.M.; Gad, A.E. Rabbits' productive, reproductive and physiological performance traits as affected by heat stress: A review. *Livest. Prod. Sci.* **2002**, *78*, 71–90. [CrossRef]
132. Hassan, R.A.; Ebeid, T.A.; Abd El-Lateif, A.I.; Ismail, N.B. Effect of dietary betaine supplementation on growth, carcass and immunity of New Zealand White rabbits under high ambient temperature. *Livest. Sci.* **2011**, *135*, 103–109. [CrossRef]
133. Morsy, W.A.; Hassan, R.A.; Abd El-Lateif, A.I. Effect of Dietary Ascorbic Acid and Betaine Supplementation on Productivity of Rabbit Does Under High Ambient Temperature. In Proceedings of the 10th World Rabbit Congress, Sharm El-Sheikh, Egypt, 3–6 September 2012; pp. 279–283.
134. Lian, P.; Braber, S.; Garssen, J.; Wichers, H.J.; Folkerts, G.; Fink-Gremmels, J.; Varasteh, S. Beyond heat stress: Intestinal integrity disruption and mechanism-based intervention strategies. *Nutrients* **2020**, *12*, 734. [CrossRef]
135. Chen, X.; Li, L.; Pu, J.; Cai, J.; Zhao, H.; Jia, G.; Liu, G.; Tian, G. Dietary Betaine improves the intestinal health and growth performance of heat-stressed growing rabbits in summer. *J. Anim. Sci.* **2023**, *101*, skad363. [CrossRef]
136. Li, Z.; Pu, J.; Zeng, T.; Cai, J.; Jia, G.; Zhao, H.; Liu, G.; Zeng, Q.; Luo, Y.; Tian, G. Effects of betaine on growth performance and intestinal health of rabbits fed different digestible energy diets. *J. Anim. Sci.* **2024**, *102*, skae029. [CrossRef]
137. Li, Z.; Pu, J.; Chen, X.; Chen, Y.; Peng, X.; Cai, J.; Jia, G.; Zhao, H.; Tian, G. Betaine addition to the diet alleviates intestinal injury in growing rabbits during the summer heat through the AAT/mTOR pathway. *J. Anim. Sci. Biotechnol.* **2024**, *15*, 41. [CrossRef]

138. Abu Hafsa, S.H.; Centoducati, G.; Hassan, A.A.; Maggiolino, A.; Elghandour, M.M.M.Y.; Salem, A.Z.M. Effects of Dietary Supplementations of Vitamin C, Organic Selenium, Betaine, and Pomegranate Peel on Alleviating the Effect of Heat Stress on Growing Rabbits. *Animals* **2024**, *14*, 950. [CrossRef] [PubMed]
139. Koskinen, E.; Junnila, M.; Katila, T.; Soini, H. A Preliminary Study on the Use of Betaine as a Cryoprotective Agent in Deep Freezing of Stallion Semen. *J. Vet. Med. Ser. A* **1989**, *36*, 110–114. [CrossRef] [PubMed]
140. Trimeche, A.; Yvon, J.M.; Vidament, M.; Palmer, E.; Magistrini, M. Effects of glutamine, proline, histidine and betaine on post-thaw motility of stallion spermatozoa. *Theriogenology* **1999**, *52*, 181–191. [CrossRef]
141. Warren, L.K.; Lawrence, L.M.; Thompson, K.N. The influence of betaine on untrained and trained horses exercising to fatigue. *J. Anim. Sci.* **1999**, *77*, 677–684. [CrossRef]
142. Rose, R.J.; Schlierf, H.A.; Knight, P.K.; Plummer, C.; Davis, M.; Ray, S.P. Effects of N,N-dimethylglycine on cardiorespiratory function and lactate production in thoroughbred horses performing incremental treadmill exercise. *Vet. Rec.* **1989**, *2*, 268–271. [CrossRef]
143. Zhang, W.; Li, X.; Liu, L.; Ma, Y.; Lu, H.; Yang, H.; Ouyang, W.; Yang, K. Effects of supplementary feeding of guanidine acetic acid and different methyl donors on athletic performance, glucose metabolism and antioxidative capacity in Yili horses. *Chin. J. Anim. Nutr.* **2021**, *33*, 6387–6397. [CrossRef]
144. de Oliveira, K.; Fachiolli, D.F.; Watanabe, M.J.; Tsuzukibashi, D.; Bittar, C.M.M.; Costa, C.; Poiatti, M.L.; de Meirelles, P.R.L. Dimethylglycine supplementation in horses performing incremental treadmill exercise. *Comp. Exerc. Physiol.* **2015**, *11*, 167–172. [CrossRef]
145. Lever, M.; Slow, S. The clinical significance of betaine, an osmolyte with a key role in methyl group metabolism. *Clin. Biochem.* **2010**, *43*, 732–744. [CrossRef] [PubMed]
146. Verbrugghe, A.; Bakovic, M. Peculiarities of one-carbon metabolism in the strict carnivorous cat and the role in feline hepatic lipidosis. *Nutrients* **2013**, *5*, 2811–2828. [CrossRef]
147. Jewell, D.E.; Jackson, M.I. Dietary Fatty Acids Change Circulating Fatty Acids, Microbial Putrefactive Postbiotics and Betaine Status in the Cat. *Animals* **2020**, *10*, 2310. [CrossRef] [PubMed]
148. Jewell, D.E.; Jackson, M.I. Dietary betaine and fatty acids change circulating single-carbon metabolites and fatty acids in the dog. *Animals* **2022**, *12*, 768. [CrossRef]
149. Jewell, D.E.; Jackson, M.I. Dietary betaine interacts with very long chain n-3 polyunsaturated fatty acids to influence fat metabolism and circulating single carbon status in the cat. *Animals* **2022**, *12*, 2837. [CrossRef]
150. Horio, M.; Ito, A.; Matsuoka, Y.; Moriyama, T.; Orita, Y.; Takenaka, M.; Imai, E. Apoptosis induced by hypertonicity in Madin Darley canine kidney cells: Protective effect of betaine. *Nephrol. Dial. Transplant.* **2001**, *16*, 483–490. [CrossRef] [PubMed]
151. Ephraim, E.; Jewell, D.E. Effect of Added Dietary Betaine and Soluble Fiber on Metabolites and Fecal Microbiome in Dogs with Early Renal Disease. *Metabolites* **2020**, *10*, 370. [CrossRef]
152. Hall, J.A.; Jewell, D.E.; Ephraim, E. Changes in the faecal metabolome are associated with feeding fiber not health status in cats with chronic kidney disease. *Metabolites* **2020**, *10*, 281. [CrossRef]
153. Mugwanya, M.; Dawood, M.A.; Kimera, F.; Sewilam, H. A meta-analysis on the influence of dietary betaine on the growth performance and feed utilization in aquatic animals. *Aquac. Rep.* **2024**, *37*, 102200. [CrossRef]
154. Adjoumani, J.J.Y.; Wang, K.; Zhou, M.; Liu, W.; Zhang, D. Effect of dietary betaine on growth performance, antioxidant capacity and lipid metabolism in blunt snout bream fed a high-fat diet. *Fish Physiol. Biochem.* **2017**, *43*, 1733–1745. [CrossRef]
155. Barry, K.J.; McClure, R.L.; Trushenski, J.T. Sea Clam-Derived Feeding Stimulants Enhance Acceptability and Intake of Reduced Fish Meal, Soy-Based Sunshine Bass Feeds. *North Am. J. Aquac.* **2017**, *79*, 115–122. [CrossRef]
156. Pinedo-Gil, J.; Martín-Diana, A.B.; Bertotto, D.; Sanz-Calvo, M.A.; Jover-Cerdá, M.; Tomás-Vidal, A. Effects of dietary inclusions of red beet and betaine on the acute stress response and muscle lipid peroxidation in rainbow trout. *Fish. Physiol. Biochem.* **2018**, *44*, 939–948. [CrossRef]
157. Dong, X.; Wang, J.; Ji, P.; Gao, X.; Sun, L.; Miao, S.; Lei, Y.; Du, X.; Zhang, X. Dietary betaine supplementation promotes growth, n-3 LC-PUFA content and innate immunity in Macrobrachium rosenbergii. *Aquaculture* **2020**, *525*, 735308. [CrossRef]
158. Ismail, T.; Hegazi, E.; Dawood, M.A.; Nassef, E.; Bakr, A.; Paray, B.A.; Van Doan, H. Using of betaine to replace fish meal with soybean or/and corn gluten meal in Nile tilapia (Oreochromis niloticus) diets: Histomorphology, growth, fatty acid, and glucose-related gene expression traits. *Aquac. Rep.* **2020**, *17*, 100376. [CrossRef]
159. La Muhamad, I.; Setiawati, M.; Wiyoto, W.; Ekasari, J. Dietary supplementation of betain to improve the growth and feed utilization in hybrid grouper (*Epinephelus lanceolatus*♂ × *Epinephelus fuscoguttatus*♀) juvenile. *J. Akuakultur. Indones.* **2021**, *20*, 24–33. [CrossRef]
160. Lu, J.F.; Luo, S.; Jin, T.C.; Wang, L.C.; Yang, G.J.; Lu, X.J.; Chen, J. Betaine protects ayu (*Plecoglossus altivelis*) against Vibrio anguillarum infection in salinity by regulating the immunomodulatory activity of monocytes/macrophages. *Aquaculture* **2021**, *536*, 736482. [CrossRef]

161. Mohseni, M.; Saltanat, N.L.; Rastravan, M.E.; Golalipour, Y. Effects of betaine supplementation in plant-protein-based diets on growth performance, haemato-immunological parameters, antioxidant status and digestive enzyme activities of juvenile Caspian trout (Salmo trutta, Kessler, 1877). *Aquac. Nutr.* **2021**, *27*, 2132–2141. [CrossRef]
162. Yeşilayer, N.; Kaymak, I.E. Effect of partial replacement of dietary fish meal by soybean meal with betaine attractant supplementation on growth performance and fatty acid profiles of juvenile rainbow trout (Oncorhynchus mykiss). *Aquac. Res.* **2020**, *51*, 1533–1541. [CrossRef]
163. Ismail, T.; Nassef, E.; Hegazi, E.; Bakr, A.; Moustafa, E.M.; Abdo, W.; Elbialy, Z.I. The modulatory effect of dietary betaine on intestinal absorptive capacity, lipogenesis and expression of lipid metabolism- and growth-related genes in Nile tilapia fed on soybean meal-based diet. *Slov. Vet. Res.* **2019**, *56*, 25–38. [CrossRef]
164. Ali, H.H.A.; Al-Faragi, J.K.A. Efficiency of betaine and β-glucan as feed additives on the growth performance and survival rate of common carp (*Cyprinus carpio* L.) fingerlings. *J. Entomol. Zool. Stud.* **2017**, *5*, 27–31.
165. Tiril, S.U.; Alagil, F.; Yagci, F.B. Effects of betaine supplementation in plant protein-based diets on feed intake and growth performance in rainbow trout (Oncorhynchus mykiss). *Isr. J. Aquac.* **2008**, *60*, 57–64. [CrossRef]
166. Tusche, K.; Berends, K.; Wuertz, S.; Susenbeth, A.; Schulz, C. Evaluation of feed attractants in potato protein concentrate based diets for rainbow trout (Oncorhynchus mykiss). *Aquaculture* **2011**, *321*, 54–60. [CrossRef]
167. Lin, J.; Zhang, C.; Lu, K.; Song, K.; Wang, L. Effect of guanidinoacetic acid and betaine supplementation in soybean meal-based diets on growth performance, muscle energy metabolism and methionine utilization in the bullfrog Lithobates catesbeianus. *Aquaculture* **2021**, *533*, 736167. [CrossRef]
168. Kolkovski, S.; Arieli, A.; Tandler, A. Visual and chemical cues stimulate microdiet ingestion in sea bream larvae. *Aquac. Int.* **1997**, *5*, 527–536. [CrossRef]
169. Koven, W.; Kolkovski, S.; Hadas, E.; Gamsiz, K.; Tandler, A. Advances in the development of microdiets for gilthead seabream, Sparus aurata: A review. *Aquaculture* **2001**, *194*, 107–121. [CrossRef]
170. Ghosh, T.K.; Chauhan, Y.H.; Mandal, R.N. Growth performance of Labeo bata (Hamilton, 1822) in freshwater and its acclimatization in brackish water with betaine as feed additive. *Aquaculture* **2019**, *501*, 128–134. [CrossRef]
171. Altuntaş, E.; Ileri, C.; Yapar, F.; Karadaş, A.; Kibar, U. Betaine osmoregulation functioning against salt stress in methicillin-resistant Staphylococcus aureus (MRSA) Metisiline dirençli Staphylococcus aureus (MRSA)'da tuz stresine karşı betain ozmoregülasyon fonksiyonu. *Turk Hij. Ve Deney. Biyol. Dergisi* **2024**, *81*, 121–134. [CrossRef]
172. Dong, X.; Xue, W.; Hua, J.; Hang, Y.; Sun, L.; Miao, S.; Wei, W.; Wu, X.; Du, X. Effects of dietary betaine in allogynogenetic gibel carp (*Carassius auratus* gibelio): Enhanced growth, reduced lipid deposition and depressed lipogenic gene expression. *Aquac. Res.* **2018**, *49*, 1967–1972. [CrossRef]
173. Jin, M.; Shen, Y.; Pan, T.; Zhu, T.; Li, X.; Xu, F.; Betancor, M.B.; Jiao, L.; Tocher, D.R.; Zhou, Q. Dietary betaine mitigates hepatic steatosis and inflammation induced by a high-fat-diet by modulating the Sirt1/Srebp-1/Pparα pathway in juvenile black seabream (*Acanthopagrus schlegelii*). *Front. Immunol.* **2021**, *12*, 694720. [CrossRef]
174. Risha, E.; Ahmed, F.; Khaled, A.A.; Hossain, F.M.A.; Akhtar, N.; Zahran, E. Interactive effects of dietary betaine and chromium picolinate on the immunomodulation, antioxidative response and disease resistance of Nile tilapia (*Oreochromis niloticus*). *Aquac. Res.* **2022**, *53*, 3464–3477. [CrossRef]
175. Shapawi, R.; Abdullah, F.C.; Senoo, S.; Mustafa, S. Nutrition, growth and resilience of tiger grouper (*Epinephelus fuscoguttatus*) × giant Grouper (*Epinephelus lanceolatus*) hybrid–A review. *Rev. Aquac.* **2019**, *11*, 1285–1296. [CrossRef]
176. Niroomand, M.; Sajadi, M.M.; Yahyavi, M.; Asadi, M. Effects of dietary Betaine on growth, survival, body composition and resistance of fry rainbow trout (*Oncorhynchus mykiss*) under environmental stress. *Iran. Sci. Fish. J.* **2011**, *20*, 135–146. [CrossRef]
177. Ali, M.E.; Abubakar, K.A.; Sogbesan, O.A. Growth performance and nutrient utilization of *Clarias gariepinus* (Burchell, 1822) fingerlings fed dietary betaine additive diets. In Proceedings of the 37th Annual Conference of the Fisheries Society of Nigeria (FISON), Yola, Nigeria, 30 October–4 November 2022; pp. 133–137.
178. Safari, R.; Hoseinifar, S.H.; Imanpour, M.R.; Hajibegloo, A.; Sanchouli, H.; Homayouni, M.; Siddik, M.A. The effects of multi-enzyme and betaine on growth performance, body composition haemato-immunological parameters and expression of growth-related genes in beluga (Huso huso). *Aquaculture* **2022**, *549*, 737784. [CrossRef]
179. Hajirezaee, S.; Ramezani, S.; Ahani, S. Betaine and the probiotic, *Lactobacillus rhamnosus* in the diet of the Common carp, *Cyprinus carpio*: Effects on growth, digestive enzyme activities, antioxidant system, humoral and mucosal immunity and resistance to Streptococcus iniae. *Aquac. Rep.* **2024**, *38*, 102282. [CrossRef]
180. Coman, G.J.; Sarac, H.Z.; Fielder, D.; Thorne, M. Evaluation of crystalline amino acids, betaine and AMP as food attractants of the giant tiger prawn (*Penaeus monodon*). *Comp. Biochem. Physiol. Part A Physiol.* **1996**, *113*, 247–253. [CrossRef]
181. He, G.L.; Shi, M.L.; Liu, Y.C.; Chen, L.T.; Chen, X.; Zhu, W.B.; Chen, Z.B.; Tan, B.P.; Xie, S.W. Effects of dietary betaine supplementation on growth performance, feed intake, intestinal histology, lipid metabolism, and immune response of black tiger shrimp (*Penaeus monodon*) fed diets containing two levels of raw feed attractants. *Aquac. Int.* **2024**, *32*, 653–673. [CrossRef]

182. Zakipour, R.E.; Akbari, M.; Arshadi, A.; Effatpanah, E. Effect of different levels of dietary Betaine on growth performance, food efficiency and survival rate of pike perch (*Sander lucioperca*) fingerlings. *Iran. J. Fish. Sci.* **2012**, *11*, 902–910.
183. Azimirad, M.; Farhangi, M.; Amiri, B.; Effatpanah Komaee, I.; Mansouri Taee, H. Effects of different levels of betaine supplementation in diet on growth, feed efficiency indices and survival rate of pikeperch (*Sander luciopreca*) larvae. *J. Fish.* **2013**, *66*, 347–358. [CrossRef]
184. Löest, C.A.; Titgemeyer, E.C.; Drouillard, J.S.; Coetzer, C.M.; Hunter, R.D.; Bindel, D.J.; Lambert, B.D. Supplemental Betaine and Peroxide-Treated Feather Meal for Finishing Cattle. *J. Anim. Sci.* **2002**, *80*, 2234–2240. [CrossRef]
185. Dunshea, F.R.; Oluboyede, K.; DiGiacomo, K.; Leury, B.J.; Cottrell, J.J. Betaine Improves Milk Yield in Grazing Dairy Cows Supplemented with Concentrates at High Temperatures. *Animals* **2019**, *9*, 57. [CrossRef]
186. Abhijith, A.; Dunshea, F.R.; Chauhan, S.S.; Sejian, V.; DiGiacomo, K. A Meta-Analysis of the Effects of Dietary Betaine on Milk Production, Growth Performance, and Carcass Traits of Ruminants. *Animals* **2024**, *14*, 1756. [CrossRef]
187. Pinotti, L.; Baldi, A.; Dell'Orto, V. Comparative mammalian choline metabolism with emphasis on the high-yielding dairy cow. *Nutr. Res. Rev.* **2002**, *15*, 315–332. [CrossRef] [PubMed]
188. Firincioglu, S.Y.F.Y.; Irshad, N. Role of Choline in Ruminant Nutrition: A Detailed Review. *J. Hell. Vet. Med. Soc.* **2022**, *73*, 3921–3928. [CrossRef]
189. Hall, L.W.; Dunshea, F.R.; Allen, J.D.; Rungruang, S.; Collier, J.L.; Long, N.M.; Collier, R.J. Evaluation of Dietary Betaine in Lactating Holstein Cows Subjected to Heat Stress. *J. Dairy Sci.* **2016**, *99*, 9745–9753. [CrossRef] [PubMed]
190. Cheng, K.F.; Wang, C.; Zhang, G.W.; Du, H.S.; Wu, Z.Z.; Liu, Q.; Guo, G.; Huo, W.J.; Zhang, J.; Chen, L.; et al. Effects of Betaine and Rumen-Protected Folic Acid Supplementation on Lactation Performance, Nutrient Digestion, Rumen Fermentation and Blood Metabolites in Dairy Cows. *Anim. Feed Sci. Technol.* **2020**, *262*, 114445. [CrossRef]
191. Mahmood, M.; Khiaosa-Ard, R.; Zebeli, Q.; Petri, R.M. Betaine Modulates Rumen Archaeal Community and Functioning during Heat and Osmotic Stress Conditions In Vitro. *Archaea* **2020**, *2020*, 8875773. [CrossRef]
192. Tapio, I.; Snelling, T.J.; Strozzi, F.; Wallace, R.J. The Ruminal Microbiome Associated with Methane Emissions from Ruminant Livestock. *J. Anim. Sci. Biotechnol.* **2017**, *8*, 7. [CrossRef]
193. Poulsen, M.; Schwab, C.; Borg Jensen, B.; Engberg, R.M.; Spang, A.; Canibe, N.; Højberg, O.; Milinovich, G.; Fragner, L.; Schleper, C.; et al. Methylotrophic Methanogenic Thermoplasmata Implicated in Reduced Methane Emissions from Bovine Rumen. *Nat. Commun.* **2013**, *4*, 1428. [CrossRef]
194. Liu, C.; Wang, C.; Zhang, J.; Liu, Q.; Guo, G.; Huo, W.J.; Pei, C.X.; Chen, L.; Zhang, Y.L. Guanidinoacetic Acid and Betaine Supplementation Have Positive Effects on Growth Performance, Nutrient Digestion and Rumen Fermentation in Angus Bulls. *Anim. Feed Sci. Technol* **2021**, *276*, 114923. [CrossRef]
195. Wang, C.; Liu, C.; Zhang, G.W.; Du, H.S.; Wu, Z.Z.; Liu, Q.; Guo, G.; Huo, W.J.; Zhang, J.; Pei, C.X.; et al. Effects of Rumen-Protected Folic Acid and Betaine Supplementation on Growth Performance, Nutrient Digestion, Rumen Fermentation and Blood Metabolites in Angus Bulls. *Br. J. Nutr.* **2020**, *123*, 1109–1116. [CrossRef]
196. Elhendawy, W.; Zeineldin, M.; Nassif, M.; Seboussi, R.; Gomaa, N.; Donia, M.; Abdelmegeid, M. Effects of Betaine Supplementation on Heat Stress-Associated Blood Metabolites, Antioxidant Profiles and Milk Production in Dairy Cows. *Egypt. J. Vet. Sci.* **2025**, *56*, 779–789. [CrossRef]
197. Peterson, S.E.; Rezamand, P.; Williams, J.E.; Price, W.; Chahine, M.; McGuire, M.A. Effects of Dietary Betaine on Milk Yield and Milk Composition of Mid-Lactation Holstein Dairy Cows. *J. Dairy Sci.* **2012**, *95*, 6557–6562. [CrossRef] [PubMed]
198. Malik, M.I.; Bilal, M.; Anwar, M.Z.; Hassan, T.; Rashid, M.A.; Tarla, D.; Dunshea, F.R.; Cheng, L. Effects of Betaine Supplementation on Dry Matter Intake, Milk Characteristics, Plasma Non-Esterified Fatty Acids, and β-Hydroxybutyric Acid in Dairy Cattle: A Meta-Analysis. *J. Anim. Sci.* **2024**, *102*, skae241. [CrossRef]
199. Wang, B.; Wang, C.; Guan, R.; Shi, K.; Wei, Z.; Liu, J.; Liu, H. Effects of Dietary Rumen-Protected Betaine Supplementation on Performance of Postpartum Dairy Cows and Immunity of Newborn Calves. *Animals* **2019**, *9*, 167. [CrossRef] [PubMed]
200. Löest, C.A.; Drouillard, J.; Titgemeyer, E.; Hunter, R.D.; Wessels, R.H. Betaine as a Dietary Supplement for Finishing Cattle. *Kans. Agric. Exp. Stn. Res. Rep.* **1998**, *1*, 76–78. [CrossRef]
201. Bock, B.J.; Brethour, J.R.; Harmoney, K.R.; Goodall, S.R. Influence of Betaine on Pasture, Finishing, and Carcass Performance in Steers. *Prof. Anim. Sci.* **2004**, *20*, 53–57. [CrossRef]
202. Parker, A.J.; Dobson, G.P.; Fitzpatrick, L.A. Physiological and Metabolic Effects of Prophylactic Treatment with the Osmolytes Glycerol and Betaine on Bos Indicus Steers during Long Duration Transportation. *J. Anim. Sci.* **2007**, *85*, 2916–2923. [CrossRef]
203. Davidson, S.; Hopkins, B.A.; Odle, J.; Brownie, C.; Fellner, V.; Whitlow, L.W. Supplementing Limited Methionine Diets with Rumen-Protected Methionine, Betaine, and Choline in Early Lactation Holstein Cows. *J. Dairy Sci.* **2008**, *91*, 1552–1559. [CrossRef]
204. Wang, C.; Liu, Q.; Yang, W.Z.; Wu, J.; Zhang, W.W.; Zhang, P.; Dong, K.H.; Huang, Y.X. Effects of Betaine Supplementation on Rumen Fermentation, Lactation Performance, Feed Digestibilities and Plasma Characteristics in Dairy Cows. *J. Agric. Sci.* **2010**, *148*, 487–495. [CrossRef]

205. Williams, S.R.O.; Milner, T.C.; Garner, J.B.; Moate, P.J.; Jacobs, J.L.; Hannah, M.C.; Wales, W.J.; Marett, L.C. Dietary Fat and Betaine Supplements Offered to Lactating Cows Affect Dry Matter Intake, Milk Production and Body Temperature Responses to an Acute Heat Challenge. *Animals* **2021**, *11*, 3110. [CrossRef]
206. Zhang, L.; An, W.J.; Lian, H.; Zhou, G.B.; Han, Z.Y.; Ying, S.J. Effects of Dietary Betaine Supplementation Subjected to Heat Stress on Milk Performances and Physiology Indices in Dairy Cow. *Genet. Mol. Res.* **2014**, *13*, 7577–7586. [CrossRef]
207. Monteiro, A.P.A.; Bernard, J.K.; Guo, J.-R.; Weng, X.-S.; Emanuele, S.; Davis, R.; Dahl, G.E.; Tao, S. Effects of Feeding Betaine-Containing Liquid Supplement to Transition Dairy Cows. *J. Dairy Sci.* **2017**, *100*, 1063–1071. [CrossRef] [PubMed]
208. Raheja, N.; Kumar, N.; Lathwal, S.S. Dietary Betaine Reduces Incidence of Follicular Cyst in Post-Partum Karan Fries Cows during Hot-Humid Season. *Indian J. Anim. Sci.* **2019**, *89*, 1332–1337. [CrossRef]
209. Shah, A.M.; Ma, J.; Wang, Z.; Zou, H.; Hu, R.; Peng, Q. Betaine Supplementation Improves the Production Performance, Rumen Fermentation, and Antioxidant Profile of Dairy Cows in Heat Stress. *Animals* **2020**, *10*, 634. [CrossRef] [PubMed]
210. Wang, C.; Liu, H.; Wang, C.; Liu, J.; Liu, H. Effects of Dietary Rumen-Protected Betaine on Lactation Performance and Serum Metabolites of Mid-Lactation Holstein Dairy Cows. *J. Agric. Food Chem.* **2020**, *68*, 13154–13159. [CrossRef]
211. Kassab, A.; Hamdon, H.; Daghash, H.; Soliman, A. Impact of Betaine Supplementation as Anti Stress on Some Hematological Parameters and Thermoregulatory Responses of Aberdeen Angus Cows in Arid Subtropical Regions. *New Val. J. Agric. Sci.* **2021**, *1*, 89–97. [CrossRef]
212. Cronje, P.B. Essential Role of Methyl Donors in Animal Productivity. *Anim. Prod. Sci.* **2016**, *58*, 655–665. [CrossRef]
213. Löest, C.A.; Bindel, D.J.; Titgemeyer, E.C.; Drouillard, J.S.; Blasi, D.A. Soybean hulls in roughage-free diets for limit-fed growing cattle. *Kans. Agric. Exp. Stat. Res.* **2017**, 60–62. [CrossRef]
214. Li, C.; Wang, Y.; Li, L.; Han, Z.; Mao, S.; Wang, G. Betaine Protects against Heat Exposure–Induced Oxidative Stress and Apoptosis in Bovine Mammary Epithelial Cells via Regulation of ROS Production. *Cell Stress Chaperones* **2019**, *24*, 453–460. [CrossRef]
215. Zhao, N.; Yang, Y.; Xu, H.; Li, L.; Hu, Y.; Liu, E.; Cui, J. Betaine Protects Bovine Mammary Epithelial Cells against LPS-Induced Inflammatory Response and Oxidative Damage via Modulating NF-κB and Nrf2 Signalling Pathway. *Ital. J. Anim. Sci.* **2022**, *21*, 859–869. [CrossRef]
216. Kim, M.J. Betaine Enhances the Cellular Survival via Mitochondrial Fusion and Fission Factors, MFN2 and DRP1. *Anim. Cells Syst.* **2018**, *22*, 289–298. [CrossRef]
217. Lee, I. Betaine Is a Positive Regulator of Mitochondrial Respiration. *Biochem. Biophys. Res. Commun.* **2015**, *456*, 621–625. [CrossRef] [PubMed]
218. Nakai, T.; Sato, T.; Teramura, M.; Sadoya, H.; Ohtani, M.; Takahashi, T.; Kida, K.; Hidaka, S. The Effect of a Continuous Supply of Betaine on the Degradation of Betaine in the Rumen of Dairy Cows. *Biosci. Biotechnol. Biochem.* **2013**, *77*, 666–669. [CrossRef] [PubMed]
219. Fernández, C.; López-Saez, A.; Gallego, L.; de la Fuente, J.M. Effect of Source of Betaine on Growth Performance and Carcass Traits in Lambs. *Anim. Feed Sci. Technol.* **2000**, *86*, 71–82. [CrossRef]
220. Fernández, C.; Sánchez-Seiquer, P.; Sánchez, A.; Contreras, A.; de la Fuente, J.M. Influence of Betaine on Milk Yield and Composition in Primiparous Lactating Dairy Goats. *Small Rumin. Res.* **2004**, *52*, 37–43. [CrossRef]
221. Ghoneem, W.M.A.; El-Tanany, R.R.A. Impact of Natural Betaine Supplementation on Rumen Fermentation and Productive Performance of Lactating Damascus Goats. *Trop. Anim. Health Prod.* **2023**, *55*, 123. [CrossRef]
222. Sánchez, P.; Muelas, R.; Sánchez, A.; Rubert, J.; Montanel, J.; Luengo, C.; Contreras, A.; Fernández, C. Influence of betaine on somatic cell count and goat milk composition. In Proceedings of the Annual Meeting of European Association for Animal Production, Budapest, Hungary, 26–29 August 2001.
223. Fernández, C.; Gallego, L.; Lopez-Bote, C.J. Effect of Betaine on Fat Content in Growing Lambs. *Anim. Feed Sci. Technol.* **1998**, *73*, 329–338. [CrossRef]
224. Banskalieva, V.; Puchala, R.; Goetsch, A.L.; Luo, J.; Sahlu, T. Effects of Ruminally Protected Betaine and Choline on Net Flux of Nutrients across the Portal-Drained Viscera and Liver of Meat Goat Wethers Consuming Diets Differing in Protein Concentration. *Small Rumin. Res.* **2005**, *57*, 193–202. [CrossRef]
225. Dong, L.; Zhong, Z.X.; Cui, H.H.; Wang, S.N.; Luo, Y.; Yu, L.H.; Loor, J.J.; Wang, H.R. Effects of Rumen-Protected Betaine Supplementation on Meat Quality and the Composition of Fatty and Amino Acids in Growing Lambs. *Animal* **2020**, *14*, 435–444. [CrossRef]
226. Brougham, B.-J.; Weaver, A.C.; Swinbourne, A.M.; Cottrell, J.J.; Kelly, J.M.; Kleemann, D.O.; van Wettere, W.H.E.J. Supplementing Dietary Betaine during Late Gestation Increases Plasma Betaine and Methionine Concentrations in Pregnant Merino Ewes but Not Neonatal Lambs. *Small Rumin. Res.* **2024**, *232*, 107226. [CrossRef]
227. Sahraei, H.R.; Kiani, A.; Azarfar, A.; Khamisabadi, H. Effect of Late Gestational Betaine Supplementation on Intermediate Metabolites, Homocysteine and Lipid Peroxidation in Pregnant Ewes and Their Offspring. *Iran. J. Appl. Anim. Sci.* **2020**, *10*, 483–489.

228. Yu, L.; Jin, Y.; Cui, H.; Luo, Y.; Dong, L.; Wang, H. Effects of Dietary Rumen-Protected Betaine Supplementation on the Antioxidant Status of Lambs. *Livest. Sci.* **2020**, *237*, 104026. [CrossRef]
229. Cai, Y.; Deng, M.; Zhang, Q.; Liu, Z.; Wang, L.; Sheng, W.; Zhang, Y.; You, P.; Wang, Z.; Wang, F. Effects of Dietary Betaine Supplementation on Biochemical Parameters of Blood and Testicular Oxidative Stress in Hu Sheep. *Theriogenology* **2021**, *164*, 65–73. [CrossRef]
230. Liu, K.; Li, F.; Tang, D.F.; Li, F.D.; Liang, Y.S.; Li, G.Z.; Deng, Y. Effects of Betaine and Rumen-Protected Fat on Finishing Production Performance and Regulation of Digestion and Metabolism in Hu Sheep. *Prataculturd Sci.* **2016**, *33*, 2565–2575.
231. Dangi, S.S.; Dangi, S.K.; Chouhan, V.S.; Verma, M.R.; Kumar, P.; Singh, G.; Sarkar, M. Modulatory Effect of Betaine on Expression Dynamics of HSPs during Heat Stress Acclimation in Goat (Capra Hircus). *Gene* **2016**, *575*, 543–550. [CrossRef] [PubMed]
232. DiGiacomo, K.; Simpson, S.; Leury, B.J.; Dunshea, F.R. Dietary Betaine Impacts the Physiological Responses to Moderate Heat Conditions in a Dose Dependent Manner in Sheep. *Animals* **2016**, *6*, 51. [CrossRef]
233. Brougham, B.-J.; Weaver, A.C.; Swinbourne, A.M.F.; Lewis Baida, B.E.; Kelly, J.M.; Walker, S.K.; Kleemann, D.O.; van Wettere, W.H.E.J. Maternal Supplementation with Dietary Betaine during Gestation to Improve Twin Lamb Survival. *Animals* **2020**, *10*, 1749. [CrossRef]
234. DiGiacomo, K.; Simpson, S.; Leury, B.J.; Dunshea, F.R. Dietary Betaine Impacts Metabolic Responses to Moderate Heat Exposure in Sheep. *Animals* **2023**, *13*, 1691. [CrossRef]
235. Brougham, B.-J.; Weaver, A.C.; Swinbourne, A.M.; Tscharke, M.R.; Munn, A.L.; Kelly, J.M.; Kleemann, D.O.; van Wettere, W.H.E.J. Maternal Supplementation with Dietary Betaine during Late Gestation Increased Ewe Plasma Creatine and Lamb Thermoregulation under Field Conditions. *Animals* **2024**, *14*, 2605. [CrossRef]

antioxidants

MDPI

Review

Dietary Supplements Derived from Food By-Products for the Management of Diabetes Mellitus

Ezgi Nur Yuksek [1], Antia G. Pereira [1,2,*] and Miguel A. Prieto [1,*]

1 Nutrition and Food Group (NuFoG), Instituto de Agroecoloxía e Alimentación (IAA), Universidade de Vigo, Campus Auga, 32004 Ourense, Spain; ezginur.yuksek@uvigo.gal

2 Investigaciones Agroalimentarias Research Group, Galicia Sur Health Research Institute (IIS Galicia Sur), SERGAS-UVIGO, 36312 Vigo, Spain

* Correspondence: antia.gonzalez.pereira@uvigo.es (A.G.P.); mprieto@uvigo.es (M.A.P.)

Abstract

The rising global incidence of diabetes has heightened the demand for prevention strategies that are both effective and environmentally sustainable. In this context, dietary supplements obtained from food processing by-products have emerged as promising candidates, combining high nutritional value with the potential to reduce food waste. These by-products contain abundant bioactive compounds, such as antioxidants, dietary fiber, vitamins, and minerals, that have been associated with improved glycemic regulation. Their beneficial effects are mediated through several interconnected biological mechanisms, including enhanced insulin sensitivity, attenuation of oxidative stress, and modulation of inflammatory pathways. The efficacy of these supplements is supported by findings from preclinical models, clinical trials, and meta-analyses, which also allow comparison with standard antidiabetic interventions. Alongside these findings, considerations related to safety, toxicity, and the regulatory framework are critical for their integration into preventive health strategies. Furthermore, market trends, technological challenges in supplement formulation, and ethical issues surrounding the valorization of food waste are key factors influencing their development and acceptance. Together, these insights underscore the dual therapeutic and ecological potential of food by-product-derived supplements in diabetes prevention, while identifying essential avenues for future research and innovation.

Keywords: diabetes prevention; insulin resistance; oxidative stress; inflammation modulation; bioactive compounds; clinical efficacy; regulatory framework; functional ingredients

1. Introduction

Diabetes mellitus (DM) is a complex and multifactorial metabolic disease characterized by high blood glucose levels, resulting from an alteration in the secretion or action of insulin, the hormone responsible for regulating the metabolism of carbohydrates, fats, and proteins [1]. DM remains a major global public health challenge, owing to its steadily increasing prevalence in both industrialized and developing countries (Figure 1). Recent estimations from the International Diabetes Federation indicate that nearly 589 million adults aged 20 to 79 are currently living with diabetes worldwide, representing approximately one in nine individuals. This number is projected to rise to 700 million by 2045 and 853 million by 2050, with the increase expected to be more pronounced in urban areas and high-income countries [2,3].

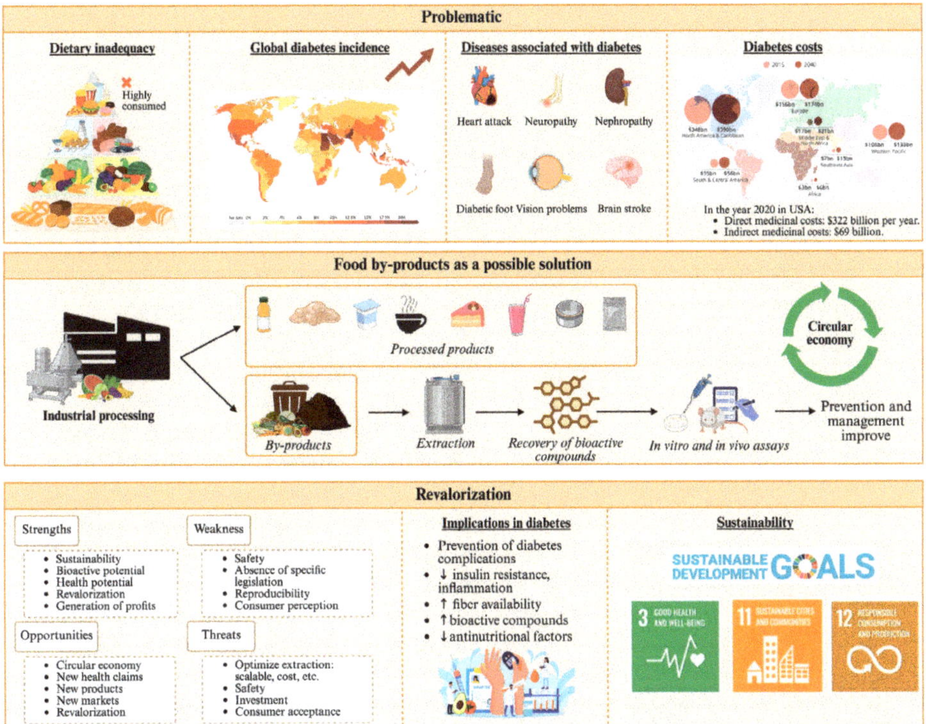

Figure 1. Exploiting food by-products for the development of supplements with antidiabetic activity. Abbreviations. ↓: decrease; ↑: increase.

The most frequent forms of this chronic disease are categorized as type 1 diabetes, type 2 diabetes (T2DM) and gestational diabetes. All these types are associated with significant health risks and complications, particularly when not adequately controlled. Such complications include, among others, kidney disease, cardiovascular disorders, neuropathy, and retinopathy, which can substantially impact patients' quality of life and life expectancy [4]. Under certain conditions, the incidence of these complications is disproportionately high. For example, in T2DM, chronic kidney disease is one of the most prevalent complications, affecting approximately 40% of individuals living with DM and contributing significantly to the global health burden associated with the disease [5].

Therefore, the growing prevalence of diabetes and its complications underscores the necessity of implementing effective prevention and management strategies. Such actions are vital to contain healthcare costs and to address the negative impact on patients' quality of life [6]. Although conventional pharmacological treatments are effective for glycemic control and prevention of complications, there are inherent limitations, such as side effects, variable adherence, and high costs, which motivate the exploration of alternatives based on functional nutrition and dietary supplements, which have garnered increasing attention due to their modifiable nature and wide accessibility. Specifically, dietary supplements containing bioactive compounds with antioxidant, anti-inflammatory, and insulin-sensitizing effects show significant promise as adjunctive measures in the prevention and management of T2DM [7]. For example, some polyphenols have been shown to improve insulin sensitivity, reduce oxidative stress, and modulate chronic inflammation, key factors in the pathophysiology of T2DM [8]. Furthermore, dietary fiber and other components can positively influence glucose metabolism and the gut microbiota, contributing to improved metabolic regulation [9]. Therefore, dietary interventions with these compounds may help mitigate oxidative stress, reduce chronic inflammation, and improve glucose metabolism, thereby potentially lowering the risk of diabetes onset and slowing disease progression [10,11].

However, despite these promising findings, there remains a lack of data for many dietary supplements regarding critical aspects such as mechanisms of action, pharmacokinetics, and potential toxicity [12]. In fact, current scientific evidence indicates that the efficacy and long-term effects of dietary supplements in diabetes prevention and management remain insufficiently characterized and require further rigorous investigation [13]. It is essential that these investigations consider that diabetes can be caused by a variety of factors, including genetic factors, alterations in insulin production or action, as well as environmental influences and lifestyles that contribute to metabolic dysfunction [1]. This implies that, in some types of diabetes, such as T1DM, which is an autoimmune disease with a fundamentally genetic etiology, the use of nutritional supplements would not have a significant impact on disease prevention, although it could contribute to reducing the risk of developing associated complications. In contrast, in T2DM, the development of which is closely linked to lifestyle, these supplements could play a relevant role in prevention by supporting the control of metabolic risk factors [14,15]. Moreover, diabetes constitutes a heterogeneous metabolic disease, in which metabolic characteristics vary significantly between individuals [16]. Therefore, the metabolism, absorption, and excretion of dietary supplements and bioactive compounds of interest can differ considerably, directly impacting their efficacy and bioavailability [17]. This interindividual variability underscores the importance of adopting personalized approaches to optimize the design and implementation of such nutritional interventions [18]. Furthermore, despite the potential preventive effect attributed to the consumption of these types of supplements, it is essential to consider that these supplements should never replace medical treatment in individuals diagnosed with diabetes. Rather, they should be understood as a complement to conventional therapies, with the potential interaction between supplements and pharmacological treatments carefully evaluated to ensure the safety and efficacy of the therapeutic approach [12].

Integrating such nutrition-based approaches into comprehensive DM prevention programs could contribute substantially to reducing the global burden of diabetes and its related complications [19]. In fact, several studies report that individuals with diabetes are 1.6 times more likely to use this kind of products than the general population [17]. Furthermore, these nutritional alternatives could be accepted by consumers, given their preference and demand for natural products, considered healthier and with a lower risk of side effects [20]. The sustainability of these supplements would be enhanced by the possibility of formulating them from extracts obtained from agro-industrial byproducts, raw material rich in bioactive compounds (e.g., polyunsaturated fatty acids, polyphenols, biopeptides, vitamins, dietary fiber, etc.), thus promoting waste recovery and minimizing environmental impact [21]. This approach contributes to the implementation of circular economy models, in which resources are efficiently reused, reducing waste generation and fostering a more responsible and sustainable production system [22]. In addition, this approach is cost-effective by generating added value to materials that would otherwise represent a cost for the food industry [23].

In this review, the potential contribution of dietary supplements derived from food waste to the prevention and management of diabetes is examined, with particular focus on their mechanisms of action, efficacy, safety profiles, toxicological considerations, and regulatory frameworks. Sustainability, ethical issues, and market dynamics are also explored, offering a multidisciplinary perspective on how the valorization of food waste may support both public health and environmental objectives.

2. Review Methodology

This review aims to provide a comprehensive overview of dietary supplements derived from food by-products for the management of diabetes mellitus, with particular attention to studies published since the year 2015. Given the limited availability of clinical studies in this area, earlier experimental research was also considered when relevant. A systematic approach was employed to ensure a thorough and unbiased synthesis of the literature. An extensive search was conducted across major academic databases, including PubMed, Scopus, Google Scholar, and Web of Science, to identify peer-reviewed articles, reviews, books, and conference proceedings relevant to the topic. The search strategy included keywords and combinations such as: "food by-products", "dietary supplements", "bioactive compounds", "diabetes", "glycemic control", "in vivo", "in vitro", and "clinical trial".

Studies were included based on their scientific relevance, methodological quality, and alignment with the objectives of the review. Exclusion criteria comprised non-peer-reviewed sources, studies lacking sufficient methodological detail, works not readily accessible, and those unrelated to by-product-derived preparations or their potential antidiabetic effects. A thematic synthesis was performed to identify consistent patterns, highlight research gaps, and provide a critical perspective on the current state of knowledge and directions for future investigation.

3. Food By-Products as Sources of Dietary Supplements for Diabetes Prevention

Each year, the food industry produces a significant quantity of materials historically regarded as waste or residues, but now commonly referred to as by-products. These are generated across all phases of the food supply chain, including harvesting, industrial processing, storage, and distribution [24]. Among the most common by-products are fruit and vegetable by-products (e.g., peels, seeds, pulp, bagasse, stems, leaves), products from the meat or fish industry (e.g., bones, spines, viscera), flour and wastewater from food industries, and other remnants that are not part of the final product [25]. All these food by-products are characterized by their richness in bioactive compounds with different biological properties, including their antioxidant, anti-inflammatory, and antitumor capacities [24]. Furthermore, scientific evidence supports the idea that many bioactive compounds exert beneficial effects in the prevention and management of chronic non-communicable diseases, especially T2DM [26]. Among the main compounds identified with antidiabetic properties are polyphenols, dietary fibers, vitamins, minerals, and other phytonutrients of functional interest [27].

Of all bioactive compounds, polyphenols have emerged as the primary focus of scientific research over the past few decades. This class of compounds can be categorized into various subclasses, including flavonoids, anthocyanins, tannins, and phenolic acids [28,29]. These molecules are characterized by reducing oxidative stress, a mechanism that significantly contributes to pancreatic β-cell dysfunction and the development of insulin resistance in individuals with T2DM. They also reduce inflammation through nuclear factor erythroid 2–related factor 2 (Nrf2) activation and nuclear factor κB (NF-Kb) inhibition and inhibit enzymes involved in carbohydrate digestion [30,31]. The bioactive compounds underlying these effects are derived from a diverse array of agro-industrial by-products (Table 1). For instance, grape seeds are an important source of proanthocyanidins and flavonoids [32,33]. Coffee grounds are particularly rich in chlorogenic acids, which have demonstrated hypoglycemic effects through mechanisms such as the inhibition of glucose absorption and the modulation of glucose metabolism [34]. Tea residues, especially from green and black tea, are a relevant source of flavonoids (primarily catechins, theaflavins, and flavanols) which contribute to antidiabetic activity by enhancing insulin sensitivity, regulating glucose uptake in peripheral tissues, and inhibiting key enzymes involved in carbohydrate digestion, such as α-amylase and α-glucosidase [35,36]. In addition to these sources, various underutilized and regionally common plants have also been identified as significant sources of phenolic compounds, including those with antidiabetic activity. These plants include both ornamental species, such as *Camellia japonica*, and wild species, such as *Verbascum* spp. [37–39].

Beyond phenolic compounds, dietary fiber also plays a crucial role in metabolic regulation and has been widely studied for its contribution to glycemic control and diabetes management. Dietary fiber with antidiabetic properties is primarily extracted from by-products of fruits, vegetables, cereals, and legumes. This fiber is characterized by its ability to regulate the rate of sugar absorption in the digestive tract, contributing to post-prandial glucose control and improving insulin sensitivity [40]. For example, citrus fruit pulp and peel, mango bark and peel, apple peel, wheat bran, onion skin, rice and wheat bran, and corn residue contain pectin, celluloses, hemicelluloses, and lignin, which have demonstrated beneficial effects in regulating glycemia [41–44]. Other by-products, such as banana peels (Table 1), are able to enhance the activity of key enzymes in the development of T2DM, such as lactase and sucrase [45]. Additionally, certain byproducts rich in fermentable fibers, such as citrus pectin or fructooligosaccharides from roots and tubers, can be used as prebiotics, improving intestinal health and energy metabolism [46].

Table 1. Case studies of by-products as a source of antidiabetic compounds.

By-Product	Compounds	Extraction	Concentration	AD Mechanism	Bioactivity	Applications	Ref.
			Polyphenols				
Orange peel	Naringin, naringenin	MA: EtOH 70%, 72 h	100 mg/kg bw, 4 weeks	↑ insulin, C-peptide, glycogen	AD, AO	PR, DS	[47]
Grapefruit peel	Polyphenols	MA: Acet 80%	80–240 μg/mL	AG	AD, AO, AM, AH	PR, DS	[48]
Lemon peel	Limonoids	MA: EtOH, 24 h	16.53 mg/mL	Join protein	AD, AO	PR, DS	[49]
Pear peel	Rutin, catechin, epicatechin	UAE: MetOH 60%	25–75%	↑ insulin SE, GU	AD, AO, AI	PR, DS	[50,51]
Pomegranate peel	Flavonoids, tannins	SE: MetOH 80%	1 mg/mL	↓ GL	AD, AO, OB	PR, DS	[52]
Orange peel	Flavonoids, phenolics	SE: MetOH 80%	1 mg/mL	↓ GL	AD, AO, OB	PR, DS	[52]
Mango peel	Flavonoids, anthocyanins	Powder	5–10%	↑ insulin SE, GU	AD, AO	PR, DS	[53]
Mango leaves	Polyphenols	SE	80%	AG	AD, AO	CF	[54]
Grape seeds	Flavonoids, procyanidins	SE: EtOH 70%, 3 h	52.01 and 152.18 mg/g dw	AG	AD, AO, OB	PR, DS	[55]
Apple pomace	Quercetin derivates	SE	20 mg/kg	AG	AD, AO	PR, FI	[56]
Carrot pomace	Polyphenols	UAE: 30 °C, 80 min, 500 W	150.6 mg/L	AG	AD, AO	PR, FI	[57]
Walnut husk	Polyphenols	SE	74.08–166.44 mg/g	AG	AD, AO, AM	PR, FI	[58]
			Fiber				
Banana peel	Dietary fiber	DE, W	nd	↓ FI, GU; ↑ insulin, GLP-1	AD, AO	PR, DS	[59]
Mango peel	Dietary fiber	SE	12.8–23%	↓ FI, GU	AD, AO	PR, FI	[60,61]
Papaya peel	Pectin, lignin	SE: EtOH	nd	↓ FI, GU	AD, MM, SA	PR, FI	[62]
Orange albedo	Pectin	SE: EtOH, 75 °C, 1 h	18.73%	AG	AD, TP	PR, FI	[63]
			Vitamins				
Orange, grapefruit, and lemon peel	Vitamin C	MA: EtOH, 24 h	110.4, 113.3, and 58.6 mg/100 g	↑ glycemic control	AD, AO	PR, FI	[64]

Table 1. *Cont.*

By-Product	Compounds	Extraction	Concentration	AD Mechanism	Bioactivity	Applications	Ref.
			Fatty acids				
Walnut oil cake	Omega-3	SE	52–70%	AG	AD, AO, AI, CP	PR, FI	[65]
Fish waste	Omega-3	UAE: hex, 60 °C, 80 min	45.1%	AG	AD, AI, CP	PR, FI	[66,67]
			Proteins				
Fish processing waste	Bioactive peptides	EAE: papain, bromelain	18.49%	↑ GLP-1, ↑ insulin SE	AD, AO	PR, FI	[68,69]
Salmon co-products	Bioactive peptides	EAE: FoodPro, alcalase	57–74%	AG	AD, AO	PR, FI	[70]
Cheese by-product	Whey proteins	SE	85%	↑ insulin SE, lipid metabolism	AD, TP	PR, FI	[71]
			Other				
Tomato waste	Lycopene	UEA: EtOH, 220 W, 25 min	12–19 μg/g dw	↓ GL	AD, AO	PR, FI	[72]
Tomato waste	Lycopene	SE: Acet:hex 1:3	5.32 mg/100 g fw	↓ GL	AD, AO, AI	PR, FI	[73]
Carrot pomace	β-Carotene	UAE: 30 °C, 80 min, 500 W	150.58 mg/L	↓ GL	AD, AO	PR, FI	[57]
Carrot pomace	Carotenoids	UAE: EtOH 50%, 40 °C, 750 W	51%	↓ GL	AD, AO, AI	PR, FI	[74]
Orange albedo	Hesperidin	SE: MetOH, RT	1.0–2.8%	↓ GL	AD, AI, CP	PR, FI	[75]

Abbreviations. Extraction. MA: maceration; SE: solvent extraction; DE: decoction; UAE: ultrasound-assisted extraction; EAE: enzyme-assisted extraction; Solvents. MetOH: methanol; Acet: acetone; W: deionized water; hex: hexane. Mechanism. GU: glucose uptake; SE: sensivity; FI: food intake; GL: glucose levels; Bioactivities. AO: antioxidant activity; AI: anti-inflammatory; AM: antimicrobial; AH: anti-hypertensive; OB: anti-obesity; AA: anti-aging; AD: anti-diabetic; AG: α-amylase and α-glucosidase inhibition; CP: cardiovascular protection; MM: microbiota modulation; SA: satiating. Applications. TP: technological properties; PR: preservative; DS: dietary supplementation; CF: commercial formulations; FI: food ingredient; Symbols. nd: not determined; ↑ improve; ↓: decrease; dw: dry weight; fw: fresh weight; RT: room temperature.

Food by-products are also a significant source of essential micronutrients such as vitamins (e.g., D, C, E, B complex) and minerals (e.g., selenium, zinc, chromium, manganese, and potassium) with antidiabetic potential [76]. For instance, vitamin D, which can be effectively extracted from cocoa bean shell or anchovy leftovers [77,78], play functional role on glucose tolerance through its effects on insulin secretion and insulin sensitivity [17]. However, these effects are not consistently observed across all vitamin D-enriched food formulations; the most significant outcomes are achieved when such formulations also include calcium, proteins, and amino acids [16,79]. In the case of vitamins C and E, which can be effectively extracted from by-products of wild apples, citrus fruits, prickly pears, and broccoli [80–83], their intake has been associated with a decreased risk of developing T2DM by up to 25%. This effect is primarily attributed to their role in enhancing insulin sensitivity and protecting pancreatic β-cell function through the mitigation of oxidative stress [84–86]. Conversely, B vitamins, which are predominantly found in significant concentrations in animal-based by-products (e.g., salmon processing residues) [87,88], contribute to glycemic control by reducing oxidative stress, improving endothelial function, and modulating homocysteine levels. In fact, deficiencies in these vitamins have been linked to insulin resistance and increased risk of T2DM and its complications [89,90]. Moreover, these vitamins are also recommended during the use of treatments for T2DM, such as metformin, whose prolonged administration has been associated with vitamin B12 deficiency [91].

Regarding minerals with antidiabetic potential, zinc and selenium are predominantly extracted from by-products generated during the processing of meat, offal, fish, and dairy products [92–94]. In addition, whole wheat processing residues represent a valuable source of trace elements such as zinc and chromium, further highlighting the potential of plant-based by-products in the recovery of essential micronutrients [92–94]. Manganese is particularly abundant in vegetable and cereal processing wastes, with recovery potentials estimated to provide 25–50% of the recommended daily intake [95]. Potassium is normally extracted for chestnuts by-products [96]. However, it is unclear whether all of potassium formulations are linked to clinically significant increases in dietary potassium [97]. Dietary formulations often incorporate several of these minerals to enhance the product's potential and benefits. However, although multivitamin and mineral-rich products are used by a large portion of the population in the form of dietary supplements, the effects of these multivitamin and mineral supplements on people with DM have barely been studied. For instance, in studies conducted with rats, it has been observed that diabetes progression in male rats is attenuated by these types of supplements. However, the consumption of these supplements in female rats did not protect against disease progression. This may suggest a gender difference in the pathogenesis of diabetes [98].

In addition to polyphenols, vitamins, minerals, and dietary fiber, which have been widely studied for their antidiabetic effects, there is growing interest in other, less-explored nutrients and bioactive compounds derived from food by-products that show promise in the prevention and management of T2DM. Preliminary studies have demonstrated the potential of protein hydrolysates obtained from animal-based food processing by-products, such as fish skins, bones, and meat trimmings, which may exert insulinotropic and glucose-lowering effects through modulation of incretin hormones and inhibition of dipeptidyl peptidase-IV activity [70,99]. Similarly, bioactive lipids and omega-3 fatty acids extracted from fish and vegetables by-products have shown beneficial effects on insulin sensitivity and inflammatory markers associated with metabolic dysfunction [67,100,101]. Furthermore, certain underutilized plant food wastes, such as fruit peels, seed cakes, and vegetable stalks, contain a variety of phytochemicals and secondary metabolites with antidiabetic potential, including alkaloids or saponins, which may contribute to glycemic control through mechanisms such as glucose uptake enhancement, α-glucosidase inhibition, and antioxidant activity.

In conclusion, food by-products represent an underutilized but extremely promising source of bioactive compounds and nutrients with therapeutic potential. Their use in the formulation of functional dietary supplements could significantly contribute to the prevention and management of T2DM, improving metabolic health through diverse and complementary molecular mechanisms. However, further scientific studies are needed, as well as consumer education courses.

4. Mechanisms of Action of Dietary Supplements Derived from Food By-Products Involved in Diabetes Prevention

Building upon the extraction and characterization of bioactive compounds from agro-industrial by-products detailed in the preceding section, it is imperative to investigate the molecular mechanisms underlying their antidiabetic properties. These bioactive constituents have been extensively documented for their ability to enhance glucose metabolism, improve insulin sensitivity, and modulate hormones and enzymes involved in glycemic regulation [102]. Their modes of action commonly encompass the stimulation of insulin secretion, attenuation of insulin resistance, and facilitation of glycogen synthesis in the liver [103]. They also modulate carbohydrate-digesting enzymes, such as α-amylase and α-glucosidase, thereby impeding glucose absorption and attenuating postprandial blood glucose spikes (Figure 2) [104]. Additionally, most of these bioactives exhibit antioxidant and anti-inflammatory properties that protect pancreatic cells from oxidative stress (OS) and mitigate chronic inflammation linked to the progression of diabetes [105]. Some substances, such as fish protein hydrolysates, can stimulate the secretion of gut hormones and suppress appetite. These effects further support weight management and glycemic control [106]. Certain supplements can also contribute to improved metabolic health by modulating gut microbiota [107]. Growing evidence highlights that supplements derived from food by-products offer a promising and multifaceted approach for the prevention and management of diabetes. A comprehensive understanding of the underlying mechanisms is vital not only to validate the therapeutic potential of these bioactive compounds but also to guide their rational incorporation into functional foods, nutraceuticals, or pharmaceutical products. Therefore, this section will systematically explore the principal mechanisms responsible for their antidiabetic effects.

4.1. Regulation of Blood Glucose Levels

The onset of DM results in impaired regulation of blood glucose levels, which in untreated individuals typically remain elevated above the normal physiological range. Maintaining glucose within this narrow range is essential for anabolic processes, as anabolism requires tightly regulated blood glucose levels [108]. In healthy individuals, the pancreas serves as a key regulator by secreting insulin to lower blood glucose and glucagon to increase it, thereby maintaining glucose homeostasis. Insulin controls glucose balance by modulating hepatic and renal glucose production, as well as promoting glucose uptake in peripheral tissues, primarily skeletal muscle [109]. However, in individuals with DM, pancreatic function is compromised; specifically, in T2DM, the pancreas exhibits impaired insulin secretion and/or beta-cell dysfunction, leading to inadequate insulin production relative to the body's needs [110].

Various strategies exist to reduce elevated blood glucose levels in T2DM. One such approach involves inhibiting the action of digestive enzymes responsible for breaking down starches into simple sugars during digestion, such as α-amylase [111]. By inhibiting this enzyme, carbohydrate absorption is limited, thereby reducing postprandial glucose spikes [112]. Consequently, insulin demand decreases, leading to more effective blood glucose control. These mechanisms contribute to protecting against insulin resistance and facilitate reductions in body weight and adiposity, two critical factors implicated in the pathogenesis of T2DM. Importantly, these effects can be achieved through the use of various bioactive compounds, as demonstrated in sources such as white kidney bean extract extracted from the *Phaseolus vulgaris* plant [111]. Another example include mangiferin, a xanthone found in mango fruit and its byproducts, that has an inhibitory effect on α-glucosidase, the enzyme responsible for hydrolyzing shorter starch chains and disaccharides [113,114].

Another complementary approach focuses on modulating hormone secretion, particularly gut hormones like glucagon-like peptide-1 (GLP-1). GLP-1 increases insulin secretion and suppresses hepatic glucose production; therapies that enhance GLP-1 activity, including GLP-1 mimetics and dipeptidyl peptidase inhibitors, are effective in lowering blood glucose and are inspired by the natural response of the body to nutrient intake and gut microbial signals [115]. Furthermore, regulating the composition and activity of the gut microbiota has emerged as a promising strategy, as a balanced microbiome can influence metabolic pathways, improve insulin sensitivity, and modulate inflammatory responses associated with T2DM [116]. For example, supplementation with prebiotics or specific probiotics has been shown to beneficially alter gut microbiota composition, leading to improved glycemic control in clinical studies [117]. Among the prebiotics evaluated, the use of cinnamon (*Cinnamomum cassia*) stands out as a dietary supplement that has demonstrated efficacy in modulating the intestinal microbiota, with the consequent significant reduction in blood glucose levels (BGL) and HbA1c levels in T2DM patients, particularly those with suboptimal glycemic control (glycosylated hemoglobin (HbA1c) >8%). Significant improvements in HbA1c levels were observed following daily supplementation with 1–6 g of cinnamon for 2 to 4 months. However, these values may vary depending on the populations and starting conditions [12,118].

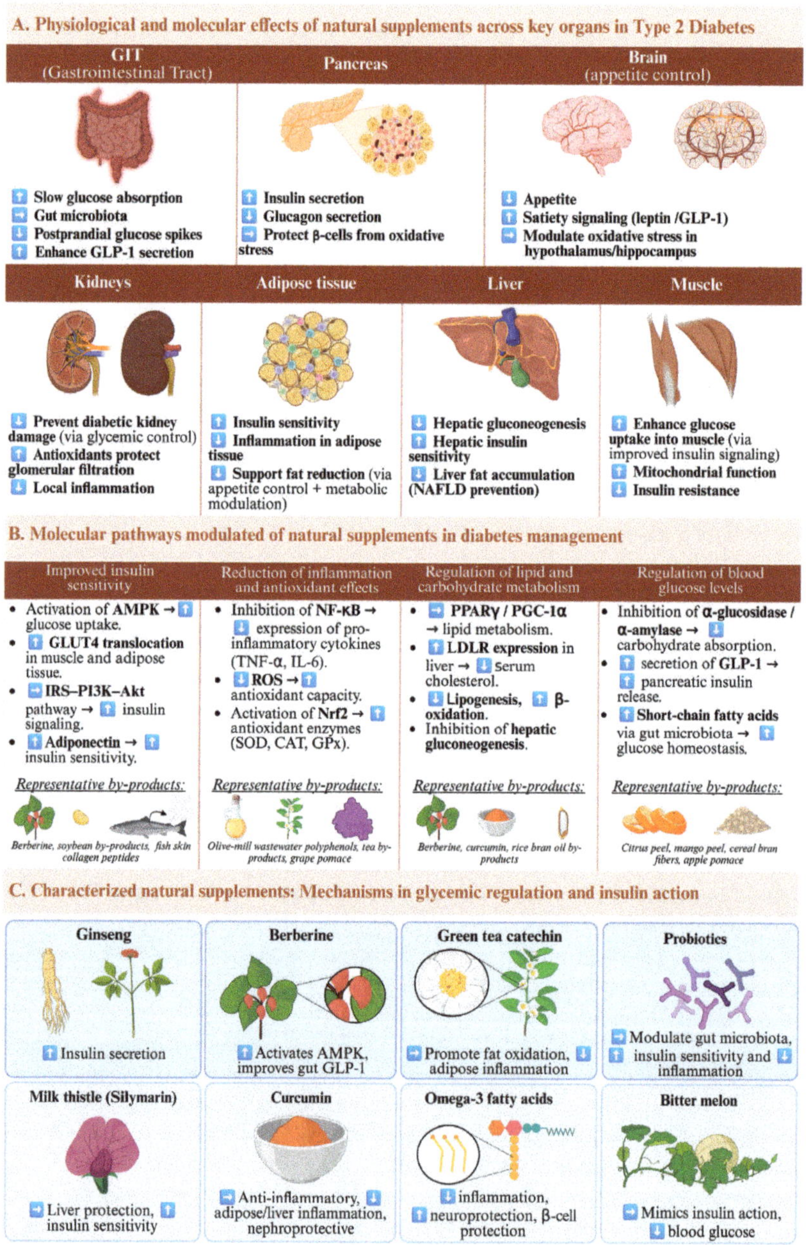

Figure 2. Main mechanism of action of antidiabetic compounds. Abbreviations. ↓: decrease; ↑: increase; →: modulate.

Another supplement that has been shown to be effective in the treatment of diabetes is chromium-based supplements, which help maintain normal blood glucose levels. However, these effects are only observed in individuals whose T2DM is due to insulin resistance and severe metabolic dysregulation, and are not effective when the condition is due to other physiological causes [119–121]. Examples of such supplements include chromium picolinate, an insulin sensitizer that promotes the activation of insulin receptor kinases at the plasma membrane. This activation initiates a signal transduction cascade that results in the translocation of glucose transporter type 4 to the cell surface, thereby amplifying insulin signaling and enhancing glucose uptake [122,123].

4.2. Improved Insulin Sensitivity

Insulin resistance is defined as one of the most critical factors in the development of DM [124]. Several studies have highlighted the potential of phenolic compounds to enhance insulin sensitivity. Consequently, vegan, and vegetarian diets, which are abundant in plant-based foods rich in phenolic compounds, have been shown to improve insulin secretion and sensitivity. This improvement is correlated with reductions in HbA1c levels and enhanced β-cell function, thereby mitigating insulin resistance [125]. Additionally, they provide protection against reactive oxygen species (ROS), which is critical for managing the inflammatory and they inhibit gluconeogenic enzymes, thereby suppressing gluconeogenesis, processes associated with DM [126].

Fatty acids have also been linked to better insulin sensitivity in both diabetic animals and patients [127]. Omega-3 fatty acids exert beneficial effects on metabolic function, improving insulin sensitivity primarily through anti-inflammatory pathways and the modulation of adiponectin levels. Specifically, α-linolenic has been reported to protect pancreatic β-cells and ameliorate diabetic neuropathy, while linoleic acid intake has been associated with a reduced risk of T2DM and improved renal outcomes in some studies [128].

Insulin sensitivity can also be improved through supplementation with certain alkaloids. For example, berberine, an alkaloid extracted from various plant sources such as *Coptis chinensis* and *Verbascum sinuatum*, help regulate blood glucose levels in vitro and in vivo by boosting insulin sensitivity, offering a comparable effect to the drug metformin. At the molecular level, berberine activates AMP-activated protein kinase, which increases GLUT4 translocation to the cell membrane in muscle and adipose tissue, enhancing glucose uptake. It also influences insulin receptor substrate signaling, improving downstream PI3K–Akt pathway activation, which contributes to better insulin signaling. In the liver, berberine upregulates low-density lipoprotein receptor expression at both mRNA and protein levels, promoting cholesterol clearance and linking lipid metabolism to glycemic control. These coordinated molecular actions support its insulin-sensitizing effects, which in some studies have been shown to be comparable to those of metformin. It also lowers HbA1c, as well as fasting and postprandial blood glucose levels [39,129,130].

4.3. Reduction in Inflammation and Antioxidant Effects

Diet plays a fundamental role in influencing cytokine concentrations and the inflammatory process, acting as a key mechanism for controlling inflammation. Following a Mediterranean style eating pattern, which includes abundant consumption of olive oil, nuts, fruits, and vegetables, has been shown to reduce levels of inflammatory cytokines [131]. On the other hand, diets that contain high amounts of trans-unsaturated fatty acids and sugars are linked to increased production of these inflammatory molecules. Therefore, dietary recommendations aimed at reducing inflammation emphasize the consumption of fats rich in polyunsaturated fatty acids, particularly omega-3 fatty acids as these nutrients help to lower cytokine levels and thereby decrease inflammation [132,133]. Nutraceuticals containing omega-3 fatty acids act as agonists of peroxisome proliferator-activated receptors (PPAR)-γ and PPAR-α, whose activation may alleviate inflammatory responses in

DM [128,134]. For instance, the effect of docosahexaenoic acid (DHA)-enriched fish oil on gene expression levels of p53, NF-κB, and (PPAR)-γ was studied in T2DM patients. In this double-blind study, results show how moderate doses of DHA-enriched fish oil supplementation increased PPAR-γ activity. Concretely, authors suggested that the substitution of polyunsaturated fatty acids (PUFAs) as natural ligands for PPAR-γ instead of chemical medicines can lead to the protection of cardiovascular system against atherosclerosis lesion formation [135].

Diet also plays a fundamental role in reducing the production of reactive oxygen species (ROS), which is vital in the management of diabetes, as the disease is associated with a compromised antioxidant defense system [136]. Elevated glucose concentrations provoke mitochondrial dysfunction, which further amplifies ROS generation. Moreover, multiple pathways contribute to ROS formation in DM, including the activation of the polyol pathway, accumulation of advanced glycation end products (AGEs), engagement of AGEs receptors, activation of protein kinase C, and heightened flux through the hexosamine biosynthetic pathway. OS also adversely affects pancreatic β-cell performance, diminishing both insulin secretion capacity and insulin quality [133,136–138]. Alpha-lipoic acid (ALA) is a well-studied antioxidant that exemplifies the therapeutic potential of such compounds in the management of diabetes. In particular, ALA has been shown to mitigate elevated oxidative stress associated with distal symmetric polyneuropathy, one of the most common complications in diabetic patients [139]. Overall, OS is a major contributing factor of diabetes, as it leads to a decrease in antioxidant enzyme activity and an increase in oxidative DNA damage [140].

4.4. Regulation of Lipid and Carbohydrate Metabolism

Various naturally occurring compounds have been investigated for their potential to modulate glucose transporters and carbohydrate metabolism, which play crucial roles in postprandial glucose and insulin responses [141]. These findings have contributed to the development of several patented formulations aimed at regulating lipid and carbohydrate metabolism. For example, mangiferin inhibits two key enzymes involved in glucose metabolism and energy homeostasis: phosphoinositide 3-kinase (PI3K) and serine/threonine protein kinase (Akt) [114]. Another example includes berberine, an alkaloid that targets key regulators of cellular energy homeostasis. It activates AMP-activated protein kinase and modulates the PPARγ co-activator (PGC)-1α, pathways involved in both lipid and carbohydrate metabolism. Through these actions, berberine promotes fatty acid oxidation, improves glucose uptake, and regulates hepatic gluconeogenesis, contributing to a balanced metabolic profile and supporting overall energy homeostasis [114]. Curcumin has also been described as a lipoxygenase inhibitor, although its precise mechanism of action is not fully defined. It also exerts lipid-lowering effects by increasing low-density lipoprotein receptor expression at the mRNA and protein levels in the liver [114]. Animal studies indicate that curcumin may improve glucose tolerance partly by stimulating glucagon-like peptide 1 (GLP-1) secretion, thereby enhancing insulin release and glycemic control [142]. At the molecular level, curcumin modulates several signaling pathways relevant to metabolic regulation. It inhibits the NF-κB pathway, reducing pro-inflammatory cytokine production, and activates Nrf2, which promotes antioxidant defense. In addition, curcumin has been reported to activate AMP-activated protein kinase, a central regulator of energy metabolism, thereby promoting fatty acid oxidation and improving insulin sensitivity. Through these combined actions, curcumin exerts effects on both lipid metabolism and carbohydrate utilization, while also protecting β-cells against oxidative and inflammatory stress [143]. Another compound whose effectiveness on lipid metabolism has been investigated in rat studies is 6-gingerol. This study suggests that 6-gingerol may positively influence lipid metabolism by hinder the movement of CD36 (which is a fatty acid transporter involved in lipid homeostasis) from the cytoplasm to the cell membrane, so that normal skeletal muscle would not take up excess fatty acids [144].

5. Efficacy of Dietary Supplements Derived from Food By-Products in Diabetes Prevention

The traditional treatment of diabetes has been widely studied in different in vitro, in vivo, and clinical trials, with a special focus on drugs such as metformin and thiazolidinedione pioglitazone (Table 2). Metformin, a guanidine derivative, is one of the primary oral therapeutic options for lowering blood glucose in T2DM as supported by numerous clinical studies [145,146]. Its mechanism of action is based on the induction of glucose uptake to the enterocyte layer of the intestine, which is associated with improved glycemic control [147]. On the other hand, thiazolidinedione pioglitazone acts as an agonist of the peroxisome proliferator-activated receptor gamma (PPARγ), leading to reduced blood glucose levels, improved insulin sensitivity, decreased vascular inflammation, and enhanced lipid profile and endothelial function [148]. In recent decades, the use of these drugs has declined due to potential adverse effects such as weight gain, bladder cancer risk, and decreased bone density. However, they are still widely used in combination with other T2DM therapeutic agents because they are cost-effective, and have regained popularity due to new positive evidence, including cardioprotective effects [149]. However, using pharmacological therapies alone presents several limitations, especially in individuals with T2DM, such as the difficulty in maintaining prolonged behavioral changes, medication side effects, and associated costs that can limit accessibility and adherence to treatment. These limitations underscore the need to explore and develop new complementary strategies that are safe, accessible, and sustainable [150]. In this context, dietary supplements derived from food by-products emerge as a promising alternative, offering potential metabolic benefits along with an ecological and economic approach, warranting further research to effectively integrate them into diabetes prevention programs.

These supplements are typically derived from natural raw materials, often utilizing by-products rich in bioactive compounds [151–153]. Among the primary bioactive constituents incorporated into supplement formulations are polyphenolic compounds, dietary fiber, vitamins, minerals, polyunsaturated fatty acids, and amino acids [154]. These compounds have demonstrated the ability to modulate key metabolic pathways, including glucose uptake, insulin sensitivity, oxidative stress, and inflammation. While multiple molecular targets underlying these effects have been proposed, it is essential to characterize the specific mechanisms of action for each individual compound. Moreover, their efficacy, bioavailability, and safety must be confirmed through in vivo studies and rigorously compared to those of conventional antidiabetic therapies. Such evidence is necessary to support their integration into functional foods, nutraceuticals, or pharmaceutical formulations aimed at diabetes prevention.

5.1. Polyphenolic Compounds

Different polyphenols have been studied as supplements for the treatment of diabetic related pathologies, showing improvements in different biomarkers mainly associated with anti-inflammatory pathways, although studies focused on these samples remain limited. For instance, a randomized controlled trial with 35 diabetic and overweight patients was assessed (Table 2). After 6 weeks of supplementation with 205 mg/day of a mixture of epigallocatechin gallate, epigallocatechin, epicatechin gallate, and epicatechin extracted from green tea, a significant effect in the weight regulation during a positive energy imbalance was observed [155]. Similar conclusions were obtained in a single-center, placebo controlled, double-blind study, where the supplementation with 530 mg/day of the same catechin mixture to 65 subjects was performed for 8 weeks. Results showed a reduction in the circulation of serum amyloid plasma, that could be explained by the stimulation of fat oxidation via activation of the LKB1/AMP-activated protein kinase pathway or through the modulation of COMT activity [156]. The effect of curcumin supplementation was also assessed in another randomized, double-blind placebo-controlled study, with doses of 1500 mg/day for 10 weeks. Results showed the positive effect of curcumin supplementation in T2DM patients, by reducing fasting blood glucose and weight [157]. In another study, a polyphenol-rich antioxidant supplement containing pomegranate extract, green tea extract, and ascorbic acid was administered to 114 male and female non-smokers (56 in the treatment group and 58 in the placebo group) with T2DM and no complications. The results demonstrated significant antagonistic effects against oxidative stress and lipid peroxidation in these patients, suggesting potential benefits in preventing cardiovascular complications associated with T2DM [158]. Therefore, polyphenols are promising candidates for developing effective adjunct therapies in diabetes management.

5.2. Fiber

Randomized controlled trials have shown that viscous dietary fibers can contribute to improved glycemic control and may reduce cardiovascular risk in individuals with diabetes. Clinical studies have reported beneficial effects following supplementation with various fibers such as insoluble fiber, galacto-oligosaccharides, chicory inulin, and beta-glucan. Similarly, animal research has demonstrated positive outcomes on blood glucose levels, HbA1c, and gut microbiota diversity when supplemented with soluble fibers derived from sources like wheat bran, barley, and beta-glucan. However, the effects on insulin sensitivity remain variable. Reviews suggest that prebiotic fibers may support glycemic regulation and gut health, while soluble fibers could also aid in managing insulin and body weight [13,159]. Despite these promising findings, further rigorous studies are necessary to determine the specific benefits of different fiber types and how best to incorporate them into dietary interventions for diabetes.

5.3. Vitamins and Minerals

Vitamins and minerals are micronutrients required in small quantities which role is to act as essential coenzymes and cofactors for metabolic and cellular reactions [160]. Their deficiency contributes to the worsening of diabetic pathology [153]. For instance, chromium (Cr) deficiency is linked to insulin signaling impairment, which is a crucial step in glucose regulation [161]. This effect was examined in several clinical studies. In one of them, 32 patients with T2DM received 200 μg/day of Cr, while the other 32 received a placebo. After 12 weeks of administration, results showed that Cr intake leads to lower fasting glucose and insulin levels and improved HOMA-IR levels, suggesting reduced insulin resistance and increased insulin sensitivity. Therefore, Cr supplementation can be considered as an accurate complementation to improve crucial parameters of metabolism, inflammation, oxidative stress, and blood pressure in T2DM patients [162]. Similar results were obtained in another study, where 600 μg/day of Cr supplementation for 4 months showed a reduction in fasting and postprandial glucose in T2DM patients [123].

A similar pattern is observed in the presence of zinc (Zn) deficiency, which has been associated with insulin secretion impairment, insulin sensitivity decrease, and inflammatory biomarkers increment, with studies showing that patients with low Zn blood concentrations are more likely to have had diabetes for a longer time, have poorer glucose control, and have reduced pancreatic β cell function [163]. For instance, in a double-blind placebo-controlled study, T2DM patients were administrated with 50 mg/day of Zn for 12 weeks, showing an improvement in fasting glucose and insulin resistance that may be explained due to the effect of this micronutrient in insulin signaling and β cells function [164]. Although these studies did not focus on vitamins and minerals present in agricultural byproducts, the presence of these micronutrients in industrial residues makes them a promising strategy to be studied for the development of supplements for diabetic prevention.

5.4. Amino Acids and Proteins

The potential benefits of amino acids (AAs) and proteins as supplements for T2DM patients have been supported by various mechanisms. AAs are not only essential substrates for protein synthesis but also play a key role in cellular energy production [165]. Furthermore, dietary protein has been shown to positively influence insulin sensitivity, adiposity, and glucose metabolism. Evidence suggests that protein intakes exceeding 1.2 g/kg/day can lead to improved glycemic control. One contributing factor is that protein metabolism requires relatively low amounts of insulin, whether endogenous or exogenous. In addition, protein consumption promotes greater satiety, which may aid in weight management. Specific AAs, particularly the branched-chain amino acids such as leucine, have been implicated in enhancing both insulin sensitivity and insulin secretion [152]. This body of scientific evidence has led to the development of various AAs-based supplements aimed at managing T2DM. For instance, the effect of AAs supplementation on T2DM patients was assessed in a randomized, single-blind, crossover study involving 65 patients evaluated the effects of essential AAs supplementation versus placebo over a 12-week period. The results demonstrated a positive impact of EAA treatment on glycemic control, with a significant reduction in HbA1c levels compared to placebo [166]. In another study, AAs supplementation of 12 g/day for 42 weeks 65–85-year-old T2DM patients showed a significant decrease on fasting and postprandial glycemic with a reduction in HbA1c. Moreover, it was observed a decrease in fasting insulin levels and HOMA-IR index during AAs supplementation. This study supports that long-term oral amino acid supplementation represents a promising complementary strategy focused on muscle and metabolic health for T2DM elderly patients [167]. These previous studies have used pure commercial AAs for supplementation. Nevertheless, agri-food by-products are known to be rich sources of these AAs. By optimizing extraction techniques, it is possible to obtain a high degree of purification of the AAs of interest [168,169]. As such, utilizing agri-food by-products as a source of AAs may offer a cost-effective and sustainable alternative for nutritional supplementation in diabetic patients, and merits further research.

5.5. Polyunsaturated Fatty Acids

PUFAs are integral components of cell membranes, having specific functional, metabolic and signaling roles. Moreover, these compounds are categorized as vital nutrients due to their role in the treatment of non-alcoholic fatty liver, autoimmune reactions, and different chronic diseases [151,170]. Since humans do not possess enzymes to place double bonds in position 3 and 6, these fatty acids (FA) are considered essential. Beyond the most relevant *n*-3 FA, alpha-linolenic acid, eicosapentaenoic acid (EPA), and docosahexaenoic acid (DHA), are specially highlighted, while linoleic and arachidonic acid are the most important *n*-6 FA. *n*-3 essential FA have shown anti-inflammatory, immunomodulatory, and antioxidant effects [171]. Concretely, supplementation of EPA and/or DHA have shown positive modulatory and anti-inflammatory effects, also suggesting providing health benefits to obese, diabetic, or those suffering from cardiovascular diseases [172], but also showing benefits to treat mental disorders such as dementia and depression [173]. However, age, health, and physiological conditions are factors affecting the adequate dose of these compounds, with a daily recommended intake that ranges from 40 to 250 mg [172]. One of the most relevant food byproducts matrices containing an excellent composition of FA are fish by-products, with high amounts of EPA and DHA [67]. Despite the lack of specific studies on the effect of these compounds extracted from fish by-products, there are studies establishing the relationship between the consumption of these compounds and benefits for the prevention or improvement of diabetes symptoms (Table 2). A double-blinded randomized study was carried out to establish the effect of EPA, and DHA supplementation in T2DM patients. This clinical trial was performed for 12 weeks in 100 patients, with doses of 500 mg of EPA and 200 mg of DHA. Results showed how FA supplementation in T2DM patients led to improvements in insulin levels, inflammation markers, antioxidant status, and serum cholesterol levels. However, results also showed no significant effect in homeostatic model assessment of insulin resistance index (HOMA-IR) and quantitative insulin sensitivity check index (QUICKI). This suggests that the potential beneficial effect of these FA supplementation is effective although limited depending on the dose [174]. For instance, in a double-blind controlled study, 56 participants with T2DM were submitted to fish oil supplementation with high content of EPA and DHA Individuals were divided in two groups: fish oil supplementation with 1.8 g EPA, 3.0 g DHA, and 5.9 g total *n*-3 of FA per day for 9 weeks and corn oil with 8.5 g of linoleic acid per day for 9 weeks. Results revealed that high doses of fish oil intake led to a moderate increase in blood glucose and decrease in insulin sensitivity. Moreover, the alteration of fat and carbohydrate utilization was identified in a time-dependent manner [175]. In a different double-blind controlled study, the administration of EPA and DHA in T2DM patients was assessed for 10 weeks. Results showed how 310 mg of EPA and 210 mg of DHA led to a decrease in HOMA-IR levels while QUICK levels were increases, suggesting that the consumption of these FA has a positive metabolic impact by improving insulin sensitivity while decreasing insulin resistance. As a result, T2DM patients had a better glucose metabolism and a reduced burden on pancreatic insulin production [176].

Table 2. Compilation of cases of studies regarding the effects of supplements in diabetes and related disorders.

Study	Treatment	Mechanism Suggested	Therapeutic Effect	Ref.
		Traditional treatments		
DBPC RCT (n = 41)	Met, 1 g, b.i.d., 26 weeks	N.S.	↑ intestinal glucose uptake, glycemic control	[147]
DBPC RCT (n = 366)	Pio, Met, Dapa, 15–30 mg/day, 24 weeks	N.S.	↓ in HbA1c of −0.38 to −0.83%	[177]
		PUFAs		
DBPC RCT (n = 170)	DHA, 350 mg/day, 2 years	N.S.	Not slowing action in NPDR progression	[178]
RCT (n = 84)	EPA, DHA, 60 + 308 mg/day, 9 months	↓ AA	↑ maternal and fetal FA status (T1DM)	[179]
Mice (n = 56)	EPA, DHA, 1%, 15 days	↓ OS, AI, ↓ β-cell apoptosis	Prevent pancreatic injury	[180]
PS (n = 47,663) **	N.S.	AI, ↓ VLDL-C	↓ CHD	[181]
DBPC RCT (n = 26) **	EPA, DHA, PUFA, 1.8, 3.0 and 5.9 g/day, 9 weeks	Change energy use, ↓ glucose uptake	↑ blood glucose; ↓ insulin sensitivity	[175]
DB (n = 100)	EPA, DHA, 500 and 200 mg/day, 12 weeks	AI	↑ insulin levels; ↓ inflammatory markers, Chl	[174]
DBPC RCT (n = 44)	EPA and DHA, 310 and 210 mg/day, 10 weeks	N.S.	↑ QUICKI levels; ↓ insulin, HOMA-IR levels (T2DM)	[176]
		Amino acids		
RCT (n = 65)	EAA 11.7 g/day//Tyr 0.1 g/day//Cys 0.4 g/day, 12 weeks	N.S.	↑ myocardial dysfunction (T2DM)	[166]
RCT (n = 34)	EAA 7.6 g/day//Tyr 0.1 g/day//Cys 0.3 g/day, 42 weeks	↑ insulin secretion, postprandial glucose	↑ insulin sensitivity, glycemic control (T2DM)	[167]

Table 2. *Cont.*

Study	Treatment	Mechanism Suggested	Therapeutic Effect	Ref.
		Vitamins and minerals		
DBPC RCT (n = 64)	Cr, 200 µg/day, 12 weeks	N.S.	↑ insulin sensitivity; ↓ insulin resistance (T2DM)	[162]
RCT (n = 71)	Cr, 600 µg/day, 4 months	↑ insulin signaling, GLUT4 translocation	↓ fasting and postprandial glucose; ≈ lipid profile	[123]
DBPC RCT (n = 44)	Zn, 50 mg/day, 12 weeks	↑ insulin signaling; ↓ insulin resistance	↓ fasting glucose and HOMA-IR (T2DM)	[164]
RCT (n = 100)	Vitamin D, 125 µg/day (5000 IU/day), 12 weeks	N.S.	≈ inflammatory and OS (T2DM)	[182]
DBPC RCT (n = 127)	Vitamin D, 100 µg/day (4000 IU/day), 48 weeks	N.S.	≈ Met (T2DM)	[183]
		Polyphenols		
RCT (n = 35) *	EGCG 120 mg/EGC 60 mg/ECG 25 mg EC; 2 caps./day; 8 weeks	AI	Regulation of weight gain	[155]
DBPC RCT (n = 64) *	EGCG, EGC, EGC, EC, 2 caps 530 mg/day, 6 weeks	↑ FO: LKB1/AMP, modulation COMT	↓ circulating SAA	[156]
DB (n = 100)	Cur, 80 mg nanocurcumin/day, 12 weeks	AI	↑ insulin levels; ↓ inflammatory, Chl	[174]
DBPC RCT (n = 53)	Cur, 1500 mg, 10 weeks	⊣protein replication, AI cytokines; ↑ RMR	↓ fasting blood glucose and weight (T2DM)	[157]

Abbreviations. Study. DBPC RCT: double-blind placebo-controlled randomized controlled trial; RCT: single-blind, placebo-controlled trial; DB: double-blind controlled study; RCT: randomized controlled trial; PS: prospective study; n: number of individuals. Compounds. Met: metformin; Pio: pioglitazone; Dapa: dapagliflozin; DHA: docosahexaenoic acid; EPA: eicosapentaenoic acid; AA: arachidonic acids; Tyr: tyrosine; Cys: cystine; Cr: chromium; EGCG: epigallocatechin gallate; EGC: epigallocatechin; ECG: epicatechin gallate; EC: epicatechin; FA: fatty acids; Cr: chromium; EAA: essential amino acids; Cur: curcumin; N.S.: not specified; Chl: cholesterol. Mechanism: VLDL-C: very low-density lipoprotein cholesterol; SAA: serum amyloid plasma; LKB1/AMP: liver kinase B1/adenosine monophosphate activated protein kinase; COMT: catechol-O-methyltransferase; HOMA-IR: Homeostatic Model Assessment for Insulin Resistance; QUICKI: Quantitative Insulin Sensitivity Check Index; RMR: resting metabolic rate. OS: oxidative stress; AI: anti-inflammatory; FO: fat oxidation; GLUT4: glucose transporter 4. Diseases. CHD: coronary heart disease; NPDR: non-proliferative diabetic retinopathy; T2DM: type 2 diabetic mellitus. Symbols. n: number of individuals in the study; b.i.d.: twice a day; *: extracts form *Camellia sinensis*; **: fish extracts; ↓ reduction; ⊣: inhibition; ↑: improve; ≈: not affected.

6. Safety and Toxicity of Food By-Product Derived Supplements

The development of dietary supplements from agri-food by-products presents a key challenge: the high heterogeneity of the extracts used as raw materials [184]. Although the final product must meet defined composition standards (Appendix A), extracts from agricultural by-products can exhibit great variability in their chemical profile [185]. This variability is determined by factors such as the type and origin of the by-product, the variety of the base food, the growing conditions, the industrial processing, and the extraction methods used [186,187]. For example, pomegranate peel extracts, rich in ellagitannins, can differ significantly in their toxicological profile depending on the drying method applied (freeze-drying vs. hot-air drying) [188]. Likewise, waste from the wine industry can concentrate contaminants such as heavy metals (lead, cadmium) if their source is not properly controlled [189]. This heterogeneity in raw materials complicates the supplement standardization process, especially in the early stages of development, and hampers toxicological and pharmacokinetic evaluation. Furthermore, the lack of standardization in the preparation of these dietary supplements complicates comparisons across studies and systematic risk assessment, creating uncertainty about their safety for long-term use or in vulnerable populations [190].

To address these problems, it is essential to conduct a detailed phytochemical characterization of the by-products used. However, most preclinical studies reported in the literature do not perform this characterization, which limits replicability and the ability to extrapolate reliable results on the effectiveness and safety of incorporating these extracts into food products. Furthermore, most of the available studies were conducted using in vitro studies or animal models (especially mice) [24,191]. In these models, some of the tested extracts, especially those rich in phenolic compounds or fiber, have been shown to reduce insulin resistance, improve glucose uptake, and modulate inflammatory pathways involved in the pathogenesis of the disease. Therefore, it is necessary to further explore this knowledge, and it is essential to conduct a greater number of in vivo studies in humans, which should focus on analyzing long-term effects.

Another factor to consider when assessing the safety of these products is the significant differences in absorption, metabolism, and excretion between species and individuals, which in some studies could explain the variability of the effects observed after taking dietary supplements [192]. For example, some studies have detected adverse effects on the liver and kidneys in animals at doses that could be achieved in humans with chronic use of certain supplements, such as mango peel extracts rich in mangiferin [193]. In this context, dose customization and individualized monitoring are key aspects not yet adequately addressed in the scientific literature. Therefore, to ensure public health, all these new food products must undergo safety assessments regulated by regional authorities, such as the European Food Safety Authority (EFSA) in the European Union (EU) or the Food and Drug Administration (FDA) in the United States (US). The EFSA approach evaluates toxicokinetics, genotoxicity, toxicity, and reproductive effects. Studies must meet ethical, methodological, and regulatory standards, including GLP and OECD guidelines. Generally, Oral administration is preferred, and botanical ingredients with a long history of safe use may be presumed safe with sufficient data [194,195].

However, the extracts used in the development of these dietary supplements may contain molecules uncharacteristic of the by-product used such as significant amounts of chemical contaminants such as pesticides and potential pathogens [196]. The primary contaminants typically identified in by-products include pesticides, mycotoxins, microbial agents, heavy metals, alkaloids, and biogenic amines (Figure 3) [194]. In the case of fruits and vegetables, proper processing is essential to prevent the appearance of mycotoxins such as patulin or ochratoxin A, highly nephrotoxic and carcinogenic compounds. Other risks associated with the use of by-products that can compromise product safety include biological instability, potential pathogenic contamination, high water activity, potential for rapid auto-oxidation, and high levels of active enzymes [197]. Most of these problems can be resolved with proper design of the byproduct collection, storage, or processing process to prevent improper handling of byproducts [198].

These biological hazards are associated with serious health risks; consequently, multiple factors must be considered when assessing the suitability of a process for extracting valuable compounds. Accordingly, it is recommended to integrate various types of analyses, such as physicochemical characterization, microbiological assessment, and contaminant detection, on the same food by-product targeted for valorization [199]. In fact, ensuring the quality and safety of these by-products is crucial to fully exploit this valorization strategy [196]. Regulatory bodies such as the EFSA stipulate that foodstuffs derived from valorized by-products must meet established safety and quality standards, ensuring that the resulting products are non-hazardous to health and suitable for human consumption. Similar requirements apply to by-products intended for use in animal feed [200,201]. To meet these requirements, in many cases the extracts are subjected to sterilization, filtration or treatment with absorbent materials before being incorporated into the food matrix [202–204]. In other studies, different extraction techniques are used, with the choice of solvent playing a key role. This is because many solvents are toxic, so only those considered GRAS can be used for food purposes. Furthermore, some solvents are characterized by increasing the concentration of the contaminants initially present in the raw material in the extracts [203].

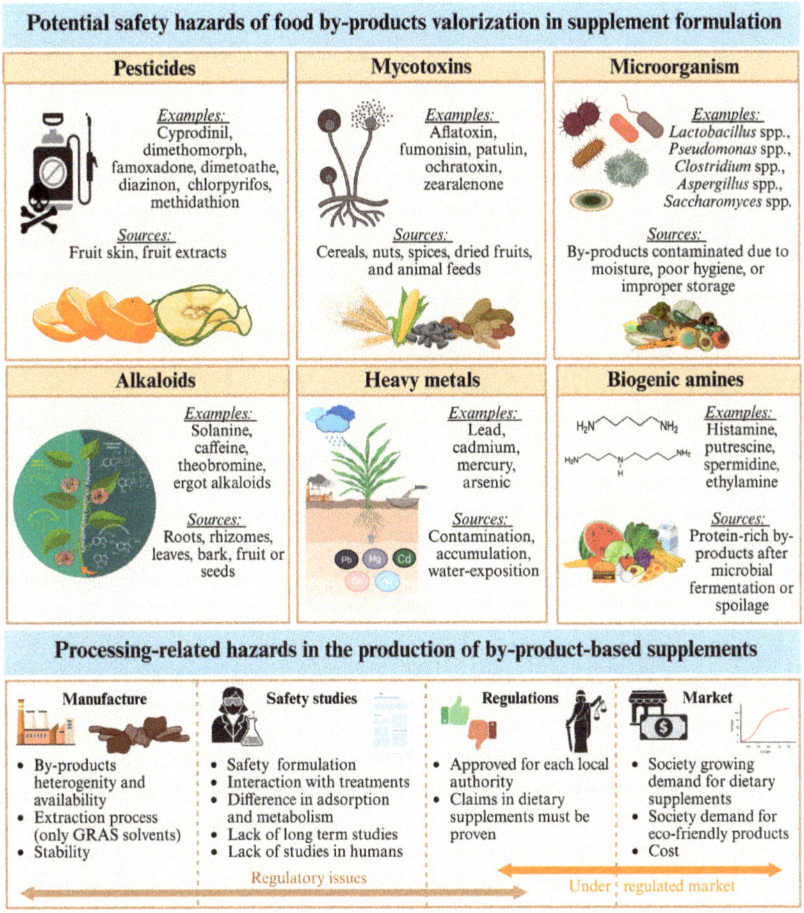

Figure 3. Biological and processing hazards associated with the production of by-product-based supplements.

Furthermore, it should be noted that these dietary supplements contain multiple bioactive compounds that can act synergistically to exert a preventive effect against diabetes mellitus type 2 (T2DM). However, these interactions can also increase the risk of toxicity or interference with other pharmacological treatments. For example, extracts rich in flavonoids can inhibit cytochrome P450 enzymes, altering the metabolism of oral antidiabetics such as metformin or sulfonylureas [205]. Likewise, supplements with a high antioxidant content can attenuate the oxidative stress necessary for insulin signaling, paradoxically negatively affecting glycemic homeostasis [206].

Likewise, prolonged use of dietary supplements, even in healthy individuals, could induce cumulative effects. This is because some polyphenols are characterized by low bioavailability but tend to accumulate in tissues, where their long-term function is not yet fully characterized. Furthermore, some studies show that some phenolic compounds exert pro-oxidant effects in cells chronically exposed to these compounds, raising questions about their safety when consumed for months or years without supervision [207,208]. Additionally, although less widespread, the cytotoxicity and mutagenic activity of by-products have also been recently evaluated by in vitro analysis in order to complement physicochemical studies [199], being necessary further investigations. Therefore, it is essential to consider subchronic and chronic toxicity studies that evaluate markers of liver and kidney damage, systemic inflammation, and endocrine disruption in relevant models, as well as pharmacokinetic studies that define parameters such as half-life, metabolism, and excretion of the main components. In this regard, novel strategies such as Food omics in combination with cellular models could be very useful to advance the knowledge of the possible toxicity or risks associated with the consumption of this type of products, since they can provide information about the compound (or compounds) responsible of toxicity and at which level of expression is impacting by applying transcriptomics, proteomics and/or metabolomics techniques [209].

Therefore, although dietary supplements developed from food by-products represent a promising and potentially sustainable strategy for the prevention of T2DM, their practical application requires rigorous evaluation from both toxicological and pharmacological perspectives. Variability in extract composition, the potential presence of contaminants, the lack of standardization of raw materials, and the limited clinical evidence available constitute significant barriers to their validation as safe and effective interventions. Furthermore, it is crucial to evaluate their effects in populations with preexisting conditions or greater biological vulnerability.

7. Regulatory Framework

Current legislation on dietary supplements varies significantly across countries. These legislative differences are primarily due to different approaches to risk assessment, the level of state intervention in the market, and the focus on preventive health [210]. This heterogeneity constitutes one of the main challenges for regulating this type of product, especially in a context of globalized trade and the increasing use of digital platforms for distribution [211]. Generally, dietary supplements are regulated as a separate category within food law, although their specific legal treatment depends on the country in question.

At the European level (Figure 4), most food legislation is based on studies and opinions issued by the European Food Safety Authority (EFSA), the body responsible for providing independent scientific advice on food safety [212,213]. Dietary supplements are covered by Directive 2002/46/EC of the European Parliament and of the Council, which considers them to be products intended to supplement the normal diet, composed of concentrated sources of nutrients or other substances with a nutritional or physiological effect, marketed in measured form [214]. This regulation applies in conjunction with other general food law provisions, such as Regulation (EC) No. 178/2002, which establishes the general principles of food law, and Regulation (EC) No. 1924/2006 on nutrition and health claims made on foods [215]. These regulations also establish the permitted nutritional claims for each food formulation. Only claims approved by the European Commission based on evaluations carried out by the various EFSA panels may be included on product labels. Furthermore, before marketing any of these products, the marketing company must notify the competent authorities in its Member State, which allows for monitoring of these types of products [216,217].

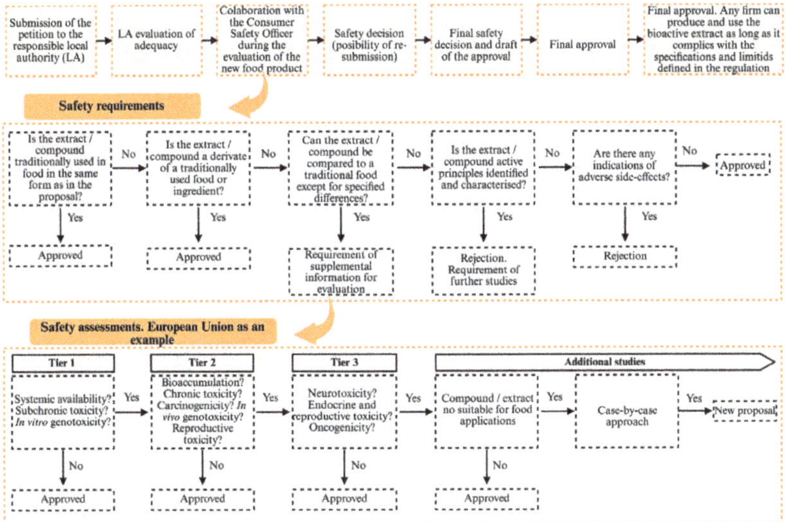

Figure 4. Regulatory framework governing the commercialization of dietary supplements derived from food by-products.

In the United States, another major market for this type of food product, these products are regulated by the Dietary Supplement Health and Education Act (DSHEA) of 1994. This law categorizes dietary supplements as a subcategory of foods, and prior approval by the US Food and Drug Administration (FDA) is not required, except in the case of new, previously unevaluated dietary ingredients [218,219]. Therefore, US legislation is less restrictive, with liability falling on the manufacturer in most cases. However, the FDA, like the EFSA, retains powers in post-market surveillance, including collecting adverse event reports and conducting random inspections. The Federal Trade Commission (FTC), for its part, oversees truthfulness in advertising, introducing a functional division of powers that can complicate the comprehensive oversight of these products [220].

In other countries, different intermediate approaches have been adopted. For example, in Canada, these types of products are classified as natural health products and are regulated by the Natural and Non-prescription Health Products Directorate [221]. This legislation is stricter than that of the United States, as it requires a pre-marketing evaluation phase. As in Europe, this is based on independent reports on the scientific evidence supporting the product's use, safety, quality, and efficacy. In Japan, dietary supplements are divided into two broad groups: foods for specific health purposes and foods with health claims. Both types of products have different legislation and requirements for approval and marketing.

Therefore, there is a significant legislative gap for these products depending on the country of sale. These regulatory divergences represent one of the major challenges for marketing this type of product due to the trade barriers they create and the increased risk of marketing dietary supplements between countries whose efficacy or safety has not been adequately tested. Various initiatives have been implemented to address this problem, including the Codex Alimentarius, a book that compiles a series of general guidelines in the field of food, which are voluntarily adopted by countries and whose application is not binding [222,223].

Another critical aspect of dietary supplement legislation is labeling control, including nutritional claims. While organizations like the EFSA strictly monitor claims about the beneficial properties of dietary supplements, in practice, it is possible to find products on the market that claim to have health benefits without any solid scientific evidence supporting them. This problem is even more acute in the digital environment, which presents a higher level of food fraud [224–226].

In this context, progress in legislation for these products must focus on increasing international cooperation and the development of legal and technological instruments that allow for greater monitoring of quality control and food safety. Likewise, it is essential to develop strategies to improve public awareness of these types of products, so that citizens can critically evaluate the information they receive. This is because the regulation of dietary supplements, far from being a merely administrative aspect, constitutes a key public health tool, which must balance consumers' right to access safe and effective products with the need to avoid unnecessary risks and deceptive commercial practices.

8. Conclusions

The increasing incidence of diabetes demands new, sustainable strategies for prevention and management. Dietary supplements derived from food by-products represent a promising source of bioactive compounds with potential antidiabetic effects. These effects are primarily attributed to improvements in insulin sensitivity, reductions in inflammation, and regulation of glucose metabolism. However, most current evidence is derived from laboratory and animal studies, with relatively few well-controlled human trials. This gap is largely due to variability in bioactive compound content and the lack of standardized dosing.

Among the supplements tested in humans, certain compounds, such as standardized extracts of polyphenols and omega-3 fatty acids, have reached a more advanced stage of development, with evidence from phase 2 clinical trials supporting their potential efficacy and safety. In contrast, many other bioactive compounds reported as antidiabetic agents remain at an earlier stage of investigation and require comprehensive toxicological assessment, pharmacokinetic evaluation, and dose optimization before clinical application can be considered. Additionally, clearer regulatory guidelines are needed to ensure the safety of these supplements. Future research should focus on conducting rigorous clinical trials to verify the benefits of these supplements in humans, while also investigating their interactions with metabolism and the gut microbiota. Improvements in extraction and delivery techniques are expected to enhance their efficacy. Furthermore, close collaboration among researchers, industry stakeholders, and policymakers will be crucial to translate these findings into practical solutions and successfully incorporate these supplements into diabetes management strategies.

Author Contributions: Conceptualization, A.G.P.; methodology, E.N.Y. and A.G.P.; software, M.A.P. and A.G.P.; validation, E.N.Y., M.A.P. and A.G.P.; formal analysis, M.A.P. and A.G.P.; investigation, E.N.Y. and A.G.P.; resources, E.N.Y., M.A.P. and A.G.P.; data curation, M.A.P. and A.G.P.; writing—original draft preparation, E.N.Y. and A.G.P.; writing—review and editing, M.A.P. and A.G.P.; visualization, M.A.P. and A.G.P.; supervision, M.A.P. and A.G.P.; project administration, M.A.P. and A.G.P.; funding acquisition, M.A.P. and A.G.P. All authors have read and agreed to the published version of the manuscript.

Funding: The research leading to these results was supported by Proyectos de Generación de Conocimiento 2023 (PID2023-148814OA-C22) and by Xunta de Galicia regarding the post-doctoral grant of A.G. Pereira (IN606B-2024/011).

Institutional Review Board Statement: Not applicable.

Informed Consent Statement: Not applicable.

Data Availability Statement: Not applicable.

Conflicts of Interest: The authors declare no conflicts of interest.

Abbreviations

The following abbreviations are used in this manuscript:

Generic.	
Cr	Chromium
DHA	Docosahexaenoic acid
DM	Diabetes mellitus
GLP-1	Glucagon-like peptide-1
HbA1c	Glycosylated hemoglobin
PUFAs	Polyunsaturated fatty acids
BGL	Blood glucose levels
COMT	Catechol-O-methyltransferase
FI	Food intake
FO	Fat oxidation
GL	Glucose levels
GLUT4	Glucose transporter 4
GU	Glucose uptake
HOMA-IR	Homeostatic model assessment of insulin resistance index
LKB1/AMP	Liver kinase B1/adenosine monophosphate activated protein kinase
NF-Kb	Nuclear factor κB
Nrf2	Nuclear factor erythroid 2–related factor 2
OS	Oxidative stress
PPAR	Peroxisome proliferator-activated receptors
QUICKI	Quantitative insulin sensitivity check index
RMR	Resting metabolic rate
ROS	Reactive oxygen species
SAA	Serum amyloid plasma
SE	Sensivity
VLDL-C	Very low-density lipoprotein cholesterol
Applications.	
CF	Commercial formulations
DS	Dietary supplementation
FI	Food ingredient
PR	Preservative
TP	Technological properties
Bioactivities.	
AA	Anti-aging
AD	Anti-diabetic
AG	α-amylase and α-glucosidase inhibition
AH	Anti-hypertensive
AI	Anti-inflammatory
AM	Antimicrobial
AO	Antioxidant activity
CP	Cardiovascular protection
MM	Microbiota modulation
OB	Anti-obesity
SA	Satiating
Compounds.	
AA	Arachidonic acids
AAs	Amino acids
AGEs	Advanced glycation end products
ALA	Alpha-lipoic acid

Chl	Cholesterol
Cr	Chromium
Cur	Curcumin
Cys	Cystine
Dapa	Dapagliflozin
DHA	Docosahexaenoic acid
EAA	Essential amino acids
EC	Epicatechin
ECG	Epicatechin gallate
EGC	Epigallocatechin
EGCG	Epigallocatechin gallate
EPA	Eicosapentaenoic acid
FA	Fatty acids
GLP-1	Glucagon-like peptide-1
HbA1c	Glycosylated hemoglobin
Met	Metformin
N.S.	Not specified
Pio	Pioglitazone
PUFAs	Polyunsaturated fatty acids
Tyr	Tyrosine
Diseases.	
CHD	Coronary heart disease
NPDR	Non-proliferative diabetic retinopathy
T2DM	Type 2 diabetic mellitus
Extraction.	
DE	Decoction
EAE	Enzyme-assisted extraction
MA	Maceration
SE	Solvent extraction
UAE	Ultrasound-assisted extraction
Solvents.	
Acet	Acetone
hex	Hexane
MetOH	Methanol
W	Deionized water
Study.	
DB	Double-blind controlled study
DBPC RCT	Double-blind placebo-controlled randomized controlled trial
n	Number of individuals
PS	Prospective study
RCT	Single-blind, placebo-controlled trial
RCT	Randomized controlled trial

Appendix A

This table summarizes the minimum quality specifications recommended for extracts derived from food by-products that have demonstrated potential antidiabetic activity. It provides guidance on source identification, extraction methods, composition, dosing considerations, purity and safety parameters, variability, and preliminary toxicity assessment. The aim is to ensure transparency, reproducibility, and safety in the preparation and use of these extracts for research purposes. The checklist is intended as a practical reference to support the standardized reporting and evaluation of bioactive by-product extracts.

Table A1. Minimum Quality Specifications for By-product-derived Extracts with Antidiabetic Potential.

Category	Parameter	Specification/Guideline	Notes
Source and Identity	Botanical/by-product identification	Include scientific name, plant part or food by-product, and origin	Ensures traceability and correct identification.
	Extraction method	Detail solvents, temperature, duration, and yield	Extraction method influences chemical composition and safety.
Composition	Marker compounds	Quantify key bioactive compounds (e.g., total polyphenols, fiber)	Allows standardization and comparison between batches or studies.
	Additional constituents	Identify other relevant compounds (e.g., flavonoids, saponins, tannins)	Supports understanding of efficacy and potential toxicity.
Dose Information	Human-relevant dose range	Express as mg/kg body weight or mg/day equivalent	Facilitates translation of experimental data to practical use.
	Unit standardization	Consistently report per extract weight or per active compound	Avoids ambiguity in dosing and reporting.
Purity and Safety	Residual solvents	Follow international guidelines	Prevents exposure to harmful solvents.
	Heavy metals	Below internationally accepted limits	Ensures safety and regulatory compliance.
	Pesticides/Mycotoxins	Below internationally accepted limits	Minimizes health risks from contamination.
	Microbiological quality	Absence of pathogens; total microbial counts within accepted limits	Ensures hygienic quality of extracts.
Variability and Reproducibility	Raw material variability	Document seasonal or batch variations	Accounts for differences in composition between raw material lots.
	Batch-to-batch consistency	Use analytical profiling (e.g., HPLC, LC-MS)	Confirms reproducibility of results.
Toxicity Data	Screening	Include in vitro cytotoxicity and, if available, in vivo safety data	Provides initial safety assessment for potential use.
Regulatory Compliance	Alignment with guidelines	All extracts should comply with the relevant legislation and regulatory requirements of the country in which they are intended to be marketed	Supports safe and regulated use of extracts.

References

1. Banday, M.Z.; Sameer, A.S.; Nissar, S. Pathophysiology of diabetes: An overview. *Avicenna J. Med.* **2020**, *10*, 174–188. [CrossRef] [PubMed]
2. Saeedi, P.; Petersohn, I.; Salpea, P.; Malanda, B.; Karuranga, S.; Unwin, N.; Colagiuri, S.; Guariguata, L.; Motala, A.A.; Ogurtsova, K.; et al. Global and regional diabetes prevalence estimates for 2019 and projections for 2030 and 2045: Results from the International Diabetes Federation Diabetes Atlas, 9th edition. *Diabetes Res. Clin. Pract.* **2019**, *157*, 107843. [CrossRef] [PubMed]
3. Hossain, M.J.; Al-Mamun, M.; Islam, M.R. Diabetes mellitus, the fastest growing global public health concern: Early detection should be focused. *Health Sci. Rep.* **2024**, *7*, e2004. [CrossRef] [PubMed]
4. Popović-Djordjević, J.B.; Katanić Stanković, J.S.; Mihailović, V.; Pereira, A.G.; Garcia-Oliveira, P.; Prieto, M.A.; Simal-Gandara, J. Algae as a Source of Bioactive Compounds to Prevent the Development of Type 2 Diabetes Mellitus. *Curr. Med. Chem.* **2021**, *28*, 4592–4615. [CrossRef]
5. Silva-Tinoco, R.; Cuatecontzi-Xochitiotzi, T.; Morales-Buenrostro, L.E.; Gracia-Ramos, A.E.; Aguilar-Salinas, C.A.; Castillo-Martínez, L. Prevalence of Chronic Kidney Disease in Individuals with Type 2 Diabetes Within Primary Care: A Cross-Sectional Study. *J. Prim. Care Community Health* **2024**, *15*, 21501319241259324. [CrossRef]
6. Galaviz, K.I.; Narayan, K.M.V.; Lobelo, F.; Weber, M.B. Lifestyle and the Prevention of Type 2 Diabetes: A Status Report. *Am. J. Lifestyle Med.* **2018**, *12*, 4–20. [CrossRef]
7. Clemente-Suárez, V.J.; Martín-Rodríguez, A.; Beltrán-Velasco, A.I.; Rubio-Zarapuz, A.; Martínez-Guardado, I.; Valcárcel-Martín, R.; Tornero-Aguilera, J.F. Functional and Therapeutic Roles of Plant-Derived Antioxidants in Type 2 Diabetes Mellitus: Mechanisms, Challenges, and Considerations for Special Populations. *Antioxidants* **2025**, *14*, 725. [CrossRef]
8. Shahwan, M.; Alhumaydhi, F.; Ashraf, G.M.; Hasan, P.M.Z.; Shamsi, A. Role of polyphenols in combating Type 2 Diabetes and insulin resistance. *Int. J. Biol. Macromol.* **2022**, *206*, 567–579. [CrossRef]

9. Myhrstad, M.C.W.; Tunsjø, H.; Charnock, C.; Telle-Hansen, V.H. Dietary Fiber, Gut Microbiota, and Metabolic Regulation—Current Status in Human Randomized Trials. *Nutrients* **2020**, *12*, 859. [CrossRef]
10. Dawi, J.; Misakyan, Y.; Affa, S.; Kades, S.; Narasimhan, A.; Hajjar, F.; Besser, M.; Tumanyan, K.; Venketaraman, V. Oxidative Stress, Glutathione Insufficiency, and Inflammatory Pathways in Type 2 Diabetes Mellitus: Implications for Therapeutic Interventions. *Biomedicines* **2024**, *13*, 18. [CrossRef]
11. Viana, M.D.M.; Santos, S.S.; Cruz, A.B.O.; de Jesus, M.V.A.C.; Lauria, P.S.S.; Lins, M.P.; Villarreal, C.F. Probiotics as Antioxidant Strategy for Managing Diabetes Mellitus and Its Complications. *Antioxidants* **2025**, *14*, 767. [CrossRef]
12. Cross, L.V.; Thomas, J.R. Safety and Efficacy of Dietary Supplements for Diabetes. *Diabetes Spectr.* **2021**, *34*, 67–72. [CrossRef]
13. Hannon, B.A.; Fairfield, W.D.; Adams, B.; Kyle, T.; Crow, M.; Thomas, D.M. Use and abuse of dietary supplements in persons with diabetes. *Nutr. Diabetes* **2020**, *10*, 14. [CrossRef]
14. Fagherazzi, G.; Ravaud, P. Digital diabetes: Perspectives for diabetes prevention, management and research. *Diabetes Metab.* **2019**, *45*, 322–329. [CrossRef]
15. Schulze, M.B.; Hu, F.B. Primary prevention of diabetes: What Can Be Done and How Much Can Be Prevented? *Annu. Rev. Public Health* **2005**, *26*, 445–467. [CrossRef]
16. Paschou, S.A.; Marina, L.V.; Spartalis, E.; Anagnostis, P.; Alexandrou, A.; Goulis, D.G.; Lambrinoudaki, I. Therapeutic strategies for type 2 diabetes mellitus in women after menopause. *Maturitas* **2019**, *126*, 69–72. [CrossRef] [PubMed]
17. Yeung, S.; Soliternik, J.; Mazzola, N. Nutritional supplements for the prevention of diabetes mellitus and its complications. *J. Nutr. Intermed. Metab.* **2018**, *14*, 16–21. [CrossRef]
18. Reynolds, A.; Mitri, J. Dietary Advice For Individuals with Diabetes. *Feingold* **2024**.
19. Bergman, M.; Buysschaert, M.; Schwarz, P.E.; Albright, A.; Narayan, K.V.; Yach, D. Diabetes prevention: Global health policy and perspectives from the ground. *Diabetes Manag. Lond. Engl.* **2012**, *2*, 309–321. [CrossRef]
20. Ekor, M. The growing use of herbal medicines: Issues relating to adverse reactions and challenges in monitoring safety. *Front. Pharmacol.* **2014**, *4*, 177. [CrossRef]
21. Pereira, A.G.; Echave, J.; Jorge, A.O.S.; Nogueira-Marques, R.; Yuksek, E.N.; Barciela, P.; Perez-Vazquez, A.; Chamorro, F.; Oliveira, M.B.P.P.; Carpena, M.; et al. Therapeutic and Preventive Potential of Plant-Derived Antioxidant Nutraceuticals. *Foods* **2025**, *14*, 1749. [CrossRef] [PubMed]
22. Maqsood, S.; Khalid, W.; Kumar, P.; Benmebarek, I.E.; Rasool, I.F.U.; Trif, M.; Moreno, A.; Esatbeyoglu, T. Valorization of plant-based agro-industrial waste and by-products for the production of polysaccharides: Towards a more circular economy. *Appl. Food Res.* **2025**, *5*, 100954. [CrossRef]
23. Kumar Gupta, R.; AE Ali, E.; Abd El Gawad, F.; Mecheal Daood, V.; Sabry, H.; Karunanithi, S.; Prakash Srivastav, P. Valorization of fruits and vegetables waste byproducts for development of sustainable food packaging applications. *Waste Manag. Bull.* **2024**, *2*, 21–40. [CrossRef]
24. Jimenez-Lopez, C.; Fraga-Corral, M.; Carpena, M.; García-Oliveira, P.; Echave, J.; Pereira, A.G.; Lourenço-Lopes, C.; Prieto, M.A.; Simal-Gandara, J. Agriculture waste valorisation as a source of antioxidant phenolic compounds within a circular and sustainable bioeconomy. *Food Funct.* **2020**, *11*, 4853–4877. [CrossRef]
25. Skwarek, P.; Karwowska, M. Fruit and vegetable processing by-products as functional meat product ingredients -a chance to improve the nutritional value. *LWT* **2023**, *189*, 115442. [CrossRef]
26. Egbuna, C.; Awuchi, C.G.; Kushwaha, G.; Rudrapal, M.; Patrick-Iwuanyanwu, K.C.; Singh, O.; Odoh, U.E.; Khan, J.; Jeevanandam, J.; Kumarasamy, S.; et al. Bioactive Compounds Effective Against Type 2 Diabetes Mellitus: A Systematic Review. *Curr. Top. Med. Chem.* **2021**, *21*, 1067–1095. [CrossRef]
27. Malik, F.; Iqbal, A.; Zia, S.; Ranjha, M.M.A.N.; Khalid, W.; Nadeem, M.; Selim, S.; Hadidi, M.; Moreno, A.; Manzoor, M.F.; et al. Role and mechanism of fruit waste polyphenols in diabetes management. *Open Chem.* **2023**, *21*, 20220272. [CrossRef]
28. Jimenez-Lopez, C.; Pereira, A.G.; Lourenço-Lopes, C.; Garcia-Oliveira, P.; Cassani, L.; Fraga-Corral, M.; Prieto, M.A.; Simal-Gandara, J. Main bioactive phenolic compounds in marine algae and their mechanisms of action supporting potential health benefits. *Food Chem.* **2021**, *341*, 128262. [CrossRef]
29. Shahidi, F.; Vamadevan, V.; Oh, W.Y.; Peng, H. Phenolic compounds in agri-food by-products, their bioavailability and health effects. *J. Food Bioact.* **2019**, *5*, 57–119. [CrossRef]
30. Fischer, M.; Timper, K.; Radimerski, T.; Dembinski, K.; Frey, D.M.; Zulewski, H.; Keller, U.; Müller, B.; Christ-Crain, M.; Grisouard, J. Metformin induces glucose uptake in human preadipocyte-derived adipocytes from various fat depots. *Diabetes Obes. Metab.* **2010**, *12*, 356–359. [CrossRef]
31. Ruan, Y.; Sun, J.; He, J.; Chen, F.; Chen, R.; Chen, H. Effect of Probiotics on Glycemic Control: A Systematic Review and Meta-Analysis of Randomized, Controlled Trials. *PLoS ONE* **2015**, *10*, e0132121. [CrossRef]
32. Echave, J.; Pereira, A.G.; Jorge, A.O.S.; Barciela, P.; Nogueira-Marques, R.; Yuksek, E.N.; Oliveira, M.B.P.P.; Barros, L.; Prieto, M.A. Grape Winemaking By-Products: Current Valorization Strategies and Their Value as Source of Tannins with Applications in Food and Feed. *Molecules* **2025**, *30*, 2726. [CrossRef]

33. Yilmaz, Y.; Toledo, R.T. Major Flavonoids in Grape Seeds and Skins: Antioxidant Capacity of Catechin, Epicatechin, and Gallic Acid. *J. Agric. Food Chem.* **2004**, *52*, 255–260. [CrossRef] [PubMed]
34. Tajik, N.; Tajik, M.; Mack, I.; Enck, P. The potential effects of chlorogenic acid, the main phenolic components in coffee, on health: A comprehensive review of the literature. *Eur. J. Nutr.* **2017**, *56*, 2215–2244. [CrossRef]
35. Wen, L.; Wu, D.; Tan, X.; Zhong, M.; Xing, J.; Li, W.; Li, D.; Cao, F. The Role of Catechins in Regulating Diabetes: An Update Review. *Nutrients* **2022**, *14*, 4681. [CrossRef]
36. Erukainure, O.L.; Chukwuma, C.I.; Nambooze, J.; Tripathy, S.; Salau, V.F.; Olofinsan, K.; Ogunlakin, A.D.; Ebuehi, O.A.T.; Unuofin, J.O. Tea Consumption and Diabetes: A Comprehensive Pharmacological Review of Black, White, Green, Oolong, and Pu-erh Teas. *Plants* **2025**, *14*, 1898. [CrossRef] [PubMed]
37. Pereira, A.G.; Garcia-Perez, P.; Cassani, L.; Chamorro, F.; Cao, H.; Barba, F.J.; Simal-Gandara, J.; Prieto, M.A. Camellia japonica: A phytochemical perspective and current applications facing its industrial exploitation. *Food Chem. X* **2022**, *13*, 100258. [CrossRef] [PubMed]
38. Pereira, A.G.; Fraga-Corral, M.; Garciá-Oliveira, P.; Jimenez-Lopez, C.; Lourenço-Lopes, C.; Carpena, M.; Otero, P.; Gullón, P.; Prieto, M.A.; Simal-Gandara, J. Culinary and nutritional value of edible wild plants from northern Spain rich in phenolic compounds with potential health benefits. *Food Funct.* **2020**, *11*, 8493–8515. [CrossRef]
39. Donn, P.; Barciela, P.; Perez-Vazquez, A.; Cassani, L.; Simal-Gandara, J.; Prieto, M.A. Bioactive Compounds of *Verbascum sinuatum L.*: Health Benefits and Potential as New Ingredients for Industrial Applications. *Biomolecules* **2023**, *13*, 427. [CrossRef]
40. Plakantonaki, S.; Roussis, I.; Bilalis, D.; Priniotakis, G. Dietary Fiber from Plant-Based Food Wastes: A Comprehensive Approach to Cereal, Fruit, and Vegetable Waste Valorization. *Processes* **2023**, *11*, 1580. [CrossRef]
41. Zhang, W.; Zeng, G.; Pan, Y.; Chen, W.; Huang, W.; Chen, H.; Li, Y. Properties of soluble dietary fiber-polysaccharide from papaya peel obtained through alkaline or ultrasound-assisted alkaline extraction. *Carbohydr. Polym.* **2017**, *172*, 102–112. [CrossRef]
42. Benítez, V.; Mollá, E.; Martín-Cabrejas, M.A.; Aguilera, Y.; Esteban, R.M. Physicochemical properties and in vitro antidiabetic potential of fibre concentrates from onion by-products. *J. Funct. Foods* **2017**, *36*, 34–42. [CrossRef]
43. Budhwar, S.; Chakraborty, M.; Sethi, K.; Chatterjee, A. Antidiabetic properties of rice and wheat bran—A review. *J. Food Biochem.* **2020**, *44*, e13424. [CrossRef]
44. Kučuk, N.; Primožič, M.; Kotnik, P.; Knez, Ž.; Leitgeb, M. Mango Peels as an Industrial By-Product: A Sustainable Source of Compounds with Antioxidant, Enzymatic, and Antimicrobial Activity. *Foods* **2024**, *13*, 553. [CrossRef] [PubMed]
45. Zou, F.; Tan, C.; Zhang, B.; Wu, W.; Shang, N. The Valorization of Banana By-Products: Nutritional Composition, Bioactivities, Applications, and Future Development. *Foods* **2022**, *11*, 3170. [CrossRef] [PubMed]
46. Holscher, H.D. Dietary fiber and prebiotics and the gastrointestinal microbiota. *Gut Microbes* **2017**, *8*, 172–184. [CrossRef]
47. Ahmed, O.M.; Hassan, M.A.; Abdel-Twab, S.M.; Abdel Azeem, M.N. Navel orange peel hydroethanolic extract, naringin and naringenin have anti-diabetic potentials in type 2 diabetic rats. *Biomed. Pharmacother.* **2017**, *94*, 197–205. [CrossRef]
48. Oboh, G.; Ademosum, A.O. Phenolic extracts from grapefruit peels (*Citrus paradisi*) inhibit key enzymes linked with type 2 diabetes and hypertension. *J. Food Biochem.* **2011**, *35*, 1703–1709. [CrossRef]
49. El-Feky, A.M.; Aboulthana, W.M.; El-Rashedy, A.A. Assessment of the in vitro anti-diabetic activity with molecular dynamic simulations of limonoids isolated from Adalia lemon peels. *Sci. Rep.* **2024**, *14*, 21478. [CrossRef]
50. Mechchate, H.; Es-safi, I.; Haddad, H.; Bekkari, H.; Grafov, A.; Bousta, D. Combination of Catechin, Epicatechin, and Rutin: Optimization of a novel complete antidiabetic formulation using a mixture design approach. *J. Nutr. Biochem.* **2021**, *88*, 108520. [CrossRef]
51. Li, X.; Wang, T.; Zhou, B.; Gao, W.; Cao, J.; Huang, L. Chemical composition and antioxidant and anti-inflammatory potential of peels and flesh from 10 different pear varieties (*Pyrus* spp.). *Food Chem.* **2014**, *152*, 531–538. [CrossRef]
52. Smaniotto, F.A.; Dluzniewski, L.T.; Bortolazzo, P.C.; Monteiro, C.S.; Baranzelli, J.; da Silva, D.T.; Somacal, S.; Conterato, G.M.M.; Emanuelli, T. In vitro assessment of antidiabetic, anti-obesogenic, and antioxidant potential of pulp and seed extracts from *Eugenia involucrata* fruits. *Food Res. Int.* **2025**, *202*, 115693. [CrossRef]
53. Gondi, M.; Basha, S.A.; Bhaskar, J.J.; Salimath, P.V.; Prasada Rao, U.J.S. Anti-diabetic effect of dietary mango (*Mangifera indica* L.) peel in streptozotocin-induced diabetic rats. *J. Sci. Food Agric.* **2015**, *95*, 991–999. [CrossRef]
54. Yousefifar, H. Development of a Food Supplement for Diabetes Type II Management with Enhanced Antidiabetic and Antioxidant Properties from Selected Local Foods. Ph.D. Thesis, University of Nairobi, Nairobi, Kenya, 2019.
55. Ivanov, Y.; Atanasova, M.; Godjevargova, T. Nutritional and Functional Values of Grape Seed Flour and Extract for Production of Antioxidative Dietary Supplements and Functional Foods. *Molecules* **2025**, *30*, 2029. [CrossRef]
56. Kammerer, D.R.; Kammerer, J.; Valet, R.; Carle, R. Recovery of polyphenols from the by-products of plant food processing and application as valuable food ingredients. *Food Res. Int.* **2014**, *65*, 2–12. [CrossRef]
57. Salehi, L.; Dinani, S.T. Application of electrohydrodynamic—Ultrasonic procedure for extraction of β-carotene from carrot pomace. *J. Food Meas. Charact.* **2020**, *14*, 3031–3039. [CrossRef]

58. Jahanban-Esfahlan, A.; Ostadrahimi, A.; Tabibiazar, M.; Amarowicz, R. A Comprehensive Review on the Chemical Constituents and Functional Uses of Walnut (*Juglans* spp.) Husk. *Int. J. Mol. Sci.* **2019**, *20*, 3920. [CrossRef]
59. Wang, M.; Yang, F.; Yan, X.; Chao, X.; Zhang, W.; Yuan, C.; Zeng, Q. Anti-diabetic effect of banana peel dietary fibers on type 2 diabetic mellitus mice induced by streptozotocin and high-sugar and high-fat diet. *J. Food Biochem.* **2022**, *46*, e14275. [CrossRef]
60. Tariq, A.; Sahar, A.; Usman, M.; Sameen, A.; Azhar, M.; Tahir, R.; Younas, R.; Issa Khan, M. Extraction of dietary fiber and polyphenols from mango peel and its therapeutic potential to improve gut health. *Food Biosci.* **2023**, *53*, 102669. [CrossRef]
61. Ajila, C.M.; Prasada Rao, U.J.S. Mango peel dietary fibre: Composition and associated bound phenolics. *J. Funct. Foods* **2013**, *5*, 444–450. [CrossRef]
62. Nieto-Calvache, J.E.; de Escalada Pla, M.; Gerschenson, L.N. Dietary fibre concentrates produced from papaya by—Products for agroindustrial waste valorisation. *Int. J. Food Sci. Technol.* **2019**, *54*, 1074–1080. [CrossRef]
63. Figueira, O.; Pereira, V.; Castilho, P.C. A Two-Step Approach to Orange Peel Waste Valorization: Consecutive Extraction of Pectin and Hesperidin. *Foods* **2023**, *12*, 3834. [CrossRef]
64. Sir Elkhatim, K.A.; Elagib, R.A.A.; Hassan, A.B. Content of phenolic compounds and vitamin C and antioxidant activity in wasted parts of Sudanese citrus fruits. *Food Sci. Nutr.* **2018**, *6*, 1214–1219. [CrossRef]
65. Song, H.; Cong, Z.; Wang, C.; He, M.; Liu, C.; Gao, P. Research progress on Walnut oil: Bioactive compounds, health benefits, extraction methods, and medicinal uses. *J. Food Biochem.* **2022**, *46*, e14504. [CrossRef]
66. Ciftci, O.; Keskin Cavdar, H. Ultrasound—Assisted Extraction Improved the Extraction Rate, Omega-3, and Polyunsaturated Fatty Acid Content of Fish Oil From Atlantic Bonito Waste. *Eur. J. Lipid Sci. Technol.* **2025**, *127*, e202400131. [CrossRef]
67. Rodrigues, M.; Rosa, A.; Almeida, A.; Martins, R.; Ribeiro, T.; Pintado, M.; Gonçalves, R.F.S.; Pinheiro, A.C.; Fonseca, A.J.M.; Maia, M.R.G.; et al. Omega-3 fatty acids from fish by-products: Innovative extraction and application in food and feed. *Food Bioprod. Process.* **2024**, *145*, 32–41. [CrossRef]
68. Ishak, N.H.; Sarbon, N.M. A Review of Protein Hydrolysates and Bioactive Peptides Deriving from Wastes Generated by Fish Processing. *Food Bioprocess Technol.* **2018**, *11*, 2–16. [CrossRef]
69. Gajanan, P.G.; Elavarasan, K.; Shamasundar, B.A. Bioactive and functional properties of protein hydrolysates from fish frame processing waste using plant proteases. *Environ. Sci. Pollut. Res.* **2016**, *23*, 24901–24911. [CrossRef] [PubMed]
70. Harnedy, P.A.; Parthsarathy, V.; McLaughlin, C.M.; O'Keeffe, M.B.; Allsopp, P.J.; McSorley, E.M.; O'Harte, F.P.M.; FitzGerald, R.J. Atlantic salmon (*Salmo salar*) co-product-derived protein hydrolysates: A source of antidiabetic peptides. *Food Res. Int.* **2018**, *106*, 598–606. [CrossRef]
71. Ostojić, S.; Pavlović, M.; Živić, M.; Filipović, Z.; Gorjanović, S.; Hranisavljević, S.; Dojčinović, M. Processing of whey from dairy industry waste. *Environ. Chem. Lett.* **2005**, *3*, 29–32. [CrossRef]
72. Mavridis, K.; Todas, N.; Kalompatsios, D.; Athanasiadis, V.; Lalas, S.I. Lycopene and Other Bioactive Compounds' Extraction from Tomato Processing Industry Waste: A Comparison of Ultrasonication Versus a Conventional Stirring Method. *Horticulturae* **2025**, *11*, 71. [CrossRef]
73. Poojary, M.M.; Passamonti, P. Extraction of lycopene from tomato processing waste: Kinetics and modelling. *Food Chem.* **2015**, *173*, 943–950. [CrossRef]
74. Umair, M.; Jabbar, S.; Nasiru, M.M.; Lu, Z.; Zhang, J.; Abid, M.; Murtaza, M.A.; Kieliszek, M.; Zhao, L. Ultrasound-Assisted Extraction of Carotenoids from Carrot Pomace and Their Optimization through Response Surface Methodology. *Molecules* **2021**, *26*, 6763. [CrossRef]
75. Victor, M.M.; David, J.M.; Cortez, M.V.M.; Leite, J.L.; da Silva, G.S.B. A High-Yield Process for Extraction of Hesperidin from Orange (*Citrus sinensis* L. osbeck) Peels Waste, and Its Transformation to Diosmetin, A Valuable and Bioactive Flavonoid. *Waste Biomass Valorization* **2021**, *12*, 313–320. [CrossRef]
76. Afolayan, A.; Wintola, O. Dietary Supplements in the Management of Hypertension and Diabetes—A Review. *African J. Tradit. Complement. Altern. Med.* **2014**, *11*, 248. [CrossRef]
77. Soares, T.F.; Oliveira, M.B.P.P. Cocoa By-Products: Characterization of Bioactive Compounds and Beneficial Health Effects. *Molecules* **2022**, *27*, 1625. [CrossRef]
78. Scurria, A.; Lino, C.; Pitonzo, R.; Pagliaro, M.; Avellone, G.; Ciriminna, R. Vitamin D3 in fish oil extracted with limonene from anchovy leftovers. *Chem. Data Collect.* **2020**, *25*, 100311. [CrossRef]
79. Lewgood, J.; Oliveira, B.; Korzepa, M.; Forbes, S.C.; Little, J.P.; Breen, L.; Bailie, R.; Candow, D.G. Efficacy of Dietary and Supplementation Interventions for Individuals with Type 2 Diabetes. *Nutrients* **2021**, *13*, 2378. [CrossRef] [PubMed]
80. Radenkovs, V.; Püssa, T.; Juhnevica-Radenkova, K.; Kviesis, J.; Salar, F.J.; Moreno, D.A.; Drudze, I. Wild apple (*Malus* spp.) by-products as a source of phenolic compounds and vitamin C for food applications. *Food Biosci.* **2020**, *38*, 100744. [CrossRef]
81. Domínguez-Perles, R.; Martínez-Ballesta, M.C.; Carvajal, M.; García-Viguera, C.; Moreno, D.A. Broccoli-Derived By-Products—A Promising Source of Bioactive Ingredients. *J. Food Sci.* **2010**, *75*, C383–C392. [CrossRef]
82. Fernández-López, J.; Fernández-Ginés, J.M.; Aleson-Carbonell, L.; Sendra, E.; Sayas-Barberá, E.; Pérez-Alvarez, J.A. Application of functional citrus by-products to meat products. *Trends Food Sci. Technol.* **2004**, *15*, 176–185. [CrossRef]

83. Barba, F.J.; Putnik, P.; Bursać Kovačević, D.; Poojary, M.M.; Roohinejad, S.; Lorenzo, J.M.; Koubaa, M. Impact of conventional and non-conventional processing on prickly pear (*Opuntia* spp.) and their derived products: From preservation of beverages to valorization of by-products. *Trends Food Sci. Technol.* **2017**, *67*, 260–270. [CrossRef]
84. Zhou, C.; Na, L.; Shan, R.; Cheng, Y.; Li, Y.; Wu, X.; Sun, C. Dietary Vitamin C Intake Reduces the Risk of Type 2 Diabetes in Chinese Adults: HOMA-IR and T-AOC as Potential Mediators. *PLoS ONE* **2016**, *11*, e0163571. [CrossRef]
85. Montonen, J.; Knekt, P.; Jarvinen, R.; Reunanen, A. Dietary Antioxidant Intake and Risk of Type 2 Diabetes. *Diabetes Care* **2004**, *27*, 362–366. [CrossRef]
86. Sluijs, I.; Cadier, E.; Beulens, J.W.J.; Spijkerman, A.M.W.; Van der Schouw, Y.T. Dietary intake of carotenoids and risk of type 2 diabetes. *Nutr. Metab. Cardiovasc. Dis.* **2015**, *25*, 376–381. [CrossRef]
87. Watanabe, F.; Bito, T.; Koseki, K. Salmon meats and by-products as excellent sources of vitamin B12. *Fish. Sci.* **2025**, *91*, 405–415. [CrossRef]
88. Zhou, Y.; He, A.; Xu, B. Natural resources, quantification, microbial bioconversion, and bioactivities of vitamin B12 for vegetarian diet. *Food Chem.* **2025**, *463*, 140849. [CrossRef] [PubMed]
89. Wang, W.; Zhang, J.; Wang, H.; Wang, X.; Liu, S. Vitamin D deficiency enhances insulin resistance by promoting inflammation in type 2 diabetes. *Int. J. Clin. Exp. Pathol.* **2019**, *12*, 1859–1867.
90. Mazur-Bialy, A.I.; Pocheć, E. Vitamin B2 deficiency enhances the pro-inflammatory activity of adipocyte, consequences for insulin resistance and metabolic syndrome development. *Life Sci.* **2017**, *178*, 9–16. [CrossRef]
91. Yao, L.; Wang, Y.; Qin, S.; Zhu, S.; Wu, L. The antidiabetic drug metformin aids bacteria in hijacking vitamin B12 from the environment through RcdA. *Commun. Biol.* **2023**, *6*, 96. [CrossRef]
92. Cabrera, C.; Lorenzo, M.L.; De Mena, C.; Lopez, M.C. Chromium, copper, iron, manganese, selenium and zinc levels in dairy products: In vitro study of absorbable fractions. *Int. J. Food Sci. Nutr.* **1996**, *47*, 331–339. [CrossRef]
93. Vasilyeva, V.T.; Sleptsova, T.V.; Lebedeva, U.M.; Abramov, A.F.; Bappagai, E.V. The content of zinc and selenium in local food products of Yakutia. *Probl. Nutr.* **2023**, *92*, 93–99. [CrossRef]
94. Filippini, T.; Cilloni, S.; Malavolti, M.; Violi, F.; Malagoli, C.; Tesauro, M.; Bottecchi, I.; Ferrari, A.; Vescovi, L.; Vinceti, M. Dietary intake of cadmium, chromium, copper, manganese, selenium and zinc in a Northern Italy community. *J. Trace Elem. Med. Biol.* **2018**, *50*, 508–517. [CrossRef]
95. Chen, C.; Chaudhary, A.; Mathys, A. Nutritional and environmental losses embedded in global food waste. *Resour. Conserv. Recycl.* **2020**, *160*, 104912. [CrossRef]
96. Torres, M.D.; Seijo, J. By-Products from the Chestnut Industry Used to Produce Natural Potassium Soaps: Physicochemical Properties. *J. Surfactants Deterg.* **2016**, *19*, 381–387. [CrossRef]
97. Picard, K.; Morris, A. Potassium Food Additives and Dietary Management of Serum Potassium: Proposed Best-Practice Recommendations. *J. Ren. Nutr.* **2025**, *35*, 221–228. [CrossRef]
98. Sárközy, M.; Fekete, V.; Szűcs, G.; Török, S.; Szűcs, C.; Bárkányi, J.; Varga, Z.V.; Földesi, I.; Csonka, C.; Kónya, C.; et al. Anti-diabetic effect of a preparation of vitamins, minerals and trace elements in diabetic rats: A gender difference. *BMC Endocr. Disord.* **2014**, *14*, 72. [CrossRef] [PubMed]
99. Kehinde, B.A.; Sharma, P. Recently isolated antidiabetic hydrolysates and peptides from multiple food sources: A review. *Crit. Rev. Food Sci. Nutr.* **2020**, *60*, 322–340. [CrossRef]
100. Phadtare, I.; Vaidya, H.; Hawboldt, K.; Cheema, S.K. Shrimp Oil Extracted from Shrimp Processing By-Product Is a Rich Source of Omega-3 Fatty Acids and Astaxanthin-Esters, and Reveals Potential Anti-Adipogenic Effects in 3T3-L1 Adipocytes. *Mar. Drugs* **2021**, *19*, 259.
101. Alfio, V.G.; Manzo, C.; Micillo, R. From Fish Waste to Value: An Overview of the Sustainable Recovery of Omega-3 for Food Supplements. *Molecules* **2021**, *26*, 1002. [CrossRef]
102. Alkhatib, A.; Tsang, C.; Tiss, A.; Bahorun, T.; Arefanian, H.; Barake, R.; Khadir, A.; Tuomilehto, J. Functional foods and lifestyle approaches for diabetes prevention and management. *Nutrients* **2017**, *9*, 1310. [CrossRef]
103. Vinayagam, R.; Xiao, J.; Xu, B. An insight into anti-diabetic properties of dietary phytochemicals. *Phytochem. Rev.* **2017**, *16*, 535–553. [CrossRef]
104. Lacroix, I.M.E.; Li-Chan, E.C.Y. Overview of food products and dietary constituents with antidiabetic properties and their putative mechanisms of action: A natural approach to complement pharmacotherapy in the management of diabetes. *Mol. Nutr. Food Res.* **2014**, *58*, 61–78. [CrossRef]
105. Jin, Y.; Arroo, R.R.J. Application of dietary supplements in the prevention of type 2 diabetes-related cardiovascular complications. *Phytochem. Rev.* **2021**, *20*, 181–209. [CrossRef]
106. Sharkey, S.J.; Harnedy-Rothwell, P.A.; Allsopp, P.J.; Hollywood, L.E.; FitzGerald, R.J.; O'Harte, F.P.M. A Narrative Review of the Anti-Hyperglycemic and Satiating Effects of Fish Protein Hydrolysates and Their Bioactive Peptides. *Mol. Nutr. Food Res.* **2020**, *64*, 2000403. [CrossRef] [PubMed]

107. Meng, X.; Li, Q.; Shi, R.; Chang, J.; Chang, H.; Li, M. Food supplements could be an effective improvement of diabetes mellitus: A review. *J. Futur. Foods* **2021**, *1*, 67–81. [CrossRef]
108. Röder, P.V.; Wu, B.; Liu, Y.; Han, W. Pancreatic regulation of glucose homeostasis. *Exp. Mol. Med.* **2016**, *48*, 19. [CrossRef]
109. Dimitriadis, G.D.; Maratou, E.; Kountouri, A.; Board, M.; Lambadiari, V. Regulation of postabsorptive and postprandial glucose metabolism by insulin-dependent and insulin-independent mechanisms: An integrative approach. *Nutrients* **2021**, *13*, 159. [CrossRef]
110. Rorsman, P.; Ashcroft, F.M. Pancreatic β-cell electrical activity and insulin secretion: Of mice and men. *Physiol. Rev.* **2018**, *98*, 117–214. [CrossRef] [PubMed]
111. Barrett, M.L.; Udani, J.K. A proprietary alpha-amylase inhibitor from white bean (*Phaseolus vulgaris*): A review of clinical studies on weight loss and glycemic control. *Nutr. J.* **2011**, *10*, 24. [CrossRef] [PubMed]
112. Farazi, M.; Houghton, M.J.; Nicolotti, L.; Murray, M.; Cardoso, B.R.; Williamson, G. Inhibition of human starch digesting enzymes and intestinal glucose transport by walnut polyphenols. *Food Res. Int.* **2024**, *189*, 114572. [CrossRef]
113. Imran, M.; Arshad, M.S.; Butt, M.S.; Kwon, J.H.; Arshad, M.U.; Sultan, M.T. Mangiferin: A natural miracle bioactive compound against lifestyle related disorders. *Lipids Health Dis.* **2017**, *16*, 84. [CrossRef]
114. DasNandy, A.; Virge, R.; Hegde, H.V.; Chattopadhyay, D. A review of patent literature on the regulation of glucose metabolism by six phytocompounds in the management of diabetes mellitus and its complications. *J. Integr. Med.* **2023**, *21*, 226–235. [CrossRef]
115. Gribble, F.M.; Reimann, F. Function and mechanisms of enteroendocrine cells and gut hormones in metabolism. *Nat. Rev. Endocrinol.* **2019**, *15*, 226–237. [CrossRef]
116. Wang, Y.; Wu, Y.; Sailike, J.; Sun, X.; Abuduwaili, N.; Tuoliuhan, H.; Yusufu, M.; Nabi, X. hua Fourteen composite probiotics alleviate type 2 diabetes through modulating gut microbiota and modifying M1/M2 phenotype macrophage in db/db mice. *Pharmacol. Res.* **2020**, *161*, 105150.
117. Chandrasekaran, P.; Weiskirchen, S.; Weiskirchen, R. Effects of Probiotics on Gut Microbiota: An Overview. *Int. J. Mol. Sci.* **2024**, *25*, 6022. [CrossRef]
118. Akilen, R.; Tsiami, A.; Devendra, D.; Robinson, N. Cinnamon in glycaemic control: Systematic review and meta analysis. *Clin. Nutr.* **2012**, *31*, 609–615. [CrossRef]
119. Wang, Z.Q.; Cefalu, W.T. Current concepts about chromium supplementation in type 2 diabetes and insulin resistance. *Curr. Diab. Rep.* **2010**, *10*, 145–151. [PubMed]
120. Asbaghi, O.; Fatemeh, N.; Mahnaz, R.K.; Ehsan, G.; Elham, E.; Behzad, N.; Damoon, A.L.; Amirmansour, A.N. Effects of chromium supplementation on glycemic control in patients with type 2 diabetes: A systematic review and meta-analysis of randomized controlled trials. *Pharmacol. Res.* **2020**, *161*, 105098. [CrossRef] [PubMed]
121. Costello, R.B.; Dwyer, J.T.; Bailey, R.L. Chromium supplements for glycemic control in type 2 diabetes: Limited evidence of effectiveness. *Nutr. Rev.* **2016**, *74*, 455–468. [CrossRef] [PubMed]
122. Lewicki, S.; Zdanowski, R.; Krzyzowska, M.; Lewicka, A.; Debski, B.; Niemcewicz, M.; Goniewicz, M. The role of chromium III in the organism and its possible use in diabetes and obesity treatment. *Ann. Agric. Environ. Med.* **2014**, *21*, 331–335. [CrossRef]
123. Paiva, A.N.; de Lima, J.G.; de Medeiros, A.C.; Figueiredo, H.A.; de Andrade, R.L.; Ururahy, M.A.; Rezende, A.A.; Brandão-Neto, J.; Almeida, M.d.G. Beneficial effects of oral chromium picolinate supplementation on glycemic control in patients with type 2 diabetes: A randomized clinical study. *J. Trace Elem. Med. Biol.* **2015**, *32*, 66–72. [CrossRef]
124. Sun, C.; Zhang, F.; Ge, X.; Yan, T.; Chen, X.; Shi, X.; Zhai, Q. SIRT1 Improves Insulin Sensitivity under Insulin-Resistant Conditions by Repressing PTP1B. *Cell Metab.* **2007**, *6*, 307–319. [CrossRef]
125. Jardine, M.A.; Kahleova, H.; Levin, S.M.; Ali, Z.; Trapp, C.B.; Barnard, N.D. Perspective: Plant-Based Eating Pattern for Type 2 Diabetes Prevention and Treatment: Efficacy, Mechanisms, and Practical Considerations. *Adv. Nutr.* **2021**, *12*, 2045–2055. [CrossRef] [PubMed]
126. Le, Y.; Wang, B.; Xue, M. Nutraceuticals use and type 2 diabetes mellitus. *Curr. Opin. Pharmacol.* **2022**, *62*, 168–176. [CrossRef] [PubMed]
127. Nguyen, H.; Pellegrini, M.V.; Gupta, V. Alpha-lipoic Acid. In *StatPearls*; StatPearls Publishing: Treasure Island, FL, USA, 2024.
128. Onaolapo, A.Y.; Onaolapo, O.J. Nutraceuticals and Diet-based Phytochemicals in Type 2 Diabetes Mellitus: From Whole Food to Components with Defined Roles and Mechanisms. *Curr. Diabetes Rev.* **2018**, *16*, 12–25. [CrossRef] [PubMed]
129. Yin, J.; Xing, H.; Ye, J. Efficacy of berberine in patients with type 2 diabetes mellitus. *Metabolism* **2008**, *57*, 712–717. [CrossRef]
130. Kong, W.; Wei, J.; Abidi, P.; Lin, M.; Inaba, S.; Li, C.; Wang, Y.; Wang, Z.; Si, S.; Pan, H.; et al. Berberine is a novel cholesterol-lowering drug working through a unique mechanism distinct from statins. *Nat. Med.* **2004**, *10*, 1344–1351. [CrossRef]
131. D'Esposito, V.; Di Tolla, M.F.; Lecce, M.; Cavalli, F.; Libutti, M.; Misso, S.; Cabaro, S.; Ambrosio, M.R.; Parascandolo, A.; Covelli, B.; et al. Lifestyle and Dietary Habits Affect Plasma Levels of Specific Cytokines in Healthy Subjects. *Front. Nutr.* **2022**, *9*, 913176. [CrossRef]

132. Zwickey, H.; Horgan, A.; Hanes, D.; Schiffke, H.; Moore, A.; Wahbeh, H.; Jordan, J.; Ojeda, L.; McMurry, M.; Elmer, P.; et al. Effect of the Anti-Inflammatory Diet in People with Diabetes and Pre-Diabetes: A Randomized Controlled Feeding Study. *J. Restor. Med.* **2019**, *8*, e20190107. [CrossRef]
133. Adefegha, S.A. Functional Foods and Nutraceuticals as Dietary Intervention in Chronic Diseases; Novel Perspectives for Health Promotion and Disease Prevention. *J. Diet. Suppl.* **2018**, *15*, 977–1009. [CrossRef]
134. Sears, B.; Perry, M. The role of fatty acids in insulin resistance. *Lipids Health Dis.* **2015**, *14*, 121. [CrossRef]
135. Naeini, Z.; Toupchian, O.; Vatannejad, A.; Sotoudeh, G.; Teimouri, M.; Ghorbani, M.; Nasli-Esfahani, E.; Koohdani, F. Effects of DHA-enriched fish oil on gene expression levels of p53 and NF-κB and PPAR-γ activity in PBMCs of patients with T2DM: A randomized, double-blind, clinical trial. *Nutr. Metab. Cardiovasc. Dis.* **2020**, *30*, 441–447. [CrossRef]
136. Shafras, M.; Sabaragamuwa, R.; Suwair, M. Role of dietary antioxidants in diabetes: An overview. *Food Chem. Adv.* **2024**, *4*, 100666. [CrossRef]
137. Yousef, H.; Khandoker, A.H.; Feng, S.F.; Helf, C.; Jelinek, H.F. Inflammation, oxidative stress and mitochondrial dysfunction in the progression of type II diabetes mellitus with coexisting hypertension. *Front. Endocrinol.* **2023**, *14*, 1173402. [CrossRef]
138. Obeme-Nmom, J.I.; Abioye, R.O.; Reyes Flores, S.S.; Udenigwe, C.C. Regulation of redox enzymes by nutraceuticals: A review of the roles of antioxidant polyphenols and peptides. *Food Funct.* **2024**, *15*, 10956–10980. [CrossRef] [PubMed]
139. Cameron, N.E.; Eaton, S.E.M.; Cotter, M.A.; Tesfaye, S. Vascular factors and metabolic interactions in the pathogenesis of diabetic neuropathy. *Diabetologia* **2001**, *44*, 1973–1988. [CrossRef]
140. Kasznicki, J.; Kosmalski, M.; Sliwinska, A.; Mrowicka, M.; Stanczyk, M.; Majsterek, I.; Drzewoski, J. Evaluation of oxidative stress markers in pathogenesis of diabetic neuropathy. *Mol. Biol. Rep.* **2012**, *39*, 8669–8678. [CrossRef]
141. Mhawish, R.; Komarnytsky, S. Small Phenolic Metabolites at the Nexus of Nutrient Transport and Energy Metabolism. *Molecules* **2025**, *30*, 1026. [CrossRef]
142. Kato, M.; Nishikawa, S.; Ikehata, A.; Dochi, K.; Tani, T.; Takahashi, T.; Imaizumi, A.; Tsuda, T. Curcumin improves glucose tolerance via stimulation of glucagon-like peptide-1 secretion. *Mol. Nutr. Food Res.* **2017**, *61*, 1600471. [CrossRef] [PubMed]
143. Buhrmann, C.; Mobasheri, A.; Busch, F.; Aldinger, C.; Stahlmann, R.; Montaseri, A.; Shakibaei, M. Curcumin modulates nuclear factor kappaB (NF-kappaB)-mediated inflammation in human tenocytes in vitro: Role of the phosphatidylinositol 3-kinase/Akt pathway. *J. Biol. Chem.* **2011**, *286*, 28556–28566. [CrossRef]
144. Peng, Z.; Zeng, Y.; Tan, Q.; He, Q.; Wang, S.; Wang, J. 6-Gingerol alleviates ectopic lipid deposition in skeletal muscle by regulating CD36 translocation and mitochondrial function. *Biochem. Biophys. Res. Commun.* **2024**, *708*, 149786. [CrossRef]
145. Piera-Mardemootoo, C.; Lambert, P.; Faillie, J.L. Efficacy of metformin on glycemic control and weight in drug-naive type 2 diabetes mellitus patients: A systematic review and meta-analysis of placebo-controlled randomized trials. *Therapies* **2021**, *76*, 647–656. [CrossRef]
146. Bailey, C.J. Metformin: Historical overview. *Diabetologia* **2017**, *60*, 1566–1576. [CrossRef] [PubMed]
147. Koffert, J.P.; Mikkola, K.; Virtanen, K.A.; Andersson, A.M.D.; Faxius, L.; Hällsten, K.; Heglind, M.; Guiducci, L.; Pham, T.; Silvola, J.M.U.; et al. Metformin treatment significantly enhances intestinal glucose uptake in patients with type 2 diabetes: Results from a randomized clinical trial. *Diabetes Res. Clin. Pract.* **2017**, *131*, 208–216. [CrossRef] [PubMed]
148. De Pablos-Velasco, P. Pioglitazone: Beyond glucose control. *Expert Rev. Cardiovasc. Ther.* **2010**, *8*, 1057–1067. [CrossRef] [PubMed]
149. Nesti, L.; Tricò, D.; Mengozzi, A.; Natali, A. Rethinking pioglitazone as a cardioprotective agent: A new perspective on an overlooked drug. *Cardiovasc. Diabetol.* **2021**, *20*, 109. [CrossRef]
150. Mittermayer, F.; Caveney, E.; De Oliveira, C.; Gourgiotis, L.; Puri, M.; Tai, L.-J.; Turner, J.R. Addressing Unmet Medical Needs in Type 2 Diabetes: A Narrative Review of Drugs under Development. *Curr. Diabetes Rev.* **2015**, *11*, 17. [CrossRef]
151. Kapoor, B.; Kapoor, D.; Gautam, S.; Singh, R.; Bhardwaj, S. Dietary Polyunsaturated Fatty Acids (PUFAs): Uses and Potential Health Benefits. *Curr. Nutr. Rep.* **2021**, *10*, 232–242. [CrossRef]
152. Flores-Hernández, M.N.; Martínez-Coria, H.; López-Valdés, H.E.; Arteaga-Silva, M.; Arrieta-Cruz, I.; Gutiérrez-Juárez, R. Efficacy of a High-Protein Diet to Lower Glycemic Levels in Type 2 Diabetes Mellitus: A Systematic Review. *Int. J. Mol. Sci.* **2024**, *25*, 10959. [CrossRef]
153. Suksomboon, N.; Poolsup, N.; Yuwanakorn, A. Systematic review and meta-analysis of the efficacy and safety of chromium supplementation in diabetes. *J. Clin. Pharm. Ther.* **2014**, *39*, 292–306. [CrossRef] [PubMed]
154. Meuffels, F.M.; Isenmann, E.; Strube, M.; Lesch, A.; Oberste, M.; Brinkmann, C. Exercise Interventions Combined with Dietary Supplements in Type 2 Diabetes Mellitus Patients—A Systematic Review of Relevant Health Outcomes. *Front. Nutr.* **2022**, *9*, 817724. [CrossRef]
155. Basu, A.; Du, M.; Sanchez, K.; Leyva, M.J.; Betts, N.M.; Blevins, S.; Wu, M.; Aston, C.E.; Lyons, T.J. Green tea minimally affects biomarkers of inflammation in obese subjects with metabolic syndrome. *Nutrition* **2011**, *27*, 206–213. [CrossRef] [PubMed]
156. Brown, A.L.; Lane, J.; Holyoak, C.; Nicol, B.; Mayes, A.E.; Dadd, T. Health effects of green tea catechins in overweight and obese men: A randomised controlled cross-over trial. *Br. J. Nutr.* **2011**, *106*, 1880–1889. [CrossRef]

157. Hodaei, H.; Adibian, M.; Nikpayam, O.; Hedayati, M.; Sohrab, G. The effect of curcumin supplementation on anthropometric indices, insulin resistance and oxidative stress in patients with type 2 diabetes: A randomized, double-blind clinical trial. *Diabetol. Metab. Syndr.* **2019**, *11*, 41. [CrossRef]
158. Kutan Fenercioglu, A.; Saler, T.; Genc, E.; Sabuncu, H.; Altuntas, Y. The effects of polyphenol-containing antioxidants on oxidative stress and lipid peroxidation in Type 2 diabetes mellitus without complications. *J. Endocrinol. Investig.* **2010**, *33*, 118–124. [CrossRef]
159. Jovanovski, E.; Khayyat, R.; Zurbau, A.; Komishon, A.; Mazhar, N.; Sievenpiper, J.L.; Blanco Mejia, S.; Ho, H.V.T.; Li, D.; Jenkins, A.L.; et al. Should Viscous Fiber Supplements Be Considered in Diabetes Control? Results From a Systematic Review and Meta-analysis of Randomized Controlled Trials. *Diabetes Care* **2019**, *42*, 755–766. [CrossRef]
160. O'Connell, B.S. Select Vitamins and Minerals in the Management of Diabetes. *Diabetes Spectr.* **2001**, *14*, 133–148. [CrossRef]
161. Bartlett, H.E.; Eperjesi, F. Nutritional supplementation for type 2 diabetes: A systematic review. *Ophthalmic Physiol. Opt.* **2008**, *28*, 503–523. [CrossRef]
162. Farrokhian, A.; Mahmoodian, M.; Bahmani, F.; Amirani, E.; Shafabakhsh, R.; Asemi, Z. The Influences of Chromium Supplementation on Metabolic Status in Patients with Type 2 Diabetes Mellitus and Coronary Heart Disease. *Biol. Trace Elem. Res.* **2020**, *194*, 313–320. [CrossRef] [PubMed]
163. Wang, X.; Wu, W.; Zheng, W.; Fang, X.; Chen, L.; Rink, L.; Min, J.; Wang, F. Zinc supplementation improves glycemic control for diabetes prevention and management: A systematic review and meta-analysis of randomized controlled trials. *Am. J. Clin. Nutr.* **2019**, *110*, 76–90. [CrossRef]
164. Khan, M.I.; Siddique, K.U.; Ashfaq, F.; Ali, W.; Reddy, H.D.; Mishra, A. Effect of high-dose zinc supplementation with oral hypoglycemic agents on glycemic control and inflammation in type-2 diabetic nephropathy patients. *J. Nat. Sci. Biol. Med.* **2013**, *4*, 336. [CrossRef]
165. Aquilani, R. Oral amino acid administration in patients with diabetes mellitus: Supplementation or metabolic therapy? *Am. J. Cardiol.* **2004**, *93*, 21–22. [CrossRef]
166. Scognamiglio, R.; Negut, C.; Piccolotto, R.; Dioguardi, F.S.; Tiengo, A.; Avogaro, A. Effects of oral amino acid supplementation on myocardial function in patients with type 2 diabetes mellitus. *Am. Heart J.* **2004**, *147*, 1106–1112. [CrossRef]
167. Solerte, S.B.; Fioravanti, M.; Locatelli, E.; Bonacasa, R.; Zamboni, M.; Basso, C.; Mazzoleni, A.; Mansi, V.; Geroutis, N.; Gazzaruso, C. Improvement of Blood Glucose Control and Insulin Sensitivity During a Long-Term (60 Weeks) Randomized Study with Amino Acid Dietary Supplements in Elderly Subjects with Type 2 Diabetes Mellitus. *Am. J. Cardiol.* **2008**, *101*, S82–S88. [CrossRef] [PubMed]
168. Görgüç, A.; Gençdağ, E.; Yılmaz, F.M. Bioactive peptides derived from plant origin by-products: Biological activities and techno-functional utilizations in food developments—A review. *Food Res. Int.* **2020**, *136*, 109504. [CrossRef] [PubMed]
169. Abish, Z.A.; Alibekov, R.S.; Tarapoulouzi, M.; Bakhtybekova, A.R.; Kobzhasarova, Z.I. Review in deep processing of whey. *Cogent Food Agric.* **2024**, *10*, 2415380. [CrossRef]
170. Djuricic, I.; Calder, P.C. Beneficial Outcomes of Omega-6 and Omega-3 Polyunsaturated Fatty Acids on Human Health: An Update for 2021. *Nutrients* **2021**, *13*, 2421. [CrossRef]
171. Zaloga, G.P. Narrative Review of n-3 Polyunsaturated Fatty Acid Supplementation upon Immune Functions, Resolution Molecules and Lipid Peroxidation. *Nutrients* **2021**, *13*, 662. [CrossRef]
172. Banaszak, M.; Dobrzyńska, M.; Kawka, A.; Górna, I.; Woźniak, D.; Przysławski, J.; Drzymała-Czyż, S. Role of Omega-3 fatty acids eicosapentaenoic (EPA) and docosahexaenoic (DHA) as modulatory and anti-inflammatory agents in noncommunicable diet-related diseases—Reports from the last 10 years. *Clin. Nutr. ESPEN* **2024**, *63*, 240–258. [CrossRef] [PubMed]
173. Chang, J.P.C.; Tseng, P.T.; Zeng, B.S.; Chang, C.H.; Su, H.; Chou, P.H.; Su, K.P. Safety of Supplementation of Omega-3 Polyunsaturated Fatty Acids: A Systematic Review and Meta-Analysis of Randomized Controlled Trials. *Adv. Nutr.* **2023**, *14*, 1326–1336. [CrossRef] [PubMed]
174. Asghari, K.M.; Saleh, P.; Salekzamani, Y.; Dolatkhah, N.; Aghamohammadzadeh, N.; Hashemian, M. The effect of curcumin and high-content eicosapentaenoic acid supplementations in type 2 diabetes mellitus patients: A double-blinded randomized clinical trial. *Nutr. Diabetes* **2024**, *14*, 14. [CrossRef] [PubMed]
175. Mostad, I.L.; Bjerve, K.S.; Bjorgaas, M.R.; Lydersen, S.; Grill, V. Effects of n−3 fatty acids in subjects with type 2 diabetes: Reduction of insulin sensitivity and time-dependent alteration from carbohydrate to fat oxidation. *Am. J. Clin. Nutr.* **2006**, *84*, 540–550. [CrossRef]
176. Farsi, P.F.; Djazayery, A.; Eshraghian, M.R.; Koohdani, F.; Saboor-Yaraghi, A.A.; Derakhshanian, H.; Zarei, M.; Javanbakht, M.H.; Djalali, M. Effects of supplementation with omega-3 on insulin sensitivity and non-esterified free fatty acid (NEFA) in type 2 diabetic patients. *Arq. Bras. Endocrinol. Metabol.* **2014**, *58*, 335–340. [CrossRef]
177. Cho, Y.K.; Kim, K.S.; Lee, B.W.; Hong, J.H.; Yu, J.M.; Lim, S.; Kim, Y.A.; Lee, C.B.; Kim, S.S.; Kwak, S.H.; et al. Efficacy and Safety of Pioglitazone Add-on in Patients with Type 2 Diabetes Mellitus Inadequately Controlled with Metformin and Dapagliflozin: A Multicenter, Randomized, Double-blind, and Placebo-controlled Study. *Clin. Ther.* **2024**, *46*, 662–669. [CrossRef]

178. García, P.P.; Martínez, F.J.H.; López, N.A.; Torre, L.C.; Sempere, M.E.T. Supplementation with a Highly Concentrated Docosahexaenoic Acid (DHA) in Non-Proliferative Diabetic Retinopathy: A 2-Year Randomized Double-Blind Placebo-Controlled Study. *Antioxidants* **2022**, *11*, 116. [CrossRef] [PubMed]
179. Ivanisevic, M.; Horvaticek, M.; Delmis, K.; Delmis, J. Supplementation of EPA and DHA in pregnant women with type 1 diabetes mellitus. *Ann. Med.* **2021**, *53*, 848–859. [CrossRef]
180. Zou, H.Y.; Zhang, H.J.; Zhao, Y.C.; Li, X.Y.; Wang, Y.M.; Zhang, T.T.; Xue, C.H. N-3 PUFA Deficiency Aggravates Streptozotocin-Induced Pancreatic Injury in Mice but Dietary Supplementation with DHA/EPA Protects the Pancreas via Suppressing Inflammation, Oxidative Stress and Apoptosis. *Mar. Drugs* **2023**, *21*, 39. [CrossRef]
181. Liu, X.; Li, Y.; Wan, X.; Zhuang, P.; Wu, Y.; Zhang, L.; Ao, Y.; Yao, J.; Zhang, Y.; Jiao, J. Association of Fish Oil Supplementation with Risk of Coronary Heart Disease in Individuals with Diabetes and Prediabetes: A Prospective Study in the UK Biobank. *Nutrients* **2023**, *15*, 3176. [CrossRef]
182. Yiu, Y.F.; Yiu, K.H.; Siu, C.W.; Chan, Y.H.; Li, S.W.; Wong, L.Y.; Lee, S.W.L.; Tam, S.; Wong, E.W.K.; Lau, C.P.; et al. Randomized controlled trial of vitamin D supplement on endothelial function in patients with type 2 diabetes. *Atherosclerosis* **2013**, *227*, 140–146. [CrossRef]
183. Angellotti, E.; D'Alessio, D.; Dawson-Hughes, B.; Nelson, J.; Cohen, R.M.; Gastaldelli, A.; Pittas, A.G. Vitamin D Supplementation in Patients with Type 2 Diabetes: The Vitamin D for Established Type 2 Diabetes (DDM2) Study. *J. Endocr. Soc.* **2018**, *2*, 310–321. [CrossRef] [PubMed]
184. Yadav, S.; Malik, K.; Moore, J.M.; Kamboj, B.R.; Malik, S.; Malik, V.K.; Arya, S.; Singh, K.; Mahanta, S.; Bishnoi, D.K. Valorisation of Agri-Food Waste for Bioactive Compounds: Recent Trends and Future Sustainable Challenges. *Molecules* **2024**, *29*, 2055. [CrossRef] [PubMed]
185. Ben-Othman, S.; Joudu, I.; Bhat, R. Bioactives from Agri-Food Wastes: Present Insights. *Molecules* **2020**, *25*, 510. [CrossRef] [PubMed]
186. Wijngaard, H.; Hossain, M.B.; Rai, D.K.; Brunton, N. Techniques to extract bioactive compounds from food by-products of plant origin. *Food Res. Int.* **2012**, *46*, 505–513. [CrossRef]
187. Kowalska, H.; Czajkowska, K.; Cichowska, J.; Lenart, A. What's new in biopotential of fruit and vegetable by-products applied in the food processing industry. *Trends Food Sci. Technol.* **2017**, *67*, 150–159. [CrossRef]
188. Mphahlele, R.R.; Fawole, O.A.; Makunga, N.P.; Opara, U.L. Effect of drying on the bioactive compounds, antioxidant, antibacterial and antityrosinase activities of pomegranate peel. *BMC Complement. Altern. Med.* **2016**, *16*, 143. [CrossRef]
189. López-Santiago, J.; García García, A.I.; Villarino, A.G.; Som, A.M.; Gómez-Villarino, M.T. Assessing wineries' performance in managing critical control points for arsenic, lead, and cadmium contamination risk in the wine-making industry: A survey-based analysis utilizing performance indicators as a results tool. *Heliyon* **2024**, *10*, e22962. [CrossRef]
190. Schilter, B.; Andersson, C.; Anton, R.; Constable, A.; Kleiner, J.; O'Brien, J.; Renwick, A.G.; Korver, O.; Smit, F.; Walker, R. Guidance for the safety assessment of botanicals and botanical preparations for use in food and food supplements. *Food Chem. Toxicol.* **2003**, *41*, 1625–1649. [CrossRef]
191. Schiebel, C.S.; Bueno, L.R.; Pargas, R.B.; de Mello Braga, L.L.V.; da Silva, K.S.; Fernandes, A.C.V.U.; dos Santos Maia, M.H.; de Oliveira, N.M.T.; Bach, C.; Maria-Ferreira, D. Exploring the biological activities and potential therapeutic applications of agro-industrial waste products through non-clinical studies: A systematic review. *Sci. Total Environ.* **2024**, *950*, 175317. [CrossRef]
192. Gurley, B.J. Pharmacokinetic Herb-Drug Interactions (Part 1): Origins, Mechanisms, and the Impact of Botanical Dietary Supplements. *Planta Medica* **2012**, *78*, 1478–1489. [CrossRef]
193. Mei, S.; Perumal, M.; Battino, M.; Kitts, D.D.; Xiao, J.; Ma, H.; Chen, X. Mangiferin: A review of dietary sources, absorption, metabolism, bioavailability, and safety. *Crit. Rev. Food Sci. Nutr.* **2023**, *63*, 3046–3064. [CrossRef]
194. Vilas-Boas, A.A.; Pintado, M.; Oliveira, A.L.S. Natural Bioactive Compounds from Food Waste: Toxicity and Safety Concerns. *Foods* **2021**, *10*, 1564. [CrossRef] [PubMed]
195. Magnuson, B.; Munro, I.; Abbot, P.; Baldwin, N.; Lopez-Garcia, R.; Ly, K.; McGirr, L.; Roberts, A.; Socolovsky, S. Review of the regulation and safety assessment of food substances in various countries and jurisdictions. *Food Addit. Contam. Part A* **2013**, *30*, 1147–1220. [CrossRef] [PubMed]
196. Socas-Rodríguez, B.; Álvarez-Rivera, G.; Valdés, A.; Ibáñez, E.; Cifuentes, A. Food by-products and food wastes: Are they safe enough for their valorization? *Trends Food Sci. Technol.* **2021**, *114*, 133–147. [CrossRef]
197. Devkota, L.; Montet, D.; Anal, A.K. Regulatory and Legislative Issues for Food Waste Utilization. In *Food Processing by—Products and their Utilization*; John Wiley & Sons, Ltd.: Hoboken, NJ, USA, 2017; pp. 535–548, ISBN 9781118432921.
198. Raak, N.; Symmank, C.; Zahn, S.; Aschemann-Witzel, J.; Rohm, H. Processing- and product-related causes for food waste and implications for the food supply chain. *Waste Manag.* **2017**, *61*, 461–472. [CrossRef] [PubMed]
199. Ribeiro, T.B.; Oliveira, A.; Coelho, M.; Veiga, M.; Costa, E.M.; Silva, S.; Nunes, J.; Vicente, A.A.; Pintado, M. Are olive pomace powders a safe source of bioactives and nutrients? *J. Sci. Food Agric.* **2021**, *101*, 1963–1978. [CrossRef]

200. European Commission. European Parliament and the Council of the EU Directive of The European Parliament and of the Council of 7 May 2002 on undesirable substances in animal feed 2002/32. *Off. J. Eur. Communities* **2002**, *L140*, 1–15.
201. European Commission. European Parliament Regulation (EC) No 178/2002 of the European Parliament and of the Council of 28 January 2002 laying down the general principles and requirements of food law, establishing the European Food Safety Authority and laying down procedures in matters of food saf. *Off. J. Eur. Communities* **2002**, *L031*, 1–24.
202. Castro-Muñoz, R.; Yáñez-Fernández, J.; Fíla, V. Phenolic compounds recovered from agro-food by-products using membrane technologies: An overview. *Food Chem.* **2016**, *213*, 753–762. [CrossRef]
203. Rebitzer, G.; Ekvall, T.; Frischknecht, R.; Hunkeler, D.; Norris, G.; Rydberg, T.; Schmidt, W.-P.; Suh, S.; Weidema, B.P.; Pennington, D.W. Life cycle assessment: Part 1: Framework, goal and scope definition, inventory analysis, and applications. *Environ. Int.* **2004**, *30*, 701–720. [CrossRef]
204. Chemat, F.; Abert-Vian, M.; Fabiano-Tixier, A.S.; Strube, J.; Uhlenbrock, L.; Gunjevic, V.; Cravotto, G. Green extraction of natural products. Origins, current status, and future challenges. *TrAC Trends Anal. Chem.* **2019**, *118*, 248–263. [CrossRef]
205. Onyeji, C.O.; Igbinoba, S.I.; Olayiwola, G. Therapeutic Potentials and Cytochrome P450-Mediated Interactions Involving Herbal Products Indicated for Diabetes Mellitus. *Drug Metab. Lett.* **2017**, *11*, 74–85. [CrossRef]
206. Martemucci, G.; Portincasa, P.; Centonze, V.; Mariano, M.; Khalil, M.; D'ALessandro, A.G. Prevention of Oxidative Stress and Diseases by Antioxidant Supplementation. *Med. Chem.* **2023**, *19*, 509–537. [CrossRef]
207. Manach, C.; Scalbert, A.; Morand, C.; Rémésy, C.; Jiménez, L. Polyphenols: Food sources and bioavailability. *Am. J. Clin. Nutr.* **2004**, *79*, 727–747. [CrossRef]
208. Stoeva, S.; Hvarchanova, N.; Georgiev, K.D.; Radeva-Ilieva, M. Green Tea: Antioxidant vs. Pro-Oxidant Activity. *Beverages* **2025**, *11*, 64. [CrossRef]
209. Acunha, T.; García-Cañas, V.; Valdés, A.; Cifuentes, A.; Simó, C. Metabolomics study of early metabolic changes in hepatic HepaRG cells in response to rosemary diterpenes exposure. *Anal. Chim. Acta* **2018**, *1037*, 140–151. [CrossRef] [PubMed]
210. Dwyer, J.T.; Coates, P.M.; Smith, M.J. Dietary Supplements: Regulatory Challenges and Research Resources. *Nutrients* **2018**, *10*, 41. [CrossRef]
211. Djaoudene, O.; Romano, A.; Bradai, Y.D.; Zebiri, F.; Ouchene, A.; Yousfi, Y.; Amrane-Abider, M.; Sahraoui-Remini, Y.; Madani, K. A Global Overview of Dietary Supplements: Regulation, Market Trends, Usage during the COVID-19 Pandemic, and Health Effects. *Nutrients* **2023**, *15*, 3320. [CrossRef] [PubMed]
212. Deluyker, H. Is scientific assessment a scientific discipline? *EFSA J.* **2017**, *15*, e15111. [CrossRef]
213. Chatzopoulou, S.; Eriksson, N.L.; Eriksson, D. Improving Risk Assessment in the European Food Safety Authority: Lessons From the European Medicines Agency. *Front. Plant Sci.* **2020**, *11*, 349. [CrossRef]
214. Coppens, P.; da Silva, M.F.; Pettman, S. European regulations on nutraceuticals, dietary supplements and functional foods: A framework based on safety. *Toxicology* **2006**, *221*, 59–74. [CrossRef]
215. Silano, V.; Coppens, P.; Larrañaga-Guetaria, A.; Minghetti, P.; Roth-Ehrang, R. Regulations applicable to plant food supplements and related products in the European Union. *Food Funct.* **2011**, *2*, 710–719. [CrossRef]
216. Anadón, A.; Ares, I.; Martínez-Larrañaga, M.-R.; Martínez, M.-A. Evaluation and regulation of food supplements: European perspective. In *Nutraceuticals*, 2nd ed.; Gupta, R.C., Lall, R., Srivastava, A., Eds.; Academic Press: Cambridge, MA, USA, 2021; pp. 1241–1271, ISBN 978-0-12-821038-3.
217. EFSA Panel on Nutrition, Novel Foods and Food Allergens (NDA); Turck, D.; Bohn, T.; Castenmiller, J.; de Henauw, S.; Hirsch-Ernst, K.I.; Knutsen, H.K.; Maciuk, A.; Mangelsdorf, I.; McArdle, H.J.; et al. Scientific advice related to nutrient profiling for the development of harmonised mandatory front-of-pack nutrition labelling and the setting of nutrient profiles for restricting nutrition and health claims on foods. *EFSA J.* **2022**, *20*, e07259.
218. Hoffman, F.A.N.N. Regulation of Dietary Supplements in the United States: Understanding the Dietary Supplement Health and Education Act. *Clin. Obstet. Gynecol.* **2001**, *44*, 780–788. [CrossRef]
219. Wallace, T.C. Twenty Years of the Dietary Supplement Health and Education Act—How Should Dietary Supplements Be Regulated? *J. Nutr.* **2015**, *145*, 1683–1686. [CrossRef]
220. Soyombo, D.A. Navigating regulatory challenges and safety considerations in lactation supplement development: A Nigeria and US Comparison. *World J. Biol. Pharm. Health Sci.* **2024**, *18*, 1–11. [CrossRef]
221. Smith, A.; Jogalekar, S.; Gibson, A. Regulation of natural health products in Canada. *J. Ethnopharmacol.* **2014**, *158*, 507–510. [CrossRef] [PubMed]
222. Petrenko, A.; Tutelyan, V. Codex Alimentarius and transparency of the global food safety system. In *Food System Transparency*; Steier, G., Friedlander, A., Eds.; CRC Press: Boca Raton, FL, USA, 2021; p. 28.
223. van der Meulen, B.; Card, M.M.; Din, A.; Fortin, N.D.; Mahmudova, A.; Maister, B.; Türkoğlu, H.G.; Bilgin, F.K.; Lederman, J.; Poto, M.P.; et al. Food regulation around the world. In *Ensuring Global Food Safety*, 2nd ed.; Martinović, A., Oh, S., Lelieveld, H., Eds.; Academic Press: Cambridge, MA, USA, 2022; pp. 11–137, ISBN 978-0-12-816011-4.

224. Miyazaki, Y.; Omori, T.; Fujitani, K.; Fujita, J.; Kawabata, R.; Imamura, H.; Okada, K.; Moon, J.-H.; Hirao, M.; Matsuyama, J.; et al. Oral nutritional supplements versus a regular diet alone for body weight loss after gastrectomy: A phase 3, multicenter, open-label randomized controlled trial. *Gastric Cancer* **2021**, *24*, 1150–1159. [CrossRef] [PubMed]
225. Esposito, G.; Sciuto, S.; Martello, E.; Pezzolato, M.; Bozzetta, E. Disclosing Frauds in Herbal Food Supplements Labeling: A Simple LC-MS/MS Approach to Detect Alkaloids and Biogenic Amines. *J. Food Prot.* **2023**, *86*, 100152. [CrossRef] [PubMed]
226. Young, A.; Woodlee, J.; Marriott, R.; Meyer, C. Dietary supplements. In *An Overview of FDA Regulated Products*; Pacifici, E., Bain, S., Eds.; Academic Press: Cambridge, MA, USA, 2025; pp. 237–256, ISBN 978-0-443-23780-5.

antioxidants

MDPI

Review

Polyphenols from Byproducts: Their Applications and Health Effects

Ranya Demir [1], Sümeyye Sarıtaş [1], Mikhael Bechelany [2,*] and Sercan Karav [1,*]

[1] Department of Molecular Biology and Genetics, Çanakkale Onsekiz Mart University, Çanakkale 17000, Türkiye; ranyaddemir@gmail.com (R.D.); sumeyyesaritas@stu.comu.edu.tr (S.S.)

[2] European Institute for Membranes (IEM), UMR-5635, University of Montpellier, ENSCM, CNRS, Place Eugène Bataillon, CEDEX 5, F-34095 Montpellier, France

* Correspondence: mikhael.bechelany@umontpellier.fr (M.B.); sercankarav@comu.edu.tr (S.K.)

Abstract

Plant byproducts represent a valuable and underutilized source of bioactive compounds. Among these, phenolic compounds have attracted growing interest from the agricultural, cosmetic, and food industries due to their diverse biological activities. These naturally occurring compounds are derived from various plant species, and they exhibit strong antioxidant, antimicrobial, and antiviral properties. Their yield, as well as quality and bioavailability, has improved with more recent advancements within green extraction, as well as purification and characterization techniques. Several phenolic compounds exhibit strong antiviral and antioxidant activities, which are highlighting their value as bioactive compounds. It is essential to evaluate extraction methods for high-yield phenolic compounds from plant byproducts so that they can contribute to the circular bioeconomy, reduction in environmental waste, and development of biomedical and food industrial applications. Their physicochemical characteristics and potential applications may lead to a determination by contributing to promising fields through expanded in vitro, in vivo, and in silico experiments. This review summarizes current research on the extraction, recovery, and applications of phenolic compounds derived from plant byproducts, providing new insights into their sustainable utilization and bioactive potential.

Keywords: agricultural biomass; sustainability; secondary metabolites; antiviral; antimicrobial; antioxidant; eco-friendly; green method

1. Introduction

Polyphenols, also known as polyhydroxyphenols and a subcategory of phenolic compounds, are natural compounds and secondary metabolites that are synthesized solely by plants [1]. They generally occur as glycosidases, which are combined with divergent organic acids that help to combat pathogenic organisms [2]. They are abundant in several types of fruits, such as grapes, berries, and apples, and vegetables, such as soybeans and onions. Other than fruits and vegetables, seeds, nuts, flowers, tree barks, and several common beverages like coffee and red wine are also known for their notability for this content [3]. They have pivotal roles in many metabolic functions, such as contributing to pollinator attraction, structural functions, and several types of defense mechanisms in plants [1,4]. Their unique bioactive characteristics, including modulating oxidative stress and inflammatory response, free radical scavenging activity, and antioxidant, antimicrobial, and prebiotic properties, are widely recognized and considered significant bioactive compounds [1,5]. Polyphenols can be divided into five primary classes based on their structural characteristics and biological functions: flavonoids, tannins, stilbenes, phenolic acids, and lignans. Each category divides into several subcategories, and each polyphenol has a unique characteristic [6].

In recent years, there has been a growing number of studies, particularly in vitro assays, investigating the physiological activities of various sources of phenolic compounds, including food-derived bioactive and nutritionally functional components [3]. Furthermore, many existing techniques have been improved after their limitations, such as instability and inefficiency in freeze-drying processes, were identified [7,8]. Recently, emerging green extraction methods have been increasingly investigated to enhance phenolic compound recovery in a more energy-efficient and sustainable way. These methods aim to provide high-yield recovery, preserve thermolabile phenolics, reduce contamination, and enhance selectivity while minimizing the use of toxic organic solvents, as well as water and energy consumption. While green extraction technologies offer a more efficient and sustainable approach to recovering phenolic compounds, their effects on bioavailability depend on the extraction agents used and processing conditions. These properties make them a promising approach for pharmaceutical, cosmetic, and environmental applications, with potential applicability in food systems when food-grade extraction agents are employed. Detailed information of green methods, including their advantages and current limitations, have been discussed [9]. Although the importance of green methods has been widely recognized, only a limited number of studies were published before the 2010s, and byproduct polyphenols received little scientific attention. In recent years, however, research interest in phenolic compounds from agro-industrial byproducts has increased markedly.

To emphasize more of this content, this review article aimed to attach importance to over 100 recently published articles about the phenolic content of byproducts. To evaluate the content of the article, bioactive properties and various plant sources of polyphenols have been included. To highlight the importance of phenolic compounds from byproducts, such as widely known pomegranate, ginger, coffee, avocado, and recent findings such as *Posidonia oceanica* (seagrass) and water caltrop shells have been outlined. Their broad-spectrum sources and bioactive attributes with potential application areas such as the food and meat industry, agriculture, cosmetics, biomedicine, and health-related applications have been highlighted. Extraction techniques for phenolic compounds from plant byproducts using advanced green technologies have been compared. Limitations and challenges, with future recommendations, have also been included to facilitate the evolutionary development in the bioavailability of plant-derived polyphenols from byproducts and their application areas.

2. Plant Derived Phenolic Compounds and Their Bioactivities

As mentioned, polyphenols are bioactive components of versatile plant species [10]. Over the past few years, they have remarked on their various interactions, such as their binding and ridging ability or specificity, and advantageous properties, including antioxidant, anti-inflammatory, therapeutic, and antimicrobial effects [11,12]. It has also been reported that they have the ability to interfere with biochemical homeostasis and affect the epigenetic modifications of chromatin [13,14]. Stilbenes possess a 1,2-diphenylethylene backbone, which defines their chemical class, and they are known for their therapeutic properties and protective effects against plant pathogens and pests. Several examples can be given, such as resveratrol, pinosylvin, etc. [15]. Resveratrol, a non-flavonoid polyphenol, is one of the most studied stilbenoids due to the exhibition of potent antioxidant, immunomodulatory, anti-inflammatory, and antiangiogenic effects [16]. A significant number of preclinical and clinical trials have reported that it can overcome the multidrug resistance in cancer cells and

https://doi.org/10.3390/antiox15010087

has the potential to enhance the sensitivity of cancer cells to chemotherapeutic agents when it is used with other clinically used drugs [17,18]. Flavonoids, which have several subcategories named anthocyanins, flavonols, flavones, flavanones, flavan-3-ols, and isoflavones, are natural phytochemicals that are widely found in fruits and vegetables and usually identified as flower pigments [19,20]. These phytochemicals are known for their antiviral properties, and they exhibit their potent activity in various stages, including viral entrance, replication phase, or translation of proteins [21,22]. Many of these flavonoids are studied against a variety of DNA and RNA viruses due to their versatile activities [23]. It has been indicated that some of the glycoside parts of flavonoids have increased their solubility, and therefore, their antiviral activity, when they are compared to their agylcone form [24]. Furthermore, several flavonoids exhibit higher inhibitory activities without toxicity and with enhanced cell proliferation [20,25]. The divergent properties of these phytochemicals increased the amount of studies and widened as in vivo [26], in vitro [27], and in silico [28] studies. Some flavonoids, such as quercetin 3-rhamnoside (Q3R), quercetin, epigallocatechin (EGC), and baicalein, are widely used for these types of studies against various types of viruses, including influenza A and B viruses [29], Japanese encephalitis virus (JEV) [30], dengue virus (DENV) [31], hepatitis C virus (HCV) [32], and human immunodeficiency viruses (HIV) [33]. Tannins are another class of polyphenols that usually have a role in regulating plant growth and protecting the plants from predators. Condensed, hydrolyzable, and phlorotannins are three subclasses of the tannin category [34]. They have potent antioxidant, antibacterial, antiviral, and anticancer properties. Generally, tannic acid, a member of tannins, is known as a flavoring and adjuvant agent [35]. Both Gram-positive and negative bacteria have been used to investigate the activity of tannins, including (*E. coli*) *Escherichia coli*, *Staphylococcus aureus* (*S. aureus*), *Yersinia enterocolitica* (*Y. enterocolitica*), *Enterococcus faecalis* (*E. faecalis*), *Streptococcus pyogenes* (*S. pyogenes*), *Listeria innocua* (*L. innocua*), and *Pseudomonas aeruginosa* (*P. aeruginosa*) [36]. As for their antiviral properties, several trials against HIV, influenza A virus, noroviruses, papillomavirus, and herpes simplex virus types 1 and 2 (HSV-1 and HSV-2) have been investigated [37]. Furthermore, they have been used in various studies in the food industry, both in vivo [38] and in vitro [39], such as in the beef industry and as preservative agents in food [40,41], making them a potential nutraceutical. Another class of polyphenol scaffold is phenolic acids, also referred to as phenol carboxylic acids, which consist of a phenolic ring and an organic carboxylic acid function [42]. These bioactive compounds are implicated in influencing the flavor, aroma, and overall sensory profile of foods and are classified into two main subclasses: hydroxycinnamic acids and hydroxybenzoic acids [43]. Phenolic acids have gained momentum owing to their strong neuroprotective effects and potential as therapeutic agents in combating cognitive and chronic diseases, including cystic fibrosis, Alzheimer's disease, and non-C or non-B hepatocellular carcinoma [44–46]. Furthermore, they exhibit antioxidant activity by preventing metal catalysis and free radical formation [47]. Finally, lignans, classified as classical lignans and neolignans [48], are commonly derived from the dimerization through oxidative reactions involving phenylpropanoid units and are known for their antiviral, anti-inflammatory, antioxidant, and anticancer activity [49]. Sesame and flax seeds consist of the highest lignan content among food groups [50]. Their anti-inflammatory activity has been tested in vitro on cyclooxygenase 1 and 2 (COX-1 and COX-2), 15-lipoxygenase (15-LOX), and phospholipases A2 (sPLA2) enzyme activities to see their inhibitory activities. Results of various assays revealed that a stronger inhibitory activity of lignans from *Schisandra rubriflora* extract exhibited anti-inflammatory activity for 15-LOX, COX-1, and COX-2 enzymes [51]. As discussed above, phenolic compounds have been investigated with various types of studies, including in vitro, in silico, and in vivo, to be able to understand the activity and expand the application areas of these bioactive compounds. One

https://doi.org/10.3390/antiox15010087

of the expanded studies has indicated that polyphenol-containing nanoparticle synthesis can be a new therapeutic approach for biomedical applications, such as in biodetection, gene delivery, bone repair, and cancer theranostics [52]. Polyphenols are highly consistent in plant-based food groups and beverages, and over 80% of polyphenol intake can be absorbed, influenced by gut microbiota. Their consumption as part of a healthy diet may help reduce the risk of chronic diseases, such as cardiovascular disease and type 2 diabetes [53]. Since they can improve endothelial function, they have also been shown to exhibit antihypertensive potential. One of the studies indicates that the improvement in endothelial function through activation of vascular eNOS and Akt signaling pathways may have an association with these effects [54]. To enhance their intake biological value and protect their structure, a research article has reviewed that the sample sizes and drying time are directly affecting their composition [55]. Moreover, according to a review article, small samples with shorter periods of drying processes conserved both the structure and nutritional value of phenolic compounds [8].

2.1. Antioxidant Activities of Phenolic Compounds

Phenolic compounds represent one of the major sources of daily antioxidant intake. In a healthy diet, their average consumption is approximately 1 g per day. This level is nearly ten times higher than the typical daily intake of vitamin C. Phenolic compounds can act in antioxidant action both directly and indirectly. Essentially, they can neutralize ROS directly due to their hydroxyl groups. The phenolic compound uses its hydroxyl group and donates one hydrogen atom to stabilize free radicals [56]. Phenolic compounds can also chelate metals to prevent radical formation at the source [57]. To increase antioxidant systems, several phenolic compounds have the ability to activate transcription factor Nrf2 to enhance cellular defense [58]. Moreover, enzymes that generate oxidative molecules can be inhibited by phenolic compounds to reduce cellular production of new radicals [59]. Furthermore, to protect lipid radicals and cell membrane integrity, several phenolic compounds can interrupt lipid peroxidation to prevent damage from free radicals [60]. Principal dietary contributors of polyphenols as antioxidants are fruits, vegetables, cereals, legumes, plant-based foods, and beverages, including coffee and green tea [61]. Dietary phenolic compounds have the ability to act as antioxidants due to their structure, including multiple hydroxyl groups [62]. They scavenge free radicals and reactive oxygen species (ROS) to interfere with oxidative stress, signaling pathways, and signaling transduction mechanisms of pro-inflammatory mediators [63,64]. On the other hand, some phenolic antioxidants may increase carbonyl stress and cause glycation-mediated protein damage. Therefore, the selection of specific phenolic compounds for any health-related application should be made with careful consideration [65]. In a 2012 review article, it has been pointed out that although most of the pharma and food industries have been fortifying whole foods and selling them with adequate and proven in vitro studies, more in vivo studies and clinical trials are required [66].

One of the recent research areas for the antioxidant activity of phenolic compounds derived from olives. Phenolic compounds from olives are rich in phytoalexins [67,68]. Olive oils, especially extra virgin olive oil, contain over 30 phenolic compounds, namely oleuropein, in both its glycated and aglycone (OLE) forms, verbascoside, oleocanthal, hydroxytyrosol (HT), tyrosol, etc. [69]. Recent studies indicate that bioactive components of extra virgin oil enhance cognitive function and reduce breast cancer risk. Moreover, they prevent oxidative damage, especially oxidative damage to low-density lipoproteins, by bonding with free radicals, breaking peroxidative chain reactions, enhancing cellular antioxidant defenses, and acting as metal chelators [70,71]. In addition, they also have the ability to modulate signaling pathways, including AMPK and JAG/SAT [72,73]. Several

https://doi.org/10.3390/antiox15010087

studies have also found that they are like adjuvant agents that maintain mitochondrial function and reduce inflammation, which shows that they have potential as neuroprotective agents for Alzheimer's disease [74,75]. According to a 2023 research article, oxidative reactions that are caused by free radicals create an unpleasant odor, and synthetic antioxidants are very effective at suppressing this kind of oxidative damage by neutralizing free radicals. Nevertheless, research also indicates that more in vivo studies are needed since it can cause a harmful effect on human health [76]. In contrast, in a 2021 research article, it has been indicated that natural phenolic antioxidants usually outperform the synthetic antioxidants when it comes to food preservation from protein and lipid oxidation [77]. Due to their diverse antioxidant activity they have also been tested on animal diets. Results showed that phenolic compounds, especially flavonoids, improve animal immunity and overall gut health, contributing to healthier animal products with better fatty acid profiles [78]. Blueberry is enriched with high anthocyanins, phenolic acids, and flavonols [79]. Similar phenolic content is also found in strawberries, while it is higher in flavan-3-ols, dihydrochalcones, and flavanones [80]. When these two berries are compared, higher phenolic content has been found in blueberries with higher antioxidant activity [81]. The high antioxidant activity of blueberries has been tested using different processes. In a 2021 research article, it is proposed to investigate two different techniques of puree processing methods: high hydrostatic pressure (HHP) and thermal processing (TP). According to the results, HPP showed a better effect for maintaining the color of puree, while TP exhibited higher antioxidant activity than HPP. Furthermore, it is also indicated that a proper HHP treatment may have the ability to activate polyphenol oxidase and β-glucosidase activity [82]. To expand the phenolic content in blueberry cultivars, 14 different types of blueberry cultivars were compared by their antioxidant and phenolic content using ABTS and ferric reducing antioxidant ability (FRAP) assays [83]. According to FRAP assay results, the highest value was $\leq$15 mmol Trolox/100 g dw, while the highest ABTS assay result was $\leq$26 mmol Trolox/100 g dw. Principal component analysis results demonstrated that the Bonifacy cultivar exhibited the highest anthocyanin content, while the Aurora cultivar exhibited the highest free amino acid content. Aronia fruits, also known as chokeberries, are another berry type that is enriched with high chlorogenic acid, flavonoids (namely quercetin, catechin, and proanthocyanidins), and anthocyanin content. Generally, black chokeberry (*Aronia melanocarpa* L.) is being used to perform biological activity studies, and different parts and extract types of the fruit, including fruit, leaves, and pomace, have been used [84]. In a research article that has included different ripeness stages of aronia fruits (unripe, ripe, and overripe), the antioxidant capacity was assessed using DPPH and FRAP assays. According to results, green, unripe aronia fruits exhibited the strongest antioxidant activity, with values of 7.59 ± 0.62 in DPPH and 15.46 ± 1.34 in FRAP [85]. Similarly, blackberries are also abundant with polyphenols, namely, tannins, anthocyanins, and flavonoids. When different parts of the blackberry plant are analyzed, immature leaves of blackberry are found to have the highest phenolic content, including catechol, gallic acid, epicatechin, and rutin [86]. Varzuru et al. [87] compared the antioxidant activities of blackberry and raspberry leaves in vitro. Results demonstrated that blackberry leaves are enriched with liposoluble antioxidants, while raspberry leaves are abundant in water-soluble antioxidants, both having the capability to mitigate oxidative stress by neutralizing free radicals and inhibiting lipid peroxidation. Individually, raspberry leaves exhibited potent scavenging activity against hydroxyl radicals, while blackberry leaves exhibited high efficacy in neutralizing superoxide radicals. In another in vitro study, the objective was to determine the antioxidant activity of commercial and wild blackberries during gastrointestinal digestion [88]. Results indicated that wild blackberry digestate exhibited more than 50% antioxidant activity after gastrointestinal digestion, while commercial blackberry

https://doi.org/10.3390/antiox15010087

digestate exhibited less than 50%. Other than antioxidant activity, anticancer properties are also studied on blackberries.

Grapes are abundant with flavonoids, especially myricetin and hesperidin, and with hydroxycinnamic acid and anthocyanins, which give the darker color to the grapes. Antioxidant activity of grapes indicates that there is a strong correlation between phenolic content and fruit coloration, meaning that darker colors indicate potentially higher antioxidant activity [89]. To elaborate further, antioxidant activities of different extracts of a grape cultivar traditionally grown in Turkey 'Karaerik,' including peel, seed, and pulp, were investigated. DPPH analysis showed that peel extracts exhibited the highest DPPH scavenging activity, and pulp extracts had the lowest score of all three [90]. When white and red grapes were compared, different levels of bioactive compounds and bioactivities, particularly antioxidant and anti-inflammatory effects, were observed. In vitro and in vivo studies indicate that grape pomace can improve cardiac function and reduce atherosclerotic lesions by reducing ROS damage and inflammation [91]. Similarly to grapes, plums are also rich in phenolic content, with more than 20 polyphenols in both free and bound forms, including catechin, epicatechin, quercetin, neochlorogenic acid, and procyanidin B2 [92]. In a comprehensive study, the antioxidant activities of six polyphenols from the Burkedin plum, catechin, ellagic acid, gallic acid, quercetin-3-glucoside, cyanidin-3-glucoside, and vanillic acid were investigated [93]. CUPRAC (CUPric Reducing Antioxidant Capacity), TPC (total phenolic content), and FRAP assay analysis showed that ellagic acid, catechin, and gallic acid exhibited higher antioxidant activity than cyanidin-3-glucoside. Moreover, flesh extracts exhibited higher antioxidant activity than seed extracts. In a research study, apple pomace samples and their both antioxidant and prebiotic activities were investigated. According to the results, apple pomace exhibited prebiotic activity against *Lacticaseibacillus paracasei* and *Bacillus subtilis* via cell proliferation and viability. However, no significant effect on *Lactobacillus acidophilus* and *Lactobacillus plantarum* was observed. Antioxidant activity of apple pomace was tested via FRAP, ABTS, and DPPH assays. Results demonstrated a high value of trolox equivalent antioxidant capacity (TEAC), indicating that water-soluble (hydrophilic) and fat-soluble (lipophilic) antioxidants were detected in the sample. Acetylcholinesterase (AChE) is a crucial enzyme that plays a role in the nervous system by breaking down the acetylcholine (ACh) neurotransmitter [94]. This essential enzyme also remains naturally in artichokes, offering maintenance of ACh levels and antioxidant protection due to its polyphenolic content [95,96]. Yellow onion peels have antimicrobial activity against *Bacillus subtilis* (*B. subtilis*), *E. coli*, *Bacillus cereus* (*B. cereus*), *S. aureus*, *P. aeruginosa*, and *Salmonella typhimurium* with MIC values between 150 and 200 μg/mL. Moreover, when it was tested on catfish fillets, it significantly reduced lipid peroxidation and oxidative spoilage under refrigerated (4 °C) and frozen (−18 °C) storage conditions, making them a potential candidate for fish preservative application [97,98]. Lactoferrin is a multifunctional milk protein that has the ability to chelate iron, and this activity accompanies various properties, including antioxidant, antimicrobial, antiviral, prebiotic, and therapeutic effects [99]. Goat lactoferrin has been used to evaluate its effect on the solubility of phenolic antioxidants. Results demonstrated that goat lactoferrin pterostilbene complexes enhanced the solubility of pterostilbene polyphenols and became a protective layer to protect the phenolic content profile. Furthermore, improvement in the emulsification and foaming properties of goat lactoferrin has also been observed [100]. Overall, articles generally discuss the main antioxidant properties of phenolic compounds, especially radical scavenging, metal chelation, and modulation of antioxidant enzymes. Antioxidant capacity of these compounds is generally analyzed with chemical assays like DDPH, ABTS, FRAP, ORAC, which mimic in vivo or in vitro effects subtly. Moreover, it is mentioned that phenolic compounds can act as antioxidants at moderate doses and

https://doi.org/10.3390/antiox15010087

pro-oxidants at high doses. However, long-term effects of intakes are mostly poorly defined in vivo studies.

2.2. Antiviral Activities of Phenolic Compounds

Previous studies demonstrated that phenolic compounds exhibit immense antiviral properties due to their hydroxyl groups attached to aromatic rings, giving them the ability to scavenge ROS and chelate iron to catalyze lipid peroxidation [101]. Phenolic compounds can directly inhibit viral entry by blocking viruses from attaching or entering host cells. This mechanism can be provided by binding viral surface proteins, altering host cell membrane receptors, or preventing membrane fusion to stop the release of viral genome. EGCG binding to targeted surface protein of influenza A virus to prevent interaction of sialic acid receptor on host cells, curcumin binding envelope protein of Dengue virus to disrupt viral entry, and quercetin binding spike protein of SARS-CoV-2 to block attachment to the host cells are examples for some of these mechanisms [102,103].

Moreover, phenolic compounds such as resveratrol and quercetin, can enhance or balance immune responses by boosting interferon production like IFN-α, IFN-β, enhancing natural killer cell activity and modulating adaptive immunity to help defense mechanism [104,105]. Several phenolic compounds, especially curcumin, can interfere with host cell signaling pathways and inhibit translation factors to disrupt viral protein synthesis [106,107]. Furthermore, several phenolic compounds, such as tannins and proanthocyanidins, have the ability to inactivate viral particles and destabilize lipid envelopes of viruses [108,109]. In a recent study, over 30 polyphenols have been tested against Porcine Epidemic Diarrhea Virus (PEDV) in vitro [110]. Results indicated that four of them exhibited anti-PENDV activity and have the potential to be candidates for anti-PENDV drugs. Phenolic compounds, including resveratrol, cyanidin, ellagic acid, and luteolin, can act as inhibitors against hepatitis B, influenza, HCV, HIV, severe acute respiratory syndrome coronavirus 2 (SARS-CoV-2), DENV, zika virus (ZIKV), and chikungunya virus by hindering viral growth and replication [101,111,112]. In a recent study, over a hundred phenolic acids, with nearly sixty abundant polyphenols, were investigated in nearly twenty Indian heritage pigmented rice varieties (IHPRV) via LC-Q-TOF-MS (a high-resolution metabolomic profiling technique) to examine their antiviral activity against SARS-CoV-2 [113]. Results indicated that nearly twenty of the abundant polyphenols exhibited antiviral activity. To date, many phenolic compounds have been studied for their diverse antiviral activity. One such compound that has been studied for its various benefits, including antiviral activity, is epigallocatechin gallate (EGCG), which is abundantly found in green tea [114]. It has been indicated that EGCG has a potent activity against DENV, HIV, chikungunya virus, SARS-CoV-2, and ZIKV viruses [115]. To elaborate, in silico studies reported that EGCG can inhibit ZIKV growth by suppressing viral entry. Further study by Kumar et al. indicated that EGCG can also interrupt various stages of replication in certain viruses, giving an exemplification of EGCG activity against ZIKV on NTPase enzyme inhibition with an IC50 value (half of the maximal inhibitory concentration) of 295.7 nM and a Ki value (binding affinity of an inhibitor to its target receptor or enzyme) of 0.387 ± 0.034 μM [116]. A recent study aimed to investigate EGCG activity against Mayaro virus, using BHK-21 and Vero E6 cell lines in vitro and in silico [117]. Results demonstrated that EGCG suppressed the Mayaro virus replication and exhibited antiviral activity in the entry and post-entry stages. Similar studies include the antiviral activity of curcumin against HCV via preventing virulence attachment [118], human herpesviruses (HHV) by hindering replication or blocking viral entry [119], and Nipah virus by hindering RNA synthesis [120], and aronia extracts against influenza virus types (type 1 and 3) by inhibiting surface proteins of the virus [121].

https://doi.org/10.3390/antiox15010087

In a recent in vitro study, several plant extracts, essential oils, and their phenolic compounds, especially quercetin and coumarins, have been investigated for their antiviral activity against certain virus types [122]. According to an investigation, quercetin can inhibit replication of several virus types, especially influenza virus types, HSV, SARS-CoV-2, porcine viruses, and HCV [120,123], while coumarins show potent activity against HIV-1 (HIV type 1) [124]. Moreover, essential oils demonstrated immense activity against different virus types, such as tea tree oil against HSV, influenza, and human papillomavirus (HPV) [125]; eucalyptus oil against HSV, influenza, and respiratory syncytial virus (RSV) [126]; peppermint oil against HSV, influenza, and adenovirus [127]; lavender oil against HSV and influenza; oregano oil against HSV and norovirus [128]; and lemon balm oil against enterovirus and HSV [122,129]. It has also been indicated that studies of essential oils exhibited different types of antiviral activity, including direct inhibition of viral growth, suppressing viral replication, inactivation of viruses, and inhibition of viral attachment [122]. Kreiser et al. investigated a group of structurally related phenolic compounds, including EGCG, quercetin, taxifolin, and naringenin. These compounds were evaluated in combination with zinc picolinate and copper sulfate. The study aimed to assess their antiviral efficacy against respiratory RNA viruses, specifically influenza A, human metapneumovirus, and human coronavirus OC43 [130]. These combinations were tested on two groups, one with naringenin and another without naringenin. Results demonstrated that the combination with naringenin reduced viral replication of human metapneumovirus, human coronavirus OC43, and influenza A virus via potentially acting as ionophores to enhance intracellular zinc levels. *Morus alba*, known as white mulberry, consists of highly flavonoids such as quercetin, kaempferol, and catechins, and phenolic acids like caffeic acid and chlorogenic acid. The antiviral activity of white mulberry was tested against pseudorabies virus (PRV), and significant inhibition of viral replication was observed [131]. The polyphenols in turmeric and red grapes, such as catechin, resveratrol, curcumin, and eriodictyol, have a strong affinity for ACE2 ligands [121,132]. These polyphenols block the replication cycle by suppressing the activity of the protease to hinder the ssRNA replication. Additionally, polyphenols, especially quercetagetin, myricetin, and EGCG, exhibit high affinity for SARS-CoV-2 RdRp (RNA polymerase of SARS-CoV-2) [133,134].

Polyphenol-rich sugarcane extract (PRSE) is another compound that is widely studied for antioxidant, antimicrobial, and antiviral activity research due to its high phenolic content. Tang et al. studied PRSE against influenza A to investigate its antiviral activity in vitro [135]. For this purpose, A549 (human lung epithelial) and MDCK (Madin-Darby Canine Kidney) cell lines were used against H3N2 and H1N1 strains. Results showed that PRSE acts as an inhibitor in the early stages of the viral replication cycle and can suppress the replication of both strains. To expand the knowledge, a year after this research, Tang et al. also studied PRSE on respiratory virus types, namely RSV, Parainfluenza Virus Type 3 (PIV-3), Adenovirus 5 (AdV5), and SARS-CoV-2 [136]. In this study, the A549, Calu-3, Vero E6 (for SARS-CoV-2), and HEp-2 (for RSV and PIV-3) cell lines were used. Results showed that the strongest inhibition was determined against SARS-CoV-2 and influenza A virus by affecting the early stages of viral replication. In contrast, little to no inhibitory effect was determined against RSV, PIV-3, and AdV5. Monkeypox virus (MPXV) is a zoonotic disease that is predicted to cause a potential global pandemic in the near future [137]. The F13L gene of MPXV encodes an envelope protein named F13, which is crucial for its viral replication cycle, and since phenolic compounds exhibit immense antiviral activities, an in silico study investigated over five polyphenols and their activity against F13 [138]. According to molecular docking results, myricetin, demethoxycurcumin, piceatannol, curcumin, matairesinol, and ellagic acid exhibited strong binding activity to F13, and more stable complexes were observed with demethoxycurcumin and myricetin, making them powerful

https://doi.org/10.3390/antiox15010087

candidates for inhibiting the F13 protein of MPVX. Due to their diverse antiviral activities, Manzoor et al. [139] investigated several compounds, including flavanols, flavonoids, and coumarins, against various fish viruses, namely viral hemorrhagic septicemia virus (VHSV), infectious hematopoietic necrosis virus (IHNV), grass carp reovirus (GCRV), singapore grouper iridovirus (SGIV), infectious pancreatic necrosis virus (IPNV), and spring viremia of carp virus (SVCV). Results showed that EGCG blocks the adhesion entry of GCRV, quercetin exhibited potent antiviral activity against SGIV and IPNV, and flavonoids such as fisetin, fustin, and sulfuretin were able to induce apoptosis of the infected cells of IHNV and VHSV. Even though the studies have expanded their knowledge and findings by demonstrating antiviral activities against several viruses and being potential antiviral candidates, further research is needed on the variability in dosage, effectiveness, and safety of phenolic compounds [140]. Recent antiviral articles highlight mechanism clearly and discuss strong in vitro signals. Nevertheless, comparative mechanism work between enveloped and non-enveloped viruses, RNA and DNA viruses are only just emerging without clear knowledge about action of phenolic compounds, whether viral-specific targets or not.

2.3. *Antimicrobial Activities of Phenolic Compounds*

Antimicrobial agents and therapies are globally crucial to overcome health challenges. To date, phytochemicals are widely studied for this purpose due to their ability to disrupt bacterial cell membranes, generate ROS, interfere with RNA synthesis, and block signaling pathways [141]. Phenolic compounds act against bacteria through multiple direct and indirect mechanisms involving cell wall, membrane, enzymes, metabolism, and pathways. They can damage structural component of bacteria and cause rapid cell lysis by disrupting cell permeability, increasing membrane rigidity, causing leakage of ions or metabolites, and damaging peptidoglycan layers [142,143]. By inhibiting essential enzymes like DNA gyrase or ATP synthase, they can cause impaired bacterial growth and inhibition of replication [144]. Other than enzymes, phenolic compounds can also directly bind bacterial proteins for denaturation, formation of irreversible complexes, and alteration in functional domains [145]. Moreover, several phenolic compounds can interfere with metabolic pathways to reduce bacterial viability by disrupting electron transportation, reducing ATP generation, and inhibiting fatty acid synthesis [146]. Furthermore, phenolic compounds can inhibit the biofilm that protects bacteria against antibiotics and immune defenses [147]. In addition to antioxidant activities of blueberry cultivars, their antimicrobial activities against Gram-positive bacteria, including *Bifidobacterium longum* (*B. longum*), *Faecalibacterium prausnitzii, Enterococcus hirae, E. faecalis, S. aureus, B. subtilis,* and *B. cereus,* and Gram-negative bacteria, including *P. aeruginosa, Vibrio harveyi, Salmonella enterica* subsp. (*Salmonella* ssp.), and *E. coli* ATCC 10536, were tested. Results showed that the Aurora and Nelson cultivars exhibited the highest antimicrobial activity for *B. cereus, P. aeruginosa,* and *Vibrio harveyi,* while the Duke and Bonus cultivars showed no antimicrobial activity. Similarly, another study proposed to investigate two different berry extracts, *Prunus spinosa* L. and *Vaccinium corymbosum,* and their antimicrobial activities via performing the minimum inhibitory concentration (MIC) method on *Shigella flexneri* ATCC 10708, *Salmonella choleraesuis* ATCC 12022, and *E. coli* ATCC 12792 [148]. According to results, *Prunus spinosa* L. extracts showed the same MIC of 250 μg/mL for *E. coli* ATCC 25922 and *Shigella sonnei* ATCC 25931, while *Vaccinium corymbosum* extracts showed different MIC results for *Salmonella enteritidis* ATCC 13076 with 2.16 mg/mL and *E. coli* ATCC 25922 with 3.74 mg/mL. Similarly, apples are also one of the studied fruits in this aspect, due to their myricetin, catechin, t-ferrulic acid, cinnamic acid, etc. [149]. When it is studied in different formulations, namely three different extract forms—hydroethanolic, microwave-assisted, and purified polypheno-

https://doi.org/10.3390/antiox15010087

lic extracts—different MIC values were observed against three pathogens [150]. Results showed that apple pomace exhibited antimicrobial efficiency against *S. aureus* with an MIC value of 2.5 mg/mL and *Shigella flexneri* (*S. flexneri*) with an MIC value of 0.625 mg/mL. Norway spruce bark is a pyramidal tree that has weeping branches. Recent studies highly investigated these trees due to their phenolic content, including stilbenes, tannins, resin acids, etc., and bioactivity features. A recent study examined two methods to compare their extraction efficiency: hot-water extraction (HW) and hydrodynamic cavitation (HC). Results showed that the HW method yielded higher total dissolved solids when it was compared to the HC method. Antimicrobial activity was tested against one Gram-positive, *S. aureus*, and one Gram-negative bacterium, *E. coli*. Antimicrobial activity tests indicated that HC extracts showed higher activity against HW extracts [151]. Pomegranate is another polyphenol-rich fruit that has high punicalagin, ellagic acid, and gallic acid [152]. Biological activities of pomegranate use various extraction types, including peel, flower, juice, pomace, etc. Predominantly, 18 polyphenols and 11 flavonoids have been identified in pomegranate peels [153,154]. This high polyphenolic content accompanies antioxidant, antibacterial, and anti-inflammatory properties [155,156]. According to studies, tannins play a crucial role in antimicrobial and anti-inflammatory activities, especially antimicrobial activity against methicillin-resistant *S. aureus* (*MRSA*) [157]. Additionally, its antioxidant activity helps to mitigate inflammation, making it a powerful candidate for antibiotics in managing respiratory infections [158]. In an in vivo study, Yassin et al. reported that pomegranate peels exhibit antimicrobial properties against *S. aureus* with 0.125 mg/mL and *MRSA* with a 0.250 mg/mL minimum inhibitory concentration (MIC) value. Besides the peel extracts, the antioxidant capacity and antimicrobial properties of pomegranate flower extracts were also investigated. Zhang et al. investigated the antimicrobial activity of pomegranate flowers against *Streptococcus* mutans. Results indicated that flower extracts inhibited biofilm formation, and as the concentration increased, inhibition zones grew larger [159].

Rosemary hydroalcoholic macerates were studied for the purpose of comparison between creams and hydrogels and evaluation of their antimicrobial, anti-inflammatory, and antioxidant properties [160]. Two types of rosemary, from the Romanian and Bulgarian coastal regions, were tested in different ethanol concentrations, which were supplied to also compare geographic origin. Results showed that rosemary, especially the Romanian species with 70% ethanol extracts, showed potential to evaluate bioactivity properties with high polyphenol content, namely, phenolic acids and flavonoids, which significantly exhibited antimicrobial activity against *Candida albicans* (*C. albicans*), *S. aureus*, and *E. coli*. In contrast, Bulgarian extracts showed no significant antimicrobial activity, indicating that geographic regions are also important for their phenolic content and bioactivities. As mentioned previously, catechins are widely studied and involved in various plant sources, namely catechin, epicatechin (EC), EGC, EGCG, epicatechin gallate (ECG), etc. [161]. They are considered antimicrobial agents due to their diverse action against different pathogens, including disrupting cell walls and membranes of the bacteria, hindering viral expression, suppressing DNA damage, ROS mechanisms, and their synergistic activities with antibiotics [162]. It has been demonstrated that their MIC values are around 1–2 mg/mL, and their minimum bactericidal concentration (MBC) values are around 2–4 mg/mL, depending on the bacteria. When it is compared to MIC, it shows the lowest concentration of an antimicrobial agent that inhibits visible bacterial growth in a subcultured antibiotic-free media [163]. Furthermore, the combination of catechins with tetracycline, erythromycin, clindamycin, azithromycin, vancomycin, and gentamicin exhibited synergistic effects with catechins [164]. To elaborate further, catechins have the ability to produce hydrogen peroxide, which is a compound that causes oxidative damage to the cells and causes an indirect kill. However, if a bacterium

https://doi.org/10.3390/antiox15010087

has a catalase enzyme, it will break down hydrogen peroxide into water and oxygen to evaluate the survival of bacteria [165,166]. As for hindering biofilm formation, catechin presence blocks the acrA gene, which is resistant to many drugs and causes the formation of biofilm in *E. coli* [164,167]. In a recent study, the polyphenolic content of *Osmanthus fragrans* was analyzed [168]. Results reported that two types of phenolic compounds were found in major quantities: phenolic acids that exhibit both antimicrobial and ROS-generating activity, and flavonoids that exhibit antifungal activity. Further study is needed to investigate its potential in food packaging by studying its activity against a fungal pathogen named *Alternaria alternata*. The results indicated that phenolic compounds from *Osmanthus fragrans* blocked spore germination, making it a candidate as an antimicrobial agent in food packaging. In a comparative review article about in vivo and in vitro studies, flavonoids, lignans, stilbenes, and phenolic acids were involved. Due to their unique structures and binding conditions, it has been demonstrated that they exhibit antimicrobial activity against not only *S. aureus* but also other foodborne Gram-positive pathogens, including *Clostridium botulinum* (*C. botulinum*), *Clostridium perfringens* (*C. perfringens*), and *Listeria monocytogenes* (*L. monocytogenes*). Other than Gram-positive, they also show potent activity against several Gram-negative bacteria, such as *Campylobacter* spp., *Salmonella* spp., *Vibrio* spp., and *E. coli*. Since they are widely studied in both in vitro and in vivo studies in a dose-dependent manner, the European Food Safety Authority (EFSA) has started to analyze their potential as antimicrobial resistance activity by investigating them on various bacterial strains and conditions [169,170]. Lund et al. studied the glycosylation process and its effect on the antimicrobial activity of quercetin. Results of the study demonstrated that it significantly affects the antimicrobial activity of quercetin, and when it is compared with glycosylated quercetin, non-glycosylated quercetin exhibited stronger antimicrobial activity against *E. coli* [171]. Other than plant sources, marine sponge species *Dysidea etheria* (*D. etheria*), *Echinodictyum dendroides* (*E. dendroides*), and *Desmapsamma anchorata* (*D. anchorata*) are also studied for their potential polyphenol content. Unexpectedly, extracts contained a low amount of polyphenol content. When their antimicrobial activity was tested, an MIC value of nearly 1024 µg/mL was obtained, indicating weak direct antimicrobial activity against the tested pathogens, and they exhibited significant activity against the *multi-resistant bacterium Pseudomonas aeruginosa* [172]. To overcome biomedical challenges like low bioavailability of bioactive compounds, microbial infection risks, and poor mechanical strength, multifunctional hydrogel systems were aimed to be developed [173]. For this purpose, Germi et al. used polyphenolic nanosheets that are derived from propolis and placed them into a gelatin methacrylate (GelMA) matrix. High polyphenolic content from flavonoid and phenolic acid levels from propolis helped the gel to involve characteristics like antimicrobial, antioxidant, and anti-inflammatory activity. To enhance their antimicrobial activity via cationic interactions, bis-imidazolium ionic liquid (BIm-IL), and to evaluate mechanical strength in hydrogel, 3-methacryloxypropyltrimethoxysilane (MPS) was used. These functionalities developed this hydrogel as a multifunctional system, making it a strong candidate for biomedical applications. To elaborate further, a recent review article compared both animal- and plant-derived polyphenol-rich extracts and their diverse effects on gel systems, including hydrogels, oleogels, nanogels, films, and hybrid gels [174]. According to the studies, combinations of gels and bioactive compounds from diverse developments include higher antioxidant activity with higher 3-ethylbenzothiazoline-6-sulfonic acid (ABTS) and 2,2-diphenyl-1-picrylhydrazyl (DPPH) scavenging activity and antimicrobial activity against *S. aureus*, *E. coli*, *C. albicans*, etc. Even though their combination opened a new perspective on gel systems, it has been crucially indicated that variability, stability, release control, and bioavailability challenges, as well as interaction or scale-up problems, are still yet to be determined. Generally, the literature identifies clearly that several phenolic compounds

https://doi.org/10.3390/antiox15010087

can disrupt membranes, chelate metals, and inhibit enzymes. However, specific binding sites or molecular targets and structure–activity relationships are generally poorly defined. Each study used appropriate extraction methods, solvents, organism strains, and MIC protocols, depending on their purpose. Although this gives variability to literature, it makes direct comparison, generation of high variable MIC values difficult. Moreover, in real foods, wound dressings, or formulations, phenolic compounds interact with proteins, lipids, and polysaccharides. These interactions can reduce free active concentration or change how they reach microbial cells, but most mechanistic work is performed in simple buffer systems, presenting a huge knowledge gap for application areas.

2.4. Therapuetic Potential of Phenolic Compounds

Antimicrobial, antiviral, anti-inflammatory, and antioxidant properties of phenolic compounds led them to be studied in diverse fields and made them study in various global health challenges, as well as the food industry, including food preservation and packaging. In a review article, it has been summarized that EGCG reduces ROS damage, curcumin supports synaptic function, resveratrol acts as a protectant for mitochondrial function, and anthocyanidins support synaptic plasticity, showing a potential for neurodegenerative disease treatments, including Parkinson's disease, Alzheimer's disease, Huntington's disease, and multiple sclerosis [175]. In another review article, it has been mentioned that several phenolic compounds, including resveratrol, piceatannol, oleuropein, ferulic acid, propolis, green tea polyphenols, hesperidin, etc., can act as neuro-, photo-, and gastroprotectants with significant anticancer activities [176]. However, poor bioavailability and clinical efficiency need further experiments to evaluate the delivery efficiency, including nano-formulations, microencapsulation, and dietary strategies. Moreover, it has been demonstrated with various studies that phenolic compounds prevent cardiovascular diseases, neurological diseases, liver diseases, and diabetes [177,178]. To enhance their bioavailability and nutritional value, bio-based polymers like proteins and polysaccharides are used for nanocarriers [179]. A life-threatening inflammatory disorder, acute pancreatitis, has been studied with phenolic compounds in preclinical trials [180]. According to results, it has been demonstrated that they have the ability to manage acute pancreatitis, but due to a lack of clinical trials, bioavailability, dose dependency, and therapeutic efficiency for acute pancreatitis are still unexplored. Fortunately, Dong et al. studied the formation of a polyphenol-rich nanocarrier to enhance the effectiveness of gemcitabine (Gem) and the bioavailability of polyphenol in pancreatic cancer [181]. For this study, a natural polyphenol, 1,2,3,4,6-pentagalloyl glucose (PGG) is used. For in vivo studies, mouse and PDX models (patient-derived xenografts) were utilized, whereas PANC-1 and Panc02 cancer cells were used for in vitro studies. According to results, the combination of this PGG nanocarrier increased the bioavailability and stability of PGG while increasing apoptosis (a regulated form of cell death) and inducing immunogenic cell death. Plant-derived vesicle-like nanoparticles are also studied for determining their diverse anticancer activities [182,183]. Accordingly, phenolic compounds from tea flowers or bitter lemon exhibit anticancer activity by causing apoptosis via ROS damage or cycle arrest and supporting the immune system [184]. Moreover, phenolic compounds from cannabis with high cannabidiol (a chemotype) content demonstrated a significant preventative effect against hepatocellular carcinoma, which is a type of liver cancer [185]. To effectively examine the bioavailability, functionality, and stability of phenolic compounds for acne therapy, they have been encapsulated into nanocarriers [186]. According to results, polyphenols, including flavonoids, showed significant anti-inflammatory activity by inhibiting COX-2, interleukin-6 (IL-6), nuclear factor kappa B (NF-κB), interleukin-1 beta (IL-1β), tumor necrosis factor-alpha (TNF-α), and prostaglandin E2 (PGE2) pathways; antibacterial activity by targeting *Propionibacterium*

https://doi.org/10.3390/antiox15010087

acnes; and antioxidant activity by scavenging ROS. Furthermore, the prevention of sebum oxidation was also observed. Phenolic compounds were also studied in delivery methods, including hydrogels, scaffolds, nanoparticles, and nanofibers, to compare their efficiency for acute and chronic wound healing [187]. Hydrogels induced a sustained drug release, while scaffolds supported cell proliferation. Moreover, nanoparticles protected the unstable polyphenols from unstable forms, while nanofibers were able to mimic the extracellular matrix structure, becoming more prevalent in diabetic wounds. Froldi et al. studied several plant-derived phenolic compounds, namely curcumin, baicalein, mangiferin, resveratrol, and quercetin, to investigate their efficiency for the prevention and treatment of a vascular disease, peripheral artery disease [188]. Results showed that baicalein and quercetin reduced inflammation by inhibiting NF-κB and TLR4 signaling pathways, as well as oxidative stress with nearly no toxicity, while curcumin was reported with liver toxicity in high doses. Mangiferin was also able to reduce ROS damage and inflammation by inhibiting the same signaling pathways dose-dependently, with several skin side effects from the mango extract. Resveratrol exhibited strong endothelial, anti-inflammatory, and antioxidant effects, but it has been reported that high doses (more than 150 mg/day) may cause thyroid issues. Bioactive compounds, namely flavonoids, alkaloids, and terpenes, exhibit unique multi-target activity by reducing neuroinflammation, improving neuroplasticity, and enhancing mitochondrial function and cognition. These properties make them a promising candidate for potential in Alzheimer's disease prevention and therapy. However, it is noted that, similarly to other clinical trials with phenolic compounds, bioavailability, stability, and delivery challenges still remain [189]. To study mitochondrial function and its effect on barrier integrity, polyphenols of kiwifruit were tested with lipopolysaccharide (LPS) in both in vivo studies that included LPS-stressed rat models and in vitro studies that included Caco-2 cell lines [190]. According to the results, it has been demonstrated that kiwifruit polyphenols improve mitochondrial function and inhibit mitophagy that is driven by AMPK/ULK2, making them a candidate food natural nutraceutical as a gut barrier protectant and for intestinal disorders that are linked to mitochondrial function. In a recent study, the pro-apoptotic activity of a blackberry species (*Rubus adenotrichos*) in liver (HepG2), stomach (AGS), skin (SK-Mel-28), and colon (SW-620) cancer cell lines were investigated. Results demonstrated that fruit and pulp extracts had strong cytotoxic effects, with $IC_{50} < 200$ μg/mL [191]. In another recent in vitro study, anti-inflammatory, antioxidant, and autophagy-modulating effects of Andean blackberry polyphenols were proposed [192]. In this study, a dose of 80 μg/mL Andean blackberry extract was used, and the results demonstrated an inhibition of the NLRP3 inflammasome, with reduced inflammation and oxidative stress. Hypoglycemic polyphenols that have been indicated as having an essential role in the inhibition of carbohydrate digestion and insulin stimulation are abundant in walnuts [193,194]. Hu et al. analyzed these hypoglycemic polyphenols derived from walnut green husk. Results reported that they enhance glucose uptake and glycogen accumulation in insulin-resistant HepG2 and 3T3-L1 cells. Moreover, their inhibition of α-glucosidase and α-amylase has also been observed [193,195]. As previously mentioned, pomegranates are rich in phenolic content and its anticancer properties were reported by exhibiting antiproliferative activity against MCF7 breast cancer with an IC50 value of 8.15 μg/m. [196]. In addition to these studies, various types of nanoparticles, especially silver nanoparticles, have been used with extracts like pomegranate peel to develop potential antimicrobial agents and antioxidants. These trials are also crucial for being green synthesis studies, meaning eco-friendly methods are being used in this research [197]. Onion is also rich in quercetin and other phenolic acids, giving antifungal and antioxidant activities, particularly against *Ascosphaera apis*, a bee pathogen [198]. A comparison between five different onion cultivars (Tank, Cobra, Bomul, Hongbanjang, and Gujji) and their different parts (root,

https://doi.org/10.3390/antiox15010087

peel, and bulb) showed that peel had the highest phenolic content among all cultivars with abundant quercetin, caffeic acid, and protocatechuic acid content [199]. Bozinou et al. [200] investigated onion skins to enhance recognition of their therapeutic potential by studying their antioxidant, antimicrobial, anti-inflammatory, and anticancer properties. In vitro studies on U-87 MG (human glioblastoma) and MCF-7 (human breast cancer) demonstrated that extracts with more than 25 μg/mL concentrations significantly suppressed cancer cell viability. Moreover, all tested onion skin samples exhibited dose-dependent cytotoxicity, making them potential adjuvants for cancer therapy and nutraceutical development [200]. Green tea is a widely studied plant for its diverse properties due to its various contents, including phenolic compounds [201]. In a study where it was compared with black tea, green tea showed the highest antioxidant activity due to its higher amount of EGCG, EC, EGC, and catechin content. Moreover, it is indicated that EGCG in green tea corresponds with apoptosis via p53 and caspase pathways. Additionally, when it is tested on MDA-MB-231 breast cancer cells via MTT (3-(4,5-dimethylthiazolyl-2)-2,5-diphenyltetrazolium bromide) assay, green tea caused high cell death at 100–200 μg/mL, compared to black tea samples [202,203]. Nevertheless, it is notable to say that synergy, additive effects, or interference with drug metabolism of polyphenol interaction against standard antiviral drugs are limited for therapeutical applications.

3. Application Areas of Phenolic Compounds from Plant Byproducts

Until recently, phenolic compounds from byproducts were associated with low production efficiency, even though these bioactive compounds show immense potential for practical applications. Industries increasingly face pressure to minimize waste, create circular economic systems, and use byproducts, especially from peels, seeds, pomace, husks, into value-added ingredients for food, pharmaceuticals, and cosmetics (Table 1).

3.1. Extraction Methods

3.1.1. Conventional Extraction Technologies

Traditional extraction methods such as maceration, Soxhlet extraction, and reflux extraction have been widely applied due to their simplicity and low equipment requirements. However, their long extraction times and extensive solvent consumption limit their efficiency and sustainability compared to emerging green technologies. In addition, they rely on large amounts of toxic organic solvents, high energy use, and harsh processing conditions that can damage sensitive bioactive compounds and generate environmental burdens.

https://doi.org/10.3390/antiox15010087

Table 1. Overview of industrial application areas and experimental contexts for phenolic compound byproducts, including their sources and application outcomes.

Industrial Application Area	Experiment Area	Phenolic Compound Byproduct Source	Type of Phenolic Compound	Type of Extraction Method	Bioactivity and Results of Applications	Reference
Food Industry	Syrup production Food additives in biscuits	Date palm and seed extract Date seed powder	Flavonoids Anthocyanins	Microwave-assisted extraction method	Increase in antioxidant activity The dose of sample affected the texture as hardness Balance in taste and texture Enhancement in nutritional value	[204,205]
	Fortification in yogurt (0.5%, 1.0%, and 1.5%)	Ginger peel extracts	Gingerol Anthocyanins Flavonoids	Ultrasound-assisted extraction Conventional solvent extraction using hot water, 80% ethanol, and 100% ethanol	Yogurt samples that are fortified with 1.5% ginger peel extracts enhanced the antioxidant activity Increase in water-holding capacity Increase in stability of pH Fortified yogurts exhibited darker colors due to higher phenolic content Microbiological tests that were carried out with *S. thermophilus* and *L. bulgaricus* resulted in longer shelf life potential with suppressed yeast and mold formation	[206,207]
	Jelly gummy	Pomegranate peel extracts	Punicalagin (both α and β form) Ellagic acid Gallic acid Coumaric acid Catechin Protocatechuic acid Chlorogenic acid Epicatechin Caffeic acid 4-Hydroxybenzaldehyde Benzoic acid Ferulic acid Quercetin	Conventional solid–liquid extraction	Bioavailability enhancement Increased firmness in texture Similar hardness between samples	[208]

https://doi.org/10.3390/antiox15010087

Table 1. *Cont.*

Industrial Application Area	Experiment Area	Phenolic Compound Byproduct Source	Type of Phenolic Compound	Type of Extraction Method	Bioactivity and Results of Applications	Reference
	Investigation of bioavailability enhancement using plasma-activated water method	Sweet orange peels	Benzoic acid Ferulic acid Caffeic acid Myricetin 3,5,6,7,4′-pentamethoxyflavone 3,5,6,7,3′,4′-hexamethoxyflavone Naringin Protocatechuic acid Apigenin Naringenin Hesperidin Hesperetin Flavonoids Narirutin Tannins Sinapinic acid Isovanillic acid 4-coumaric acid Tangeretin Neohesperidin	Plasma-activated water method	Decrease in unwanted compounds, which increased the bioavailability	[209,210]
	Raw ground beef patties	Acacia bark and leaves	Flavonols, Proanthocyanidins Bioflavonoids/polyflavonoids Hydrolyzable tannins Hydroxybenzoic acids Hydroxycinnamic acids Dihydrochalcones Abundantly: Epicatechin p-coumaroyltrifolin B Procyanidin B5 (−)-epigallocatechin Procyanidin C1	Conventional solid–liquid extraction	Both samples exhibited an increase in shelf life and stability Higher antioxidant capacity Reduction in color deterioration Reduction in lipid peroxidation	[211]

https://doi.org/10.3390/antiox15010087

Table 1. *Cont.*

Industrial Application Area	Experiment Area	Phenolic Compound Byproduct Source	Type of Phenolic Compound	Type of Extraction Method	Bioactivity and Results of Applications	Reference
	Investigation of encapsulation efficiency with lactose crystals	Cheese whey and maize inflorescences	Chlorogenic acid Maysin	Hot-aqueous extraction and encapsulation	Increasement in antioxidant capacity, stability, and bioavailability Protection from degradation	[212]
	Investigation of effect in bioavailability	*Posidonia oceanica* (seagrass)	Flavonoids Stilbenes Coumarins Tannins Lignans	Solid–liquid extraction	Antifungal activity against *Penicillium italicum*, *Botrytis cinerea*, *Penicillium digitatum*, *Penicillium expansum*, *Aspergillus niger*, and *Geotrichum candidum* Antiviral activity against *murine norovirus* and *feline calicivirus* Potential in food preservation	[213]
	Human keratinocytes HaCaT cells	Blueberry pomace	Cyanidin-3-glucoside Quercetin-3-glucoside	Ultrasound-assisted extraction	High antioxidant activity Preservation of HaCaT cells from oxidative stress Developmental potential as food additives	[214]
Cosmetic Applications	Investigation with NADES extraction process to enhance bioavailability	Hazelnut skin and shell	(−)-epicatechin (+)-catechin Phloridzin Quercetin-3-*O*-rhamnoside Myricetin-3-*O*-rhamnoside	NADES extraction	Protection against UV-induced damage Enhancement in photostability Antioxidant and skin-healing properties Promoter in process of inhibition of lipid peroxidation and the aging process	[215]

https://doi.org/10.3390/antiox15010087

Table 1. *Cont.*

Industrial Application Area	Experiment Area	Phenolic Compound Byproduct Source	Type of Phenolic Compound	Type of Extraction Method	Bioactivity and Results of Applications	Reference
	Investigation with antibiotics (β-lactams, amphenicols, tetracyclines, and quinolones)	Grape pomace	Catechins Resveratrol Gallic acid	Solid–liquid extraction Supercritical Fluid extraction Ultrasound-assisted extraction Microwave-assisted extraction Enzyme-assisted High-Voltage Electric Discharge Ohmic Heating High-Pressure Processing Deep Eutectic Solvents (DESs)	Inhibition of antibacterial enzymes, including peroxidase, lactase, pectinases, xylanases, and cellulases Antimicrobial properties against *E. coli*, *S. Typhimurium*, *S. aureus*, *P. aeruginosa*, and *L. monocytogenes* Synergistic effect on the inhibition of *S. aureus* and *E. coli* Potential as cosmeceuticals and nutraceuticals by inhibiting *Cutibacterium acnes*, *as well as reducing inflammation and redness*	[216–219]
	Investigation of their effect on food quality and methodology efficiency	Date seed powder	Protocatechuic acid p-hydroxybenzoic acid Vanillic acid Syringic acid Caffeic acid Coumaric acid Ferulic acid p-hydroxycinnamic acid Chlorogenic acid Quercetin-3-*O*-glucoside Catechin Procyanidin Anthocyanins Proanthocyanidins	Conventional solvent extraction	Strong antioxidant activity Potential to enhance food quality as food additives and functional oils	[220]

https://doi.org/10.3390/antiox15010087

Table 1. *Cont.*

Industrial Application Area	Experiment Area	Phenolic Compound Byproduct Source	Type of Phenolic Compound	Type of Extraction Method	Bioactivity and Results of Applications	Reference
	Investigation into bioavailability and potential applications	Olive byproducts	Hydroxytyrosol Tyrosol Oleuropein Caffeic acid Rutin Resveratrol Catechins	Ultrasound-assisted extraction Microwave-assisted extraction Enzyme-assisted extraction Supercritical Fluid extraction Pressurized Liquid extraction Ohmic Heating Pulsed Electrical Field	Potential as natural antioxidants in food, oil, and beverages Sustainable biomass waste sources	[221,222]
	Investigation into bioavailability and potential applications	Wine byproducts	Hydroxytyrosol Tyrosol Oleuropein Caffeic acid Rutin Resveratrol Catechins	Ultrasound-assisted extraction Microwave-assisted extraction Enzyme-assisted extraction Supercritical Fluid extraction Pressurized Liquid extraction Ohmic Heating Pulsed Electrical Field	UV-protectant and anti-aging properties Sustainable biomass waste sources Natural antioxidants	[221,223]
Pharmaceutical and Health Applications	Investigation on bioactivity	Cacao bean shells	Quercetin Kaempferol Proanthocyanidins	Microwave-assisted extraction Ultrasound-assisted extraction	Rich phenolic content Antimicrobial activity against *S. mutans* 7-methylxanthine metabolite did not exhibit inhibitory effect against *S. mutans* alone	[224,225]

https://doi.org/10.3390/antiox15010087

Table 1. *Cont.*

Industrial Application Area	Experiment Area	Phenolic Compound Byproduct Source	Type of Phenolic Compound	Type of Extraction Method	Bioactivity and Results of Applications	Reference
	Investigation into bioactivity and potential health applications	Persimmon byproducts	Gallic acid Quercetin	Solid-assisted extraction	Antimicrobial activities: Inhibition of *E. coli* and biofilm activity Synergistic effect with gentamicin against *S. aureus* Potential candidate by inhibiting bacterial adhesion to HaCaT and Caco-2 cells Antiviral activity by inhibiting biofilm formation, blocking β-lactamase	[226,227]
	Investigation of bioactivity and effect of using a DES extraction–resin adsorption system	Grape pomace	Quercetin-3-*O*-rutinoside Quercetin-*O*-pentoside Myricetin-*O*-rutinoside Gallic acid Caffeic acid Syringic acid Vanillic acid Chlorogenic acid *p*-coumaric acid Petunidin-rutinoside Malvidin-rutinoside β-Type (epi)catechin tetramer Quercetin-glucuronide *p*-coumaric acid Malvidin-3-*O*-glucoside Malvidin-3-*O*-(*p*-coumaroyl)glucoside (+)-catechin (−)-epicatechin Syringic acid (−)-gallocatechin	Solid–liquid extraction Supercritical Fluid extraction Ultrasound-assisted extraction Microwave-assisted extraction Enzyme-assisted High-Voltage Electric Discharge Ohmic Heating High-Pressure Processing Deep Eutectic Solvents (DESs)	Improving lipid profiles that are effective for cardiovascular diseases Being able to preserve their antioxidant and α-glucosidase with DES extraction–resin adsorption system	[228]

https://doi.org/10.3390/antiox15010087

Table 1. *Cont.*

Industrial Application Area	Experiment Area	Phenolic Compound Byproduct Source	Type of Phenolic Compound	Type of Extraction Method	Bioactivity and Results of Applications	Reference
	In vitro study with Vero cells and HepG-2 liver cells	Mango peel extract	Gallic acid Chlorogenic acid	*	Antimicrobial activity against *S. aureus and S. mutans* Antiviral activity adenovirus-7 and HSV-1 Cytotoxicity effect against HepG-2 liver cancer cells	[229]
	Investigation into bioactivity	Avocado seed powder	Quercetin Gallic acid	Conventional solid–liquid extraction	Antimicrobial activity against *E. coli, S. aureus, Salmonella* spp., *P. aeruginosa*, and *C. albicans* Anti-inflammatory activity against TNF-α, IL-6, and IL-1β pathways and COX-2 and iNOS enzymes	[230]
	Investigation into bioactivity	Cherry seed	Caffeic acid Ferulic acid Gallic acid Protocatechuic acid	Conventional solid–liquid extraction Ultrasound-assisted extraction Enzyme-assisted extraction Supercritical Fluid extraction	Modulation of lipid metabolism and insulin sensitivity Promote glucose utilization Contribution to glucose intake	[231–233]
	Investigation into bioactivity	Longan fruit	Proanthocyanidins Phenolic acids	Ultrasound-assisted extraction Enzyme-assisted extraction Microwave-assisted extraction	High antioxidant and anti-inflammatory activities Potential candidates against NAFD and obesity	[234]

https://doi.org/10.3390/antiox15010087

Table 1. *Cont.*

Industrial Application Area	Experiment Area	Phenolic Compound Byproduct Source	Type of Phenolic Compound	Type of Extraction Method	Bioactivity and Results of Applications	Reference
	In vivo experiment with d-galactose-induced aging mice	Walnut oil process byproducts	Procyanidin dimer B2 isomer I Procyanidin dimer B1 isomer II Malvidin hexoside Malvidin Coumaroyl-hexoside Petunidin hexoside (−)-epicatechin Quercetin Glucuronide Myricetin Rhamnoside (+)-catechin (iso)-rhamnetin Myricetin	Conventional solid–liquid extraction	Restore in antioxidant and SOD activity with WKP hydrolysates Improvement in liver and kidney tissue morphology Reduction in MDA concentration Reduction in expression of SREBP-1c by pretreatment of malvidin hexosides	[235]
	In vitro study to investigate the potential in joint disorders using macrophage- and chondrocyte-like cell lines	Hemp leaf	Cannabinoids Flavonoids	Heat-reflux extraction	Reduction in symptoms of osteoarthritis Improvement in joint function	[236]
	In vivo study for investigation of their antidiabetic activity	Pomegranate tree (*Punica granatum*) bark	Flavonoids Phenolic acids Tannins	Soxhlet extraction	Inhibition of α-glucosidase and α-amylase	[237]
	Investigation into sustainable methods to evaluate bioactivity	Garlic byproducts	Flavonoids	Supercritical Fluid extraction	No significant improvement observed Lack of standardization and material assessment	[238,239]

https://doi.org/10.3390/antiox15010087

Table 1. *Cont.*

Industrial Application Area	Experiment Area	Phenolic Compound Byproduct Source	Type of Phenolic Compound	Type of Extraction Method	Bioactivity and Results of Applications	Reference
Sustainable Materials and Packaging	Investigation into their bioactivities in cooked chicken noodle	Mango, papaya and loquat byproducts	- In mango: Mangiferin Catechin Quercetin Kaempferol Rhamnetin Anthocyanins Gallic acid Ellagic acid Propyl gallate Methyl gallate Benzoic acid Protocatechuic acid - In loquat: 3-p-coumaroylquinic acid Caffeoylquinic acid 5-feruloylquinic acid Quercetin Kaempferol derivatives, including galactoside, glucoside, and rhamnoside - In papaya: Ferulic acid Mandelic acid Syringic acid Vanillic acid Myricetin Conifer aldehyde	Conventional solid–liquid extraction	Strongest antioxidant activity in mango seed extracts Higher phenolic content in seed extracts than peel extracts Better color maintenance in mango seed extracts Reduction in lipid and protein oxidation in mango peel extracts Partial effect of papaya and loquat extracts in reduction in lipid and protein oxidation	[240]

https://doi.org/10.3390/antiox15010087

Table 1. *Cont.*

Industrial Application Area	Experiment Area	Phenolic Compound Byproduct Source	Type of Phenolic Compound	Type of Extraction Method	Bioactivity and Results of Applications	Reference
	Investigation into their potential in food-packaging industry using polyphenols and citric acid	Poplar bark and veneer	*	Chemical-assisted solid–liquid extraction	Resulted in high antioxidant activity, mechanical resistance, and bonding properties	[241]

* Not identified.

https://doi.org/10.3390/antiox15010087

3.1.2. Green Extraction Technologies

Green extraction methods use safer solvents, lower temperatures, and reduced energy inputs, allowing higher recovery of stable, high-quality phenolic compounds. Green extraction methods provide this valorization by making the process cleaner, scalable, compliant with regulatory expectations, and more aligned with consumer demand for natural, sustainable products (Figure 1) [9,242,243]. Upon analysis of plant materials, hydrolytic treatment gives an easier polyphenol recovery of non-extractable phenolic compounds. Bound phenolic compounds such as proanthocyanidins and phenolic acids like ferulic acids, are chemically attached to cell wall components in plant tissues that cannot be extracted easily without hydrolysis. Once released, they exhibit higher bioactivities compared to free ones [244].

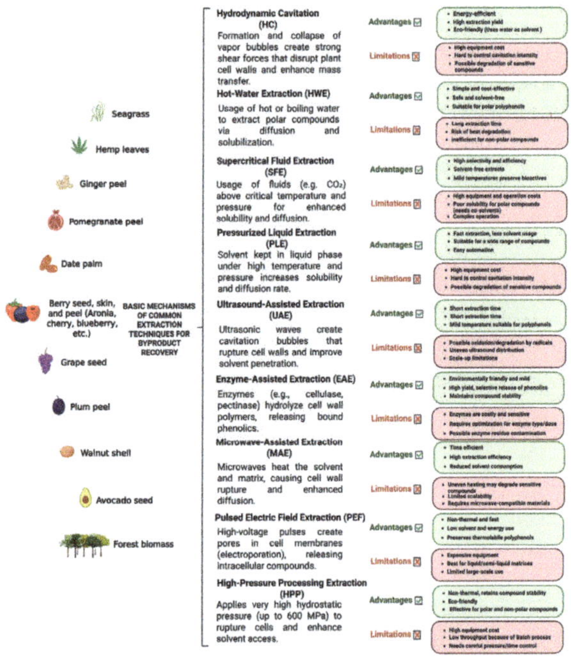

Figure 1. A comparative overview of different extraction techniques for polyphenol recovery, highlighting their main principles, advantages, and limitations.

High-Pressure-Assisted Extraction (HPAE), a related technique, offers an eco-friendly alternative to conventional heat treatments by reducing extraction time, solvent use, and energy consumption while increasing yield [245]. In comprehensive research, optimum conditions of HPAE for phenolic compounds from olive pomace has been investigated [246]. Results showed that HPAE emerged as a more efficient and environmentally friendly extraction method than traditional solvent extraction for recovering phenolic compounds from olive pomace. Overall, HPAE method enhanced the yield and antioxidant activity of phenolic compounds from olive pomace without altering chemical structures, making it a promising green technology for valorizing olive industry byproducts.

HC is a green and energy-efficient technology that uses controlled formation, growth, and collapse of vapor bubbles in a liquid flow to enhance physical and chemical processes. It has been used particularly for extraction, mixing, cell disruption, and biomass valorization [247]. An extensive review article aimed to summarize and evaluate HC for extracting and valorizing valuable compounds from plants and biomass [248]. The summary suggests

https://doi.org/10.3390/antiox15010087

that HC significantly enhances mass transfer, cell wall rupture, and matrix disintegration, resulting in higher extraction yields of bioactive compounds compared to conventional or ultrasonic extraction. Moreover, it demands less energy than UAE and can be easily scaled up for industrial-level operation with simple configurations and reduced solvent use. It represents a promising tool for green chemistry, biofuel production, and circular bioeconomy applications, as well as a powerful technique for eco-friendly valorization of plant and biomass materials. Nevertheless, current limitations include equipment erosion, process optimization complexity, variability among feedstocks, and limited industrial-scale data. Overcoming these requires better reactor design, process modeling, and integration with complementary green technologies.

Supercritical Fluid extraction (SFE) is a green extraction technique that utilizes fluids in conditions above their critical temperature and pressure, where they exhibit both gas-like diffusivity and liquid-like solvating power. SFE enables efficient extraction under mild conditions, minimizing the degradation of heat-sensitive phenolic compounds. The process occurs in a closed, oxygen-free system, reducing oxidation and light-induced degradation. Additionally, solvent recovery is simple, as CO_2 reverts to gas upon depressurization, allowing for reuse and reduced solvent waste. Because CO_2 is a non-polar solvent, its ability to extract polar compounds such as polyphenols is limited. To overcome this, small amounts of polar co-solvents, especially ethanol, are often added, which slightly increases complexity and cost [249,250]. In a comprehensive study, recovery of tannins from various biomass resources using SPE was investigated. Results highlighted that SFE enables highly selective extraction of specific tannin fractions (hydrolyzable vs. condensed tannins) by adjusting pressure, temperature, and co-solvent composition. When reactivity and byproducts are analyzed, it has been suggested that under supercritical or subcritical conditions solvent reactivity (especially water) can cause secondary reactions such as hydrolysis and depolymerization, generating gallic acid, ellagic acid, and sugars from complex tannins. These reactions can either lower yield or create valuable derivatives, depending on the application [251].

Pressurized Liquid extraction (PLE), also known as Accelerated Solvent Extraction (ASE), is an automated and environmentally friendly extraction technique that employs high pressure and temperature to enhance the efficiency of compound recovery. Elevated pressures allow solvents to remain in the liquid phase above their boiling points, which increases solubility and diffusion rates of target compounds while reducing solvent viscosity and surface tension, thereby facilitating efficient extraction from the matrix. PLE is recognized as a rapid, economical, and efficient alternative to conventional extraction methods, often yielding equal or higher concentrations of phenolic compounds. The method can be operated in dynamic or static modes, providing flexibility and reproducibility. Despite its efficiency, the high temperature and pressure conditions may risk degradation of thermolabile compounds, and the requirement for specialized equipment increases initial operational costs [252]. A large-scale study was conducted to develop and optimize a sustainable PRE process using ethanol–water mixtures as green solvents for recovering bioavailable phenolic antioxidants from grape seed byproducts [253]. Evaluation of the bioavailability and stability of phenolic compounds during in vitro digestion and Caco-2 cell transport, linking extraction conditions with biological relevance, was also investigated. Results highlighted that optimized low-temperature PLE (75% ethanol, 20 °C, 11 min) efficiently extracts bioavailable antioxidant phenolics from grape seed byproducts with minimal degradation and strong sustainability advantages. This process supports circular economy valorization of winery waste and offers a scalable green-technology route for functional-ingredient production. Nevertheless, to achieve industrial viability and biologi-

https://doi.org/10.3390/antiox15010087

cal relevance, future research must focus on optimizing bioavailability, enhancing energy and solvent efficiency, and integrating PLE into holistic circular bioeconomy frameworks.

Ultrasound-assisted extraction (UAE) is a modern extraction technique that uses acoustic cavitation, which refers to the formation and collapse of microbubbles caused by ultrasonic waves, to disrupt plant cell walls and enhance the release of intracellular compounds. UAE enables efficient recovery of compounds bound within the plant matrix and prevents thermal degradation of bioactive molecules due to its operation at relatively low temperatures. The efficiency depends on ultrasonic parameters (frequency, intensity, and duration), and overexposure can lead to partial degradation or oxidation of sensitive compounds. Additionally, scalability for industrial applications remains a technical challenge [254]. In a research article where two potato peel varieties (Lady Claire and Lady Rosetta) are used, UAE with traditional solid–liquid extraction (SLE) tested different ultrasonic frequencies (33 and 42 kHz) have been investigated [255]. Results showed that UAE significantly improved phenolic recovery and antioxidant activity compared to SLE. Moreover, research proved that UAE is a faster, more effective, and environmentally friendly method under 33 kHz than 42 kHz for obtaining antioxidant-rich phenolic compounds from potato processing waste.

Microwave-Assisted Extraction (MAE) has emerged as a highly efficient technique for isolating bioactive compounds from food matrices. This method offers significant advantages over conventional extraction processes, including reduced extraction time, lower solvent consumption, and enhanced automation. The integration of ultrasonication with MAE further improves extraction efficiency by increasing yield and minimizing component degradation. Modified MAE systems address the limitations of traditional approaches, offering solvent-free or low-solvent operation, higher extraction efficiency, and greater energy savings. Overall, MAE provides a rapid, uniform, and energy-efficient means of extracting bioactive compounds under optimized conditions [256,257]. A research article aimed to optimize UAE of phenolic compounds from Annona muricata (soursop) leaves to evaluate their antioxidant and antimicrobial properties [258]. The parameters that are included in this study were extraction time, temperature, and solvent concentration to optimize the conditions and investigate their influence on yield of bioactive compounds. Results showed that UAE exhibits as a rapid, sustainable, and effective method for extracting antioxidant and antimicrobial compounds from Annona muricata leaves.

Enzyme-Assisted Extraction (EAE) utilizes the catalytic activity of specific enzymes to hydrolyze the components of the cellular matrix, thereby breaking down cell walls and facilitating the release of intracellular metabolites into the surrounding medium. The process generally involves the use of a solvent, which may be either organic or aqueous in nature, to enhance the penetration and recovery of target compounds. EAE offers several advantages, including effective cell wall disruption, selective extraction of metabolites, mild operating conditions, and improved extraction yield and rate. However, despite these benefits, large-scale application of EAE remains limited due to several drawbacks. These include the high cost of enzymes, incomplete cell wall degradation that necessitates additional purification steps, enzyme instability under certain processing conditions, limited recyclability and reusability, and loss of catalytic activity after repeated use [259]. To improve the release of insoluble-bound phenolic compounds from winemaking byproducts (grape skins and seeds), a study investigated the influence of EAE method [260]. To determine their efficiency in improving phenolic yield, antioxidant activity, and enzyme inhibition, two enzymes Pronase (a protease) and Viscozyme (a carbohydrase complex) have compared. Results showed that both enzymes increased the ratio of soluble to insoluble-bound phenolics, meaning more phenolics were converted into bioavailable forms. Moreover, viscozyme was more effective than Pronase, extracting higher amounts of gallic acid, cate-

https://doi.org/10.3390/antiox15010087

chin, and prodelphinidin dimer A, and uniquely releasing p-coumaric, caffeic acids, and procyanidin dimer B, which Pronase failed to extract. Furthermore, soluble phenolics from enzyme-treated samples significantly inhibited α-glucosidase and lipase, suggesting potential in managing hyperglycemia and obesity. Nevertheless, it has been reported that even after enzymatic treatment, some phenolics remained bound. Although EAE was effective, alkaline hydrolysis (NaOH) still achieved higher phenolic yields and antioxidant activity, indicating that EAE might need optimization to match chemical extraction efficiency.

Pulsed Electric Field (PEF) extraction is an emerging green technology that applies short, high-voltage pulses to create pores in cell membranes, a process known as electroporation. This increases cell permeability and facilitates the release of intracellular bioactive compounds. PEF enables efficient extraction at ambient temperatures, reducing both extraction time and energy consumption while preserving heat-sensitive compounds. It also requires smaller amounts of non-toxic solvents, minimizing environmental impact. However, optimization of field strength and process parameters is essential to balance extraction efficiency and prevent excessive cell damage [261]. In an extensive study, optimization of PEP parameters for recovering phenolic compounds from white grape pomace have been investigated [262]. Selected parameters including electric field strength, energy input, solvent composition, temperature, and extraction time were investigated. Study showed that optimal PEF parameters were 3.8 kV/cm, 10 kJ/kg, 50% ethanol, 50 °C, 190 min for recovering phenolic compounds of white grape pomace. Although, PEP provides higher yields, reduced solvent use, and shorter extraction times, further research on industrial validation, complete phenolic characterization, bioavailability testing, and sustainability assessment to unlock its full potential in the circular bioeconomy. Moreover, hybridizing green extraction methods such as integrating PEF with UAE, EAE, or MAE, can improve yields while lowering energy and solvent inputs. Overall, it has been highlighted that UAE, NADES, and UA-NADES offer better results in efficiency and green chemistry, while PLE provides higher yields in industrial application areas. Moreover, PEF and EAE are suggested as pretreatments to enhance other methods, while SFE provides pure extracts, especially in non-polar phenolic compounds. When used synergistically, combination of these methods has potential to provide high-value compounds for medical, pharmaceutical, cosmetic, and functional food applications [263–265].

Versatile wine production wastes, including grape pomace, seeds, and peels, have been investigated to analyze byproduct phenolic content, such as phenolic acids, stilbenes, and flavonoids [266]. They were studied to compare various techniques, namely, PEF, high-pressure processing, SFE, pressurized liquid, EAE, UAE, and MEA extraction methods. Results showed that MAE and UAE methods yielded the highest extraction and stronger antioxidant activity, while EAE method gave more purified extractions when it was combined with other methods. Moreover, the combination of UAE and PEF extraction methods preserved the total phenolic content by more than 40%. Since wine byproducts are not fully used, this study suggests that they can support zero-waste production and evaluate value-added recovery mostly in agriculture.

Natural deep eutectic solvent (NADES) extraction is a green extraction method that uses mixtures of natural compounds such as sugars, amino acids, and organic acids as eco-friendly solvents to extract bioactive compounds from plants or other materials. These solvents are formed by strong hydrogen bonding between components, creating a liquid with high solubility power. The method is efficient, safe, and biodegradable, making it suitable for food, cosmetic, and pharmaceutical applications. However, NADES are often viscous and difficult to recover, which can limit large-scale use [263,267]. In pursuit of a similar goal, Liu et al. [268] investigated the phenolic content of water caltrop shells, aquatic plants that are mainly distributed across Europe, Asia, and Africa, with a green

https://doi.org/10.3390/antiox15010087

extraction method named ultrasound-assisted natural deep eutectic solvents (UA-NADESs) to enhance the efficiency of extracting bioactive compounds from natural materials. Results showed a high yield phenolic compounds, which can be further investigated for applications in food preservation, cosmetic formulations, and pharmaceutical products. Correspondingly, rose oil production causes a lot of waste of phenolic compounds with high-yield byproducts. To recover them, the ultrafiltration method was studied using rose distillation wastewater [269]. Two membranes, 1 and 10 kDa, were tested, and 1 kDa showed higher purification and yield of phenolic compounds with 88.5%. Ultrafiltration pledges to a more isolated and stabilized recovery. Hence, these recovered compounds preserved their antioxidant activity and suggested that they can be further studied as cosmeceuticals, functional foods, natural preservatives, and nutraceuticals. Broccoli leaves are rich in oligosaccharides and bioactive compounds, including non-extractable phenolic compounds, giving them the potential for having functional properties [270,271]. To investigate bioactive properties and evaluate the functional dietary fiber SFE technology, which provides a higher concentration of oligosaccharides, improving water solubility and extraction yield of phenolic compounds to enhance antioxidant activity [272]. Results demonstrated that supercritical technology effectively valorizes broccoli leaf byproducts, evaluating their bioactive properties with optimal conditions of 191 bar, 40 °C, and 1 h. Citrus processing causes a wide range of citrus peel and waste products that are rich in various bioactive compounds, including phenolic content [273,274]. In a research article where lignocellulosic byproducts like STF231 that are derived from the medicinal plant extract industry have been used, different extraction techniques are compared to see the impact on their total phenolic content, as well as antioxidant activity [275]. In this study, conventional hydroethanolic extraction (CE), NADES, UAE, and their combinations are compared. CE is a classic method used to pull bioactive compounds out of plant material using a mixture of water and ethanol as solvent [276]. Results showed that CE exhibited the strongest antioxidant activity, while NADES and UAE combination produced the highest polyphenol recovery. Further analysis has been conducted using tools that focuses on evaluating the sustainability profile of the sample preparation phase, especially Green Analytical Procedure Index (GAPI), AGREE (Analytical Green Chemistry Evaluation), and AGREEprep (Analytical Green Chemistry Evaluation of Extraction Preparation) [277]. Analytical results showed that the combination of NADES and UAE methods were the most sustainable method, with low toxicity, low cost, and high efficiency, suggesting it as a sustainable alternative to traditional organic solvent extraction. It is also highlighted that only one specific plant byproduct was investigated in this case, indicating that results may vary across all species. Furthermore, no biological or in vivo tests were conducted to confirm health effects of the extracts.

Overall, it is notable to say that each technique offers advantages to be more suitable for specific applications while presenting divergent limitations. Conventional methods are still widely used, simple, and accessible, while suffering from higher solvent toxicity, poor economic sustainability, and lower extract quality and efficiency.

In comparison, SFE provides high extraction efficiency, better mass transfer, and is environmentally friendly while being a suitable alternative for thermolabile compounds. Nonetheless, it is a costly method with limited effectiveness for polar compounds. Unlike SFE, UAE is a cost-effective technique with a rapid and energy-efficient method with good performance at low temperatures. However, its efficiency depends on the sample, such as the plant matrix, and it may require additional steps like filtration. MAE also provides fast extraction with high yields, reduced solvent use, and good preservation of bioactive compounds. Nevertheless, it involves non-uniform heating, limited sustainability for non-polar compounds, and high equipment costs like SFE. EAE provides high selectivity and is suitable for mild operation conditions and release of bound phenolics. However, its high enzyme cost, high sensitivity to process conditions, and long extraction time are a challenge. PEF is an eco-friendly and highly efficient method that increases cell permeability and preserves heat-sensitive compounds. However, it involves high equipment costs, and it is less effective for highly rigid matrices. PLE provides high-efficiency extraction by using elevated temperature and pressure to enhance solvent solubility and penetration. This results in time-efficient extraction with reduced solvent consumption. However, it again requires high enzyme cost and has a risk of thermal degradation of thermolabile compounds. HWE, which is a safe, eco-friendly, and simpler method, is a suitable technique for polar compounds. Nonetheless, its major challenges are long extraction time, high thermal exposure, and risk of degradation of heat-sensitive compounds. Moreover, HC is an energy-efficient and solvent-saving technique that enhances mass transfer and extraction yields. Nonetheless, it has limited control over cavitation intensity and risk of degradation of sensitive bioactives. HPP is a high-efficiency method with high solvent access and minimal thermal damage, which makes this technique suitable for both polar and non-polar compounds. However, it requires high capital cost with complex process control. DESs, along with NADES, which is a type of DES, are emerging as green solvents with high extraction yields, but their high viscosity and the necessity of downstream filtration remain as challenges. Additionally, high-voltage electric discharge (HVED) is an energy-efficient method that provides high extraction yield. Nonetheless, it lacks selectivity and has a risk of causing compound degradation. Lastly, ohmic heating is an energy- and time-efficient method with reduced solvent usage, and it provides good preservation of heat-sensitive compounds. However, high equipment cost and limitation by dependence on the electrical conductivity of the material are the main challenges.

3.2. Food Industry

Date palm byproducts are produced by the processing of syrup production. To evaluate these low-cost agricultural wastes, functional biscuits have been aimed to be developed. Tests involved texture, color, proximate, sensory, and shelf life analysis [204]. Results demonstrated that total phenolic capacity and antioxidant activity have increased. Samples with more extracts showed hardness as texture compared to others, but water capacity remained stable across storage conditions. Moreover, biscuits with 10% date palm extract had the highest consumer acceptance with a balance in taste, texture, and nutritional enhancement. Hence, this study presents the date palm byproducts as functional foods with nutritional enhancement and sensory acceptability due to their higher dietary fiber, antioxidant, and anti-inflammatory properties. Phenolic compounds from date seeds and seed extracts exhibit strong antioxidant activities. Therefore, the use of date-based fibers or phenolic extracts has potential to improve the quality of foods. Moreover, they have potential for generating functional oils, natural additives. Although date seed powder is being used in food products, it has been highlighted that date seeds have not traditionally been consumed, suggesting more studies for safety profile. Food industry byproducts are rich in valuable phenolic compounds and no longer waste but strategic resources. Through green extraction techniques, including PLE, SFE, UAE, MAE, EAE, HC, and PEF, degradation of thermolabile compounds and recovery of phenolic compounds have been improved. Furthermore, through circular economy approaches, as well as less solvent usage and energy demand, these materials can be transformed into functional foods, nutraceuticals, active packaging materials, and biorefineries [278,279].

In another zero-waste valorization study, ginger peel extracts were studied for yogurt to test its efficiency for shelf life and microbiological stability [207]. Experiments were tested as fortification of yogurt with 0.5%, 1.0%, and 1.5% of ginger peel extracts for 21 days.

https://doi.org/10.3390/antiox15010087

Microbiological tests were performed as total viability count, formation of yeasts and molds, and activity of *L. bulgaricus* and *Streptococcus thermophilus* (*S. thermophilus*). Results showed that fortified yogurts with 1.5% ginger peel extracts enhanced antioxidant activity and phenolic content up to triple the amount of the control groups, which were yogurts with no extracts. Moreover, pH remained more stable, and water-holding capacity was increased. Color characteristics of fortified yogurts with more extracts exhibited darker colors due to higher phenolic content. According to results, 80% ethanol extracts exhibited highest efficiency. Moreover, freeze-drying process on samples along with 80% ethanol extraction resulted in the highest total phenolic content between samples. As for microbiological tests, both *S. thermophilus* and *L. bulgaricus* remained in the desired range, while yeast and mold formation were significantly suppressed, indicating a longer shelf life with various bioactive health benefits. In a similar aim, pomegranate peels were investigated with the microencapsulation method in vitro as jelly gummy prototypes to enhance their value as byproducts, since they are considered nearly 70% waste product from juice processing [208]. The polyphenol content of pomegranate peel shows an instability to oxygen, light, and heat, even though there is a high amount of phenolic content. To be able to eliminate this challenge, microencapsulation was performed to enhance stability and bioavailability. Results demonstrated that the microencapsulation method significantly enhanced their stability and bioavailability, being a sensory-acceptable and functional vehicle candidate as a valuable method for delivering extracts in high yields. Moreover, microencapsulated extract exhibited higher bioavailability but need for an optimization process. In another study, sweet orange peels were investigated by using plasma-activated water with a central composite rotatable design to eliminate compounds that reduce their nutritional values [209]. Experimental results showed that the total bioactive content was significantly enhanced by plasma-activated water, and there has been a reduction in unwanted compounds, including terpenoids, limonin, narginin, tannins, phytic acid, and saponins, suggesting plasma-activated water as a permissible method to eliminate undesirable compounds to evaluate nutritional value as nutraceutical or functional food candidates.

To evaluate the biological value and bioavailability of phenolic compounds of acacia bark and leaves, they have been investigated on raw ground beef patties to test their shelf-life development individually [211]. Results indicated that both phenolic content from leaves and bark significantly improved oxidative shelf life with higher antioxidant activity, reduced color deterioration, and lower lipid oxidation. In a comprehensive review about using phenolic compounds in meat products, it is suggested that nanotechnological methods, such as active films with polyphenols, improve their shelf life with more stability and antioxidant effect. Moreover, addressing challenges and limitations that affect phenolic compound efficiency has been listed as meat type, stabilization, dose optimization, standardization, and consumer accessibility [280].

In a comprehensive study, valorization of two agricultural byproducts, cheese whey and maize inflorescences, has been tested to investigate encapsulation efficiency of bioactive compounds such as phenolic compounds into lactose crystals, with the aim of enhancing their stability and antioxidant properties, as well as protecting them from degradation [212]. Results showed that lactose-polyphenol co-crystals that are developed from valorization of two byproducts provided antioxidant protection and stability under UV-light conditions with an efficient encapsulation capability, promising an industrial recovery of lactose and plant phenolic compounds. *Posidonia oceanica* is a widely studied seagrass that is endemic to the Mediterranean Sea. Billions of tons of biomass waste are obtained due to human activity and climate change [213]. To valorize these biomass wastes into sustainable byproducts, their bioactive properties have been investigated [281]. To make a comprehensive experiment, biomass wastes are prepared with three different methods: heat-assisted, UAE,

https://doi.org/10.3390/antiox15010087

and organic solvent-based. These extracts were tested on *Penicillium italicum* (*P. italicum*), *Botrytis cinerea* (*B. cinerea*), *Penicillium digitatum* (*P. digitatum*), *Penicillium expansum* (*P. expansum*), *Aspergillus niger* (*A. niger*), and *Geotrichum candidum* (*G. candidum*) to test their antifungal activity. Both UAE and heat-assisted extracts significantly inhibited *P. digitatum* and *B. cinerea*, while organic solvent-based extracts exhibited a general anti-fungal activity against all the targets. The same extract types were also tested on *murine norovirus* (MNV) and *feline calicivirus* (FCV). Results indicated that only organic-solvent-based extracts exhibited antiviral activity against FCV and MNV due to their lipid and non-glycosylated polyphenol content. Hence, it has been suggested that these biomass waste byproducts have a powerful value to be used in food preservation and biomedical applications. In an in vitro environmental sustainability study, blueberry fruit waste and its bioactivities were investigated in human keratinocytes HaCaT cells [214]. To extract the phenolic compound efficiently, optimized UAE was used with a response surface methodology. The extracted compounds exhibited high antioxidant activity due to their phenolic content, mainly anthocyanidins. Moreover, the antioxidant activity of these bioactive compounds preserved HaCaT cells from oxidative stress that is caused by hydrogen peroxide, suggesting a development for dietary supplements or functional food derivatives from blueberry pomace in the food industry.

3.3. Cosmetic Applications

Hazelnut byproducts are also abundant in bioactive compounds with a wide spectrum of properties. To evaluate the sustainability of hazelnut byproducts that were obtained from skin, shell, and defatted flour, NADES extraction has been used to compare traditional and UAE techniques [215]. Results showed that skin extracts contained a unique phenolic profile with the highest antioxidant activity compared to shell and defatted flour. Moreover, NADES with betaine and sorbitol yielded the highest efficiency for skin extracts, while the fructose and glycerol combination yielded the highest for defatted extracts and fructose and lactic acid for shell extracts. Compounds from hazelnut byproducts, such as catechins that are found in the skin extracts, protect against UV-induced damage and enhance photostability, while gallic acid and flavonols provide skin-healing properties due to their anti-inflammatory activity. Furthermore, the antioxidant activity of these extracts may prevent lipid peroxidation and aging progress, exhibiting anti-aging properties, making them valuable byproducts for health-related applications, anti-aging cosmetic products, and UV protection formulations.

Grape pomace flavonoids exhibit antimicrobial activities by chelating iron, forming hydrogen bonds between versatile microbial enzymes and proteins, protecting the structural integrity of Gram-positive bacterial cell walls, and disrupting the other membranes of Gram-negative bacteria [282]. Stilbenes from grape pomace have the potential to inhibit bacterial enzymes, namely, peroxidase, lactase, pectinases, xylanases, and cellulases, via degrading their tertiary structure [216,217]. Furthermore, antimicrobial activities against foodborne pathogens, namely *E. coli*, *S. Typhimurium*, *S. aureus*, *P. aeruginosa*, and *L. monocytogenes*, have also been demonstrated [228]. When phenolic compounds from grape pomace are combined with antibiotics, such as β-lactams, amphenicols, tetracyclines, and quinolones, a synergistic effect of inhibition against *S. aureus* and *E. coli* has been observed [216].

Phenolic compounds from date seeds and seed extracts exhibit strong antioxidant activities. Therefore, use date-based fibers or phenolic extracts to improve the quality of foods. Moreover, they have potential for generating functional oils and natural additives. Although date seed powder is being used in food products, it has been highlighted that date seeds have not traditionally been consumed, suggesting more studies for safety profile [220]. Moreover, it has been mentioned that green extraction methods can be a suitable alternative

https://doi.org/10.3390/antiox15010087

for their cost-effective and target-selective properties. Since olive and wine byproducts are rich in phenolic compounds, especially hydroxytyrosol, tyrosol, oleuropein, caffeic acid, rutin, catechin, and resveratrol, they are valuable candidates for sustainable biomass waste sources. These recovered phenolic compounds can be applied as natural antioxidants in foods, oil, and beverages, as well as nutraceuticals and supplements for health benefits due to their bioactivities. Moreover, they have been suggested as potential candidates as cosmeceuticals due to their UV-protective and anti-aging properties. Nevertheless, it has been highlighted that many green extraction methods have stability issues and the initial equipment and operating costs are high for supercritical and membrane systems [221–223]. As mentioned, grape byproducts are rich in phenolic content, dietary fiber, and other bioactive compounds [283]. Phenolic compounds, including catechins, resveratrol, and gallic acid, significantly exhibit antimicrobial activity against *S. epidermidis*, *E. coli*, and *S. aureus* via disrupting bacterial cell walls [283,284]. Moreover, evidential studies that experimented with grape byproducts in the Kirby–Bauer method demonstrated that they exhibit inhibition of biofilm formation of Gram-positive bacterial zones, significantly against *Cutibacterium acnes* [218]. These antimicrobial properties of grape byproducts provide stability for products, promising candidates for acne-prone skin, protection from outer contamination, and reduced inflammation and redness, making them suitable for integration into nutraceuticals and cosmetic products [219].

3.4. Pharmaceutical and Health Applications

Correspondingly to the previously discussed study, activities of bounded polyphenols from persimmon byproducts were investigated to contribute them as food byproducts, prebiotics, therapeutics, or nutraceutical agents [226,227]. Results showed dietary fibers selectively inhibited *E. coli* and biofilm activity, along with synergistic bactericidal effects with gentamicin (an aminoglycoside antibiotic) against *S. aureus*. Moreover, it inhibited virulence factors by inducing membrane damage, inhibiting biofilm formation, and blocking β-lactamase that induces multi-resistance to the β-lactam antibiotics. Furthermore, it has also been observed that they inhibited pathogen adhesion to HaCaT and Caco-2 cells, making these persimmon-derived phenolic compounds, significantly gallic acid and quercetin, powerful candidates for being used as nutraceuticals or functional ingredients to mitigate various bacterial infections.

Cacao bean shells are rich in phenolic content. In a recent study, cacao bean shell extracts from different origins were tested for their phytochemical profile and antimicrobial activity [224]. Results showed that extracts exhibited antimicrobial activity against *Streptococcus mutans* (*S. mutans*). Moreover, analyses confirmed one metabolite, 7-methylxanthine; however, it did not show an inhibitory effect against *S. mutans* when it was tested alone, suggesting that a synergistic effect with other extracts might give a potential activity. Grape pomace is rich in bioactive compounds, including versatile phenolic compounds, and it has been shown that these compounds improve lipid profiles by affecting oxidized low-density lipoproteins and lipid serums, affecting cardiovascular diseases [228].

Since traditional solvents like ethanol, acetone, and methanol are effective but toxic to the environment, Wang et al. studied recovering these phenolic compounds from citrus peels by using a green method of integrated extraction–adsorption system using resins and deep eutectic solvent (DES), which is less toxic and more biodegradable [274,285]. Results of the usage of the DES extraction–resin adsorption system showed a significant yield of recovery for phenolic compounds. Moreover, extracted phenolic compounds preserved their antioxidant and α-glucosidase inhibitory activity, promising to be strong candidates for natural agents for type 2 diabetes. In an in vitro study, mango peel extract was investigated for its photochemical profile in cultured cell lines, namely Vero cells and

https://doi.org/10.3390/antiox15010087

cancerous HepG-2 liver cells [229]. More than 15 phenolic compounds were identified, while gallic acid and chlorogenic acids were found abundantly. Identified compounds exhibited antimicrobial activity against *S. aureus and S. mutans*, antiviral activity against adenovirus-7 and HSV-1, and cytotoxicity against HepG-2 liver cancer cells, suggesting their use as a therapeutic agent in oncology and infectious disease management.

Avocado seed is considered one of the most highly valued fruits in the world, with various components including vitamins, enzymes, proteins, oligosaccharides, and phenolic compounds [286]. Industrial processing of avocado has resulted in a substantial amount of waste from seeds and peels due to a lack of information on recovery methods. To open a gate to eliminate this problem, many studies have been conducted, and one of them analyzed phenolic content of avocado seeds to evaluate its nutritional value in agriculture. Results showed that avocado seed powder consists of high tannins and flavonoids like quercetin and gallic acids [230]. Moreover, antimicrobial activities of avocado seed polyphenols have been documented against *E. coli, S. aureus, Salmonella* spp., *P. aeruginosa*, and *C. albicans* via disrupting bacterial cell walls, inhibiting bacterial enzymes, and targeting fungal pathogens. Furthermore, anti-inflammatory activity against pathways like TNF-α, IL-6, and IL-1β, and COX-2 and iNOS enzymes that are crucial for the inflammation process have been observed, promising a powerful candidate for functional food additives. Polyphenolic compounds of cherry byproducts from seeds are abundantly identified as caffeic, ferulic, gallic, and protocatechuic acids. These compounds modulate insulin sensitivity and lipid metabolism and enhance glucose utilization, which are key metabolic pathways of metabolic syndrome [231,232]. Additionally, they can contribute to glucose intake by modulating GLUT4 by influencing its expression and translocation [233]. Longan fruit contains valuable bioactive compounds, including proanthocyanidins, alkaloids, and phenolic acids, that provide benefits for managing non-alcoholic fatty liver disease (NAFLD) and obesity. Byproducts of longan fruit have been investigated for their antioxidant and anti-inflammatory activities. Slides demonstrated that they can regulate lipid metabolism genes and inflammatory pathways, enhance liver detoxification, and show potential for liver protection. Additionally, exhibiting potent activities makes them outperform synthetic antioxidants and makes them potential candidates for a more sustainable option for health-related applications [234].

In a recent in vivo study, both extractable and non-extractable polyphenols are investigated for their potential activity against obesity and metabolic disorders [235]. The treatment of obese rats with extractable and non-extractable polyphenols was conducted over a period of 8 weeks. Results demonstrated that non-extractable polyphenols exhibited better results by decreasing liver triglycerides by over 45%. In contrast, no reduction in insulin resistance from either type of polyphenol was observed. Walnuts have been used for their versatile benefits and put through various processing, such as oil production. The byproduct of the walnut oil process, walnut kernel with pellicle (WKP), has been investigated for its effect on gastrointestinal digestion via valorization of protein- and polyphenol-rich hydrolysates.

Further study investigated the anti-aging activity of WKP hydrolysates in vivo by testing d-galactose-induced aging mice. Experimental results demonstrated that WKP hydrolysates restored total antioxidant capacity and superoxide dismutase (SOD) activity, tissue morphology in liver and kidney, and reduced concentration of malondialdehyde (MDA), a byproduct of lipid peroxidation, in liver, kidney, and serum tissues. Thus, WKP hydrolysates are potential candidates as reusable byproducts with health benefits and sustainable properties. Hemp leaf byproducts contain a high amount of cannabinoids, flavonoids, etc., which shows a variety of these bioactive compounds in abundance in seeds, stems, hurds, and leaves themselves [287,288].

https://doi.org/10.3390/antiox15010087

An in vitro study investigated these hemp leaf byproducts with hyaluronic acid gel formulations, which are a commonly used gel type for medical applications, such as for viscous supplementation for joint disorders [236]. Experimental studies are conducted in macrophage-like (RAW 264.7) and chondrocyte-like (C-20/A4) cell lines. Results showed that hyaluronic acid gel provided a sustainable and localized delivery of cannabinoids to joint spaces, reduced symptoms of conditions like osteoarthritis, and improved joint function by restoring viscoelastic properties of synovial fluid, promising a powerful candidate for therapeutic applications. The bark of the pomegranate tree (*Punica granatum*) is enriched with versatile bioactive compounds, namely phytochemicals like flavonoids, phenolic acids, and tannins, which are considered agro-industrial waste [289]. To evaluate its agricultural value, its antidiabetic properties were investigated in vivo [237]. In this study, the Soxhlet extraction method has been used to evaluate extraction efficiency and to reduce the risk of contamination with unwanted compounds with continuous recovery of analytes of the sample matrix. The results demonstrated that extracts evaluated inhibitory effects on α-glucosidase and α-amylase, exhibiting significant enzyme inhibition with no acute toxicity in mice at 2000 mg/kg. Garlic byproducts consist of proteins, phenolic compounds, dietary fibers, and polysaccharides, and they can be generated from harvesting or processing, such as stems, skin, leaves, and root plates [290]. Qui et al. [238] investigated garlic byproducts for their antioxidant, antimicrobial, and anti-inflammatory properties to find sustainable valorization methods. The comprehensive investigation suggested that challenges like a lack of standardization and material assessment and the necessity of cost-effective green methods with analysis need to be addressed to valorize garlic byproducts for potential applications in material, health, and agriculture-based applications. Bioavailability is one of the other challenges in this application area because phenolic compounds are generally poorly absorbed, rapidly metabolized, and present at very low aglycone levels in plasma. By designing synergistic polyphenol combinations, formulations to improve oral take and consider interactions with drugs that affect pharmacokinetics, and enhanced bioavailability, as well as their role in application areas such as in food industry, pharma, medical, and cosmetics [291].

3.5. Sustainable Materials and Packaging

In a 2024 research article, antioxidant activities of peel and seed extracts from mango, papaya, and loquat using different extraction solvents have been investigated to analyze and evaluate the potential antioxidant effects of these extracts in cooked chicken models during refrigerated storage [240]. During investigation, the study focused on their ability to control lipid and protein oxidation and maintain color stability. Experimental results showed that regardless of solvent, mango seed extracts exhibited the strongest antioxidant activity in ABTS, DPPH, and FRAP assays. Furthermore, mango seed extracts had the highest levels of phenolic compounds, flavonoids, and condensed tannins. It has also been pointed out that seed extracts were richer in antioxidants than peel extracts. Applicational results highlighted that mango seed extracts maintained color better and effectively reduced lipid and protein oxidation during ten days of refrigerated storage. Papaya and loquat extract also reduced oxidation in a less effective way. The study suggested that evaluating the effects in different meat types and processing conditions can be further investigated for future approaches. A 2019 review discussed several future approaches to optimize process, assess sustainability, and improve stability [292]. For instance, combining processes to extract multiple compounds from same waste stream to maximize value, applying response surface methodology or machine learning models to optimize extraction parameters have been discussed. Moreover, it has been suggested that using nano- and microencapsulation can improve stability and controlled release of sensitive bioactives in food or nutraceutical

https://doi.org/10.3390/antiox15010087

applications. Further limitations can be listed as complication in collecting, processing, and standardizing raw byproducts that have short shelf life because of high moisture content [220,293]. Biodegradation refers to the natural breakdown of materials through enzymatic processes initiated by microorganisms. To improve functional properties, the combination of bioplastics with complementary barrier capabilities has been investigated using polymer blending and multilayer configurations [35]. Results indicate phenolic acids are a powerful candidate for food packaging and preservation applications due to their broad spectrum of antioxidant, antimicrobial, and antiviral activities. To evaluate one of the most valuable waste products into valuable byproducts, Fernandes et al. [241] investigated forest byproducts for developing sustainable particleboards to potentially contribute to the food-packaging industry. In these experiments, poplar bark and veneers with bio-adhesives containing citric acid have been used. Experimental studies indicated that citric acid enhanced the bonding properties by more than 50% but negatively stimulated the antioxidant capacity. Thus, particle boards exhibited high mechanical resistance and antioxidant activity, making them a potential candidate for eco-friendly, sustainable food packaging.

Overall, green extraction techniques have a promising potential as an alternative to conventional methods. Nonetheless, their broader industrial application areas require careful consideration of several key challenges as discussed previously. To summarize these challenges, standardization of processing parameters remains an essential aspect to ensure reproducibility, along with consistent extraction efficiency and reliable dose determination of bioactive compounds generally. Additionally, standardization of extracts is necessary for downstream applications, including encapsulation or formulation optimization or in vitro and in vivo applications. Moreover, enhancing extraction efficiency in terms of time and yield with reduced solvent consumption is essential for cost-effective and suitable production. Although most green extraction techniques demonstrate high yields with short process times, further optimization is required for energy input, solvent selection, and scalability. By addressing these limitations, their enhanced potential will expand their application areas, including the food, pharmaceutical, cosmetic, health, and sustainable food packaging industries. It is notable to say that a strategic combination of suitable green extraction methods offers a potential pathway to overcome current challenges. Byproducts are emerging for their significant economical and environmental advantages. Hence, overcoming these challenges and improving extract quality and bioavailability will enable the recovery of higher-value bioactive compounds and facilitate the improvement toward more suitable and industrially viable extraction processes.

4. Conclusions

Phenolic compounds form byproducts that possess beneficial attributes with their immense biological properties, including antioxidant, antimicrobial, antiviral, anti-inflammatory, and antifungal activities. Valorization of these phenolic compounds from waste products requires unique methods that are also acquiring eco-friendly and cost-effective properties. To overcome limitations such as standardization, bioavailability, sustainability, dose determination, delivery, stability, lack of variability management, and knowledge, innovative methods need to be developed. Several extraction techniques, including HC, hot-water extraction, SFE, PLE, UEA, EAE, MAE, PEF, high-pressure processing extraction, and NADES-based techniques have been indicated to have the potential to overcome some of the challenges during sustainability studies, but additional research is required. The main challenge remains scaling sustainable and optimized processes while ensuring economic and regulatory practicability, leading to zero-waste and high valuable food systems. Moreover, assessing synergistic or antagonistic interactions between different

https://doi.org/10.3390/antiox15010087

byproduct materials, as well as investigating encapsulation or formulation strategies, is essential to improve extract stability and application in the food industry, as well as in other application areas. Furthermore, there is insufficient standardization in phenolic extraction, identification, and quantification. These different methods produce incomparable data, reducing reliability, even though they are sustainable and powerful candidates for green extraction methods. Even though some of the byproducts, including pomegranate peel, ginger peel, and longan fruit waste extracts, are well studied, further research is needed for some of the byproducts, including avocado seed, coffee, and seagrass, to uncover their efficiency in their potential application fields. Although byproducts from phenolic compounds are risen interest, other than TPC, their activity against specific molecules or interactions such as polyphenol–polyphenol or polyphenol–protein, are merely discussed. Furthermore, in vivo studies are still limited and well-designed clinical trials using polyphenol-rich byproducts extracts as bioactives are poorly defined. Similarly, the majority of in vitro studies demonstrate only in vitro activity and perform further studies like dose-dependent modeling subtly. Moreover, even though green extraction methods provide many advantages, processes in whole extracts are still limited compared to purified standards. As previously mentioned in antimicrobial section, different processes and methods make it difficult to directly compare values, therefore translating byproduct phenolic compounds into real antimicrobial products. This gap also makes it difficult to understand about selectivity of byproduct phenolic compounds. Moreover, while phenolic compounds are generally seen as safe and natural, more studies about pro-oxidant, genotoxic, or endocrine effects at certain doses or in certain forms for byproduct extracts. The food industry and agricultural applications have been rapidly evolving, but further studies and information need to be gathered to equally grow the evolution in cosmetic and health-related applications. Warranting further research will potentially help to evaluate the food and meat industry, along with cosmetic, biomedical, health-related, and agricultural applications, and evolve the sustainable research to overcome limitations in methods, knowledge, underusage, bioavailability, and nutritional value of phenolic compounds from byproducts.

Author Contributions: Conceptualization, S.K., S.S. and R.D.; writing—original draft preparation, R.D., S.S. and M.B.; writing—review and editing, S.K., S.S., R.D. and M.B.; visualization, S.S., R.D. and S.K. All authors have read and agreed to the published version of the manuscript.

Funding: This research received no external funding.

Institutional Review Board Statement: Not applicable.

Informed Consent Statement: Not applicable.

Data Availability Statement: No new data were created or analyzed in this study. Data sharing is not applicable to this article.

Conflicts of Interest: The authors declare no conflicts of interest.

References

1. Bertelli, A.; Biagi, M.; Corsini, M.; Baini, G.; Cappellucci, G.; Miraldi, E. Polyphenols: From Theory to Practice. *Foods* **2021**, *10*, 2595. [CrossRef]
2. Mukherjee, C.; Chakraborty, S. Study of Dietary Polyphenols from Natural Herbal Sources for Providing Protection against Human Degenerative Disorders. *Biocatal. Agric. Biotechnol.* **2021**, *33*, 101956. [CrossRef]
3. Cutrim, C.S.; Cortez, M.A.S. A Review on Polyphenols: Classification, Beneficial Effects and Their Application in Dairy Products. *Int. J. Dairy Technol.* **2018**, *71*, 564–578. [CrossRef]
4. Quideau, S.; Deffieux, D.; Douat-Casassus, C.; Pouységu, L. Plant Polyphenols: Chemical Properties, Biological Activities, and Synthesis. *Angew. Chem. Int. Ed.* **2011**, *50*, 586–621. [CrossRef]
5. Lang, Y.; Gao, N.; Zang, Z.; Meng, X.; Lin, Y.; Yang, S.; Yang, Y.; Jin, Z.; Li, B. Classification and Antioxidant Assays of Polyphenols: A Review. *J. Future Foods* **2024**, *4*, 193–204. [CrossRef]

6. Grosso, G. Effects of Polyphenol-Rich Foods on Human Health. *Nutrients* **2018**, *10*, 1089. [CrossRef]
7. Cao, H.; Saroglu, O.; Karadag, A.; Diaconeasa, Z.; Zoccatelli, G.; Conte-Junior, C.A.; Gonzalez-Aguilar, G.A.; Ou, J.; Bai, W.; Zamarioli, C.M.; et al. Available Technologies on Improving the Stability of Polyphenols in Food Processing. *Food Front.* **2021**, *2*, 109–139. [CrossRef]
8. Coşkun, N.; Sarıtaş, S.; Jaouhari, Y.; Bordiga, M.; Karav, S. The Impact of Freeze Drying on Bioactivity and Physical Properties of Food Products. *Appl. Sci.* **2024**, *14*, 9183. [CrossRef]
9. Picot-Allain, C.; Mahomoodally, M.F.; Ak, G.; Zengin, G. Conventional versus Green Extraction Techniques—A Comparative Perspective. *Curr. Opin. Food Sci.* **2021**, *40*, 144–156. [CrossRef]
10. Feng, W.; Hao, Z.; Li, M. Isolation and Structure Identification of Flavonoids. In *Flavonoids—From Biosynthesis to Human Health*; Justino, G.C., Ed.; InTech: London, UK, 2017; ISBN 978-953-51-3423-7.
11. Perron, N.R.; Brumaghim, J.L. A Review of the Antioxidant Mechanisms of Polyphenol Compounds Related to Iron Binding. *Cell Biochem. Biophys.* **2009**, *53*, 75–100. [CrossRef]
12. Urquiaga, I.; Leighton, F. Plant Polyphenol Antioxidants and Oxidative Stress. *Biol. Res.* **2000**, *33*, 55–64. [CrossRef]
13. Leri, M.; Scuto, M.; Ontario, M.L.; Calabrese, V.; Calabrese, E.J.; Bucciantini, M.; Stefani, M. Healthy Effects of Plant Polyphenols: Molecular Mechanisms. *Int. J. Mol. Sci.* **2020**, *21*, 1250. [CrossRef] [PubMed]
14. Rasouli, H.; Farzaei, M.H.; Khodarahmi, R. Polyphenols and Their Benefits: A Review. *Int. J. Food Prop.* **2017**, *20*, 1700–1741. [CrossRef]
15. Dubrovina, A.S.; Kiselev, K.V. Regulation of Stilbene Biosynthesis in Plants. *Planta* **2017**, *246*, 597–623. [CrossRef] [PubMed]
16. Cucciolla, V.; Borriello, A.; Oliva, A.; Galletti, P.; Zappia, V.; Ragione, F.D. Resveratrol: From Basic Science to the Clinic. *Cell Cycle* **2007**, *6*, 2495–2510. [CrossRef]
17. Ko, J.-H.; Sethi, G.; Um, J.-Y.; Shanmugam, M.K.; Arfuso, F.; Kumar, A.P.; Bishayee, A.; Ahn, K.S. The Role of Resveratrol in Cancer Therapy. *Int. J. Mol. Sci.* **2017**, *18*, 2589. [CrossRef]
18. Mondal, A.; Bennett, L.L. Resveratrol Enhances the Efficacy of Sorafenib Mediated Apoptosis in Human Breast Cancer MCF7 Cells through ROS, Cell Cycle Inhibition, Caspase 3 and PARP Cleavage. *Biomed. Pharmacother.* **2016**, *84*, 1906–1914. [CrossRef]
19. Giampieri, F.; Cianciosi, D.; Alvarez-Suarez, J.M.; Quiles, J.L.; Forbes-Hernández, T.Y.; Navarro-Hortal, M.D.; Machì, M.; Casanova, R.D.J.P.; Espinosa, J.C.M.; Chen, X.; et al. Anthocyanins: What Do We Know until Now? *J. Berry Res.* **2023**, *13*, 1–6. [CrossRef]
20. Badshah, S.L.; Faisal, S.; Muhammad, A.; Poulson, B.G.; Emwas, A.H.; Jaremko, M. Antiviral Activities of Flavonoids. *Biomed. Pharmacother.* **2021**, *140*, 111596. [CrossRef]
21. Cushnie, T.P.T.; Lamb, A.J. Antimicrobial Activity of Flavonoids. *Int. J. Antimicrob. Agents* **2005**, *26*, 343–356, Correction in *Int. J. Antimicrob. Agents* **2006**, *27*, 181. [CrossRef]
22. Friedman, M. Overview of Antibacterial, Antitoxin, Antiviral, and Antifungal Activities of Tea Flavonoids and Teas. *Mol. Nutr. Food Res.* **2007**, *51*, 116–134. [CrossRef] [PubMed]
23. Ahmad, A.; Kaleem, M.; Ahmed, Z.; Shafiq, H. Therapeutic Potential of Flavonoids and Their Mechanism of Action against Microbial and Viral Infections—A Review. *Food Res. Int.* **2015**, *77*, 221–235. [CrossRef]
24. Imanishi, N.; Tuji, Y.; Katada, Y.; Maruhashi, M.; Konosu, S.; Mantani, N.; Terasawa, K.; Ochiai, H. Additional Inhibitory Effect of Tea Extract on the Growth of Influenza A and B Viruses in MDCK Cells. *Microbiol. Immunol.* **2002**, *46*, 491–494. [CrossRef]
25. Lani, R.; Hassandarvish, P.; Shu, M.-H.; Phoon, W.H.; Chu, J.J.H.; Higgs, S.; Vanlandingham, D.; Abu Bakar, S.; Zandi, K. Antiviral Activity of Selected Flavonoids against Chikungunya Virus. *Antivir. Res.* **2016**, *133*, 50–61. [CrossRef]
26. Papakyriakopoulou, P.; Velidakis, N.; Khattab, E.; Valsami, G.; Korakianitis, I.; Kadoglou, N.P. Potential Pharmaceutical Applications of Quercetin in Cardiovascular Diseases. *Pharmaceuticals* **2022**, *15*, 1019. [CrossRef]
27. He, W. Epigallocatechin Gallate Inhibits HBV DNA Synthesis in a Viral Replication—Inducible Cell Line. *World J. Gastroenterol.* **2011**, *17*, 1507. [CrossRef] [PubMed]
28. Ancuceanu, R.; Lascu, B.E.; Drăgănescu, D.; Dinu, M. In Silico ADME Methods Used in the Evaluation of Natural Products. *Pharmaceutics* **2025**, *17*, 1002. [CrossRef]
29. Wu, W.; Yang, L.; Zhang, S.; Hao, D.; Yang, Y.; Li, X.; Wu, G.; Zhang, L.; Luo, Z. Development of a New Characteristic Parameter—Waveform Index of Finger Blood Volume Pulse. *Comput. Assist. Surg.* **2016**, *21*, 6–10. [CrossRef]
30. Johari, J.; Kianmehr, A.; Mustafa, M.; Abubakar, S.; Zandi, K. Antiviral Activity of Baicalein and Quercetin against the Japanese Encephalitis Virus. *Int. J. Mol. Sci.* **2012**, *13*, 16785–16795. [CrossRef]
31. Phadungsombat, J.; Lin, M.Y.-C.; Srimark, N.; Yamanaka, A.; Nakayama, E.E.; Moolasart, V.; Suttha, P.; Shioda, T.; Uttayamakul, S. Emergence of Genotype Cosmopolitan of Dengue Virus Type 2 and Genotype III of Dengue Virus Type 3 in Thailand. *PLoS ONE* **2018**, *13*, e0207220. [CrossRef] [PubMed]
32. Nahmias, Y.; Goldwasser, J.; Casali, M.; Van Poll, D.; Wakita, T.; Chung, R.T.; Yarmush, M.L. Apolipoprotein B–Dependent Hepatitis C Virus Secretion Is Inhibited by the Grapefruit Flavonoid Naringenin. *Hepatology* **2008**, *47*, 1437–1445. [CrossRef]

https://doi.org/10.3390/antiox15010087

33. Li, B.Q.; Fu, T.; Dongyan, Y.; Mikovits, J.A.; Ruscetti, F.W.; Wang, J.M. Flavonoid Baicalin Inhibits HIV-1 Infection at the Level of Viral Entry. *Biochem. Biophys. Res. Commun.* **2000**, *276*, 534–538. [CrossRef]
34. Fabbrini, M.; D'Amico, F.; Barone, M.; Conti, G.; Mengoli, M.; Brigidi, P.; Turroni, S. Polyphenol and Tannin Nutraceuticals and Their Metabolites: How the Human Gut Microbiota Influences Their Properties. *Biomolecules* **2022**, *12*, 875. [CrossRef]
35. Ordoñez, R.; Atarés, L.; Chiralt, A. Biodegradable Active Materials Containing Phenolic Acids for Food Packaging Applications. *Compr. Rev. Food Sci. Food Saf.* **2022**, *21*, 3910–3930. [CrossRef]
36. Czerkas, K.; Olchowik-Grabarek, E.; Łomanowska, M.; Abdulladjanova, N.; Sękowski, S. Antibacterial Activity of Plant Polyphenols Belonging to the Tannins against Streptococcus Mutans—Potential against Dental Caries. *Molecules* **2024**, *29*, 879. [CrossRef] [PubMed]
37. Kaczmarek, B. Tannic Acid with Antiviral and Antibacterial Activity as A Promising Component of Biomaterials—A Minireview. *Materials* **2020**, *13*, 3224. [CrossRef]
38. Cardoso-Gutierrez, E.; Aranda-Aguirre, E.; Robles-Jimenez, L.E.; Castelán-Ortega, O.A.; Chay-Canul, A.J.; Foggi, G.; Angeles-Hernandez, J.C.; Vargas-Bello-Pérez, E.; González-Ronquillo, M. Effect of Tannins from Tropical Plants on Methane Production from Ruminants: A Systematic Review. *Vet. Anim. Sci.* **2021**, *14*, 100214. [CrossRef] [PubMed]
39. Roldan, M.B.; Cousins, G.; Muetzel, S.; Zeller, W.E.; Fraser, K.; Salminen, J.-P.; Blanc, A.; Kaur, R.; Richardson, K.; Maher, D.; et al. Condensed Tannins in White Clover (*Trifolium repens*) Foliar Tissues Expressing the Transcription Factor TaMYB14-1 Bind to Forage Protein and Reduce Ammonia and Methane Emissions in Vitro. *Front. Plant Sci.* **2022**, *12*, 777354. [CrossRef] [PubMed]
40. Tuyen, P.; Xuan, T.; Khang, D.; Ahmad, A.; Quan, N.; Tu Anh, T.; Anh, L.; Minh, T. Phenolic Compositions and Antioxidant Properties in Bark, Flower, Inner Skin, Kernel and Leaf Extracts of *Castanea crenata* Sieb. et Zucc. *Antioxidants* **2017**, *6*, 31. [CrossRef]
41. Molino, S.; Pilar Francino, M.; Ángel Rufián Henares, J. Why Is It Important to Understand the Nature and Chemistry of Tannins to Exploit Their Potential as Nutraceuticals? *Food Res. Int.* **2023**, *173*, 113329. [CrossRef]
42. Lafay, S.; Gil-Izquierdo, A. Bioavailability of Phenolic Acids. *Phytochem. Rev.* **2008**, *7*, 301–311. [CrossRef]
43. Clifford, M.N. Chlorogenic Acids and Other Cinnamates—Nature, Occurrence, Dietary Burden, Absorption and Metabolism. *J. Sci. Food Agric.* **2000**, *80*, 1033–1043. [CrossRef]
44. Caruso, G.; Godos, J.; Privitera, A.; Lanza, G.; Castellano, S.; Chillemi, A.; Bruni, O.; Ferri, R.; Caraci, F.; Grosso, G. Phenolic Acids and Prevention of Cognitive Decline: Polyphenols with a Neuroprotective Role in Cognitive Disorders and Alzheimer's Disease. *Nutrients* **2022**, *14*, 819. [CrossRef]
45. Kumar, N.; Goel, N. Phenolic Acids: Natural Versatile Molecules with Promising Therapeutic Applications. *Biotechnol. Rep.* **2019**, *24*, e00370. [CrossRef]
46. Rashedinia, M.; Rasti Arbabi, Z.; Sabet, R.; Emami, L.; Poustforoosh, A.; Sabahi, Z. Comparison of Protective Effects of Phenolic Acids on Protein Glycation of BSA Supported by In Vitro and Docking Studies. *Biochem. Res. Int.* **2023**, *2023*, 9984618. [CrossRef]
47. Rashmi, H.B.; Negi, P.S. Phenolic Acids from Vegetables: A Review on Processing Stability and Health Benefits. *Food Res. Int.* **2020**, *136*, 109298. [CrossRef]
48. Cui, Q.; Du, R.; Liu, M.; Rong, L. Lignans and Their Derivatives from Plants as Antivirals. *Molecules* **2020**, *25*, 183. [CrossRef]
49. Jang, W.Y.; Kim, M.-Y.; Cho, J.Y. Antioxidant, Anti-Inflammatory, Anti-Menopausal, and Anti-Cancer Effects of Lignans and Their Metabolites. *Int. J. Mol. Sci.* **2022**, *23*, 15482. [CrossRef] [PubMed]
50. Baldi, S.; Tristán Asensi, M.; Pallecchi, M.; Sofi, F.; Bartolucci, G.; Amedei, A. Interplay between Lignans and Gut Microbiota: Nutritional, Functional and Methodological Aspects. *Molecules* **2023**, *28*, 343. [CrossRef] [PubMed]
51. Barker, D. Lignans. *Molecules* **2019**, *24*, 1424. [CrossRef]
52. Guo, Y.; Sun, Q.; Wu, F.; Dai, Y.; Chen, X. Polyphenol-Containing Nanoparticles: Synthesis, Properties, and Therapeutic Delivery. *Adv. Mater.* **2021**, *33*, 2007356. [CrossRef]
53. Williamson, G. The Role of Polyphenols in Modern Nutrition. *Nutr. Bull.* **2017**, *42*, 226–235. [CrossRef]
54. Hügel, H.M.; Jackson, N.; May, B.; Zhang, A.L.; Xue, C.C. Polyphenol Protection and Treatment of Hypertension. *Phytomedicine* **2016**, *23*, 220–231. [CrossRef]
55. Ochmian, I.; Figiel-Kroczyńska, M.; Lachowicz, S. The Quality of Freeze-Dried and Rehydrated Blueberries Depending on Their Size and Preparation for Freeze-Drying. *Acta Univ. Cibiniensis Ser. E Food Technol.* **2020**, *24*, 61–78. [CrossRef]
56. Sandoval-Acuña, C.; Ferreira, J.; Speisky, H. Polyphenols and Mitochondria: An Update on Their Increasingly Emerging ROS-Scavenging Independent Actions. *Arch. Biochem. Biophys.* **2014**, *559*, 75–90. [CrossRef] [PubMed]
57. Megías, C.; Pastor-Cavada, E.; Torres-Fuentes, C.; Girón-Calle, J.; Alaiz, M.; Juan, R.; Pastor, J.; Vioque, J. Chelating, Antioxidant and Antiproliferative Activity of Vicia Sativa Polyphenol Extracts. *Eur. Food Res. Technol.* **2009**, *230*, 353–359. [CrossRef]
58. Na, H.-K.; Surh, Y.-J. Modulation of Nrf2-Mediated Antioxidant and Detoxifying Enzyme Induction by the Green Tea Polyphenol EGCG. *Food Chem. Toxicol.* **2008**, *46*, 1271–1278. [CrossRef] [PubMed]
59. Fraga, C.G.; Oteiza, P.I.; Hid, E.J.; Galleano, M. (Poly)Phenols and the Regulation of NADPH Oxidases. *Redox Biol.* **2023**, *67*, 102927. [CrossRef]

https://doi.org/10.3390/antiox15010087

60. Valgimigli, L. Lipid Peroxidation and Antioxidant Protection. *Biomolecules* **2023**, *13*, 1291. [CrossRef]
61. Scalbert, A.; Johnson, I.T.; Saltmarsh, M. Polyphenols: Antioxidants and Beyond. *Am. J. Clin. Nutr.* **2005**, *81*, 215S–217S. [CrossRef]
62. Rudrapal, M.; Khairnar, S.J.; Khan, J.; Dukhyil, A.B.; Ansari, M.A.; Alomary, M.N.; Alshabrmi, F.M.; Palai, S.; Deb, P.K.; Devi, R. Dietary Polyphenols and Their Role in Oxidative Stress-Induced Human Diseases: Insights into Protective Effects, Antioxidant Potentials and Mechanism(s) of Action. *Front. Pharmacol.* **2022**, *13*, 806470. [CrossRef]
63. Apel, K.; Hirt, H. REACTIVE OXYGEN SPECIES: Metabolism, Oxidative Stress, and Signal Transduction. *Annu. Rev. Plant Biol.* **2004**, *55*, 373–399. [CrossRef]
64. Gebicki, J.M.; Nauser, T. Fast Antioxidant Reaction of Polyphenols and Their Metabolites. *Antioxidants* **2021**, *10*, 1297. [CrossRef]
65. Obrenovich, M.E.; Nair, N.G.; Beyaz, A.; Aliev, G.; Reddy, V.P. The Role of Polyphenolic Antioxidants in Health, Disease, and Aging. *Rejuvenation Res.* **2010**, *13*, 631–643. [CrossRef] [PubMed]
66. Chiva-Blanch, G.; Visioli, F. Polyphenols and Health: Moving beyond Antioxidants. *J. Berry Res.* **2012**, *2*, 63–71. [CrossRef]
67. Bucciantini, M.; Leri, M.; Nardiello, P.; Casamenti, F.; Stefani, M. Olive Polyphenols: Antioxidant and Anti-Inflammatory Properties. *Antioxidants* **2021**, *10*, 1044. [CrossRef]
68. Mazzocchi, A.; Leone, L.; Agostoni, C.; Pali-Schöll, I. The Secrets of the Mediterranean Diet. Does [Only] Olive Oil Matter? *Nutrients* **2019**, *11*, 2941. [CrossRef] [PubMed]
69. Fabiani, R.; Rosignoli, P.; De Bartolomeo, A.; Fuccelli, R.; Servili, M.; Montedoro, G.F.; Morozzi, G. Oxidative DNA Damage Is Prevented by Extracts of Olive Oil, Hydroxytyrosol, and Other Olive Phenolic Compounds in Human Blood Mononuclear Cells and HL60 Cells. *J. Nutr.* **2008**, *138*, 1411–1416. [CrossRef]
70. Cicerale, S.; Lucas, L.; Keast, R. Biological Activities of Phenolic Compounds Present in Virgin Olive Oil. *Int. J. Mol. Sci.* **2010**, *11*, 458–479. [CrossRef] [PubMed]
71. Cicerale, S.; Lucas, L.; Keast, R. Antimicrobial, Antioxidant and Anti-Inflammatory Phenolic Activities in Extra Virgin Olive Oil. *Curr. Opin. Biotechnol.* **2012**, *23*, 129–135. [CrossRef]
72. Liu, W.; Wang, Z.; Xia, Y.; Kuang, H.; Liu, S.; Li, L.; Tang, C.; Yin, D. The Balance of Apoptosis and Autophagy via Regulation of the AMPK Signal Pathway in Aging Rat Striatum during Regular Aerobic Exercise. *Exp. Gerontol.* **2019**, *124*, 110647. [CrossRef]
73. Erdman, V.V.; Nasibullin, T.R.; Tuktarova, I.A.; Somova, R.S.; Mustafina, O.E. Association Analysis of Polymorphic Gene Variants in the JAK/STAT Signaling Pathway with Aging and Longevity. *Russ. J. Genet.* **2019**, *55*, 728–737. [CrossRef]
74. Harris, J.; Lang, T.; Thomas, J.P.W.; Sukkar, M.B.; Nabar, N.R.; Kehrl, J.H. Autophagy and Inflammasomes. *Mol. Immunol.* **2017**, *86*, 10–15. [CrossRef]
75. Bostancıklıoğlu, M. An Update on the Interactions between Alzheimer's Disease, Autophagy and Inflammation. *Gene* **2019**, *705*, 157–166. [CrossRef] [PubMed]
76. Hassanpour, S.H.; Doroudi, A. Review of the Antioxidant Potential of Flavonoids as a Subgroup of Polyphenols and Partial Substitute for Synthetic Antioxidants. *Avicenna J. Phytomed.* **2023**, *13*, 354–376. [CrossRef] [PubMed]
77. Gutiérrez-del-Río, I.; López-Ibáñez, S.; Magadán-Corpas, P.; Fernández-Calleja, L.; Pérez-Valero, Á.; Tuñón-Granda, M.; Miguélez, E.M.; Villar, C.J.; Lombó, F. Terpenoids and Polyphenols as Natural Antioxidant Agents in Food Preservation. *Antioxidants* **2021**, *10*, 1264. [CrossRef] [PubMed]
78. Bešlo, D.; Golubić, N.; Rastija, V.; Agić, D.; Karnaš, M.; Šubarić, D.; Lučić, B. Antioxidant Activity, Metabolism, and Bioavailability of Polyphenols in the Diet of Animals. *Antioxidants* **2023**, *12*, 1141. [CrossRef]
79. Cesa, S.; Carradori, S.; Bellagamba, G.; Locatelli, M.; Casadei, M.A.; Masci, A.; Paolicelli, P. Evaluation of Processing Effects on Anthocyanin Content and Colour Modifications of Blueberry (*Vaccinium* spp.) Extracts: Comparison Between HPLC-DAD and CIELAB Analyses. *Food Chem.* **2017**, *232*, 114–123. [CrossRef]
80. Da Silva, F.L.; Escribano-Bailón, M.T.; Pérez Alonso, J.J.; Rivas-Gonzalo, J.C.; Santos-Buelga, C. Anthocyanin Pigments in Strawberry. *LWT Food Sci. Technol.* **2007**, *40*, 374–382. [CrossRef]
81. Mustafa, A.M.; Angeloni, S.; Abouelenein, D.; Acquaticci, L.; Xiao, J.; Sagratini, G.; Maggi, F.; Vittori, S.; Caprioli, G. A New HPLC-MS/MS Method for the Simultaneous Determination of 36 Polyphenols in Blueberry, Strawberry and Their Commercial Products and Determination of Antioxidant Activity. *Food Chem.* **2022**, *367*, 130743. [CrossRef]
82. Zhang, W.; Shen, Y.; Li, Z.; Xie, X.; Gong, E.S.; Tian, J.; Si, X.; Wang, Y.; Gao, N.; Shu, C.; et al. Effects of High Hydrostatic Pressure and Thermal Processing on Anthocyanin Content, Polyphenol Oxidase and β-Glucosidase Activities, Color, and Antioxidant Activities of Blueberry (*Vaccinium* spp.) Puree. *Food Chem.* **2021**, *342*, 128564. [CrossRef]
83. Lachowicz-Wiśniewska, S.; Pratap-Singh, A.; Ochmian, I.; Kapusta, I.; Kotowska, A.; Pluta, S. Biodiversity in Nutrients and Biological Activities of 14 Highbush Blueberry (*Vaccinium corymbosum* L.) Cultivars. *Sci. Rep.* **2024**, *14*, 22063. [CrossRef]
84. Saracila, M.; Untea, A.E.; Oancea, A.G.; Varzaru, I.; Vlaicu, P.A. Comparative Analysis of Black Chokeberry (*Aronia melanocarpa* L.) Fruit, Leaves, and Pomace for Their Phytochemical Composition, Antioxidant Potential, and Polyphenol Bioaccessibility. *Foods* **2024**, *13*, 1856. [CrossRef]
85. Dobros, N.; Zielińska, A.; Siudem, P.; Zawada, K.D.; Paradowska, K. Profile of Bioactive Components and Antioxidant Activity of *Aronia melanocarpa* Fruits at Various Stages of Their Growth, Using Chemometric Methods. *Antioxidants* **2024**, *13*, 462. [CrossRef]

https://doi.org/10.3390/antiox15010087

86. Halim, M.A.; Kanan, K.A.; Nahar, T.; Rahman, M.J.; Ahmed, K.S.; Hossain, H.; Mozumder, N.H.M.R.; Ahmed, M. Metabolic Profiling of Phenolics of the Extracts from the Various Parts of Blackberry Plant (*Syzygium cumini* L.) and Their Antioxidant Activities. *LWT* **2022**, *167*, 113813. [CrossRef]
87. Varzaru, I.; Oancea, A.G.; Vlaicu, P.A.; Saracila, M.; Untea, A.E. Exploring the Antioxidant Potential of Blackberry and Raspberry Leaves: Phytochemical Analysis, Scavenging Activity, and In Vitro Polyphenol Bioaccessibility. *Antioxidants* **2023**, *12*, 2125. [CrossRef] [PubMed]
88. Sánchez-Velázquez, O.A.; Mulero, M.; Cuevas-Rodríguez, E.O.; Mondor, M.; Arcand, Y.; Hernández-Álvarez, A.J. *In Vitro* Gastrointestinal Digestion Impact on Stability, Bioaccessibility and Antioxidant Activity of Polyphenols from Wild and Commercial Blackberries (*Rubus* spp.). *Food Funct.* **2021**, *12*, 7358–7378. [CrossRef] [PubMed]
89. Muzolf-Panek, M.; Waśkiewicz, A. Relationship between Phenolic Compounds, Antioxidant Activity and Color Parameters of Red Table Grape Skins Using Linear Ordering Analysis. *Appl. Sci.* **2022**, *12*, 6146. [CrossRef]
90. Kupe, M.; Karatas, N.; Unal, M.S.; Ercisli, S.; Baron, M.; Sochor, J. Phenolic Composition and Antioxidant Activity of Peel, Pulp and Seed Extracts of Different Clones of the Turkish Grape Cultivar 'Karaerik'. *Plants* **2021**, *10*, 2154. [CrossRef]
91. Bocsan, I.C.; Măgureanu, D.C.; Pop, R.M.; Levai, A.M.; Macovei, Ș.O.; Pătrașca, I.M.; Chedea, V.S.; Buzoianu, A.D. Antioxidant and Anti-Inflammatory Actions of Polyphenols from Red and White Grape Pomace in Ischemic Heart Diseases. *Biomedicines* **2022**, *10*, 2337. [CrossRef]
92. Yu, J.; Li, W.; You, B.; Yang, S.; Xian, W.; Deng, Y.; Huang, W.; Yang, R. Phenolic Profiles, Bioaccessibility and Antioxidant Activity of Plum (*Prunus salicina* Lindl). *Food Res. Int.* **2021**, *143*, 110300. [CrossRef]
93. Johnson, J.B.; Hoyos, B.E.; Mani, J.S.; Reynolds, M.; Altvater, J.; Naiker, M. Identification of Phenolics Responsible for the High Antioxidant Activity in Burdekin Plum (*Pleiogynium timoriense*) Fruit. *Food Chem. Adv.* **2022**, *1*, 100081. [CrossRef]
94. Trang, A.; Khandhar, P.B. Physiology, Acetylcholinesterase. In *StatPearls*; StatPearls Publishing: Treasure Island, FL, USA, 2025.
95. Abd El-Aziz, N.M.; Eldin Awad, O.M.; Shehata, M.G.; El-Sohaimy, S.A. Antioxidant and Anti-Acetylcholinesterase Potential of Artichoke Phenolic Compounds. *Food Biosci.* **2021**, *41*, 101006. [CrossRef]
96. Iglesias-Carres, L.; Bruno, A.; D'Antuono, I.; Linsalata, V.; Cardinali, A.; Neilson, A.P. In Vitro Evidences of the Globe Artichoke Antioxidant, Cardioprotective and Neuroprotective Effects. *J. Funct. Foods* **2023**, *107*, 105674. [CrossRef]
97. Phan, T.-H.-A.; Nguyen, T.-P.; Tran, T.-T.-A. Extracting, Evaluating Biological Activities of Phenolic Compounds from Yellow Onion Peels (*Allium cepa* L.) and Their Applicability for Fish Preservation. *IOP Conf. Ser. Earth Environ. Sci.* **2021**, *947*, 012042. [CrossRef]
98. Oyawoye, O.M.; Olotu, T.M.; Nzekwe, S.C.; Idowu, J.A.; Abdullahi, T.A.; Babatunde, S.O.; Ridwan, I.A.; Batiha, G.E.; Idowu, N.; Alorabi, M.; et al. Antioxidant Potential and Antibacterial Activities of *Allium cepa* (Onion) and *Allium sativum* (Garlic) against the Multidrug Resistance Bacteria. *Bull. Natl. Res. Cent.* **2022**, *46*, 214. [CrossRef]
99. Demir, R.; Sarıtaş, S.; Bechelany, M.; Karav, S. Lactoferrin: Properties and Potential Uses in the Food Industry. *Int. J. Mol. Sci.* **2025**, *26*, 1404. [CrossRef] [PubMed]
100. Ma, H.; Zhang, H.; Abubaker, M.A.; Al-Wraikat, M.; Li, L.; Liu, Y. Goat Lactoferrin–Pterostilbene Complexes as Novel Edible Functional Proteins with Enhanced Ultraviolet Stabilization and Antioxidant Properties. *Food Hydrocoll.* **2025**, *158*, 110538. [CrossRef]
101. Davidova, S.; Galabov, A.S.; Satchanska, G. Antibacterial, Antifungal, Antiviral Activity, and Mechanisms of Action of Plant Polyphenols. *Microorganisms* **2024**, *12*, 2502. [CrossRef]
102. Song, J.-M.; Lee, K.-H.; Seong, B.-L. Antiviral Effect of Catechins in Green Tea on Influenza Virus. *Antivir. Res.* **2005**, *68*, 66–74. [CrossRef]
103. Manjunathan, R.; Periyaswami, V.; Mitra, K.; Rosita, A.S.; Pandya, M.; Selvaraj, J.; Ravi, L.; Devarajan, N.; Doble, M. Molecular Docking Analysis Reveals the Functional Inhibitory Effect of Genistein and Quercetin on TMPRSS2: SARS-COV-2 Cell Entry Facilitator Spike Protein. *BMC Bioinform.* **2022**, *23*, 180. [CrossRef] [PubMed]
104. Li, T.; Zhu, J.; Yu, Q.; Zhu, Y.; Wu, C.; Zheng, X.; Chen, N.; Pei, P.; Yang, K.; Wang, K.; et al. Dietary Flavonoid Quercetin Supplement Promotes Antiviral Innate Responses Against Vesicular Stomatitis Virus Infection by Reshaping the Bacteriome and Host Metabolome in Mice. *Mol. Nutr. Food Res.* **2024**, *68*, 2300898. [CrossRef] [PubMed]
105. Zainal, N.; Chang, C.-P.; Cheng, Y.-L.; Wu, Y.-W.; Anderson, R.; Wan, S.-W.; Chen, C.-L.; Ho, T.-S.; AbuBakar, S.; Lin, Y.-S. Resveratrol Treatment Reveals a Novel Role for HMGB1 in Regulation of the Type 1 Interferon Response in Dengue Virus Infection. *Sci. Rep.* **2017**, *7*, 42998. [CrossRef]
106. Luo, R.; Li, S.; Yang, C.; Tang, B.; Li, L.; Luo, C. Curcumin Inhibits the Development of Pancreatic Cancer by Targeting the Circ_0079440/miR-522-3p/EIF4A1 Pathway. *Cell Biochem. Biophys.* **2024**, *83*, 377–390. [CrossRef]
107. Peng, Y.; Zheng, J.; Zhou, Y.; Li, J. Characterization of a Novel Curcumin Analog P1 as Potent Inhibitor of the NF-κB Signaling Pathway with Distinct Mechanisms. *Acta Pharmacol. Sin.* **2013**, *34*, 939–950. [CrossRef] [PubMed]
108. Iqbal, N.; Poór, P. Plant Protection by Tannins Depends on Defence-Related Phytohormones. *J. Plant Growth Regul.* **2025**, *44*, 22–39. [CrossRef]

https://doi.org/10.3390/antiox15010087

109. Gescher, K.; Kühn, J.; Lorentzen, E.; Hafezi, W.; Derksen, A.; Deters, A.; Hensel, A. Proanthocyanidin-Enriched Extract from *Myrothamnus flabellifolia* Welw. Exerts Antiviral Activity against Herpes Simplex Virus Type 1 by Inhibition of Viral Adsorption and Penetration. *J. Ethnopharmacol.* **2011**, *134*, 468–474. [CrossRef]
110. Hu, Y.; Li, Y.; Zhu, H.; Wang, D.; Zhou, J.; Li, B. Screening and Comparative Study of Four Anti-PEDV Candidate Drugs in Vitro. *Anal. Biochem.* **2025**, *701*, 115804. [CrossRef]
111. De Angelis, M.; Della-Morte, D.; Buttinelli, G.; Di Martino, A.; Pacifici, F.; Checconi, P.; Ambrosio, L.; Stefanelli, P.; Palamara, A.T.; Garaci, E.; et al. Protective Role of Combined Polyphenols and Micronutrients against Influenza A Virus and SARS-CoV-2 Infection In Vitro. *Biomedicines* **2021**, *9*, 1721. [CrossRef]
112. Mandal, M.K.; Domb, A.J. Antimicrobial Activities of Natural Bioactive Polyphenols. *Pharmaceutics* **2024**, *16*, 718. [CrossRef]
113. Rasaily, S.; Haldipur, A.C.; Srividya, N. LC-Q-TOF-MS Phenolic Profiling, Data Mining for Antiviral Compounds, and in Silico Analysis against SARS-CoV-2 in Heritage-Pigmented Rice Metabolome. *Food Biosci.* **2025**, *63*, 105613. [CrossRef]
114. Capasso, L.; De Masi, L.; Sirignano, C.; Maresca, V.; Basile, A.; Nebbioso, A.; Rigano, D.; Bontempo, P. Epigallocatechin Gallate (EGCG): Pharmacological Properties, Biological Activities and Therapeutic Potential. *Molecules* **2025**, *30*, 654. [CrossRef] [PubMed]
115. Eggers, M.; Jungke, P.; Wolkinger, V.; Bauer, R.; Kessler, U.; Frank, B. Antiviral Activity of Plant Juices and Green Tea against SARS-CoV-2 and Influenza Virus. *Phytother. Res.* **2022**, *36*, 2109–2115. [CrossRef]
116. Kumar, D.; Sharma, N.; Aarthy, M.; Singh, S.K.; Giri, R. Mechanistic Insights into Zika Virus NS3 Helicase Inhibition by Epigallocatechin-3-Gallate. *ACS Omega* **2020**, *5*, 11217–11226. [CrossRef]
117. Da Conceição, P.J.P.; Ayusso, G.M.; Carvalho, T.; Duarte Lima, M.L.; Marinho, M.D.S.; Moraes, F.R.; Galán-Jurado, P.E.; González-Santamaría, J.; Bittar, C.; Zhang, B.; et al. In Vitro Evaluation of the Antiviral Activity of Polyphenol (−)-Epigallocatechin-3-Gallate (EGCG) Against Mayaro Virus. *Viruses* **2025**, *17*, 258. [CrossRef] [PubMed]
118. Bibi, S.; Nisar, M.; Rafique, S.; Waqas, M.; Zahoor, M.; Idrees, M.; Nazir, N.; Ihsan, M.; Salmen, S.H.; Alharbi, S.A.; et al. Harnessing Nature's Gifts: *Salix Nigra* and Its Potential for Combating Hepatitis C Virus (HCV). *ACS Omega* **2023**, *8*, 42987–42999. [CrossRef]
119. Hassan, S.T.S.; Šudomová, M.; Mazurakova, A.; Kubatka, P. Insights into Antiviral Properties and Molecular Mechanisms of Non-Flavonoid Polyphenols against Human Herpesviruses. *Int. J. Mol. Sci.* **2022**, *23*, 13891. [CrossRef] [PubMed]
120. Prabhu, S.; Kalaimathi, K.; Thiruvengadam, M.; Ayyanar, M.; Shine, K.; Amalraj, S.; Ceasar, S.A.; Priya, S.P.; Prakash, N. Antiviral Mechanisms of Dietary Polyphenols: Recent Developments as Antiviral Agents and Future Prospects in Combating Nipah Virus. *Phytochem. Rev.* **2024**. [CrossRef]
121. Chojnacka, K.; Skrzypczak, D.; Izydorczyk, G.; Mikula, K.; Szopa, D.; Witek-Krowiak, A. Antiviral Properties of Polyphenols from Plants. *Foods* **2021**, *10*, 2277. [CrossRef]
122. Srivastava, R.; Dubey, N.K.; Sharma, M.; Kharkwal, H.; Bajpai, R.; Srivastava, R. Boosting the Human Antiviral Response in Conjunction with Natural Plant Products. *Front. Nat. Prod.* **2025**, *3*, 1470639. [CrossRef]
123. Agrawal, P.K.; Agrawal, C.; Blunden, G. Quercetin: Antiviral Significance and Possible COVID-19 Integrative Considerations. *Nat. Prod. Commun.* **2020**, *15*, 1934578X20976293. [CrossRef]
124. Sharapov, A.D.; Fatykhov, R.F.; Khalymbadzha, I.A.; Zyryanov, G.V.; Chupakhin, O.N.; Tsurkan, M.V. Plant Coumarins with Anti-HIV Activity: Isolation and Mechanisms of Action. *Int. J. Mol. Sci.* **2023**, *24*, 2839. [CrossRef] [PubMed]
125. Garozzo, A.; Timpanaro, R.; Stivala, A.; Bisignano, G.; Castro, A. Activity of *Melaleuca alternifolia* (Tea Tree) Oil on Influenza Virus A/PR/8: Study on the Mechanism of Action. *Antivir. Res.* **2011**, *89*, 83–88. [CrossRef]
126. Mieres-Castro, D.; Ahmar, S.; Shabbir, R.; Mora-Poblete, F. Antiviral Activities of Eucalyptus Essential Oils: Their Effectiveness as Therapeutic Targets against Human Viruses. *Pharmaceuticals* **2021**, *14*, 1210. [CrossRef] [PubMed]
127. Garozzo, A.; Timpanaro, R.; Bisignano, B.; Furneri, P.M.; Bisignano, G.; Castro, A. In Vitro Antiviral Activity of *Melaleuca Alternifolia* Essential Oil. *Lett. Appl. Microbiol.* **2009**, *49*, 806–808. [CrossRef]
128. Feng, A.; Si, T.; Du, B.; Sun, Q.; Yang, Y.; Rong, R. Antiviral Potential of Essential Oils and Their Plant Sources, Delivery Methods, and Drug Development Strategies with Special Focus on Their Mechanisms: A Comprehensive Review. *Phytochem. Rev.* **2025**, *24*, 1667–1689. [CrossRef]
129. Elshafie, H.S.; Camele, I.; Mohamed, A.A. A Comprehensive Review on the Biological, Agricultural and Pharmaceutical Properties of Secondary Metabolites Based-Plant Origin. *Int. J. Mol. Sci.* **2023**, *24*, 3266. [CrossRef]
130. Kreiser, T.; Zaguri, D.; Sachdeva, S.; Zamostiano, R.; Mograbi, J.; Segal, D.; Bacharach, E.; Gazit, E. Inhibition of Respiratory RNA Viruses by a Composition of Ionophoric Polyphenols with Metal Ions. *Pharmaceuticals* **2022**, *15*, 377. [CrossRef] [PubMed]
131. Zhang, X.; Yang, J.; Liu, F.; Mo, M.; Farooq, M.; Li, J.; Yao, C.; Wei, W. Antiviral Activity of *Morus alba* L. Extract against Pseudorabies Virus. *J. Ethnopharmacol.* **2025**, *336*, 118719. [CrossRef]
132. Quirós-Fallas, M.I.; Vargas-Huertas, F.; Quesada-Mora, S.; Azofeifa-Cordero, G.; Wilhelm-Romero, K.; Vásquez-Castro, F.; Alvarado-Corella, D.; Sánchez-Kopper, A.; Navarro-Hoyos, M. Polyphenolic HRMS Characterization, Contents and Antioxidant Activity of *Curcuma longa* Rhizomes from Costa Rica. *Antioxidants* **2022**, *11*, 620. [CrossRef]
133. Lei, J.; Kusov, Y.; Hilgenfeld, R. Nsp3 of Coronaviruses: Structures and Functions of a Large Multi-Domain Protein. *Antivir. Res.* **2018**, *149*, 58–74. [CrossRef]

134. Li, C.-W.; Chao, T.-L.; Lai, C.-L.; Lin, C.-C.; Pan, M.Y.-C.; Cheng, C.-L.; Kuo, C.-J.; Wang, L.H.-C.; Chang, S.-Y.; Liang, P.-H. Systematic Studies on the Anti-SARS-CoV-2 Mechanisms of Tea Polyphenol-Related Natural Products. *ACS Omega* **2024**, *9*, 23984–23997. [CrossRef]
135. Tang, C.; Carrera Montoya, J.; Fritzlar, S.; Flavel, M.; Londrigan, S.L.; Mackenzie, J.M. Polyphenol Rich Sugarcane Extract (PRSE) Has Potential Antiviral Activity against Influenza A Virus in Vitro. *Virology* **2024**, *590*, 109969. [CrossRef] [PubMed]
136. Tang, C.; Flavel, M.; Londrigan, S.L.; Mackenzie, J.M. Polyphenol Rich Sugarcane Extract Restricts Select Respiratory Viruses Depending on Their Mode of Entry. *Virology* **2025**, *606*, 110500. [CrossRef] [PubMed]
137. Mileto, D.; Riva, A.; Cutrera, M.; Moschese, D.; Mancon, A.; Meroni, L.; Giacomelli, A.; Bestetti, G.; Rizzardini, G.; Gismondo, M.R.; et al. New Challenges in Human Monkeypox Outside Africa: A Review and Case Report from Italy. *Travel Med. Infect. Dis.* **2022**, *49*, 102386. [CrossRef]
138. Rout, M.; Mishra, S.; Dey, S.; Singh, M.K.; Dehury, B.; Pati, S. Exploiting the Potential of Natural Polyphenols as Antivirals against Monkeypox Envelope Protein F13 Using Machine Learning and All-Atoms MD Simulations. *Comput. Biol. Med.* **2023**, *162*, 107116. [CrossRef] [PubMed]
139. Manzoor, Z.; Sajad, A.; Qadiri, S.S.N.; Shah, F.A.; Dar, S.A.; Mandu, S.M. Polyphenols as Antiviral Agents: Assessing Their Potential Usage and Benefits in Aquaculture. *Aquac. Int.* **2025**, *33*, 106. [CrossRef]
140. Sharma, S.; Ghosh, P.; Kar, C.; Ghosh, R. Anticipating Viral Challenges: A Perspective on Phytochemicals against Existing and Emerging Viruses. *Anti-Infect. Agents* **2025**, *23*, e22113525312781. [CrossRef]
141. Ghosh, S.; Basu, S.; Anbarasu, A.; Ramaiah, S. A Comprehensive Review of Antimicrobial Agents Against Clinically Important Bacterial Pathogens: Prospects for Phytochemicals. *Phytother. Res.* **2025**, *39*, 138–161. [CrossRef]
142. Zhong, W.; Tang, M.; Xie, Y.; Huang, X.; Liu, Y. Tea Polyphenols Inhibit the Activity and Toxicity of *Staphylococcus aureus* by Destroying Cell Membranes and Accumulating Reactive Oxygen Species. *Foodborne Pathog. Dis.* **2023**, *20*, 294–302. [CrossRef]
143. Wang, L.; Li, T.; Wu, C.; Fan, G.; Zhou, D.; Li, X. Unlocking the Potential of Plant Polyphenols: Advances in Extraction, Antibacterial Mechanisms, and Future Applications. *Food Sci. Biotechnol.* **2025**, *34*, 1235–1259. [CrossRef]
144. Bae, J.-Y.; Seo, Y.-H.; Oh, S.-W. Antibacterial Activities of Polyphenols against Foodborne Pathogens and Their Application as Antibacterial Agents. *Food Sci. Biotechnol.* **2022**, *31*, 985–997. [CrossRef]
145. Nassarawa, S.S.; Nayik, G.A.; Gupta, S.D.; Areche, F.O.; Jagdale, Y.D.; Ansari, M.J.; Hemeg, H.A.; AL-Farga, A.; Alotaibi, S.S. Chemical Aspects of Polyphenol-Protein Interactions and Their Antibacterial Activity. *Crit. Rev. Food Sci. Nutr.* **2023**, *63*, 9482–9505. [CrossRef]
146. Makarewicz, M.; Drożdż, I.; Tarko, T.; Duda-Chodak, A. The Interactions between Polyphenols and Microorganisms, Especially Gut Microbiota. *Antioxidants* **2021**, *10*, 188. [CrossRef]
147. Slobodníková, L.; Fialová, S.; Rendeková, K.; Kováč, J.; Mučaji, P. Antibiofilm Activity of Plant Polyphenols. *Molecules* **2016**, *21*, 1717. [CrossRef]
148. Bernal-Gallardo, J.O.; Molina-Torres, J.; Angoa-Pérez, M.V.; Cárdenas-Valdovinos, J.G.; García-Ruíz, I.; Ceja-Díaz, J.A.; Mena-Violante, H.G. Phenolic Compound Content and the Antioxidant and Antimicrobial Activity of Wild Blueberries (*Vaccinium stenophyllum* Steud.) Fruits Extracts during Ripening. *Horticulturae* **2021**, *8*, 15. [CrossRef]
149. Li, W.; Yang, R.; Ying, D.; Yu, J.; Sanguansri, L.; Augustin, M.A. Analysis of Polyphenols in Apple Pomace: A Comparative Study of Different Extraction and Hydrolysis Procedures. *Ind. Crops Prod.* **2020**, *147*, 112250. [CrossRef]
150. Geana, E.-I.; Ciucure, C.T.; Niculescu, V.-C.; Marinas, I.C.; Pircalabioru, G.G.; Dutu, D.; Trusca, R.; Oprea, O.-C.; Ficai, A.; Andronescu, E. Valorization of Apple Pomace by Obtaining Some Bioactive Ingredients with Antioxidant, Antimicrobial and Prebiotic Activities. *Food Bioprod. Process.* **2025**, *150*, 182–197. [CrossRef]
151. Tienaho, J.; Liimatainen, J.; Myllymäki, L.; Kaipanen, K.; Tagliavento, L.; Ruuttunen, K.; Rudolfsson, M.; Karonen, M.; Marjomäki, V.; Hagerman, A.E.; et al. Pilot Scale Hydrodynamic Cavitation and Hot-Water Extraction of Norway Spruce Bark Yield Antimicrobial and Polyphenol-Rich Fractions. *Sep. Purif. Technol.* **2025**, *360*, 130925. [CrossRef]
152. Sun, S.; Huang, S.; Shi, Y.; Shao, Y.; Qiu, J.; Sedjoah, R.-C.A.-A.; Yan, Z.; Ding, L.; Zou, D.; Xin, Z. Extraction, Isolation, Characterization and Antimicrobial Activities of Non-Extractable Polyphenols from Pomegranate Peel. *Food Chem.* **2021**, *351*, 129232. [CrossRef] [PubMed]
153. Feng, Y.; Lin, J.; He, G.; Liang, L.; Liu, Q.; Yan, J.; Yao, Q. Compositions and Biological Activities of Pomegranate Peel Polyphenols Extracted by Different Solvents. *Molecules* **2022**, *27*, 4796. [CrossRef]
154. Ruan, J.-H.; Li, J.; Adili, G.; Sun, G.-Y.; Abuduaini, M.; Abdulla, R.; Maiwulanjiang, M.; Aisa, H.A. Phenolic Compounds and Bioactivities from Pomegranate (*Punica granatum* L.) Peels. *J. Agric. Food Chem.* **2022**, *70*, 3678–3686. [CrossRef]
155. Morgene, M.F.; Botelho-Nevers, E.; Grattard, F.; Pillet, S.; Berthelot, P.; Pozzetto, B.; Verhoeven, P.O. *Staphylococcus aureus* Colonization and Non-Influenza Respiratory Viruses: Interactions and Synergism Mechanisms. *Virulence* **2018**, *9*, 1354–1363. [CrossRef]

https://doi.org/10.3390/antiox15010087

156. Zhang, H.; Wang, M.; Yu, G.; Pu, J.; Tian, K.; Tang, X.; Du, Y.; Wu, H.; Hu, J.; Luo, X.; et al. Comparative Analysis of the Phenolic Contents and Antioxidant Activities of Different Parts of Two Pomegranate (*Punica granatum* L.) Cultivars: 'Tunisia' and 'Qingpi'. *Front. Plant Sci.* **2023**, *14*, 1265018. [CrossRef]
157. Turlewicz-Podbielska, H.; Augustyniak, A.; Pomorska-Mól, M. Viral Co-Infections of the Porcine Respiratory Tract: Insight into the Local Cytokine Response. *Med. Weter.* **2022**, *78*, 212–221. [CrossRef]
158. Kumar, N.; Pratibha; Neeraj; Sami, R.; Khojah, E.; Aljahani, A.H.; Al-Mushhin, A.A.M. Effects of Drying Methods and Solvent Extraction on Quantification of Major Bioactive Compounds in Pomegranate Peel Waste Using HPLC. *Sci. Rep.* **2022**, *12*, 8000. [CrossRef] [PubMed]
159. Wu, W.; Jiang, S.; Liu, M.; Tian, S. Simultaneous Process Optimization of Ultrasound-Assisted Extraction of Polyphenols and Ellagic Acid from Pomegranate (*Punica granatum* L.) Flowers and Its Biological Activities. *Ultrason. Sonochem.* **2021**, *80*, 105833. [CrossRef] [PubMed]
160. Sahlabgi, A.; Lupuliasa, D.; Stanciu, G.; Lupșor, S.; Vlaia, L.L.; Rotariu, R.; Predescu, N.C.; Rădulescu, C.; Olteanu, R.-L.; Stănescu, S.-G.; et al. The Development and Comparative Evaluation of Rosemary Hydroalcoholic Macerate-Based Dermatocosmetic Preparations: A Study on Antioxidant, Antimicrobial, and Anti-Inflammatory Properties. *Gels* **2025**, *11*, 149. [CrossRef] [PubMed]
161. Munggari, I.P.; Kurnia, D.; Deawati, Y.; Julaeha, E. Current Research of Phytochemical, Medicinal and Non-Medicinal Uses of *Uncaria gambir* Roxb.: A Review. *Molecules* **2022**, *27*, 6551. [CrossRef]
162. Alkufeidy, R.M.; Ameer Altuwijri, L.; Aldosari, N.S.; Alsakabi, N.; Dawoud, T.M. Antimicrobial and Synergistic Properties of Green Tea Catechins against Microbial Pathogens. *J. King Saud Univ. Sci.* **2024**, *36*, 103277. [CrossRef]
163. Rodríguez-Melcón, C.; Alonso-Calleja, C.; García-Fernández, C.; Carballo, J.; Capita, R. Minimum Inhibitory Concentration (MIC) and Minimum Bactericidal Concentration (MBC) for Twelve Antimicrobials (Biocides and Antibiotics) in Eight Strains of *Listeria monocytogenes*. *Biology* **2021**, *11*, 46. [CrossRef]
164. Jubair, N.; R., M.; Fatima, A.; Mahdi, Y.K.; Abdullah, N.H. Evaluation of Catechin Synergistic and Antibacterial Efficacy on Biofilm Formation and acrA Gene Expression of Uropathogenic *E. coli* Clinical Isolates. *Antibiotics* **2022**, *11*, 1223. [CrossRef] [PubMed]
165. Erttmann, S.F.; Gekara, N.O. Hydrogen Peroxide Release by Bacteria Suppresses Inflammasome-Dependent Innate Immunity. *Nat. Commun.* **2019**, *10*, 3493. [CrossRef]
166. Ayub, A.; Cheong, Y.K.; Castro, J.C.; Cumberlege, O.; Chrysanthou, A. Use of Hydrogen Peroxide Vapour for Microbiological Disinfection in Hospital Environments: A Review. *Bioengineering* **2024**, *11*, 205. [CrossRef]
167. Xu, W.; Lin, Z.; Cortez-Jugo, C.; Qiao, G.G.; Caruso, F. Antimicrobial Phenolic Materials: From Assembly to Function. *Angew. Chem. Int. Ed.* **2025**, *64*, e202423654. [CrossRef]
168. Wang, L.; Li, T.; Wu, C.; Shi, J.; Fan, G.; Zhou, D.; Li, X. Inhibitory Effect of Polyphenol Extract of *Osmanthus fragrans* on *Alternaria alternata* and Development of a Novel Antimicrobial Film. *Food Biosci.* **2025**, *64*, 105882. [CrossRef]
169. Prabhu, S.; Molath, A.; Choksi, H.; Kumar, S.; Mehra, R. Classifications of Polyphenols and Their Potential Application in Human Health and Diseases. *Int. J. Physiol. Nutr. Phys. Educ.* **2021**, *6*, 293–301. [CrossRef]
170. Manso, T.; Lores, M.; De Miguel, T. Antimicrobial Activity of Polyphenols and Natural Polyphenolic Extracts on Clinical Isolates. *Antibiotics* **2021**, *11*, 46. [CrossRef] [PubMed]
171. Lund, M.N. Reactions of Plant Polyphenols in Foods: Impact of Molecular Structure. *Trends Food Sci. Technol.* **2021**, *112*, 241–251. [CrossRef]
172. De Oliveira Souza, G.G.; Gonçalves Castro, J.W.; Nascimento, L.L.L.; Inácio Da Silva, M.; Duarte Leite, D.O.; Garcia Santos, G.J.; Janaine Camilo, C.; Alencar De Menezes, I.R.; Martins Da Costa, J.G. Chemical Profile, Antioxidant and Antimicrobial Activity of Marine Sponge Species Combined with Multivariate Statistical Analyses: *Desmapsamma anchorata*, *Dysidea etheria* and *Echinodictyum dendroides*. *Chem. Biodivers.* **2025**, *22*, e202402156. [CrossRef]
173. Zaki-Germi, S.; Afshar, D.; Akbari, A.; Nikfarjam, N. Multifunctional Bionanocomposite Hydrogels Based on Gelatin Methacrylate and Polyphenolic 2D Nanoparticles Decorated with Antimicrobial Bis(Imidazolium)-Based Ionic Liquids. *Compos. Sci. Technol.* **2025**, *262*, 111059. [CrossRef]
174. Kawee-ai, A. Advancing Gel Systems with Natural Extracts: Antioxidant, Antimicrobial Applications, and Sustainable Innovations. *Gels* **2025**, *11*, 125. [CrossRef]
175. Jalouli, M.; Rahman, M.A.; Biswas, P.; Rahman, H.; Harrath, A.H.; Lee, I.-S.; Kang, S.; Choi, J.; Park, M.N.; Kim, B. Targeting Natural Antioxidant Polyphenols to Protect Neuroinflammation and Neurodegenerative Diseases: A Comprehensive Review. *Front. Pharmacol.* **2025**, *16*, 1492517. [CrossRef]
176. Di Sotto, A.; Di Giacomo, S. Plant Polyphenols and Human Health: Novel Findings for Future Therapeutic Developments. *Nutrients* **2023**, *15*, 3764. [CrossRef]
177. Mehmood, A.; Usman, M.; Patil, P.; Zhao, L.; Wang, C. A Review on Management of Cardiovascular Diseases by Olive Polyphenols. *Food Sci. Nutr.* **2020**, *8*, 4639–4655. [CrossRef] [PubMed]
178. García-Conesa, M.-T.; Larrosa, M. Polyphenol-Rich Foods for Human Health and Disease. *Nutrients* **2020**, *12*, 400. [CrossRef]

https://doi.org/10.3390/antiox15010087

179. Zhang, Z.; Li, X.; Sang, S.; McClements, D.J.; Chen, L.; Long, J.; Jiao, A.; Jin, Z.; Qiu, C. Polyphenols as Plant-Based Nutraceuticals: Health Effects, Encapsulation, Nano-Delivery, and Application. *Foods* **2022**, *11*, 2189. [CrossRef] [PubMed]
180. Niu, C.; Zhang, J.; Okolo, P.I. Therapeutic Potential of Plant Polyphenols in Acute Pancreatitis. *Inflammopharmacology* **2025**, *33*, 785–798. [CrossRef]
181. Dong, Y.; Li, J.; Dai, Y.; Zhang, X.; Jiang, X.; Wang, T.; Zhao, B.; Liu, W.; Sun, H.; Du, P.; et al. A Novel Nanocarrier Based on Natural Polyphenols Enhancing Gemcitabine Sensitization Ability for Improved Pancreatic Cancer Therapy Efficiency. *Mater. Today Bio* **2025**, *30*, 101463. [CrossRef] [PubMed]
182. An, Y.; Sun, J.-X.; Ma, S.-Y.; Xu, M.-Y.; Xu, J.-Z.; Liu, C.-Q.; Wang, S.-G.; Xia, Q.-D. From Plant Based Therapy to Plant-Derived Vesicle-Like Nanoparticles for Cancer Treatment: Past, Present and Future. *Int. J. Nanomed.* **2025**, *20*, 3471–3491. [CrossRef]
183. Solanki, R.; Rawat, L.; Tabasum, S.; Pal, S.; Patel, S.; Sabarwal, A. A Comprehensive Review of Anti-Cancer Mechanisms of Polyphenol Honokiol and Nano Carrier-Based Approaches to Enhance Its Therapeutic Potential. *Phytochem. Rev.* **2025**. [CrossRef]
184. Feng, T.; Wan, Y.; Dai, B.; Liu, Y. Anticancer Activity of Bitter Melon-Derived Vesicles Extract against Breast Cancer. *Cells* **2023**, *12*, 824. [CrossRef] [PubMed]
185. Tajik, T.; Baghaei, K.; Moghadam, V.E.; Farrokhi, N.; Salami, S.A. Extracellular Vesicles of Cannabis with High CBD Content Induce Anticancer Signaling in Human Hepatocellular Carcinoma. *Biomed. Pharmacother.* **2022**, *152*, 113209. [CrossRef] [PubMed]
186. Puspadewi, R.; Milanda, T.; Muhaimin, M.; Chaerunisaa, A.Y. Nanoparticle-Encapsulated Plant Polyphenols and Flavonoids as an Enhanced Delivery System for Anti-Acne Therapy. *Pharmaceuticals* **2025**, *18*, 209. [CrossRef]
187. Kaparekar, P.S.; Anandasadagopan, S.K. The Potential Role of Bioactive Plant-Based Polyphenolic Compounds and Their Delivery Systems—As a Promising Opportunity for a New Therapeutic Solution for Acute and Chronic Wound Healing. *Curr. Pharmacol. Rep.* **2022**, *8*, 321–338. [CrossRef]
188. Froldi, G.; Ragazzi, E. Selected Plant-Derived Polyphenols as Potential Therapeutic Agents for Peripheral Artery Disease: Molecular Mechanisms, Efficacy and Safety. *Molecules* **2022**, *27*, 7110. [CrossRef]
189. Chen, D.; Sun, Y. Current Status of Plant-Based Bioactive Compounds as Therapeutics in Alzheimer's Diseases. *J. Integr. Neurosci.* **2025**, *24*, 23090. [CrossRef]
190. Xu, R.; Zhou, Y.; Niu, H.; Xue, S.; Fu, L.; Huang, Q.; Sun, Y. Kiwifruit Polyphenols Regulate Mitophagy to Promote Intestinal Barrier Function: Based on Proteomics, In Vitro and In Vivo Study. *Food Biosci.* **2025**, *64*, 105847. [CrossRef]
191. Madrigal-Gamboa, V.; Jiménez-Arias, J.; Hidalgo, O.; Quesada, S.; Pérez, A.M.; Azofeifa, G. Membrane Processing Effect of Blackberry (*Rubus adenotrichos*) on Cytotoxic and Pro-apoptotic Activities against Cancer Cell Lines. *J. Food Process. Preserv.* **2021**, *45*, e15575. [CrossRef]
192. Castejón-Vega, B.; Kyriakidis, N.C.; Alcócer-Gómez, E.; Giampieri, F.; González-Paramás, A.M.; Cordero, M.D.; Alvarez-Suarez, J.M. Modulatory Effect of Andean Blackberry Polyphenols on Genes Related to Antioxidant and Inflammatory Responses, the NLRP3 Inflammasome, and Autophagy. *J. Berry Res.* **2024**, *14*, 41–59. [CrossRef]
193. Hu, X.; Xie, J.; Bai, Y.-Y.; Hong, Z.-S.; Zhang, L.; Gong, W.-Y.; Pan, Y.-X.; Wang, W.-J.; Su, M.; Sheng, J.; et al. Extraction, In Vitro Hypoglycaemic Activity and Active Ingredient Analysis of Polyphenols from Walnut Green Husk. *J. Funct. Foods* **2024**, *122*, 106508. [CrossRef]
194. Kim, Y.; Keogh, J.; Clifton, P. Polyphenols and Glycemic Control. *Nutrients* **2016**, *8*, 17. [CrossRef]
195. Farazi, M.; Houghton, M.J.; Nicolotti, L.; Murray, M.; Cardoso, B.R.; Williamson, G. Inhibition of Human Starch Digesting Enzymes and Intestinal Glucose Transport by Walnut Polyphenols. *Food Res. Int.* **2024**, *189*, 114572. [CrossRef]
196. Yassin, M.T.; Mostafa, A.A.-F.; Al Askar, A.A. In Vitro Evaluation of Biological Activities and Phytochemical Analysis of Different Solvent Extracts of *Punica granatum* L. (Pomegranate) Peels. *Plants* **2021**, *10*, 2742. [CrossRef] [PubMed]
197. Saad, P.G.; Castelino, R.D.; Ravi, V.; Al-Amri, I.S.; Khan, S.A. Green Synthesis of Silver Nanoparticles Using Omani Pomegranate Peel Extract and Two Polyphenolic Natural Products: Characterization and Comparison of Their Antioxidant, Antibacterial, and Cytotoxic Activities. *Beni-Suef Univ. J. Basic Appl. Sci.* **2021**, *10*, 29. [CrossRef]
198. Olszowy-Tomczyk, M.; Garbaczewska, S.; Wianowska, D. Correlation Study of Biological Activity with Quercetin and Phenolics Content in Onion Extracts. *Molecules* **2022**, *27*, 8164. [CrossRef]
199. Kim, Y.; Kim, Y.-J.; Shin, Y. Comparative Analysis of Polyphenol Content and Antioxidant Activity of Different Parts of Five Onion Cultivars Harvested in Korea. *Antioxidants* **2024**, *13*, 197. [CrossRef]
200. Bozinou, E.; Pappas, I.S.; Patergiannakis, I.-S.; Chatzimitakos, T.; Palaiogiannis, D.; Athanasiadis, V.; Lalas, S.I.; Chatzilazarou, A.; Makris, D.P. Evaluation of Antioxidant, Antimicrobial, and Anticancer Properties of Onion Skin Extracts. *Sustainability* **2023**, *15*, 11599. [CrossRef]
201. Mirmojarabian, S.; Karimi, A.; Lorigooini, Z.; Javadi-Farsani, F.; Soltani, A.; Moradi, M.-T. Phytochemical Properties and Antiviral Effect of Green Tea (*Camellia sinensis*) Extract on Adenovirus In Vitro. *J. Shahrekord Univ. Med. Sci.* **2022**, *24*, 104–110. [CrossRef]
202. Teixeira Oliveira, J.; Machado Da Costa, F.; Gonçalvez Da Silva, T.; Dotto Simões, G.; Dos Santos Pereira, E.; Quevedo Da Costa, P.; Andreazza, R.; Cavalheiro Schenkel, P.; Pieniz, S. Green Tea and Kombucha Characterization: Phenolic Composition, Antioxidant Capacity and Enzymatic Inhibition Potential. *Food Chem.* **2023**, *408*, 135206. [CrossRef] [PubMed]

https://doi.org/10.3390/antiox15010087

203. Deo, A.S.; Devi, P.J.A.; Sijisha, K.S.; Anusha, R.; Mishra, T.; Mathew, S.; Abraham, K.M.; Jagadish, R.; Priya, S. Comparative Studies on the Antioxidant, Anticancer and Anti-Inflammatory Activities of Green Tea, Orthodox Black Tea and CTC Black Tea. *J. Food Sci. Technol.* **2024**, *61*, 1315–1325. [CrossRef]
204. Ranasinghe, M.; Alghaithi, M.; Mugdil, P.; Sundarakani, B.; Stathopoulos, C.; Maqsood, S. Fiber and Polyphenol Enriched Biscuits Using Date Palm Byproduct: Physiochemical Characteristics, Sensory Properties, In Vitro Digestion, and Storage Stability. *J. Food Sci.* **2025**, *90*, e17667. [CrossRef]
205. Ranasinghe, M.; Sivapragasam, N.; Mostafa, H.; Airouyuwa, J.O.; Manikas, I.; Sundarakani, B.; Maqsood, S.; Stathopoulos, C. Valorizing Date Seeds in Biscuits: A Novel Approach to Incorporate Bioactive Components Extracted from Date Seeds Using Microwave-Assisted Extraction. *Resour. Environ. Sustain.* **2024**, *15*, 100147. [CrossRef]
206. Mustafa, I.; Chin, N.L. Antioxidant Properties of Dried Ginger (*Zingiber officinale* Roscoe) Var. Bentong. *Foods* **2023**, *12*, 178. [CrossRef]
207. Almehizia, A.A.; Al-Omar, M.A.; Naglah, A.M.; Zen, A.A.; Rouf, M.A.; Nupur, A.H.; Rahamn, M.S.; Adam, A.M.A.; Mazumder, M.A.R. Incorporation of Ginger Peel Polyphenol into Yogurt Improves Its Stability: Physicochemical and Microbiological Characterizations. *Bull. Chem. Soc. Ethiop.* **2024**, *39*, 201–213. [CrossRef]
208. Gonzales, E.; Bustamante, A.; García-Díaz, D.; Sanhueza, L.; Orellana, J.F.; Fredes, C.; Jiménez, P.; Chávez, V.; Echeverría, F. Assessing Quality and Polyphenol in Vitro Bioaccessibility in Healthy Jelly Gummies with Microencapsulated and Non-Encapsulated Pomegranate Peel Extract. *Food Chem.* **2025**, *470*, 142611. [CrossRef]
209. Bansode, V.; Chandra Panda, T.; Jaddu, S.; Sahoo, S.; Subrahmanyam, K.; Vignesh, V.; Niranjan, T.; Chandra Pradhan, R.; Dwivedi, M. Enhancing Nutritional Potential: Plasma-Activated Water Treatment on Sweet Orange Peel Powder—Polyphenols, Flavonoids, Antioxidants, and Anti-Nutrients Optimization. *IEEE Trans. Plasma Sci.* **2025**, *53*, 51–62. [CrossRef]
210. Medina-Herrera, N.; Martínez-Ávila, G.C.G.; Robledo-Jiménez, C.L.; Rojas, R.; Orozco-Zamora, B.S. From Citrus Waste to Valuable Resources: A Biorefinery Approach. *Biomass* **2024**, *4*, 784–808. [CrossRef]
211. Ngongoni, K.N.; Pfukwa, T.M.; Mapiye, C. Keeping Quality of Raw Ground Beef Patties Fortified with Polyphenols Extracted from *Acacia mearnsii* Bark and Leaves. *Meat Sci.* **2025**, *219*, 109665. [CrossRef]
212. Rusindo-Rodríguez, I.M.; Ramos-Sánchez, V.H.; Chávez-Flores, D.; Delgado, E.; Chávez-Martínez, A.; Luján-Torres, V.E.; Orozco-Mena, R.E.; Sepúlveda, D.R.; Gutiérrez-Méndez, N. Co-Crystals of Lactose-Polyphenols Created from Two Food-Related By-Products: Cheese Whey and Maize Inflorescences. *Food Bioprocess Technol.* **2025**, *18*, 4513–4528. [CrossRef]
213. Telesca, L.; Belluscio, A.; Criscoli, A.; Ardizzone, G.; Apostolaki, E.T.; Fraschetti, S.; Gristina, M.; Knittweis, L.; Martin, C.S.; Pergent, G.; et al. Seagrass Meadows (*Posidonia oceanica*) Distribution and Trajectories of Change. *Sci. Rep.* **2015**, *5*, 12505. [CrossRef]
214. Piechowiak, T.; Skóra, B.; Grzelak-Błaszczyk, K.; Sójka, M. Extraction of Antioxidant Compounds from Blueberry Fruit Waste and Evaluation of Their In Vitro Biological Activity in Human Keratinocytes (HaCaT). *Food Anal. Methods* **2021**, *14*, 2317–2327. [CrossRef]
215. Bencresciuto, G.F.; Carnevale, M.; Paris, E.; Gallucci, F.; Santangelo, E.; Migliori, C.A. A Sustainable Alternative for Cosmetic Applications: NADES Extraction of Bioactive Compounds from Hazelnut By-Products. *Sustainability* **2025**, *17*, 1516. [CrossRef]
216. Sanhueza, L.; Melo, R.; Montero, R.; Maisey, K.; Mendoza, L.; Wilkens, M. Synergistic Interactions between Phenolic Compounds Identified in Grape Pomace Extract with Antibiotics of Different Classes against *Staphylococcus aureus* and *Escherichia coli*. *PLoS ONE* **2017**, *12*, e0172273. [CrossRef] [PubMed]
217. Daglia, M. Polyphenols as Antimicrobial Agents. *Curr. Opin. Biotechnol.* **2012**, *23*, 174–181. [CrossRef]
218. Armari, M.; Zavattaro, E.; Trejo, C.F.; Galeazzi, A.; Grossetti, A.; Veronese, F.; Savoia, P.; Azzimonti, B. Vitis Vinifera L. Leaf Extract, a Microbiota Green Ally against Infectious and Inflammatory Skin and Scalp Diseases: An In-Depth Update. *Antibiotics* **2024**, *13*, 697. [CrossRef]
219. Tsiapali, O.I.; Ayfantopoulou, E.; Tzourouni, A.; Ofrydopoulou, A.; Letsiou, S.; Tsoupras, A. Unveiling the Utilization of Grape and Winery By-Products in Cosmetics with Health Promoting Properties. *Appl. Sci.* **2025**, *15*, 1007. [CrossRef]
220. Salih, N.K.-E.M.; Alam, M.Z.; Haris, S.; Kamal-Eldin, A.; Al-Marzouqi, A.H. Prospective Applications of Phenolic Compounds in Processing By-Products of Date Fruits (*Phoenix dactylifera* L.). *Discov. Food* **2025**, *5*, 198. [CrossRef]
221. Tapia-Quirós, P.; Montenegro-Landívar, M.F.; Reig, M.; Vecino, X.; Cortina, J.L.; Saurina, J.; Granados, M. Recovery of Polyphenols from Agri-Food By-Products: The Olive Oil and Winery Industries Cases. *Foods* **2022**, *11*, 362. [CrossRef]
222. Galanakis, C.M.; Tsatalas, P.; Galanakis, I.M. Implementation of Phenols Recovered from Olive Mill Wastewater as UV Booster in Cosmetics. *Ind. Crops Prod.* **2018**, *111*, 30–37. [CrossRef]
223. Barba, F.J.; Zhu, Z.; Koubaa, M.; Sant'Ana, A.S.; Orlien, V. Green Alternative Methods for the Extraction of Antioxidant Bioactive Compounds from Winery Wastes and By-Products: A Review. *Trends Food Sci. Technol.* **2016**, *49*, 96–109. [CrossRef]
224. Rojo-Poveda, O.; Ribeiro, S.O.; Anton-Sales, C.; Keymeulen, F.; Barbosa-Pereira, L.; Delporte, C.; Zeppa, G.; Stévigny, C. Evaluation of Cocoa Bean Shell Antimicrobial Activity: A Tentative Assay Using a Metabolomic Approach for Active Compound Identification. *Planta Med.* **2021**, *87*, 841–849. [CrossRef]

225. Disca, V.; Travaglia, F.; Carini, C.; Coïsson, J.D.; Cravotto, G.; Arlorio, M.; Locatelli, M. Improving the Extraction of Polyphenols from Cocoa Bean Shells by Ultrasound and Microwaves: A Comparative Study. *Antioxidants* **2024**, *13*, 1097. [CrossRef] [PubMed]
226. Moreno-Chamba, B.; Salazar-Bermeo, J.; Martínez-Madrid, M.C.; Lizama, V.; Martín-Bermudo, F.; Berná, G.; Neacsu, M.; Saura, D.; Martí, N.; Valero, M. Bound Galloylated Compounds in Persimmon Upcycled Dietary Fiber Modulate Microbial Strains Associated to Human Health after in Vitro Digestion. *LWT* **2022**, *156*, 113011. [CrossRef]
227. Moreno-Chamba, B.; Salazar-Bermeo, J.; Navarro-Simarro, P.; Narváez-Asensio, M.; Martínez-Madrid, M.C.; Saura, D.; Martí, N.; Valero, M. Autoinducers Modulation as a Potential Anti-Virulence Target of Bacteria by Phenolic Compounds. *Int. J. Antimicrob. Agents* **2023**, *62*, 106937. [CrossRef]
228. Lopes, J.D.C.; Madureira, J.; Margaça, F.M.A.; Cabo Verde, S. Grape Pomace: A Review of Its Bioactive Phenolic Compounds, Health Benefits, and Applications. *Molecules* **2025**, *30*, 362. [CrossRef]
229. Albrahim, J.S.; El-Fakharany, E.M.; El-Gendi, H.; Saleh, A.K.; El-Maradny, Y.A. Therapeutic Perspectives of *Mangifera indica* L. Peel Extract: Phytochemical Profile, Antimicrobial, Anticancer, and Antiviral Efficacy. *Biomass Convers. Biorefin.* **2025**, *15*, 11371–11394. [CrossRef]
230. Siol, M.; Sadowska, A. Chemical Composition, Physicochemical and Bioactive Properties of Avocado (*Persea americana*) Seed and Its Potential Use in Functional Food Design. *Agriculture* **2023**, *13*, 316. [CrossRef]
231. Gandhi, G.R.; Jothi, G.; Antony, P.J.; Balakrishna, K.; Paulraj, M.G.; Ignacimuthu, S.; Stalin, A.; Al-Dhabi, N.A. Gallic Acid Attenuates High-Fat Diet Fed-Streptozotocin-Induced Insulin Resistance via Partial Agonism of PPARγ in Experimental Type 2 Diabetic Rats and Enhances Glucose Uptake through Translocation and Activation of GLUT4 in PI3K/p-Akt Signaling Pathway. *Eur. J. Pharmacol.* **2014**, *745*, 201–216. [CrossRef]
232. Nunes, A.R.; Alves, G.; Falcão, A.; Lopes, J.A.; Silva, L.R. Phenolic Acids from Fruit By-Products as Therapeutic Agents for Metabolic Syndrome: A Review. *Int. J. Mol. Sci.* **2025**, *26*, 3834. [CrossRef] [PubMed]
233. Santana-Gálvez, J.; Cisneros-Zevallos, L.; Jacobo-Velázquez, D. Chlorogenic Acid: Recent Advances on Its Dual Role as a Food Additive and a Nutraceutical against Metabolic Syndrome. *Molecules* **2017**, *22*, 358. [CrossRef] [PubMed]
234. Zubairu, I.K.; Rakariyatham, K.; Bai-Ngew, S.; Leksawasdi, N.; Regenstein, J.M.; Lao, F.; Hong, H.; Shin, W.-S.; Alzahrani, K.J.; Phimolsiripol, Y. Nutritional and Therapeutic Potential of Longan Fruit By-Products for Liver Diseases: Pathway to Functional Foods. *Curr. Nutr. Rep.* **2025**, *14*, 28. [CrossRef]
235. Servin-Uribe, R.I.; Castilla-Ramírez, P.; Ramírez, I.F.P.; Jiménez, J.P.; Reynoso-Camacho, R. In Vivo Differential Effects of Extractable and Non-Extractable Phenolic Compounds from Grape Pomace on the Regulation of Obesity and Associated Metabolic Alterations. *Plant Foods Hum. Nutr.* **2025**, *80*, 37. [CrossRef]
236. Frusciante, L.; Geminiani, M.; Shabab, B.; Olmastroni, T.; Roncucci, N.; Mastroeni, P.; Salvini, L.; Lamponi, S.; Trezza, A.; Santucci, A. Enhancing Industrial Hemp (*Cannabis sativa*) Leaf By-Products: Bioactive Compounds, Anti-Inflammatory Properties, and Potential Health Applications. *Int. J. Mol. Sci.* **2025**, *26*, 548. [CrossRef]
237. Laaraj, N.; Bouhrim, M.; Kharchoufa, L.; Tiji, S.; Bendaha, H.; Addi, M.; Drouet, S.; Hano, C.; Lorenzo, J.M.; Bnouham, M.; et al. Phytochemical Analysis, α-Glucosidase and α-Amylase Inhibitory Activities and Acute Toxicity Studies of Extracts from Pomegranate (*Punica granatum*) Bark, a Valuable Agro-Industrial By-Product. *Foods* **2022**, *11*, 1353. [CrossRef] [PubMed]
238. Qiu, Z.; Zheng, Z.; Xiao, H. Sustainable Valorization of Garlic Byproducts: From Waste to Resource in the Pursuit of Carbon Neutrality. *Compr. Rev. Food Sci. Food Saf.* **2025**, *24*, e70151. [CrossRef]
239. Fernandez, S.P.; González, R.E. Nutritional Quality, Bioactive Compounds, and Antioxidant Activity of Nine Clones of Fresh Garlic and Its Black Garlic Derivative: A Comparative Study. *Biol. Life Sci. Forum* **2024**, *40*, 29.
240. Cava, R.; Ladero, L. Using Polyphenol-Rich Extracts from Tropical Fruit Byproducts to Control Lipid and Protein Oxidation in Cooked Chicken Models. *Eur. Food Res. Technol.* **2024**, *250*, 2809–2820. [CrossRef]
241. Fernandes, R.A.; Ferreira, N.; Lopes, S.; Freitas, B.; Santos, J.; Martins, J.M.; Carvalho, L.H. Antioxidant Particleboards Produced from Forest By-Products with Application in the Food Packaging Industry. *Polymers* **2025**, *17*, 216. [CrossRef]
242. Panja, P. Green Extraction Methods of Food Polyphenols from Vegetable Materials. *Curr. Opin. Food Sci.* **2018**, *23*, 173–182. [CrossRef]
243. Ameer, K.; Shahbaz, H.M.; Kwon, J. Green Extraction Methods for Polyphenols from Plant Matrices and Their Byproducts: A Review. *Compr. Rev. Food Sci. Food Saf.* **2017**, *16*, 295–315. [CrossRef]
244. Dzah, C.S.; Duan, Y.; Zhang, H.; Serwah Boateng, N.A.; Ma, H. Latest Developments in Polyphenol Recovery and Purification from Plant By-Products: A Review. *Trends Food Sci. Technol.* **2020**, *99*, 375–388. [CrossRef]
245. Alexandre, E.M.C.; Araújo, P.; Duarte, M.F.; De Freitas, V.; Pintado, M.; Saraiva, J.A. High-Pressure Assisted Extraction of Bioactive Compounds from Industrial Fermented Fig by-Product. *J. Food Sci. Technol.* **2017**, *54*, 2519–2531. [CrossRef]
246. Okur, I.; Oztop, M.H.; Alpas, H. Optimization and Comparison of High-Pressure-Assisted Extraction of Phenolic Compounds from Olive Pomace. *ACS Food Sci. Technol.* **2022**, *2*, 1862–1869. [CrossRef]

https://doi.org/10.3390/antiox15010087

247. Panda, D.; Saharan, V.K.; Manickam, S. Controlled Hydrodynamic Cavitation: A Review of Recent Advances and Perspectives for Greener Processing. *Processes* **2020**, *8*, 220. [CrossRef]
248. Wu, Z.; Ferreira, D.F.; Crudo, D.; Bosco, V.; Stevanato, L.; Costale, A.; Cravotto, G. Plant and Biomass Extraction and Valorisation under Hydrodynamic Cavitation. *Processes* **2019**, *7*, 965. [CrossRef]
249. Farías-Campomanes, A.M.; Rostagno, M.A.; Coaquira-Quispe, J.J.; Meireles, M.A.A. Supercritical Fluid Extraction of Polyphenols from Lees: Overall Extraction Curve, Kinetic Data and Composition of the Extracts. *Bioresour. Bioprocess.* **2015**, *2*, 45. [CrossRef]
250. Bessa, C.; Francisco, T.; Dias, R.; Mateus, N.; Freitas, V.D.; Pérez-Gregorio, R. Use of Polyphenols as Modulators of Food Allergies. From Chemistry to Biological Implications. *Front. Sustain. Food Syst.* **2021**, *5*, 623611. [CrossRef]
251. Słota, P.; Harasym, J.; Jacukowicz-Sobala, I. Supercritical Fluid Extraction—A Sustainable and Selective Alternative for Tannin Recovery from Biomass Resources. *Appl. Sci.* **2025**, *15*, 5914. [CrossRef]
252. Dobroslavić, E.; Elez Garofulić, I.; Šeparović, J.; Zorić, Z.; Pedisić, S.; Dragović-Uzelac, V. Pressurized Liquid Extraction as a Novel Technique for the Isolation of *Laurus nobilis* L. Leaf Polyphenols. *Molecules* **2022**, *27*, 5099. [CrossRef]
253. Nieto, J.A.; Santoyo, S.; De Sá, M.; Baoshan, S.; Reglero, G.; Jaime, L. Comprehensive Study of Sustainable Pressurized Liquid Extractions to Obtain Bioavailable Antioxidant Phenolic Compounds from Grape Seed By-Products. *Processes* **2024**, *12*, 2308. [CrossRef]
254. Savic, I.M.; Savic Gajic, I.M. Optimization of Ultrasound-Assisted Extraction of Polyphenols from Wheatgrass (*Triticum aestivum* L.). *J. Food Sci. Technol.* **2020**, *57*, 2809–2818. [CrossRef]
255. Kumari, B.; Tiwari, B.K.; Hossain, M.B.; Rai, D.K.; Brunton, N.P. Ultrasound-assisted Extraction of Polyphenols from Potato Peels: Profiling and Kinetic Modelling. *Int. J. Food Sci. Technol.* **2017**, *52*, 1432–1439. [CrossRef]
256. Pogorzelska-Nowicka, E.; Hanula, M.; Pogorzelski, G. Extraction of Polyphenols and Essential Oils from Herbs with Green Extraction Methods—An Insightful Review. *Food Chem.* **2024**, *460*, 140456. [CrossRef] [PubMed]
257. López-Salazar, H.; Camacho-Díaz, B.H.; Ocampo, M.L.A.; Jiménez-Aparicio, A.R. Microwave-Assisted Extraction of Functional Compounds from Plants: A Review. *BioResources* **2023**, *18*, 6614–6638. [CrossRef]
258. Tanruean, K.; Luangkamin, S.; Srisurat, T.; Bunmusik, W.; Suttiarporn, P. Optimization of Microwave-Assisted Extraction Process for Production of Polyphenol-Rich Crude Extract from *Cinnamomum iners* Leaves. *Appl. Sci.* **2025**, *15*, 1265. [CrossRef]
259. Das, S.; Nadar, S.S.; Rathod, V.K. Integrated Strategies for Enzyme Assisted Extraction of Bioactive Molecules: A Review. *Int. J. Biol. Macromol.* **2021**, *191*, 899–917. [CrossRef]
260. De Camargo, A.C.; Regitano-d'Arce, M.A.B.; Biasoto, A.C.T.; Shahidi, F. Enzyme-Assisted Extraction of Phenolics from Winemaking by-Products: Antioxidant Potential and Inhibition of Alpha-Glucosidase and Lipase Activities. *Food Chem.* **2016**, *212*, 395–402. [CrossRef] [PubMed]
261. Lakka, A.; Bozinou, E.; Stavropoulos, G.; Samanidis, I.; Athanasiadis, V.; Dourtoglou, V.G.; Makris, D.P.; Lalas, S.I. Enhancement of Polyphenols Recovery from *Rosa canina*, *Calendula officinalis* and *Castanea sativa* Using Pulsed Electric Field. *Beverages* **2021**, *7*, 63. [CrossRef]
262. Carpentieri, S.; Ferrari, G.; Pataro, G. Optimization of Pulsed Electric Fields-Assisted Extraction of Phenolic Compounds From White Grape Pomace Using Response Surface Methodology. *Front. Sustain. Food Syst.* **2022**, *6*, 854968. [CrossRef]
263. Dai, Y.; Van Spronsen, J.; Witkamp, G.-J.; Verpoorte, R.; Choi, Y.H. Natural Deep Eutectic Solvents as New Potential Media for Green Technology. *Anal. Chim. Acta* **2013**, *766*, 61–68. [CrossRef]
264. Mustafa, A.; Turner, C. Pressurized Liquid Extraction as a Green Approach in Food and Herbal Plants Extraction: A Review. *Anal. Chim. Acta* **2011**, *703*, 8–18. [CrossRef]
265. Rosales, T.K.O.; Fabi, J.P. Valorization of Polyphenolic Compounds from Food Industry By-Products for Application in Polysaccharide-Based Nanoparticles. *Front. Nutr.* **2023**, *10*, 1144677. [CrossRef] [PubMed]
266. Drosou, C.; Kyriakopoulou, K.; Laina, K.T.; Bimpilas, A.; Tsimogiannis, D.; Krokida, M. Revolutionizing Wine Waste: Advanced Techniques for Polyphenol Recovery from White Wine Byproducts. *Agriculture* **2025**, *15*, 648. [CrossRef]
267. Paiva, A.; Craveiro, R.; Aroso, I.; Martins, M.; Reis, R.L.; Duarte, A.R.C. Natural Deep Eutectic Solvents—Solvents for the 21st Century. *ACS Sustain. Chem. Eng.* **2014**, *2*, 1063–1071. [CrossRef]
268. Liu, Y.; Li, R.; Wang, Q.; Liu, X.; Gong, Z. Polyphenols and Flavonoids in Water Caltrop Shells from Different Producing Areas of China: Identification, Green Extraction and Quantitative Analysis. *Ind. Crops Prod.* **2025**, *225*, 120606. [CrossRef]
269. Dushkova, M.; Mitova, M.; Bakardzhiyski, I.; Miteva-Petrova, M.; Menkov, N. Application of Ultrafiltration for Recovery of Bioactive Phenolic Compounds from Rose Wastewater. *Appl. Sci.* **2025**, *15*, 2040. [CrossRef]
270. Huang, J.; Liu, Y.; Han, F.; Fang, Z.; Yang, L.; Zhuang, M.; Zhang, Y.; Lv, H.; Wang, Y.; Ji, J.; et al. Genetic Diversity and Population Structure Analysis of 161 Broccoli Cultivars Based on SNP Markers. *Hortic. Plant J.* **2021**, *7*, 423–433. [CrossRef]

https://doi.org/10.3390/antiox15010087

271. Ghimire, U.; Pliakoni, E.; Yu, F.; Brecht, J.K.; Liu, T. Identifying Genes Regulated during Natural, on-Plant Senescence in Broccoli (*Brassica oleracea*) in Contrast to Postharvest Senescence. *Postharvest Biol. Technol.* **2023**, *206*, 112535. [CrossRef]
272. Rivas, M.Á.; Benito, M.J.; Martín, A.; De Guía Córdoba, M.; Gizaw, Y.; Casquete, R. Development of Supercritical Technology to Obtain Improved Functional Dietary Fiber for the Valorization of Broccoli By-product. *J. Sci. Food Agric.* **2025**, *105*, 2203–2214. [CrossRef]
273. Shah, M.A.; Tariq, S.; Abuzar, S.M.; Ilyas, K.; Qadees, I.; Alsharif, I.; Anam, K.; Almutairi, R.T.; Al-Regaiey, K.A.; Babalghith, A.O.; et al. Peel Waste of Citrus Fruits: A Valuable and Renewable Source of Polyphenols for the Treatment of Diabesity. *Curr. Res. Biotechnol.* **2024**, *7*, 100204. [CrossRef]
274. Wang, S.; Feng, Y.; Yu, X.; Yang, Z.; Jiao, P.; Niu, Q. Integrated Deep Eutectic Solvent Extraction and Resin Adsorption for Recovering Polyphenols from *Citrus reticulata* Blanco Peels: Process Optimization, Compositional Analysis, and Activity Determination. *Sep. Purif. Technol.* **2025**, *355*, 129560. [CrossRef]
275. Marchetti, F.; Gugel, I.; Costa, S.; Baldisserotto, A.; Foletto, A.; Gugel, I.; Baldini, E.; Manfredini, S.; Vertuani, S. A Sustainable Approach to Valuable Polyphenol and Iridoid Antioxidants from Medicinal Plant By-Products. *Antioxidants* **2024**, *13*, 1014. [CrossRef] [PubMed]
276. Rocchetti, G.; Blasi, F.; Montesano, D.; Ghisoni, S.; Marcotullio, M.C.; Sabatini, S.; Cossignani, L.; Lucini, L. Impact of Conventional/Non-Conventional Extraction Methods on the Untargeted Phenolic Profile of *Moringa oleifera* Leaves. *Food Res. Int.* **2019**, *115*, 319–327. [CrossRef] [PubMed]
277. Armenta, S.; Garrigues, S.; Esteve-Turrillas, F.A.; De La Guardia, M. Green Extraction Techniques in Green Analytical Chemistry. *TrAC Trends Anal. Chem.* **2019**, *116*, 248–253. [CrossRef]
278. Coşkun, N.; Sarıtaş, S.; Bechelany, M.; Karav, S. Polyphenols in Foods and Their Use in the Food Industry: Enhancing the Quality and Nutritional Value of Functional Foods. *Int. J. Mol. Sci.* **2025**, *26*, 5803. [CrossRef]
279. Kammerer, D.R.; Kammerer, J.; Valet, R.; Carle, R. Recovery of Polyphenols from the By-Products of Plant Food Processing and Application as Valuable Food Ingredients. *Food Res. Int.* **2014**, *65*, 2–12. [CrossRef]
280. Bai, T.; Wang, X.; Du, W.; Cheng, J.; Zhang, J.; Zhang, Y.; Klinjapo, R.; Asavasanti, S.; Yasurin, P. Recent Advances, Challenges, and Functional Applications of Natural Phenolic Compounds in the Meat Products Industry. *Antioxidants* **2025**, *14*, 138. [CrossRef]
281. Benito-González, I.; López-Rubio, A.; Martínez-Abad, A.; Ballester, A.-R.; Falcó, I.; González-Candelas, L.; Sánchez, G.; Lozano-Sánchez, J.; Borrás-Linares, I.; Segura-Carretero, A.; et al. In-Depth Characterization of Bioactive Extracts from Posidonia Oceanica Waste Biomass. *Mar. Drugs* **2019**, *17*, 409. [CrossRef]
282. Galante, M.; Brassesco, M.E.; Maragoni Santos, C.; Beres, C.; Fai, A.E.C.; Cabezudo, I. Grape Pomace as a Natural Source of Antimicrobial Agents for Food Preservation. *Front. Nutr.* **2025**, *12*, 1650450. [CrossRef]
283. Yilmaz, Y.; Toledo, R.T. Major Flavonoids in Grape Seeds and Skins: Antioxidant Capacity of Catechin, Epicatechin, and Gallic Acid. *J. Agric. Food Chem.* **2004**, *52*, 255–260. [CrossRef] [PubMed]
284. Di Stefano, V.; Buzzanca, C.; Melilli, M.G.; Indelicato, S.; Mauro, M.; Vazzana, M.; Arizza, V.; Lucarini, M.; Durazzo, A.; Bongiorno, D. Polyphenol Characterization and Antioxidant Activity of Grape Seeds and Skins from Sicily: A Preliminary Study. *Sustainability* **2022**, *14*, 6702. [CrossRef]
285. Rajasekharan, R.; Paul, S.A.; Madhavan, A.; Sindhu, R.; Binod, P.; Awasthi, M.K.; Pandey, A. Green Strategies for Extraction of Nanocellulose from Agricultural Wastes—Current Trends and Future Perspectives. In *Green Sustainable Process for Chemical and Environmental Engineering and Science*; Elsevier: Amsterdam, The Netherlands, 2023; pp. 269–288. ISBN 978-0-323-95156-2.
286. Duarte, P.F.; Chaves, M.A.; Borges, C.D.; Mendonça, C.R.B. Avocado: Characteristics, Health Benefits and Uses. *Ciênc. Rural* **2016**, *46*, 747–754. [CrossRef]
287. Zuo, X.; Gu, Y.; Wang, C.; Zhang, J.; Zhang, J.; Wang, G.; Wang, F. A Systematic Review of the Anti-Inflammatory and Immunomodulatory Properties of 16 Essential Oils of Herbs. *Evid. Based Complement. Alternat. Med.* **2020**, *2020*, 8878927. [CrossRef]
288. Chagas, M.D.S.S.; Behrens, M.D.; Moragas-Tellis, C.J.; Penedo, G.X.M.; Silva, A.R.; Gonçalves-de-Albuquerque, C.F. Flavonols and Flavones as Potential anti-Inflammatory, Antioxidant, and Antibacterial Compounds. *Oxid. Med. Cell. Longev.* **2022**, *2022*, 9966750. [CrossRef]
289. Belgacem, I.; Li Destri Nicosia, M.G.; Pangallo, S.; Abdelfattah, A.; Benuzzi, M.; Agosteo, G.E.; Schena, L. Pomegranate Peel Extracts as Safe Natural Treatments to Control Plant Diseases and Increase the Shelf-Life and Safety of Fresh Fruits and Vegetables. *Plants* **2021**, *10*, 453. [CrossRef]
290. Jang, H.-J.; Lee, H.-J.; Yoon, D.-K.; Ji, D.-S.; Kim, J.-H.; Lee, C.-H. Antioxidant and Antimicrobial Activities of Fresh Garlic and Aged Garlic By-Products Extracted with Different Solvents. *Food Sci. Biotechnol.* **2018**, *27*, 219–225. [CrossRef]
291. Tresserra-Rimbau, A.; Lamuela-Raventos, R.M.; Moreno, J.J. Polyphenols, Food and Pharma. Current Knowledge and Directions for Future Research. *Biochem. Pharmacol.* **2018**, *156*, 186–195. [CrossRef] [PubMed]

https://doi.org/10.3390/antiox15010087

292. Mourtzinos, I.; Goula, A. Polyphenols in Agricultural Byproducts and Food Waste. In *Polyphenols in Plants*; Elsevier: Amsterdam, The Netherlands, 2019; pp. 23–44, ISBN 978-0-12-813768-0.
293. Saini, R.K.; Khan, M.I.; Kumar, V.; Shang, X.; Lee, J.-H.; Ko, E.-Y. Bioactive Compounds of Agro-Industrial By-Products: Current Trends, Recovery, and Possible Utilization. *Antioxidants* **2025**, *14*, 650. [CrossRef]

https://doi.org/10.3390/antiox15010087

MDPI AG
Grosspeteranlage 5
4052 Basel
Switzerland
Tel.: +41 61 683 77 34

Antioxidants Editorial Office
E-mail: antioxidants@mdpi.com
www.mdpi.com/journal/antioxidants

www.ingramcontent.com/pod-product-compliance
Lightning Source LLC
Chambersburg PA
CBHW061836210926
51850CB00006B/907
* 9 7 8 3 7 2 5 8 7 4 5 2 1 *